T0376288

FAST FORDS

Published in February 2024

ISBN 978-1-910505-83-0

Published by Evro Publishing
Westrow House, Holwell, Sherborne, Dorset DT9 5LF, UK

Edited by Steve Rendle
Designed by Richard Parsons

Printed and bound in China by 1010 Printing International Ltd

www.evropublishing.com

COVER PHOTOGRAPHS

Front

Row by row, from left. Row 1: BTCC Mondeo, *Author's archive*; 'Group 1+' Capri 3.0, *Motorsport Images*; Group 2 Escort RS1600, *Author*. Row 2: Escort RS Cosworth, *McKlein*; Cortina Lotus Mk1, *Author*; Mustang, *Author*. Row 3: RS200, *Ford*. Row 4: Sierra XR4x4, *Author's archive*. Row 5: Focus WRC, *Ford*; Focus RS, *Ford*; Anglia, *Murdock-USA archive*. Row 6: GT40, *Jeff Bloxham*; Sierra RS Cosworth, *Author's archive*; Supervan 1, *Ford*.

Back

The author pictured in the USA in 2002, *Tim Crete*.

AUTHOR

Jeremy Walton is an award-winning author of 37 original motoring books and one racing-themed novel. Ford became an area of special expertise and his earlier books about the marque are regarded as benchmark works. He has worked for *Motoring News* and *Motor Sport*, as well as Ford Motor Company's Competition and Advanced Vehicle Operations departments in Britain and Germany. Walton raced Escort, Capri , Fiesta and Sierra Cosworth derivatives, and his varied Ford driving experiences since 1967 range from Fiesta to GT40, via V8 Supervans and America's Mustangs and Thunderbirds. Besides racing, his motoring activities have included journalism, TV advertising and automotive consultancy.

50 YEARS UP CLOSE AND PERSONAL WITH FORD'S FINEST

BY JEREMY WALTON

CONTENTS

INTRODUCTION

Because I once worked full-time at Ford of Britain, from 1972 to 1975, and have been a long-time consultant to the company, you might suspect that this book is a toothless promotion for all things Blue Oval. Not so. Why?

One reason is that my full-time period at Ford, albeit a massively rewarding professional boost, was largely a negative personal experience that even left a long-term health legacy. I should also highlight that I had five years of writing about all sorts of cars before I became a Ford employee, and afterwards 46 more years as a freelancer. Of my 37 books on motoring subjects, just six have been about Fords, although maybe it seems like more because a few went to multiple editions with various publishers.

I had no family or geographical links with Ford prior to joining as full-time Motorsport Public Relations Officer. When I left, I returned to my Thames Valley roots, selling all but my wife's Ford road car to buy a then-new Lancia Beta Coupé.

True, since then and into the 21st century I have had regular and profitable Ford consultancies, including one involving TV advertising. Additionally, I steered more than 200 Fords in my journalistic and racing life. I also had more than a dozen company Fords over long periods (three months to three years), and loaned competition machinery, plus cars that I bought and resold when their work was done.

However, I spent my personal cash far more often on other manufacturers' cars. Here's a vehicle record, much of it mundane as I had two kids and a wife to feed, a house from the age of 21, and I divorced and remarried twice after I was 44.

Teen years: Four motorcycles (Norton Jubilee, Triumph T120/T110, Ariel Golden Arrow and Leader); new 1965 Mini 848cc (following successful third-party insurance claim and disposal of all personal motorcycles); return to bikes with company Honda 400F to collect and deliver editorial content, 1975–77.

Growing up at Ford: 1972 Ford Escort Mexico special build (PEV 805K); 1971 MG Midget 1275.

Adult: 1972 Ford Escort 1300 GT estate; 1973 Ford Escort RS2000 special build; 1975 Lancia Beta Coupé; 1976 BMW 528; 1976 Ford RS2000 Group 1 rally car (post-season, bought/resold); 1977 Vauxhall 1.2 Chevette; 1981 Ford 2.8i Capri; 1982 VW Polo; 1980 and 1983 Ford Escort XR3/3i (bought/resold company cars); 1985 Ford Escort 1.6 L estate; 1986 and 1988 Honda CRX; 1990 VW Passat estate.

Divorced/remarried: VW Golf MkII Sport and Golf MkII GTI; Ford Sierra Cosworth three-door (bought and resold as a race car); Sierra Cosworths, rear-drive and 4x4 (bought/resold as company cars); Alfa Romeo 155 and 156.

Longest ownership: 1986 BMW 635 CSi (16 years).

Other BMWs: 318d, 120d, E36 M3 convertible and Z3 2.0 six-cylinder.

Classics/limited editions: 1958 Austin-Healey Sprite; 1998 Lotus Elise 135 Sport; 2006 Audi TT quattro Sport.

My first car was the inevitable basic 848cc Mini and equally predictably it got tuned to explosive levels. That Ford's competitions staff at Boreham kept straight faces on my first visit was a credit to their unusual degree of diplomacy that vaporised when they got to know me better.

After that 1965 Mini disappeared from my life as a partial deposit on my first house, I was fully employed, with company vans or cars, either via motoring media or Ford until 1977, when I struck out on my own. Comparative poverty was reflected by a 1,256cc Vauxhall Chevette that I stomached for a year, before I was awarded company-car status again, despite always remaining self-employed. Into the '80s, magazine work and TV advertising 'Precision Driving' jobs allowed me to buy one of the first Ford Escort XR3s, followed by an XR3i. A company 2-litre Capri subsequently inhabited my drive, alongside a loaned 3-litre. I returned to Capris for the compilation of the second of four editions of Capri books for two publishers.

The road to Ford employment

Meanwhile, I quietly switched to a BMW for long-distance miles, including turning up on Ford advertising shoots with my four-door 528, a very comfortable, swift and smooth device. It was frowned upon by Ford executives sent to monitor filming, but popular with the late Peter Harrison, a stunning character who ensured I kept this lucrative 'Precision Driving' sideshow intact despite much opposition. There were naturally bitter Ford of Britain objections to a known motoring journalist/former employee accessing pre-announcement models, all made worse when *Car* magazine ran a pre-launch story rubbishing the new front-drive Escort.

Also, my apparently perfect Blue Oval relationship on the outside – wonderful access to (nearly) all areas and hardware – had some harsh, gritty realities within as I graduated from the banter and informality of eight editorial colleagues to a Ford payroll with employee numbers measured in thousands. Initially, for me the negatives outweighed the positives of employment at the company, and when the job opportunity as a Motorsport Public Relations Officer came up, with additional responsibility for the two-year-old Ford Advanced Vehicles (FAV), I did not apply.

I was happy reporting for sister titles *Motor Sport* and *Motoring News*, as well as racing touring cars with a chance of graduating from a 1300 Escort Sport to a V8 Chevy Camaro that could win 1973 races outright, instead of less satisfying class victories. I achieved a surprise seventh overall in the 1972 Spa 24 Hours with a production-specification BMW 3.0 CS and knew that Munich had ambitions to overhaul Ford and others in European touring car racing. I had tested Broadspeed's pre-Batmobile prediction of the lightweight CSL coupé as an overawed 1972 accomplice to Niki Lauda at Silverstone.

As Ford worked through its list of over 400 applicants for the Motorsport PR job, Nick Brittan, a respected Ford consultant and former racer, advised me in the strongest terms to apply. Nick was a mentor: I had been weaned on his book *How to go Saloon Car Racing*, based around his exploits in his 'Green Bean' Ford Anglia, and he had earlier steered me away from my original Coventry competition links (Emery 998 Hillman Imp and pre-production Avenger Tiger) with brisk and relevant 'dad' advice: 'No proper cash or support with them, boy, and you've got kids and a mortgage.'

I took Nick seriously. Seriously enough to go through the rigmarole of a formal letter, although the job was no longer advertised and Ford still had plenty of worthy candidates to interview. Against the odds, I got the job, but a fractious relationship awaited. The competitions people at Boreham were the bright spot, but politics overcame me in other assignments to FAV and the main Brentford HQ.

Ford of Britain top-floor executives had been ordered to dump an ageing and prominent speechwriter who had carelessly lost American favour. Mr Unfavoured was rehoused away from the high-profile HQ and shuffled into the FAV office allotted to me. His senior status blocked my access to many overseas outings for the company, particularly the East African Safari. In fact I did much more overseas travel with Ford connections after I left the company.

▲ A personal highlight of my earlier career was to drive Roger Williamson's Ford Anglia racer. Roger was tragically killed in the 1973 Dutch Grand Prix when on the brink of Formula 1 stardom. Here we are at a wintry Mallory Park in 1970 for a *Motoring News* story. *Motorsport Images/Andrew Marriott*

Another negative as a Ford employee was the 1973–74 fuel crisis: the fuel shortages saw Ford withdraw European factory entries from motorsport in a blink, and many events were cancelled anyway. When following the 2,000-mile RAC Rally route in 1973, we were allowed only three gallons of fuel per stop, this in an uprated Capri RS3100 pre-production weapon with four occupants.

The combined efforts of that fuel crisis and stress (childhood asthma returned and resulted in some miserable sick-note days) prompted me to leave a fat salary and good career prospects at the prospering Good Ship Ford. My wife and kids had settled into Essex life better than I had. Yet, I was supremely lucky in that some key Ford of Britain people stayed in touch after I took the door marked 'EXIT'. Thus I got an assortment of freelance/consultancy opportunities that would never have come way if I hadn't served my time as an employee.

Spoiler alert: excessive name-dropping

The upsides of the Ford job covered prestige events and VIP connections that I would never have experienced as a grubby reporter. Rallies, from club level to Monte Carlo and RAC were natural to a Stuart Turner-managed sports operation, but totally new to me. Previously being responsible for lending rally-wise journalists temperamental RS1600 road cars ensured I found my way around most events.

Prestige people? I enjoyed a tank-driving day on Salisbury Plain laid on by Prince Michael of Kent's regiment for Ford-contracted lady rally driver Gillian Fortescue-Thomas. There were some wry grins to be had during the Silverstone day that saw Mark Phillips trash Tom Walkinshaw's RS2000 before Princess Anne's withering gaze.

More impressive were drives accompanied by two Formula 1 World Champions, Jackie Stewart ('JYS') and Emerson Fittipaldi ('Emmo'). They were inspiring experiences that changed some of the blinkered hardcore motor racing media attitudes I had held previously. That was especially the case with JYS, with whom there were a number of outings, stretching from 1973–74 Capri RS3100 to a Mondeo in the '90s. For Emmo, Ford deployed a flagship 3-litre Granada Ghia that I steered in ultra-gentle mode – you never want to make a total fool of yourself driving such talented goods! – when chauffeuring him from Heathrow to a Ford-Cosworth DFV function in London. He was a total gent, putting this young stranger at immediate ease.

Anyway, that's enough name-dropping. Read on through the main chapters to see if I made good use of so many faster Fords over 50 years.

In 1970 I acquired my first Ford rally jacket. The jacket shown in this 2023 'selfie' dates back to a 1997 Swedish rally spectating trip. I've kept it, despite the fact that it is several sizes too big – and very warm in the British climate, rather than in the Swedish winter conditions it was designed for! *Author*

Post-Ford working life

On leaving Ford, I obtained some contracted writing opportunities. A backbone was Italy's *Quattroruote* magazine as UK correspondent from 1986 to 2003, plus work as a regular columnist and feature writer for *Cars & Car Conversions*, and features/road tests/track tests for *Motor Sport* and *Motoring News*. The latter supplied my earliest Capris, subsequent XR3s and two four-door Cosworths. A divorce and second marriage propelled a move to life as a full freelancer or, as my colleagues dubbed me, 'media tart'.

Besides outlets in Europe and North America, there were occasional British national newspaper breaks, such as *The Guardian* for the aforementioned Jackie Stewart story. The *Daily Express* supported Nigel Mansell tales from his Unipart Formula 3 days and Gordon Spice's historic Capri victories, particularly at the last long-circuit Spa 24 Hours. *The Observer* would take items such as RAC Rally previews that featured 'Batman', as the British media dubbed Ari Vatanen in his Ford Escort era.

There were still one-off Ford jobs from loyal former colleagues, especially GT40 driving jobs (see Chapter 5). Incidentally, when the only Cosworth four-cylinder GT70 – the example that originally contested French events in BP colours – emerged from decades in hibernation under a drab matt-black paint scheme and disfigured with a slave Escort/Fiesta-era CVH engine, it proved a lucky break for me – and some hands-on sub-contractors – as it was destined to be reborn rather publicly in its original colours. Thus, financially rewarding low-profile background press-release drafts, or forensic research into the contorted tale of that GT70 in the run up to the Goodwood Festival of Speed, were welcome to both my bank account and inflated post-divorce mortgage payments. I still did occasional TV car ads, but stepped outside Ford for presenter/reporter roles. TV work came via Sky News (including Ferrari's 50th anniversary, borrowing a Henley neighbour's 355), and Meridian (the southern ITV channel) embracing varied stories from Golf tyre tests to a televised Vauxhall Vectra one-make Thruxton crashfest as a driver/reporter. I was also involved in business films for BMW dealers and others.

However, as the '90s closed and a new millennium approached, I had the reassurance of regular American work from three book publishers and *Roundel*, the BMW Car Club of North America magazine. Additionally, US and French (*Auto Hebdo*) one-off jobs arrived when my backside kissed various European high-performance or track cars.

However, as the title suggests, this book focuses on my experiences with fast Fords, before, during and after my time on the Blue Oval payroll.

AUTHOR'S ACKNOWLEDGEMENTS AND THANKS: 1967–2022

Without these folk, mentioned in chaotic chronological order, this book – and a significant part of my 54 years scribbling words about Fords – wouldn't have happened. Their guidance, or access to Fords from showroom to silly-fast, was invaluable, unlocking contacts, experiences and quotes.

Some died during the long span that I have been privileged to earn a living from Ford tales. That makes it doubly important for me to acknowledge my debts, especially the married-life self-employment support of my first wife, Patricia Ann Rolfe, who died in 2019.

I would like to acknowledge the help of the following people. *(*Deceased signified by an asterisk)*

Martyn Watkins: Editor *Cars & Car Conversions*. My mentor who facilitated generous access to many Ford steering wheels, including 1968–69 street and competition Escort Twin Cams. ***Harry Calton:** Ford of Britain PR, then Ford-era Aston Martin. Simply the best. ***Bill Meade:** Ford Competitions Centre, Boreham, Essex staff. Enabled and engineered so much. ***Mick Jones:** Ford Competitions Centre, Boreham, Essex staff. Unmatched knowledge. ***Peter Ashcroft:** Ford Competitions Centre, Boreham, Essex staff. Engine builder, became Competitions Manager. Insider knowledge and my factory access, from Capris to C100. ***Stuart McCrudden:** Ford Competitions Centre, Boreham, Essex staff. My first Ford Boreham guide, and always a formidable Ford 'fixit' presence, from media to motorsports, from Fiestas to Formula 1 retro. ***Janos Odor:** Founder/owner of Janspeed, Salisbury, Wiltshire. Enabled 1972 Ford race season, and Sierra Cosworth 1988 race share with son Kieth. Reunited May 2022. ***Keith Greene:** Racing and team-manager legend. Discount Capri components at Broadspeed, plus access to Alan Mann, Gordon Spice, Chris Craft. ***Roger Willis:** Motorsport face of Castrol. Book and Escort Sport sponsors, plus my first Capri. **Paul Harrison:** Ford of Europe. Last Capri factory access and support into the 21st century. **Ivan Bartholomeusz:** Ford Heritage UK. Delivered/rescued Lotus Cortina FGF 113C. ***Alf Belsen and Paul Wilson:** Formerly of Ford Press Garage, Brentford, Middlesex. Paul also advised on Sierra Cosworth race-track running repairs; enabled track access to RS2000 MkV at Brands Hatch and 2013 Fiesta ST, Llandow. **Andrew Marriott:** TV sports reporter and fount of motorsport knowledge, formerly my *Motoring News*/*Motor Sport* mentor. Enabled Roger Williamson's Anglia test. ***Nick Brittan:** Loved his saloon car racing book, heeded realistic advice. **Jochen Neerpasch:** Ford Cologne Competitions Manager pre-BMW Motorsport. **Mike Kranefuss:** Jochen Neerpasch's successor. ***Gerry Birrell:** Ford Competitions and Ford Advanced Vehicles-contracted driver. ***Tom Walkinshaw:** Gerry Birrell's successor. Class winner in Capri XWC 713L; founder TWR. **John Fitzpatrick:** Broadspeed and Ford-contracted driver, also supplied rare Anglia European knowledge via book *Fitz: My Life at the Wheel*. ***Chris Craft:** Broadspeed and Ford-contracted driver. Provided my first race-car passenger ride (Escort 1300 GT) and let me race his 3-litre Capri. One of nature's funniest gentlemen! ***John Miles:** Lotus legend, *Autocar* word-star. Inspiration via 1968 Brands race school to Nürburgring Capri 2.8i and Capri 3.2-litre. **David Brodie and *Tony Pond:** Race and rally heroes, multiple encounters and tuition, Brodie contacts helped with this book. **Barry Reynolds:** Ford of Britain/consultant. Enabled my remunerative 'Precision Driving' TV advertisements and more. **Charles Reynolds:** Former Ford at Boreham administration manager. Shared RS3100 on 1973 RAC Rally coverage, and assorted adventures. Cool insights into Fords on events. **Dave Cook/*Peter Clark:** CC Racing, Kirbymoorside. Access to supreme race Capri lore. ***Mike Smith:** BBC DJ and presenter. Allowed access to his race Capri II (1975) and Snetterton 24 Hours-winning Escort RS Turbo. **Thomas Ammerschläger:** Ford of Germany engineer. Zakspeed/Köln Capri insights. **John Young:** SuperSpeed, Ilford. Access/insights, V8 Capri and Colin Hawker DFVW. **Mike Hodges:** Henley, Oxfordshire. Enabled USA access and UK BF Goodrich sponsorship. **Jesse Crosse:** Editor at *Performance Car*, keeper/restorer of most original Sierra RS Cosworth (ex-press) road car D990 PVW. Also loaned me his outstanding Escort Mexico, now sold. **Peter Osborne:** Thame, Oxfordshire, business partner most of adult working life – 1968-to date. ***John Haynes (publishing) and *Clive ('CR') Richardson (colleague and friend):** Jointly enabled first JW book. 'CR' shared overseas forays, a six-hour race in an MoT failure, and a Scottish duck pond, and supplied Cologne Capri and Supervan 1 text/pictures for this book. **Michael Black:** Oxfordshire. Shared adventures, including Spa 24 Hours trips 1972–73. **Alastair Mayne:** A source of wisdom at Graham Goode Racing, Leicester, from RS1600 to Focus. Boss Graham did it all the Ford hard way too, from Anglia to Sierra. **Sue Collins:** Collins Performance Engineering. Enabled 1991 track thrills in Sierra Cosworth RS500 and RS200. **Norm Murdoch:** Team Blitz, Ohio USA. Amazing knowledge and support for Capri in America, long after it was dropped from the Ford/Mercury network. **Dave Lampitt:** Engineer, Gloucestershire. Enabled 2021 reunion with XWC 713L, 'my' Spa 24 Hours Capri. **Matthew Carter:** Former Editor of *Classic & Sportscar*/*Autocar*. Logical knowledge and retrieval of Cortina Savage published material. **Allen Patch:** XR Owners Club. Outstanding knowledge/provision of XR4i/XR links to past Ford advertising-agency personnel and Escort front-drive data.

CHAPTER 1
A POPULAR CAREER MOVE

How Ford became part of my working life and prompted this book

After a two-year NUJ (National Union of Journalists) apprenticeship, starting in 1965, my desperate goal to become a motoring journalist looked somewhat distant when I began work as a sub-editor on agricultural magazines belonging to a major publishing group. This behemoth also owned *Motor* and *Autocar*, but these weekly motoring magazines only employed people with more advanced educational credits to their names, preferably with an engineering link to the automotive industry.

INTO THE WORLD OF MOTORING JOURNALISM

Yet my two-year NUJ apprenticeship did prove a vital qualification when, aged 21, I landed a job at a pioneering monthly magazine that wanted both youth and recognised journalistic experience.

This was a rapidly transforming flagship magazine that had recently changed name and content from dull *Cars Illustrated* to clumsy but accurate *Cars & Car Conversions*, sometimes rudely nicknamed *Cars & Car Perversions*, but commonly shortened to just *CCC* or *Triple C*. It appealed to a younger audience busily spending their booming Swinging Sixties wages on transforming affordable mass-production saloons into noisier – and occasionally faster – vehicles with a personal touch. King of that heap, and the key to my new employer's rapidly accelerating circulation, was the original Mini.

There were masses of speed shops with 'go-faster' kits on offer, which meant that *CCC* swiftly became a prosperous haven for the expanding tuning business. The magazine even opened its own short-lived speed shop, West Five Speed Centre in Ealing, handy for the Queen Victoria, a large pub across the street.

As it happened, I owned a basic 848cc Mini and had bought the magazine regularly. That was fortunate, for *CCC* had a complete sub-culture of editorial and automotive language that required

an immediate handbrake turn from my formal NUJ writing style and the no-nonsense reportage and clarity that the agricultural magazines had demanded. *CCC* spoke effectively to people the established motoring media disdained.

CCC was owned by Ken Gregory and Alfred Moss, two men whose long-time connection had begun when Ken – a predecessor to Bernie Ecclestone in his business-dealing prowess – managed the career of Alfred's son, Stirling. While the parent company, Speed & Sports Publications Ltd, was run by grown-ups, *CCC*'s staff were young tearaways – mostly under 25 – with an uninhibited attitude to speed in the dawning era of the blanket 70mph limit and breathalysers. There were frequent parties featuring flamboyant females, great pop music, weak alcohol and, ooh, shared roll-ups that packed more mind-bending additives than contemporary king-size filter fags.

Lunchtime business was often conducted over a beer or three. In fact the most helpful writing advice I received came from MD Julian Berrisford at that handy Queen Victoria bar: 'Write as if you're talking to somebody your own age in a pub.' The writing style in *CCC* would be almost incomprehensible today, a strange amalgam of public-school slang and shorthand motorsport and car-component references that were pretty obscure even then. Certainly, I later had to turn my typewriter through 180° to write for the supremely establishment *Motor Sport*, which was then Britain's best-selling motoring magazine. I also learned a few harsh journalistic lessons at *Motoring News* with the requirement to deliver legible full-length event reports overnight under the editorial guidance of Michael 'M.R.G.' Greasley.

Initially, the *CCC* job focused on my NUJ-trained production skills, and I was assigned to producing spin-off tuning manuals. These skinny but uniquely informative books were originally created on inky, hand-cranked sheets, within a tiny room adjacent to the only lavatory in the building. These *CCC*-branded titles flourished, and by the early '70s there were over 20 of them, mainly covering mass-market Minis and Fords. Produced with the assistance of major tuning companies that had motorsport contracts with

▲ This view of Berpop at Brands Hatch in 1972 helps to demonstrate why this upright '50s survivor became a folk legend as it waged war with Minis and Anglias. *Jeff Bloxham*

▲ The February 1969 edition of *Cars & Car Conversions* featured my track test of Berpop – and showed an exciting new Ford on the front cover.

car manufacturers, these manuals catalogued benchmark names such as Downton and Janspeed (Mini), and SuperSpeed and Broadspeed (Ford).

Months passed before I was allowed near a road test car let alone the kind of track outings that became a major part of my working life, but gradually the fair-minded treatment of Editor Martyn Watkins, and the expansion of the magazine, allowed this long-haired youth to handle the hardware. Initially, I shared and then drove readers' cars, a terrifying experience all round. Particularly popular with the people who paid my £1,040 a year salary were regular visits to as many speed shops as possible, generating not only editorial pages but extra advertising revenue too.

CCC had been prescient in taking a regular 'partwork' series from Mini racer Clive Trickey, who not only won on track but could also recount the technical stories behind his successes, generating best-selling editions of the *Tuning a Mini* books. Fortunately for me, it became apparent that Ford-aligned readers and advertisers – as well as Ford of Britain – felt under-represented in this frenzy of Mini coverage. Even for me, a Mini owner who tuned his bumperless example to various hilariously explosive stages, it felt like the Ford contingent had a point.

I managed to capture a reader's 105E Anglia with the almost mandatory 1,650cc/ front disc brake tuning recipe from SuperSpeed in Ilford, but more about reworked Anglias later. For the magazine and me, the Blue Oval turning point came when a BMC press representative refused to allow access to the competitions department in Abingdon for a proposed colour feature. Ford weren't so shy – and had a lot more money to spend.

Dearborn dollars pushed Ford of Britain to chase hard after the apparently impossible ambition to displace the home-grown British Motor Corporation (BMC) from the top of the British sales charts. That meant utilising the halo effect of every performance and motorsport avenue that Ford could explore. From Cosworth-Ford V8s at the pinnacle of Formula 1 to an extensive list of competition components for its saloon cars, Ford was hot to trot out its performance programmes to the media. Such performance parts were sold from Boreham, the company's competitions centre in Essex.

For us at *CCC*, the results were immediate and spectacular. Ford of Britain's cooperation provided 'access all areas' at Boreham, and that open attitude swept in comparative newcomers like me. Ford in many forms sustained its welcome to me for 50 years, making this book possible.

Throughout my writing career, I have tried to deliver articles that put readers in my privileged driving seat rather than trying to set lap records or obtain a race outing in a test car. It happened that I was offered track test cars to race, but it was a side issue, and usually because I hadn't crashed, or the owner needed publicity.

POPULAR SENSATION

Based on the year of manufacture, I begin my earliest Fast Fords tale with one of the most unlikely vehicles in this book. One that has nothing to do with official Ford links and everything to do with a resourceful and generous individual, a king amongst competitive budget racers utilising durable, low-cost components.

The starting point for ingenious Mike Berman of Leicestershire in the period 1964–78 was the cheapest Ford sold in Britain in the '50s, one that traced its origins back to pre-war years. The Ford Popular 103E roamed British roads in big numbers, as the company's Dagenham factory pumped out 155,340 of these narrow, upright 'Perpendicular Pops' in the years 1953–59. I have played the pedant and provided the 103E suffix to avoid any confusion with later models called Popular. Ford's frequent habit of giving completely different designs the same name meant that Popular remained Ford's British shorthand for cheap, stripped-down offerings, such as some pretty desperate '70s Escort derivatives that fought for economy-conscious customers during slack times.

Under the skin: a roadgoing racer

With a narrow girth (56.5in) allied to a tallish roofline (64.5in), a 'sit-up-and-beg' Popular looked an unlikely basis for a racing project. Yet the toughness and affordability of many components – such as the 1,172cc side-valve engine – proved seductive to impecunious club racers and constructors. A low kerb weight ensured that many were given the hot-rod/drag-strip treatment with bigger engines and American custom styling cues.

Showroom Ford 'Pops' of the '50s proffered just 30bhp to propel a skinny 1,624lb (737kg) to a giddy 60mph maximum. They cost under £400 (a little over £10,000 today) and that modest price told punters not to expect standard items like a heater or more than three forward gears, with synchromesh only on top. Seating was resolutely vinyl-covered, bumpers were painted rather than chromed, and windscreen wipers were vacuum-operated (slower wipes the harder you pressed the accelerator). Thanks to pallid six-volt electrics, undernourished sparks could

▼ The original Berpop on racing duty during 1968 at Mallory Park, probably with owner Mike Berman at the wheel. *Motorsport Images*

▲ How Berpop looked in early 1969, at about the time I drove it, in 1,650cc/130bhp guise and complete with wildly extended steel wheels. *Jeff Bloxham*

day in my track apprenticeship. He offered experienced professional opinions on the cars, which varied from 1-litre sports racing Costin-Nathan to a substantial American 4.7-litre Ford Falcon. Steve did it all with a light touch, even when I disappeared in a cloud of rubber smoke, rotating the Martin Birrane-owned Falcon at Copse.

But Berpop was the star for *CCC* as it had such effective home-built features, all for total expenditure that Mike reckoned at only '£300 excluding labour'.

The key oily bits featured at that Silverstone encounter played supporting roles to that extraordinary appearance. That was owed to the glass-fibre front (bonded to radical aluminium wheelarch extensions), a complete assembly that lifted off for excellent access to the motor and suspension. In those comparatively relaxed days, the fabric sunroof panel remained, but the skimpy alloy-sheet floorpan was reinforced out of sight by light-gauge steel braces.

The other extreme modification to the Popular outline was a massive tweak to the girth, from 45in to 64in. Far fatter steel wheels carried treaded race tyres, the 8.5in rears and 6.5in fronts exhibiting massive rim offsets. Somehow, the plot was steered by an assortment of Standard-Triumph parts, as often featured in unique kit cars and earlier Lotus production.

A notable 2½° of negative camber was applied to the front suspension, adjusted via spacers on the bottom wishbones. That front assembly featured double wishbone A-arms and an inclined coil spring/telescopic Armstrong shock absorber, with nothing left of the original transverse-leaf layout as Mike designed and fabricated the surrounding mounts.

The Morris Minor live rear axle carried a locked differential to deliver abrupt traction instead of paying for a more progressive limited-slip differential, but had the supreme racing plus point of offering diverse final-drive ratios (10 in total) to suit British tracks ranging from twisty Brands Hatch to high-speed Silverstone. On the day of my test, the car had a Brands-type 4.2:1 final-drive, which meant that it ran out of revs at little over

demand cold-start assistance from human exertion on a starting handle.

Enough of the rewinds to post-war austerity Britain. What set Mike Berman and his renditions on a Popular base to becoming a club racing cult hero?

Mike, who nowadays runs a theatrical costume company, is an engineer of diverse talents. He built the aptly named Berpop – an abbreviated combination of his surname and 'pop' – because he could, simply 'for fun' according to friend and fellow competitor Gerry Taylor. As Gerry recalled in June 2011: 'Berpop began life as a road smoker and we used to drive down to Silverstone in it from Mike's garage in Oadby.' When, in 1964, the police objected to its hearty noise levels on public roads, Mike started to compete with JBK 280 off the public highways.

By the time I drove Berpop, for *CCC* at a dry but breezy Silverstone on Thursday 21 November 1968, it was a club-racing legend. The high roof and proudly pre-war outline of Ford's most basic economy car became notably visible amongst hordes of boxy Minis and Imps, never mind lowered Anglia and early Escort racers.

My first major track test shaped up with, according to *CCC*'s billing, 'four of the cars that entertained us and many other watchers in '68'. I was relieved that established factory Mini driver Steve Neal (owner of 100+ Wheels and father of BTCC star Matt) arrived to oversee this significant

BERPOP ENTERS THE DRAG SCENE

A few weeks before our track test, Mike Berman decided to do something very different with Berpop that involved no cornering capers at all. Together with a few other intrepid circuit racers from the special saloons category, they entered their track hardware at Santa Pod, the best-known British drag-racing venue.

Other circuit contenders included Mick Hill in 'Janglia', the Jaguar-powered Anglia that preceded his famous Boss Capri V8s, and Mike Bennion with his 4.7-litre Morris Minor V8. All were lumped into the Junior Competition Eliminator category featuring some drag-racing regulars.

Berpop was handled very competently by Nick Lees throughout a day of pitting his wits not just against many more powerfully propelled opponents but also the tricky 'Christmas Tree' of sequential start lights for the standing quarter mile. Astonishingly, Berpop with its 130bhp Berman-tweaked Cortina engine put all those rivals away! The final official shoot-out was Bennion's V8 versus our perpendicular Popular hero, Lees recording 15.23sec/90.25mph to eclipse the bestial Morris Minor's 15.64sec/92.25mph.

Bottom: Another Santa Pod visitor, from the circuits this time, in the form of Nick Lees in the 1648 Ford "Berpop". Took away Junior Competition Title as a souvenir

SANTA POD DRAG

THE FINAL drag meeting at the Santa Pod Strip on December 1 was a great success and Harold Bull (997 cc Dragster Stripduster) made a tremendous impact with four very hairy runs at over 130 mph, culminating in a fantastic top dragster final when he went everywhere but straight to record a shattering time of 10.1s. (132.1 mph). All this was achieved with a blown head gasket which was hastily repaired with a bit of "gum" between runs.

Runner-up to Bull was Alan Blount whose Chevy powered Weekend Warrior was handicapped with only five pistons functioning but managed a creditable final effort at 11.4s. John Whitmore (Obsession) took the Middle Drag title with a run at 13.81s., this beating Pete Wilson's Martini which travelled all the way from Guernsey to establish a quicker but losing time of 13.58s. (in fact 8 mph faster). Top Competition eliminator was won by Dave Collis in his newly purchased Chevy powered Opas 1 scoring with a time of 13.4s. to the Dickson Marshall Racing Team's Good Vibrations at 13.7s. The Middle Competition bracket was an all circuit racers affair with Nick Lees taking the money in the incredible but under-geared Ford Berpop at 15.23s. Runner-up was the 5.1-litre Chevy engined Morris Minor 1000 of Mike Bennion at 15.64s. In the Top Street section, the Dodge Charger of Doug Harler had an easy win over Pete Shelton's Shell-tuned Cortina which bobbed on the line. Reg Clark had no trouble in the Junior Street runs and he easy blew off Fred Bolden's Anglia with a time of 18.43s in his 3.8-litre Jaguar. Harold Bull's shattering time of 10.1s. (132.1 mph) establishes a new record for the quickest and fastest sub-1000 cc engined machine in the world. Just a quick note to tell you that plans are well under way for the BDR and HRA to present the first full-scale national dragster exhibition on January 25-26 at the Fairfield Halls, Croydon, London. It will be called "Drag Racing '69."

We had a full field of eight cars for the Junior Competition eliminator, and only one of these was a regular competitor—Geoff Pearson with his "Mini-Hater"—this went down under a first round blast from Peter Yorke (modified Triumph Spitfire) who'd driven up from Guernsey (with Pete Wilson's "Martini" on the trailer) all to no avail though for he went down under the charge of one of the visiting circuit racers—Mike Bennion in his 292 Ford powered Morris Minor—this is one of the lowest and meanest looking machines ever to hit the strip. One of our other visitors, Mick Hill in his Jaguar powered '58 Anglia "Janglia" put down PBTS man John Tarbox who'd borrowed a 1498 Anglia for the last meet of the season, only to be shown the way home by fellow circuit racer Nick Lees in Mike Berman's wild 1648 Ford "Berpop" — Bob Fisher's Ford Falcon special also fell under the "Berpop" spell, and "would be custom king" Simon "the Pieman" Eldridge met first round defeat at the hands of the movin' Minor of Mike Bennion.

For the final it was an all circuit match, with the "Berpop" getting the lead out of the chute and holding it for the win at 15.23 and 90.25 miles an hour over Mike Bennion's trailing 15.64 and 92.25 miles an hour.

A very close match race was later held between "Berpop" and "Janglia" with the Ford powered machine losing only once on the last run via the red light — these kind've machines are more than welcome at the strip, and with a bit of practice on "reading the lights" and "straight line shifting" should prove a mean match for the best of our quarter-milers.

Cuttings from *Drag Racing & Hot Rod* and *Motoring News* tell of Berpop's unlikely late-1968 success within an invitation quarter-mile challenge at Santa Pod. Nick Lees, one of many Berpop guest drivers, beat several fancied competitors, including Mick 'Capri V8' Hill in his 'Janglia'. The final run saw Berpop heroically defeat a 5.1-litre V8 Morris Minor with a 15.23sec/90.25mph performance. *Courtesy of Mike Berman*

100mph on the Grand Prix circuit. The gearbox was an amiable all-synchromesh four-speed from a Ford source, equipped with close ratios (first and second were always a chasm apart on period street Fords) and hitched to a propshaft that Berman manufactured to his specifications, including balancing.

Berpop's butt was suspended on coil springs of Cortina Lotus rating with adjustable dampers, restrained via top and bottom radius arms. Talking of restraints, Berpop at that time featured simple front disc/rear drum braking, Mike having abandoned a previous graft of Cortina Lotus A-frame and discs.

That a pre-war design could compete with front-running Minis and Anglias was down to low weight (1,456lb/660kg) and a 1,650cc stretch of the non-crossflow Cortina 1500 GT motor, a simple overhead-valve unit that pumped out 100bhp more than the original side-valve as it spun raucously to 7,500rpm. The period power boost came from a brace of Weber 40 DCOE carburettors on tubular manifolding and a hidden Cosworth A6 camshaft profile. Mike did most of the design and labour in all his steps from road use to ever more race-biased Berpop guise.

Even the intensive hands-on work needed to turn a mass-manufactured cylinder head of the '60s into a deeper-breathing competition unit was Mike's handwork. That meant ensuring efficient gas flow through precise matching and polishing of ports, plus installing larger inlet and exhaust valves, actuated by lighter and strengthened valve gear, to the precision trials of cylinder head depth shaves, achieving a final compression ratio of 11:1.

The result was a credit to Mike, creating a competitive track animal that also covered thousands of public-road miles in the 130bhp trim I experienced.

'IT BECAME MY TURN. SURE, I HAD COMPETED THE PREVIOUS SEASON IN AUTOCROSS AND RALLYCROSS, BUT I WAS A NOVICE WITH MY GRADUATION TO REGULAR RACING STILL A WHILE AWAY.'

Behind the wheel: steering a cult club racer

Even by the ingeniously resourceful standards of British club racing over 50 years ago, Mike Berman's home-brewed Berpop literally stood out from the crowd. Despite the radically lower and wider stance, this Popular – with its many mechanical transplants and implants – towered over contemporary Mini and Anglia rivals.

As the lightweight body cladding quivered and the durable motor clambered the slopes to a reverberating 7,500rpm limit, the opposition were frequently shocked into submission. Yes, even future Grand Prix challenger Roger Williamson and his mighty Anglia found the shaky Berpop apparition occupying all available rear-view mirror space and on one occasion, at Mallory Park home turf for both protagonists, Berman was the winner.

'By gawd, what an experience,' were professional Steve Neal's opening words at our Silverstone test day. Releasing his six-foot frame from Berpop's embrace, Steve expounded: 'The car was fairly safe, although it doesn't look it. It steers well and it gets round the corners well with some understeer: there was enough power to change this to tail out. After a couple of laps it felt OK, but I still wasn't keen on the restricted vision or the brakes, which seemed lousy with too much travel and no real stopping power.'

That was the esteemed opinion of a skilled and experienced racer, one used to peak performance from every aspect of his race cars. Steve was then driving on and over the limit in a factory-backed Britax Mini Cooper S against other works Minis and Fords as well as a very ambitious and talented team-mate in Gordon Spice.

It became my turn. Sure, I had competed the previous season in autocross and rallycross, but I was a novice with my graduation to regular racing still a while away. So my Silverstone knowledge at that time was limited to test days in road cars. Racing Berpop back then made strong demands of the many drivers Mike allowed into that unique cockpit. That November day we had a then regular Berpop pilot, John Wales, along for educational company and he said, 'It takes me a couple of laps to acclimatise!'

The flyweight glass-fibre driver's door, with

SILVERSTONE TEST DAY

By J. Walton Driving: Steve Neal Photos: Max Le Grand.

MIKE BERMAN'S 1650cc Berpop

THAT LITTLE DEUCE COUPE has got nothing on this club dicer, created by Mike Berman, driven by John Wales and sired by a common Ford Pop. Yes folks this is the one you've been waiting for, a track test of the supreme example of British ingenuity . . . may we present your friend and our's, Berpop. But before we launch into the turgid detail we have a special message for "Romper Noad." Peter, we are going to tell you exactly how it is, where it's at and generally kick it around, because now everyone can own a Berpop. It's the truth fella, M. Berman's firm are going to sell replicas, slightly detuned for road use. Think of that; wind up the elastic in all the gears and blow your mind at over the ton in a Pop! What's more it's got roadholding and front disc brakes to control this un-pop-like urge. Mind you there are still snags to the original which put-off even the bravest wheelmen; like a lack of visibility and poor brakes, by competition standards.

So let's tell you what has happened to bring about the transformation. Mike Berman of 5 The Morwoods, Oadby, Leics., has had the car for five years, during which time he's constantly modified it. He used it as a road car until the police objected to the hearty noise level and hounded him off the public highway. Cost up to the time of our Silverstone track test was £300, excluding labour.

The engine is giving out something like 130 b.h.p. at present from its 1650 c.c. That capacity should give you a clue as to whose engine they used as a starting point. Yep, Mr. Ford's 1500 GT Cortina lump, but there are other changes now, apart from the capacity. Like the Lotus crank, Cosworth A 6 cam and twin sidedraught 40 DCOEs. Mike gained a lot of power by headwork too, fitting larger valves: $1\frac{5}{16}$ ins. inlet and $1\frac{3}{16}$ ins. exhaust, and scraping away at the face until a compression ratio of 11.1, or so, was present. The rev limit of this big four is 7,500 r.p.m., partly made possible by the lightened valve gear. As the engine has done many thousands of road and race miles without trouble, this seems to be a good limit.

The gearbox is a close ratio, all-synchro Ford unit; from there the power goes via a Berman manufactured prop shaft that is properly balanced. The rear axle is from a Minor 1000; for the test it had a Brands final drive of 4.2. The great advantage of using a BMC axle is that you can choose from any of 10 final drives. The axle incorporates a locked differential, mainly because he can't afford a limited slip.

In the cornering department they've really gone to town, widening the rear track from 3 ft. 9 ins. to 5 ft. 4 ins.! Steel wide wheels were fitted with $8\frac{1}{2}$ ins. rims at the rear and $6\frac{1}{2}$ ins. at the front. Also up front there is double wishbone suspension with coil springs and adjustable Armstrongs, mounted on a Berman-designed front chassis section that took the place of the original. The floorpan is reinforced underneath by light gauge steel — a wise precaution when you see that the floor is all alloy panel. Rear suspension consists of top and bottom radius arms, with the axle damped by adjustable shockers and sprung by Mk 1 Lotus-Cortina coils. Mike did try A bracket Lotus suspension and discs on the rear earlier last season, but both of these had been dropped by the time we drove it. However, it still has disc brakes at the front and an adjustable ratio of front to rear stopping effort. The steering parts are from Standard-Triumph, though whether you'd be able to tell is a matter for debate. Another point about the steering is that camber angles can be changed by fitting or removing spacers from the bottom wishbones; it had $2\text{-}2\frac{1}{2}$ degrees of negative for our trial.

A lot of weight saving tricks are used to bring the car down from the standard $14\frac{1}{2}$ cwt. to its present 13 cwt. Apart from rust and the sunroof they've fitted a complete glassfibre bonnet/wings effort. The wing extensions are aluminium, as is the bootlid.

Sitting pretty in your fireside chair? Think yourself lucky fella, as we describe in mind-bending detail our laps of Silverstone GP circuit.

Neal having just returned from his stint, we climb in and take a pew. We find the high set and sturdy seat reassuring, but not the tiny windows and stark Pop steering wheel with its spindly column. Belted and fired up, we begin to wonder why everyone bar John Wales is grinning like the keeper is coming to hand round the cookies. Steve and John both warn the reporter to take it easy and watch out for bumps, while the others just lick their lips and thirst for action. The clutch has a long travel and we ease it out gently while the engine spits its disapproval at 4,000 r.p.m., more spluttering 'til it clears at 5,000 and we emerge for our first ever trip round the GP circuit. Playing chicken we stay in third and potter round looking for some clues to the way ahead. We very quickly find it's got more roadholding than can ever be possible in a Pop. After a few laps we speed up a little; on the bumpy exit to Club Corner it does bounce off line a little with the arse-end sliding slowly out. Inside, we're assaulted by a cacophony of sound as the motor winds up to 7,500 in fourth and all the body parts moan in sympathy. Still, you're doing over the ton down there and that's something that's never to be forgotten in a car designed to struggle over 60 m.p.h. During our first session we run out of petrol after four or five laps, so that when it's time for our second trembling session we've had plenty of time to watch John Wales at work in the Pop. As the car is geared for Brands we weren't astonished to find that John had been backing off in three places on the GP circuit because the engine has run out of revs in top.

Steve Neal's candid comments

Steve's comments on the car were forthright and to the point: As we watched him disentangle himself from the harness he managed a brave smile with the words, "By gawd, what an experience." Later on after he'd had time to recover we asked for a few more words on the car, after much hilarity this what he thought: "It steers well and gets round the corners well with some understeer; there's enough power to change this to tail out. The car is fairly safe, although it doesn't look it. Suspension and gearing were set up for Brands and I found the settings too harsh" (John Wales and Nick Lees, their faithful 18-year-old mechanic, slackened the dampers off for our runs). Neal frowns for a bit and then goes on, "after a couple of laps it feels OK but I still wasn't keen on the restricted vision or the brakes, which seemed lousy with too much travel and no real stopping power. There's no body whip during hard cornering and the car could cope with an extra 20 b.h.p. Engine noise and body resonance are almost painful, there's plenty of wind noise too." Summing it up he said, "Mike Berman's ideas are good, it's certainly a very different competition car."

245

▲ At far left in the big photo, the apprentice observes as Mini Cooper S British Saloon Car Championship contender Steve Neal prepares for a run in Berpop. The first track test of my writing career started a few laps later, as pictured at bottom right.

no visible handle and simple tube braces within, opened a bit like that of a Frogeye Sprite: fumble at the corner of the Perspex side 'glass' and you find a lever to release the internal lock. Inside you could use the estate agent's euphemism of 'plenty of character features'. For modern racers who compete with features like HANS neck braces, all-enveloping seat side sections and elaborate safety harnesses, the truth would be closer to 'brutal bodging'.

Back in 1968, I was surrounded by stark aluminium sheeting to cover the gearbox and propshaft. Both gearbox and simple period race seat were set well back, the seat at least a foot behind the centre pillar, which carried a roll-cage hoop with rearward support bars. A quartet of switches and a kill button for the ignition were sensibly grouped on the centre panel, ahead of the wooden-top four-speed gear lever.

Sticking out a long way from the original front bulkhead, multiple steel tubes supported a sporty leather-rim steering wheel. The resulting rearward seat location, barely above the equally basic floor, was an extremely clever engineering prediction, albeit in conjunction with an engine conventionally mounted ahead of the production bulkhead. It reflected an overwhelming need to get human ballast as far back and as low as possible. Vision was less of a priority!

All-round vision, or a lack of it, was the first aspect that struck this track-testing newcomer. The view forward through the virtually flat screen was fine, with bonnet poking out ahead, but sight lines to the side and rearwards were naturally constricted by the ancient outline of a pre-war design. There was a driver's door mirror and a central interior mirror, but the latter just monitored the minimalist period rear screen. None of this was a problem on acres of Silverstone Grand Prix circuit lacking the intense traffic of modern track days but it could have been tricky during a race.

On the plus side, the simple bucket seat felt reassuring with built-up sides and rigid block mounting complementing the embrace of a four-point safety harness. A little less comforting, the steering column and wheel were 'spindly' by 1968 standards and downright skeletal compared with today's beefy cabin kit.

Steve Neal had warmed Berpop thoroughly, so the loyal motor fired readily with barely a tickle of throttle. When I peeked out of the circular hole in the driver's door Perspex in my open-face helmet, I saw a lot of grinning faces, most reminding me of zookeepers waiting to feed the bears fresh meat, but there were also serious expressions on the faces of Neal and Wales, strenuously advising caution.

A comparatively gentle competition clutch with long travel allowed this stranger to balance forward motion against the need for 4,000rpm minimum. Too few revs and even a motor with previous public-road form spat back in protest. A few more splutters and we were left with a clear 1,650cc throat ready to flick up to 7,500rpm, once we had got over the novelty of my first laps around the full 2.9-mile circuit (hired at £7.50 for our four test cars). Yes, I was tackling the layout featuring flat-out 160mph Woodcote, where men like Jochen Rindt demonstrated the difference between regular racers and racing gods.

That first session was as much about learning the track layout as it was concerned with educating myself in Berpop lore, as I had such limited track experience. Positively, I liked the amiable four-speed gear-change (early five-speed Ford gearboxes never matched these smooth operators) and loved the sensations of speed relative to my previous performance benchmark, my modified 650cc Triumph motorcycle. I found the amount of noise within the cabin tolerable – many contemporary competition Minis resonated more.

'I WAS IMPRESSED BY THE FEROCITY WITH WHICH A VEHICLE DESIGNED TO RUN NO MORE THAN 60MPH FLAPPED ITS WAY BEYOND 100MPH FOR THREE DRIVERS THAT DAY. SO IT WASN'T TOO MUCH OF A SURPRISE WHEN IT RAN OUT OF PETROL.'

As motorcycles were then part of my perspective, I was very impressed with the grip delivered by treaded race tyres and loads of front-wheel negative camber. The only apparent vices to me were heavy loads at the steering wheel, just prior to a distinct tendency to hop over bumpy camber.

Out on the Hangar Straight, the front bobbed up and down dementedly. Here the professionals would run out of revs pretty quickly in top gear and that became a tachometer worry for me after the initial exploratory laps. I was impressed by the ferocity with which a vehicle designed to run no more than 60mph flapped its way beyond 100mph for three drivers that day. So it wasn't too much of a surprise when it ran out of petrol.

A second outing was generously allowed and regular on-event mechanic Nick Lees (then aged just 18) and John Wales recommended and executed a slightly softer damper setting. The slacker damper action protected this novice against the 'bunny hops' Berpop exhibited over bumps. It was a wise precaution as I had gained enough confidence to wonder if it was worth using the brakes at all. Driven harder, steering loads escalated to Mini competition levels before Berpop would unstick.

Now third gear was only needed for the few sharper curves on the circuit, particularly Becketts. Driven that way by a novice, Berpop still averaged over 80mph, which tells you how effective Berman's engineering had become. Possibly it also tells us why I didn't really notice the lack of decent braking. Mike had proved Berpop could stay ahead of Roger Williamson's Anglia (in its earlier format) around the tight Mallory track, which includes a hairpin. So there were retardation forces available, when desperation and close combat with an opponent were in the mix.

I totally enjoyed the Berpop experience, partly because I knew no better: motorcycles and a lot of crude off-road miles autocrossing left me with high bumpy-surface tolerance. Also, a zero appreciation of anything but demonic noise levels!

Next, the heroic tales of the now iconic Ford Anglia in the UK, continental Europe and as an airborne automotive film star!

WHAT HAPPENED TO BERPOP NEXT?

That Silverstone track day did influence Mike Berman's thinking and he went on to construct 11 radically reworked Populars with a variety of engines from basic 100E side-valves to a pedigree race V8. More of that in a minute, but first here's what happened to the original Berpop after those *CCC* features of 1969, told in Mike's words in 2021.

'At the time I was initially disappointed with the first run, but after a bit of tweaking it got better and we went away determined to improve things even more. I then carried on with more modifications especially to the brakes and suspension. Then tweaked the engine for a bit more power to around 160bhp. We continued to race at tracks all over England with many memorable dices with names such as Roger Williamson and Gerry Marshall.'

According to Gerry Taylor, Mike was such an 'easy-going bloke' that he allowed many other drivers to race Berpop, but his patience reached its limit when one of them totalled it. As Mike told me in 2021, 'A friend, Richard de la Rue, asked if he could drive my car at Silverstone and in return I could drive his car, a 750 special called 'Simplicity', in a separate handicap race. His race was first and off he went – but on the fourth lap he lost it and wrote off Berpop's body. I was very tempted to do the same to his.'

Undaunted, Mike began construction of a very special Berpop 2. Here's how he remembered the process, with some pedigree parts installed and significantly uprated bodywork: 'With the bodywork and other components ruined, I decided to build Berpop 2, which was much more of a racing car. I started from scratch with another body that featured glass-fibre for doors, boot and complete front end. I enhanced the chassis with a spaceframe that used March and Lola suspension components. I fitted a Jaguar rear differential and bought a V8 engine from Bruce McLaren's Zerex Special. This was a Tecalemit fuel-injected 4-litre Oldsmobile with about 340bhp.'

▲ Pictured in a publicity photo for Mike Berman's local Leicestershire press, this is the second generation of Berpop strutting its street cred. *Leicestershire Mercury*

▼ Mike Berman with the ever more radical Berpop 2, created after another driver wrote off the original Berpop. This was, in Mike's words, 'much more of a racing car'. *Courtesy of Mike Berman*

As before, Mike shared his Big Boy's toy generously: 'Various other drivers had a go, amongst them John Wales, Bill Cox and Martin Sellicks.' Then Mike sadly concluded Berpop 2's CV: 'In 1978, after having had a fantastic time, I decided to sell it. It was bought by someone whose name I have forgotten. I think it then got sold on to Geoff Kramer, who converted it to road use. I believe it was then stolen, never to be seen again. The theory is that it was broken up and the parts sold off. We'll never know!'

Yet, that wasn't the end of the Berpop saga, as Mike explained: 'I continued to build replicas for customers. They ranged from engine replacements featuring a Ford 100E up to Rover V8s. Altogether I built 11 such modified Populars, one of which even entered the London-to-Sydney Marathon.'

Do any of those Berpops survive? Unlikely in original guise, but if you happen to know differently, please let the publisher know.

CHAPTER 2

ANGLIA

Ford's first UK million seller

I can find justification approaching 65 years on from its debut for dubbing the 1959–67 Ford Anglia (105E) a true 'People's Car', much as lazy writers described Princess Diana as the 'People's Princess'. This fourth edition of the Anglia was always quirky with that notchback rear window and transplanted, finned, Americana topping the rear wings. Back in late-'50s Britain, being born the same year as the legendary and trendy Mini always overshadowed this Anglia, then Ford's UK bestseller.

A sea change in popularising Anglia affection to rival that for the Mini occurred decades after it was a Ford showroom resident. The underestimated Ford baby clasped a wider public to its tinny chest courtesy of J.K. Rowling's *Harry Potter*. The 2002 movie, *Harry Potter and the Chamber of Secrets*, highlighted the workaday Ford's unsuspected magical properties. These included flight at express-train velocities, heroic rescues and an ability to carry more people and payload than a contemporary coach (see 'Harry Potter's airborne Anglia' on page 23).

THE ANGLIA 105E ENTERS THE MARKET

Hardcore drivers who weren't seduced by Mini wizardry, or who had a soft spot for affordable Fords of comparative simplicity and powertrain durability, could see performance and economic virtues in the Anglia's cost-conscious specification. For motor racing and rallying enthusiasts, the introduction of the 105E engine with overhead valves and free-revving amiability meant the Anglia almost immediately attracted attention as an alternative to the more obvious hordes of Minis. This was partly due to Ford dollars permeating global motorsport during the '60s, releasing high-performance hardware that could be transplanted at amateur and professional levels.

Sadly, Anglia bodywork was no durability

match for the tough Ford powertrains. Mass-produced car bodies were notoriously prone to the tinworm's ravages until relatively recently. Ford was no exception to that rule, adding flaky paints for good measure.

I already had some personal '60s experience of the basic Anglia Deluxe, as borrowed from my then father-in-law for a West Sussex weekend, shortly after I married his eldest daughter. My regular Mini 850 was ill after some over-enthusiastic modifications resulted in my hooligan driving breaking the crankshaft, so there were strict family instructions to behave. At 20 years of age I did note that there must be no scars in evidence on the vehicle's return, but still drove it as hard as I could along favourite roads.

I was impressed by the tough engine's comparative (to a Mini) smoothness at maximum gear speeds and the pleasantly rewarding gear-change; a distinct advance over the Mini's elongated 'wand' gear-lever shifts. The cornering wasn't amazing in the Mini sense, but the aged steering system and cart-sprung rear axle did the job well enough to cut a chosen cornering path accurately. I add the caution that, when an Anglia is fitted with wider wheels and grippy fat tyres, it can corner faster, yet there is also a much better chance of tipping it up on two wheels, or the roof, as many a racer has discovered…

So what was implanted within this basic showroom Ford some 60 years ago?

Built in the period 1959–67, this was the fourth British mass-production Ford to use the Anglia badge and arguably the most significant step in the UK Ford division's ambitions to lead its home market and gather some motorsport respect, with over a million manufactured on the Dagenham and Halewood production lines, or supplied for Knock Down (KD) kit assembly overseas. Yet those achievements aren't part of the headline public perception today, for millions of memories are dominated by the Harry Potter Anglia's magical movie stardom.

The Anglia's 1959 showroom appearance was never going to match the cute appeal that the Mini defined. The lines exhibited a gawky mix of Americana (finned rear wings and that reverse-slanted rear glass) and European thinking. The

▲ The action for the author's road test of Roger Williamson's Anglia was set against the short track at Mallory Park, with the hairpin section blocked off. The Anglia was never anything less than exhilarating. *Motorsport Images/Andrew Marriott*

➤ The Anglia was an affordable mount for late '60s/early '70s TV Rallycross. This is Trevor Fox looking for a way ahead of Don Godden's Mini at Lydden Hill. Win Percy – later a British Saloon Car Champion and TWR Mazda, Rover and Jaguar driver – regularly featured in Anglia rallycross action. *Peter J. Osborne*

body had glimpsed a wind tunnel and the front end appeared rather more aero friendly than contemporary small cars. That odd aft design allowed useful boot space, but even more capacious van and estate variants were subsequent additions to the line-up. I cannot believe that customers were irresistibly lured and compelled to purchase by the official sales spiel invoking clear rear vision in the rain from the sheltered rear window!

In many respects, particularly in comparison with home-market Mini and Imp rivals, the smallest Ford's engineering was primitive. A leaf-sprung rear axle, MacPherson-strut front suspension and aged recirculating-ball steering were never going to earn media applause. Fortunately for Ford of Britain, however, there was initially none of the forthright media criticism that was to emerge in later decades.

Ford's tardy adoption of an OHV four-cylinder motor and slick four-speed gearbox soon won praise from those who worked powertrains hardest, for the massively short stroke (48mm, with 81mm bore) engine, running elevated 8.9:1 compression, just loved to work its little pistons off at elevated engine revs.

Additionally, Ford ensured that rallying, plus various endurance and record attempts (see 'GP stars endure at Goodwood' on page 33) underlined those strong points. So the most affordable British Ford was unfashionable, but it did set new production records for the UK Company – benchmarks soon overhauled by the first Cortina's subsequent and immediate commercial success.

During the 1959 debut year, Anglia road tests revealed a low kerb weight of 1,624lb (737kg). Such modesty on the scales aided acceptable contemporary performance and fuel economy, all assisted by that aerodynamically conscious body of unbelievably narrow track (46in). Those basic statistics in 1959 were reported by *Motor* magazine, along with fuel consumption of 41mpg in a harsh week of testing, plus a maximum speed averaging 75mph (indicated as a tad over 80mph on the optimistic rectangular speedo). A 0–60mph time below 27sec, coupled to a standing quarter mile completed in some 23sec, were competitive in that market sector over 60 years ago.

The UK price was £610, a bit more than an early Mini – which Ford teardown analysis famously reported as costing more to make than BMC's retail price for the car!

Time for more-agile Anglia tales...

➤ Showtime! The Anglia has recovered from years of neglect, overshadowed by the Escort and Capri at British classic car gatherings. This self-styled 'Angry Box' example, with radically revised engine bay and plum paint finish, shows the standards now achieved. *Author*

HARRY POTTER'S AIRBORNE ANGLIA

For 21st century children, a blue 1959 (some sources say 1962) Anglia Deluxe, with the distinctive rear-window recess, is pure magic. Transformed by film special effects, the flying mechanical star aided the escape of Harry Potter from the suburban horrors of Privet Drive in the 2002 epic *Harry Potter and the Chamber of Secrets*. Mistress of the best-seller lists, J.K. Rowling apparently liked the baby of the Ford range in her teenage years because, 'it reminded her of her days spent with a good friend, Sean. Sean drove a Ford Anglia and the two drove around in it on days when boredom swept over them', according to multiple Potter fan-base sources.

As you'd expect, there are plenty of myths and much outright fake news around the fictional automotive hero, including mentions of multiple registration plates. Certainly the plate reads 7990 TD in its *Harry Potter and the Chamber of Secrets* debut, but you can find other registrations in publicity material and in original Potter books. I like the rear plate COS 207 that appears on some of the model Potter Anglias and elsewhere with a credible explanation: COS translates as Chamber Of Secrets, and 207 is the page number in the book on which the car rescues Harry Potter from suburban hell!

As anyone who has been around or researched major movies will know, cars can be made to do anything cinematically, assuming that the budget and a talented special-effects crew is available. The top man in Potter's case was supervisor John Richardson, and the headline flying act centred on a stripped Anglia body mounted on a stabilised gimbal capable of near 360° rotation. The gimbal unit was attached to a crane, which gave the crew freedom to rotate the Anglia to virtually any angle for the 'flying' sequences.

For other appearances, multiple Anglias (up to 16 according to some sources) were purchased in varying condition. Some of the Potter Anglias were destined to be chopped in half to allow improved camera access.

What happened to the Potter Anglias? Famously, one fully running car got pinched during filming and was subsequently found in a castle (not a Potter edifice). Singer Liam Payne from the internationally successful music group One Direction bought another Potter filming Anglia and apparently paid 'a six-figure sum' for it. You can be sure there are numerous Potter Anglia clones around today, just as there are so many examples of *Only Fools and Horses* Reliant three-wheelers and classic James Bond Astons. Speaking of Bond, there were also James Bond film appearances for the Anglia back in the day. The 1963 Sean Connery epic *Doctor No* has a green example as the static backdrop for the murder of a lead character, and an Anglia also appears in background airport and hotel shots in both *Doctor No* and *Thunderball*. Nerd interval over...

The *Harry Potter* film franchise brought the small English Ford with American style global recognition 43 years after the Anglia was first launched. *Alamy*

THE ANGLIA GOES RACING WITH BROADSPEED

Ford of Britain got behind the Anglia as a factory race and rally car throughout much of the '60s and was rewarded with the 1966 British Saloon Car Championship drivers' title for John Fitzpatrick and Team Broadspeed, along with second place in the Entrants' Trophy. Driver and team had just switched over from the Mini establishment to grasp that useful title, and followed up in 1967 with second place in the drivers' championship, along with notable outings in France and Spain to show what the baby Ford could do.

The Anglia was a popular saloon-car racing choice for international events in its production heyday and continues to entertain within lower-cost classic-racing formulae today. I had significant personal links with Broadspeed in period, so they occupy a healthy portion of prestige race space in this book, but before I get into that, I must pay tribute to the efforts of others with the competition Anglia.

Ford supported its range-starter strongly in international rallying, but the Anglia was naturally overshadowed by bigger-capacity opponents and by the quicker racing saloons in Ford's armoury. The Blue Oval would dominate in the UK and make an impact on overseas touring-car racing with the Cortina Lotus, plus the Galaxie, Falcon and Mustang outright winners from America – massive machines to European eyes. A V8 soundtrack typified British events after Dan Gurney had shown that American litres from Chevrolet (powering his Impala SS) could defeat Jaguar on Goodwood home ground. The exception to the early-'60s rule came when Jim Clark was around, tricycling the Cortina Lotus in which he won the 1964 British tin-top title – a tale told in the next chapter.

'TAKING ELEMENTS OF COSWORTH AND HOLBAY ENGINE EXPERIENCE, KNOWLEDGE OF THE ANGLIA 105E ENGINE GAINED IN FORMULA 3, PLUS THE GROUP 5 RULE FREEDOMS, RALPH BROAD'S ANGLIAS COULD NOW PACK A WINNING 1-LITRE CLASS PUNCH IN BRITAIN AND EUROPE.'

Meanwhile, what to do with the Anglia? Supporting sub-contractors seemed to be the answer.

Pre-1965, Essex-based SuperSpeed, with driver/co-owner Mike Young and a young and flamboyant Chris Craft, demonstrated that Mini Coopers could be fought and occasionally beaten in the small-capacity classes. Yet the full 1,293cc engine of competition 1275 Cooper S Minis was too much to cope with within the rules (showroom Anglias featured only 997cc and 1,200cc motors). For 1966 the rules changed to a much freer Group 5 format, and increasingly successful Mini Cooper S entrant Ralph Broad became available for Ford to sub-contract. The energetic Ralph, with rapid-fire speech and a team of outstanding technicians, built racing Group 5 Anglias for Ford in Bristol Street, Birmingham, under the Broadspeed banner.

Taking elements of Cosworth and Holbay engine experience, knowledge of the Anglia 105E engine gained in Formula 3, plus the Group 5 rule freedoms, Ralph Broad's Anglias could now pack a winning 1-litre class punch in Britain and Europe. The motor was redeveloped to enable multiple downdraught Weber carburation or Tecalemit fuel injection to be used. These endeavours extracted over 100bhp per litre, the high-revving four-cylinder units mated to Hewland five-speed gearboxes. Broadspeed saw that the suspension received just as much attention, running a coil-over-spring and radius-rod layout that is so effective for a live axle. In short, they produced a thoroughly well-balanced Ford of class-winning power that looked similar to the showroom article.

In 1964 young Midlander John Fitzpatrick and the Mini Cooper S had been beaten in the premier British Saloon Car Championship points only by the combination of Jim Clark and Ford-funded Team Lotus with the first-edition Cortina – a performance that established 'Fitz'.

John Fitzpatrick's regional upbringing and natural competition talent had put him on the radar of then Mini Cooper S specialist Ralph Broad. Increasing Broadspeed support saw John graduate from privateer Cooper S racer to paid factory-status wheelman. Employed by Cooper Car Company and Ken Tyrrell in 1964, when both were BMC-allied to field Cooper S racers in various capacities, John had a profitable, high-profile Mini Cooper season.

By comparison, 1965 was a bit of a setback for Fitz, with too many retirements and a single class victory with Broadspeed, who were now frustrated in a queue for support behind other Mini specialists. Ralph Broad swiftly found new suitors for his engineering talents, reappearing for the 1966 season with a brace of well-funded 1-litre Anglias. In those days, the British Championship was divided into four capacity classes, so it was possible to win the outright title on points awarded for class victories. That is just what Broad and Ford aimed to do, running in the smallest capacity class, but it was a bumpy ride.

John Fitzpatrick was always part of the plan, and started the season with a class win, but the more experienced racing and rally driver hired for the Broadspeed Ford assault was Peter Procter. Tragically, the second round of the British series, at Goodwood, saw Procter tipped into a major accident by another competitor at one of the fastest sections of the track (the scene of Stirling Moss's life-altering crash). Peter managed to escape the ensuing major fire – the result of a split metal fuel tank – but suffered such serious burns that his racing career was over. A welcome by-product was that safety-bag fuel tanks became the norm in saloon car racing as well as other formulae, along with fireproof overalls and (initially rudimentary) roll cages.

For Broadspeed and Ford, the prospect of recruiting another driver at short notice loomed. Grand Prix driver Trevor Taylor's sister had already proven that she shared a slice of her brother's talent, with outings in Anglias and Minis. That she was attractive, as well as more than a match for most male drivers, brought the Ford Broadspeed equipe a publicity bonus. John reported in his 2016 book, *Fitz: My Life at the Wheel*, that Anita Taylor also 'moderated Ralph's language when things went awry'.

▲ After running Cooper S Minis, Birmingham-based Broadspeed switched to Ford for 1966 and prepared Anglias for British racing. Former SuperSpeed and Ford Competitions employee Chris Craft is seen here in typical airborne action at Thruxton chicane. *Ford*

John Fitz successfully completed the 1966 class assignment, securing six of eight qualifying race wins. One class win was lost after an encounter with Bill McGovern (Emery Imp) that provided Fitz with a puncture and demoted his Ford to fourth in class, Anita Taylor taking the 1-litre victory ahead of the factory-backed Alan Fraser Imps. It was actually a very tightly fought championship result, as John 'Smokin' Rhodes had an equal number of class wins in the Cooper Car Company Mini Cooper S, running in the class above the Anglia. John and the Broadspeed Ford finally got the nod as they had more outright victories.

Looking back to that 1966 season, John Fitzpatrick recalled: 'The Anglia days were good fun with Ralph. Henry Taylor approached Ralph at the end of the 1965 season to see if he would be interested in running a team of Anglias in the BSCC. He said he had told Peter Proctor he could win the championship if he took the drive. Ralph said he wanted me to drive as well and that was agreed. What Henry did not expect was for me to be quicker than Peter, and of course Peter then had his fiery shunt at Goodwood: that was when Anita Taylor came into the team.' In his book, John expanded on this major female driver signing: 'Anita's presence in the team certainly gained us massive publicity and she was very fast, keeping me honest throughout the year and stepping in to win when I had a problem. She was a great asset to the team. She retired at the end of 1967 to concentrate on family life.'

▲ John Fitzpatrick won the 1966 British Saloon Car Championship outright in a Broadspeed Anglia. The team and 'Fitz' were rewarded with some 1967 European Touring Car Championship outings, and here Fitzpatrick is seen in the car he shared with Trevor Taylor at Barcelona, where they finished third overall, against strong opposition. *Ford*

John and Anita were teamed together again in 1967, but Anita's brother Trevor was also no stranger to a Broadspeed Anglia or Escort cockpit. He delivered some sensational overall results when teamed with Fitzpatrick in the small-capacity Fords, magnificently demonstrated when Taylor/Fitzpatrick, running in the 1-litre Ford at Barcelona's scenic Montjuïc track, finished a tremendous third overall in a sports, GT and saloon car event. Beaten only by two Porsche 911s, the small family Ford and two talented drivers put an awful lot of larger-capacity machines to shame.

Fortune wasn't quite so kind to Broadspeed Fords in 1967, as the 1-litre Imps had matured into class acts, assisted by factory funding for the Fraser team and the presence of factory employee Bernard Unett at the wheel, supported by driving talent such as Tony Lanfranchi. Later, a privately entered Imp would dominate the British Championship when preparation guru George Bevan and driver Bill McGovern established a class-winning 1970–72 stranglehold on the series.

Nevertheless, John Fitzpatrick's 1967 Anglia took seven class victories versus two for Bernard Unett's Fraser Imp. Occasionally, the 1-litre Broadspeed Anglia would also beat the SuperSpeed Anglias in the larger 1.3-litre class, plus all but the most professional Mini Cooper Ss. John finished overall runner-up to Ford Falcon conductor and expat Australian legend Frank Gardner, who would become a good friend off-track.

In March 2021 John recalled for me his experiences of racing the earlier small Fords: 'The Anglia was a bit of a handful for me to start with, as I was used to racing front-wheel-drive cars, but I soon got the hang of it and the high-revving Holbay-based engine was like driving a sewing machine after the Minis. I seem to remember we used 9,000rpm.' Comparing the Anglia with the later Broadspeed Escorts, John added these illuminating comments: 'The Escorts were far more manageable, with their wider wheels and wheelbase, in fact the BDA was quite a rocket-ship.'

More about that later – Ford needed to promote a new small car in 1968, to replace the Anglia, but the Escort story must follow at a distance, as there are other Anglia and then Cortina tales to tell…

ROGER WILLIAMSON ANGLIA MAGIC

It wasn't all about factory money buying Anglias decent results for sales promotion. Roger Williamson, an impecunious but determined former karting champion, deployed a flyweight club-racing Anglia as a stepping stone on his ascent to Formula 1. Roger became the only British driver to make motor racing's premier league from the fiercely individual and resourceful ranks of Special Saloon Car racing. We will explore this Anglia/young ace combination thoroughly, because I was lucky enough to drive this championship-winning Ford when it had completed an astonishing season of victories, employing well-used components.

It was a dank and dreary December day when I had the chance to experience Roger's rather different kind of Anglia magic. Instead of the 21st-century special-effects airborne Harry Pottermobile in sky blue, I had the reality check of driving the very special flyweight red-and-yellow Anglia steered by the extraordinary racing talent of Roger from 1968 to 1970, with championship victory in the last of those years. It was a very important day for me, as this £1,150 Ford was being offered for sale to fund Roger's (successful) single-seater ambitions.

Responsibilities came with my drive. I was entrusted with this precious Anglia because respected *Motoring News* (*MN*) Sports Editor Andrew Marriott ('A.R.M.') had promised the close-knit Williamson family team that I would bring it back unscathed.

Appearing in the Christmas (24 December 1970) issue of *MN* guaranteed a captive audience of potential punters with time on their hands. Such timely sales publicity suited the Williamson cause – although their offer to take a road car in part exchange proved more important to clinching the deal itself, as you will learn in a separate story ('Sold to the man with the Golden Voice!' on page 28).

I had followed some of Roger's outings during 1970, including a very wet Brands Hatch finale to the Hepolite Glacier Special Saloon Championship on 29 November. Roger had finished second overall in that downpour to complement a quartet of earlier outright victories that polished his championship season of 14 class wins. However, I didn't know the Williamsons well.

Thanks to Facebook friend Dave Smith and *Autosport*, the results Roger scored with this variation on an Anglia theme are easy to admire. They included a famous outright victory at Crystal Palace that you can find on YouTube, and another non-championship outright win at Thruxton that saw the 1-litre car turn a record 1m 52.2s lap time, just weeks before that Hepolite Glacier finale.

My mentor at *Motoring News* and *Motor Sport*, the aforementioned A.R.M., facilitated this test, knowing it would be good for the Williamsons and our paper. That Anglia, plus the father-and-son Williamson team from North End Motors outside Leicester, worked hard for their results.

Andrew Marriott reminded me in 2021: 'I knew Roger had won a lot of races in a 200cc kart [with four-speed gearbox], which was faster than most kart categories back then – and I also knew that Roger was focused on getting into Formula 1, even then. Mind you, he had pedigree, as his dad – always known as "Dodge" – had been a

▼ Roger Williamson took the unlikely route of a 1-litre Ford Anglia on his way to his Formula 1 apprenticeship. Here he is seen during his 1966 Hepolite Glacier Special Saloon Car Championship season, leading at Crystal Palace. *Alamy*

SOLD TO THE MAN WITH THE GOLDEN VOICE!

Voice-over artist Gerry Taylor, who bought the Williamson Anglia after my test, said in 2021: 'I had been racing Minis for a number of years, and had raced against Roger, but was never quick enough to be placed. I concluded that I had to have a better car if I was ever going to do any good. Several Anglias later, and into 1970, I was about to pack it all in.

'Speaking to Roger, he said that he was selling his championship-winning Anglia, would I be interested? I knew that if I could get that car, then it would be such a big step forward. Roger said he loved my Volvo P1800S, and would I be interested in doing a deal with the Volvo? That red Volvo (BNT 466B) was built by Jensen at its Wednesbury works.

'We met at Mallory Park, and Roger tried the Volvo and wanted it. The deal was done at £1,000, so that was a simple swap – no money changed hands. I purchased some spares, and extra wheels with wets, at £400. I did have a 1-litre BDA built, but it was nowhere near the power of the Holbay, so I reverted to the original Williamson unit, and went on winning into 1974–75 in the well-known Swish sponsorship colours.

'I was the new owner of that famous car. The car was so different to anything I had driven before – it was all set up and needed nothing adding to it. History shows what Roger achieved in it, and I went on to set class lap records and win races with it too. The engineering and the set-up of the car was such that I added nothing new to that car from the day I acquired it from Roger, to the day I retired from racing in 1975.'

HEPOLITE GLACIER
Outright Winner
Roger Williamson's Hepolite Championship winning Anglia is now offered for sale, after Racing Car Show commitment. Genuine enquiries only.
Tel: Leicester 22808, after 7 p.m. 61721.
Road car taken in part exchange.

The advertisement that accompanied the Christmas track test in *Motoring News*. It was the final paragraph – 'Road car taken in part exchange' – that apparently clinched the deal. *Author's archive*

top speedway rider and was well known in the Leicester area.'

Roger converted to car racing from karting in 1967 at the age of 19 with an ex-Ian Mitchell Mini of minimal 848cc engine capacity. He demonstrated winning ability with four outright victories and many more class wins, setting new records over both the shorter and longer Mallory Park circuits. Andrew Marriott underlines that even the route into a club-racing Anglia wasn't easy: 'It wasn't all as simple as going from winning in Minis to a Ford Anglia. That Formula 1 ambition saw the Williamsons buy a Cooper T71 Formula 3 car [ex-Bev Bond] with a 1-litre "screamer" of a Holbay motor. Sadly for Roger, the Cooper got burnt out in the family garage – which dealt mainly in commercial vehicles – but that 1-litre engine survived.'

The author believes Roger then had five initial outings in a Ford Anglia converted from a road car, with the classic 1,650cc engine stretch (based on the Cortina 1500 GT motor), fighting with our opening chapter's Berpop hero and creator Mike Berman in local events.

Andrew Marriott continued to put Roger's story in perspective after that garage burnout: 'That pretty well wrote off 1968 for Roger, and it was 1969 before the combination of Roger, the red Anglia with a yellow stripe and that 1-litre motor arrived in British club racing.' Once that combination was established, Roger began setting new records, notably sharing the shorter-circuit 1-litre class time at Mallory with future triple British Saloon Car Champion Bill McGovern, leaving the lap record at 40.6s.

Andrew added: 'Remember that there were a lot of Anglias about back then. They ranged from Broadspeed's immaculate and fully trimmed international racers to quick bigger-engine devices with Ford engineers driving under the East Anglian Racing banner. But Roger's Anglia was different.'

Andrew went on to explain: 'It was a bit rough around the edges, but light and very well balanced. Not so many Anglia people raced with smaller engines in club racing, so the main opposition came from 1-litre Imps and Minis. There were lots of entries and the quality of racing was tremendous, so there was plenty of press coverage of these clubby events. Roger was so smooth, so naturally talented and so determined that it wasn't long before racing insiders noticed there was real talent here. Roger would be the only driver to make it straight from racing an amateur club-racing saloon to a professional Formula 3 – and finally a Formula 1 – drive.'

By the time *Motoring News* and I got to grips with the Anglia, Roger had captured the 1970 BRSCC Hepolite Glacier Special Saloon Car Championship title outright, after a tight points fight with Rob Mason's 1.3-litre Mini Cooper S. Roger's full season covered more than just the Hepolite Glacier series and ensured he appeared before millions via BBC TV cameras, twice. His tally included three class victories at Oulton Park and Brands Hatch, two wins at Mallory Park and Castle Combe, plus class honours at Mondello Park in Ireland, along with the televised races at Crystal Palace and Thruxton.

Even more impressive were five outright victories, the most important of which, to influence potential sponsors and spread his reputation as a rising racing star, was his end-of-season October outing at Crystal Palace. The BBC cameras whirred and Murray Walker uttered his quick-fire commentary.

I had attended Brands Hatch on 2 August to see Williamson win during round 11 of the Hepolite Glacier series, as had many competitors from other classes. All scrambled to watch these Special Saloon events, as the racing had the excitement to engage substantial crowds, as well as experienced competitors and mechanics.

Even hard-bitten club-racing legends like David Brodie paid tribute to Roger's extraordinary talent: 'In his little 1-litre Anglia he was the fastest driver I ever saw into and out of a corner. He had the self-belief, the pure confidence in his own ability that all great drivers have.'

In his round-up of the 1970 season for *Cars & Car Conversions Yearbook*, Mike Kettlewell paid tribute not just to Roger but also to the often-overlooked racing abilities of the Ford Anglia: 'We may have just seen the last of the great Anglias in a world that was progressively swamped by more and more Escorts. Seeing Roger go through the Esses at Mallory, demonstrating a really sorted car, was certain to leave a very stirring impression on the mind of the beholder that here indeed was an Anglia to remember. A Champion Anglia.'

▲ Happy chappy! Roger in the office of his flyweight Anglia, which brought his name to the notice of influential motorsport backers, including Les Thacker at BP and Tom Wheatcroft, father of the reborn Donington track. They supported Roger's career all the way to Formula 1. *Courtesy of Gerry Cannell*

Now we concentrate on car and driver, as I found them on that memorable Leicestershire short-circuit day.

Under the skin: a very Special Saloon

The worthy champion Anglia appeared for our Mallory encounter in its usual red and yellow war-paint, the front wheel arches and their additional eyebrow extensions filled to capacity with Firestone wet-weather tyres and 7.5x13in Minilites. These were borrowed from British championship regular and (then) Ford dealer Vince Woodman, assisting Roger in his startling wet-weather performances at Brands Hatch and Thruxton: normally the Anglia ran Dunlops, which we reverted to later on 7in rims for a drier second session.

The most obvious radical race feature on the Anglia outline was the truly featherweight panels. A mixture of L72 aluminium-alloy panels for the sides and rear, plus glass-fibre for the bonnet, boot, roof and door skins, slashed the kerb weight significantly. My contemporary guess, guided by the Williamsons, was 1,232lb (559kg), which would have been right in race Mini territory. Benefitting from hindsight and my part-time attendance at Lotus Engineering, where kerb weights were a religion, not a bar-room boast, I would add the weight of a limited-slip-differential rear axle and a separate oil cooler. Also considering the fact that the Anglia was a longer and larger vehicle than the pioneering transverse, front-drive Mini, I would revise that original guess to a still skimpy 1,320lb (599kg).

To power those pounds, the Anglia deployed an equally race-focused 123bhp motor. Yes, there was the remains of the rescued Holbay 1-litre four-cylinder from the Cooper F3 car, but it was shorn of its original carburation and exhaust, as it would now be put to work in a new saloon-car environment, with relevant regulations to meet. The carburation remained downdraught, but instead of the previous single unit, the Anglia employed a pair of Weber 40mm DCNLs on a manifold Holbay had developed for sports cars. The exhaust manifolding and side-exit pipe was based on Janspeed plumbing. One feature on these '70s racers that foretold the future was the presence of a brace bar between the engine bay's suspension top mounts. In period it was credited with limiting camber changes in the front suspension, but such braces – albeit in beefed-up form – later saw widespread use in many performance cars, improving body stiffness as well as suspension behaviour.

Internally, the motor carried hardened-steel components developed for Formula 3, but on inspection of the crankshaft and bearings, it was obvious that there was a fundamental change from the usual three-bearing layout, aimed at improving reliability. Instead of searching for the last five horsepower and a 10,000rpm rev limit, the Williamsons' hard-won experience demanded durability over erratic reliability and frequent expensive rebuilds. Thus their Anglia featured a five-bearing bottom end derived from the contemporary Escort, and a 9,400rpm limit. The high-rpm motor was soothed by a two-gallon-capacity dry-sump system, the oil tank originally in the boot for weight-balance reasons. However, the tank got so over-cooled there that they moved it forward to sit cosily adjacent to the hard-worked engine.

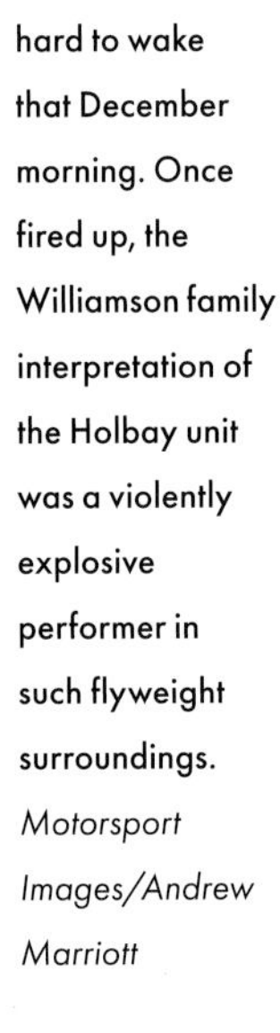

▼ The 1-litre motor was hard to wake that December morning. Once fired up, the Williamson family interpretation of the Holbay unit was a violently explosive performer in such flyweight surroundings. *Motorsport Images/Andrew Marriott*

There were five engine-related retirements during the season, ranging from a simple spark-plug protest to head gaskets that failed twice, plus a piston once, and a broken crankshaft that stopped early-season play before that rev limit was imposed. Such retirements interrupted the points-gathering mission (and hence the championship challenge), making it a tense season for the cash-conscious equipe.

By contrast, the transmission, suspension and brake systems were straightforward, proven period hardware. A four-speed close-ratio gearbox based on the 'Bullet' specification from Ford Performance Parts offered the racing luxury of synchromesh and reverse. Salisbury supplied the limited-slip differential, the final-drive ratio a relatively high 6:1 for shorter-circuit work.

The rear axle was suspended by period conventional steel leaf springs, operating alongside a Panhard rod and adjustable Armstrong dampers. Up front, the Williamsons adapted some stiffened Cortina Lotus struts, with Armstrong damping and a double-tube anti-roll bar that cut roll angles considerably.

They also experimented with a quicker steering rack to remind Roger of his karting days, but it was not a success and they reverted to modified Cortina/Anglia components. Cortina Lotus front discs were used, augmented by classic Ferodo DS11 pads, but at the rear Roger's natural late-braking style needed nothing more than plain production Anglia drums utilising standard linings. Unusual back in these Special Saloon days was the fitment of servo assistance, Roger commenting: 'I use a servo because the effort is less than normal and it is easy to feel when the wheels are locking.' A trademark cheeky grin was present as he added: 'Who wants to stand on the seat to put the brakes on, anyway?'

▲ The moment of truth, as Roger Williamson briefs the author in the small Ford that provided valuable recognition during Roger's rapid rise to the premier league. *Motorsport Images/Andrew Marriott*

Behind the wheel: a roaring success

That 1-litre engine was truly trick and not just a transplant from a Formula 3 car. In the hands of the apprentice track tester, and mindful of the need to stay on the black stuff, engine revs would be constrained to 8,500rpm, with a minimum of 6,200rpm required for splutter-free progress, although Roger normally ran a race maximum of 9,400rpm after restraint during practice lappery. Williamson senior commented: 'We treat the car like a little baby right up 'til the time it reaches the start of a race.'

Naturally, Leicestershire and the Mallory track were British December cool on test day, and the highly stressed engine took a while to wake up. Roasting the spark plugs – and a paddock tow – finally brought life to planet Anglia. Masking-tape over cooling vents was required to attain working temperatures for oil and water, and over five minutes were needed before the flyweight Ford carried enough temperature to face the track.

Even before exiting the paddock, my learning curve needed more work. Strapped in a simple Britax race harness, Roger leaning through the open driver's door to make himself heard above the Holbay cacophony, I missed the significant throttle-action advice: 'No quick blips.' In the ensuing embarrassing silence, I could hear Roger and the close-knit crew's ribald comments all too clearly. Fortunately, the engine was warm enough to restart swiftly, and I kept to the recommended regularly spaced accelerator depressions. That brought a very satisfying succession of barks, roars and shouts of sheer mechanical joy, all resonating around that stripped metallic cabin.

Functionality was the key throughout the interior, but I was lucky that Roger and I were of similar build, so the modestly contoured race seat, with cloth inserts, was a reassuringly supportive fit, and the deeply dished leather-rim steering wheel was well positioned. The flat, black dash panel carried a Smiths rev counter with period tell-tale – a feature that recorded the maximum rpm attained rather than the team relying on what the driver decided to tell them. A combined water and oil temperature gauge reported 60–70°C that day, but the vital read-out was that for oil pressure, pointedly highlighted by Roger's stubby finger with the instruction it should not read less than 80psi.

The opening lap was mildly disturbing. The

➤ I shared a stark Anglia cockpit with four-dial instrumentation, including the vital, large rev-counter dial, plus a passenger-footwell mounted brake servo that Roger found helpful for fearsome late-braking manoeuvres! *Motorsport Images/Andrew Marriott*

mile-long, initially wet Mallory Park short circuit offered the challenges of a long, fast right-hander and a tight chicane that cut out the traditional hairpin feature of the full circuit. The engine still needed time and over 6,000rpm to clear any hiccups. As I turned into the quick right at Gerard's, I could see so much daylight around the driver's door frame that I was convinced the door was ajar, and there were some shudders from the rear axle as I spent far too long on overrun during the exploratory opening laps.

The initial negatives faded fast under the pleasures of running the amazing small-capacity engine towards exhilarating higher rpm. Deafening, it reminded me why drivers find motorsport so invigorating. The wetter morning session also allowed me to enjoy the chatty steering feedback through the slower chicane section, so I knew that the consequent oversteer power slides were rewardingly easy to suppress, even for a stranger.

That session was curtailed when a Mini driver mated his machine with the Gerard's banking: car and driver were just muddy and undaunted. The driver's wife seemed to suffer most, commenting: 'There goes my washing machine money!'

Cheered by a drier track and the fact that I had brought the precious Ford back unscathed, the afternoon session became a genuine pleasure. The servo-assisted brakes were superbly progressive and the gearbox ratios seemed well suited to that high final-drive ratio, the synchromesh shifts as slick as an estate agent's patter. The Anglia's low weight and battle-proven suspension meant I could now bat through Gerard's at my allotted 8,500 shrill revs in third, changing into top on the mud-slicked exit bump. The car would skip back into line amiably, while the slower off-camber sections of the chicane could now be conquered in third, with the engine beginning to enjoy life at 7,000rpm before swallowing the brief interlude of Kirkby Straight in fourth, just beyond 100mph.

Once I had extracted myself from the now warm cockpit and comfy seat, I immediately asked how much it would cost to own such a rewarding ride?

Roger's answers were succinct and prophetic: '£1,150 and it's fully sorted to go class winning for you next season sir!' As the 'Sold to the man with the golden voice!' panel on page 28 recounts, that Williamson Anglia performance promise was fully delivered to the next owner.

Roger also delivered on his abundant driving talent when he graduated to single-seaters under the patronage of then Donington circuit owner Tom Wheatcroft. Roger made it all the way to Formula 1, where tragedy unfolded at the 1973 Dutch Grand Prix, ending his life after only two F1 outings.

GP STARS ENDURE AT GOODWOOD

Back in the '60 and '70s, a popular way to endorse the durability and speed of new cars lay in the endurance run. Ford was a big fan of such filmed and heavily promoted demonstrations, even submitting the Transit van to trial by sustained speed at Monza, Italy's cathedral of motorsport. However, the Anglia preceded all that in 1960 with a week at Goodwood's swift circuit featuring three of the 997cc production Fords.

The objective, successfully attained, was to drive a typical motorist's annual average of 10,000 miles in seven day-and-night stints.

Some world-class talent was hired to drive those miles. Heading the bill was Graham Hill, subsequently to become a double Formula 1 World Champion, backed up by McLaren founder and fellow F1 ace Bruce McLaren. Also with Grand Prix and world championship sports car-racing pedigree was Roy Salvadori. These luminaries were backed up by prominent trade figures with competition experience and leading journalists of the day, including Gordon Wilkins and Tommy Wisdom, both internationally recognised competition drivers.

The dress code featured no racing protection – perhaps a flat cap and a dapper neck scarf – and one of the motoring writers chomped contentedly on a cigar as he departed following one of the fuel stops (carried out every three hours, with tyre swaps at six hours). The Anglias – Graham Hill's in white, the other two plummy shades – did have optional radios on board and, with steel wheels fitted, looked a little more firmly suspended than showroom cars. However, the tyres looked period spindly and carried treads that appeared to be entirely standard.

The trio of cars, with three drivers apiece, experienced Britain's wintry weather, pounding through significant rain and standing water to keep lap times within the 2m 14s to 2m 19s bracket. The cars averaged around 60mph, achieving a creditable 32mpg according to the figures monitored by RAC timekeepers and observers.

Bruce McLaren, Graham Hill and Roy Salvadori pose with the Anglia they used for the endurance run at Goodwood in October 1960. *Getty Images*

CHAPTER 3

CORTINA

A '60s motorsport winner and Britain's best-seller for 20 years

In Britain, from 1962 to 1982 Cortinas were showroom stylish, underpinned by simple but rugged and durable components, and were particularly reliable when judged against underdeveloped and underfunded domestic rivals. The established best-sellers had been haphazardly assembled by Ford's rivals at British Leyland and its forebear BMC (British Motor Corporation), far from world-class opposition. Straightforward Cortina components and big-order discounts appealed to fleet buyers, providing Ford of Britain with full order books and leadership of the sales charts – a dominant position that Ford maintained from 1972 to 2020 with Cortina, Escort, Fiesta and Focus models.

THE CORTINA STORY: FAMILY FAVOURITES

A low kerb weight and versatile, tough powertrains for the Cortina also appealed to hardcore motorsport and performance-tuning specialists, and to spectators. Early Cortinas found an enthusiastic reception amongst that community, especially when backed by Ford corporate promotional cash in Britain and overseas, funding legendary '60s competitors such as Team Lotus and Alan Mann Racing.

Official Ford production figures for five editions of the Cortina total 4.2 million (over 3 million built at Dagenham) and show progressive sales increases for each generation. However, the overall picture was less encouraging. The final two (1976–82) models – officially designated Mark IV and Cortina 80 – were well beyond their sell-by date outside the UK. So the subsequent aero-conscious Sierra, featuring independent rear suspension, was sorely needed by the early '80s, especially in more demanding and engineering-conscious European markets such as Germany.

The first Cortinas (1962–66) predated my 1967 full-time employment as a motoring journalist, thus my experience of those models was

retrospective. The second generation of Cortinas (1966–70) stuck with the simple oily bits – four-cylinder engine, live back axle on leaf springs, and MacPherson-strut front suspension. As with many second-edition Fords, it got a little plumper, more comfortable and less sporty, but Ford also created a more durable version of the Cortina (Lotus) Twin Cam. I had one as a staff car at *Motoring News*, but it had been significantly modified with BRM-uprated Lotus-Ford Twin Cam motor. More of that later, and also of the legendary Cortina Savage 3-litre V6 transplants from Jeff Uren/Race Proved.

The revered, affordable – and now often overlooked – original 78bhp Cortina 1500 GT morphed during the second generation into the 1600 GT, achieved via the use of the crossflow engine introduced in September 1967. By the 1968 model year, and the 1967 Paris motor show, the 88bhp 1600 GT crossflow lay at the heart of another very clever and profitable Cortina – the prestigious 1600E mix-and-match of Cortina Twin Cam componentry and E for Executive classier cosmetics. Today, the 1600E has an auction value second only to the Lotus/Twin Cam variants. The 1600E boasted an air of luxury inside and out for £865 in 1968. Within could be found a wood-veneer dashboard, walnut door cappings, leather-rim sports steering wheel, a quartet of extra dials, and plush recliner seats with a leather look. Externally, the showroom lures included unique paint with blue or gold coach-line decals, shiny Rostyle steel wheels and twin spotlights.

The 1600E featured Cortina Lotus suspension settings alongside that tough 1.6-litre GT crossflow engine – a supremely versatile unit that also served the Escort Mexico and Capri, plus Formula Ford single-seater racers. The engines used in the latter were laboriously hand-worked to deliver 100bhp from primarily production parts using the black-magic art of blueprinting, or 'authorised cheating'.

We had a staff 1600E at *Motoring News* too: I borrowed that majestic Roman Purple example from advertising manager Hilary Weatherley for more formal occasions. These included an uncle's funeral and transporting various ladies, from my mum to more easily impressed females.

▲ The treasured and traditional wheel-lifting image of a Cortina Lotus in action, this time at Goodwood in 1964 for the second round of the British Saloon Car Championship. Jim Clark finished second to Jack Sears's Galaxie. *Motorsport Images*

▲ An interesting photograph depicting Cortina Lotus MkIs on the assembly line at Lotus's first major factory at Cheshunt. *Ford*

By contrast, the third-generation Cortina, announced in 1970, was all about showroom sales in a 'Coke bottle'-pumped transatlantic style, a reference to the rising flanks of the rear wing panels. The overall design was a mash-up of tastes, predictably created by an international team comprising US executives, plus the talented Patrick Le Quément, who went on to greater things at Renault. After a fraught, strike-ridden and quality-control-lacking start in 1970–71, during 1972 the Cortina MkIII retrieved lost ground and ripped past the equally strike-prone British Leyland establishment's 1100/1300 sales-chart regulars.

The third edition's bulbous body was wider and heavier than its Cortina ancestors, but 2.3in shorter. For motorsport and performance-orientated customers, this Cortina signalled the retreat of the model from frontline action, especially as it had put on weight and was 176lb (80kg) heavier than the second-generation car. Plus, the Ford Cologne-designed engine range was deliberately restricted to then new single-overhead-camshaft (Pinto) units, as Ford had motorsport-homologated competition versions of Escort Twin Cam/RS1600 and Capri 3.0 V6/RS types to fulfil any 1968–74 sporting ambitions. Although the first- and second-generation Escort saw extensive action in premier-league international motorsport, the major international honours for Cortinas and Capris were achieved with first-edition models, which make major money at classic-car auctions over 50 years after their motorsport heyday.

That Anglo-German-manufactured Pinto engine for the Cortina MkIII took its common nickname from the US-market Ford Pinto compact, which shared the power unit. The European Pinto units featured a heavy cast-iron cylinder block and tall cylinder head, and were rated at a maximum of 98bhp in production trim, but became popular in the performance world when lightly reworked to fit into two generations of Escort RS2000.

Returning to the Cortina MkIII, by chance I did drive a non-production 3-litre Cortina on a hilarious Belgian press launch (an Irish contingent had lunched all the four-cylinder production models). Another 3-litre Cortina MkIII featured a factory V6 with 210 turbocharged horsepower and outstanding 1,000–5,000rpm access to calm pulling power. I was fortunate enough to share some hilarious road miles in that abused Cortina (PWC 906K) with the late and great Gerry Birrell.

◀ The Cortina 1600E was a blend of ex-Lotus Twin Cam performance items alongside the practical 1600 GT crossflow engine, topped by shiny paint and Rostyle wheels. *Motorsport Images*

▼ A Cortina III (PWC 906K) in non-production 3-litre, 210bhp turbocharged format. *Ford*

Chuckling Gerry became a professional race and impressive development driver who knew how to get the best out of the least-promising automotive ingredients. The Cortina wrapped around that V6 turbo was to a shaky blancmange recipe, which was revealed when drivers tried to match the turbocharged thrust with the car's cornering ability.

When at Ford, I also contributed minimally to a PR coup masterminded by Public Relations colleagues Harry Calton and Steve Sturgess – one that saw them capture Noel Edmonds for a 1973 motor racing season in a Cortina 2000 GXL. Back then, Edmonds was a massively popular disc jockey hosting the BBC Radio 1 Breakfast Show, with an audience measured in millions.

Although the Cortina continued to dominate the British sales charts, mainly via bulk fleet sales, I had much less to do with the mass-market Ford post-1976. At that point, the ageing mechanicals of the Cortina were covered in boxier bodies that, with the aid of a cost-conscious 1979 revamp, lasted until the aerodynamically fashionable Sierra came to the Europe-wide sales rescue in 1982. However, I did use an entertaining and smooth late-model V6 Cortina Estate with Cologne production 2.3-litres to cover the November 1979 RAC Rally. The tough task was to follow the day-and-night rally route, create assorted media reports, send hastily phoned route messages, mostly for the organisers, plus retrieve stranded competitors and produce a full feature report for the now-defunct *Motor* magazine.

After that I participated in minimal Cortina press events with its successive revamps, right up to the end of the '90s. There was then a chance to drive a production Sierra and Cortina back-to-back, but I was wearing BMW goggles that day and was left underwhelmed by both obsolescent Fords in low-power, mass-sales specification.

JIM CLARK'S LEGENDARY MkI CORTINA RACER

On 4 October 1988, I had a date with the Cortina Lotus registered BJH 417B that is the benchmark Cortina Lotus to my mind – far from the most powerful or developed amongst restored classic-racing Cortinas, but the summit of Cortina motorsport aristocracy.

Try this motor-racing bloodline tick-list:

- Team Lotus creation and ran to class or outright victories in every British Championship round entered. Tick!
- Driver, Jim Clark, double F1 World Champion (1963 and '65), 1965 Indianapolis 500 winner, 1964 British Saloon Car Champion. Tick!
- Long-term owner: Triple Indianapolis 500 winner Dario Franchitti. Tick!
- Former prime exhibit at expanded Jim Clark Motorsport Museum, Duns, Scotland. Final big tick!

Back in 1988 I had no idea how significant this Cortina would become, and it's relevant that when I originally wrote this section, in 2022, BJH 417B was up for sale with a London specialist (see the Postscript at the close of this section).

I had attended the Silverstone Grand Prix track principally to track test the British Touring Car Championship (BTCC) pacesetting Andy Rouse Sierra RS500 for *Performance Car* magazine, the RS500 feature appearing in the January 1989 issue. I had a vague idea that it would be neat to compare the 140–150bhp Cortina Lotus/Twin Cam with the 500–530bhp turbocharged Sierra, back-to-back. However, that media day's constricted schedule and the commercial demands of Andy's Kaliber low-alcohol beer sponsor meant the misfiring Cortina didn't get the track time or media coverage it deserved.

Back then, this aristocrat amongst Cortinas was owned by Nissan dealer and talented saloon racer Andy Middlehurst. By 2021, Andy was well established amongst the classic-racing cognoscenti as the successful pilot of an Australian-owned Lotus 25 single-seater and his own Lotus 43. Oh, and Andy also owned a pioneering first-generation Lotus Elite to while away the road miles and salve his historic Lotus itch.

Andy remained cheerfully informative despite a 22-year gap in our Cortina conversations, and was a key player in the deals that saw BJH 417B become the property of Dario Franchitti in 2014. By then, it had passed through assorted hands following Andy's sale of the car, providing an instructive guide to Cortina Lotus values over the years.

In December 1964, at the close of the racing season, Team Lotus offered for sale a trio of race Cortinas with BJH-prefix registrations (417B, 418B and 419B) that had been guided by stars such as (Sir) Jackie Stewart and Sir John Whitmore, besides 417B's starring role with Clark. The asking price was '£1,600 each, or near offer'. By 2003, and the Brands Hatch liquidation auction of Tom Walkinshaw Racing (TWR) assets, BJH 417B reportedly made £90,000.

When it changed hands a decade later, the valuation was 'north of £200,000'. Writing in 2023, I understand you could anticipate an asking price beyond a quarter of a million pounds, possibly

➤ Hold the front page! Clark in airborne action at Oulton Park makes the cover of the 17 April 1964 issue of *Autosport*. When photographer Francis Penn asked Clark about the Cortina's spectacularly scary cornering action, he quipped: 'You want to be in it!' *Motorsport Images*

▲ An April 1964-season British Saloon Car Championship confrontation at Aintree between the agile Jim Clark 1.6-litre Team Lotus Cortina and the gargantuan 7-litre V8 Galaxie of Jack Sears. *Motorsport Images*

£300,000. That would be a record for a Cortina, but almost cheap if you value competition credentials, when you think what Ford GT40s and Shelby Cobras make without a championship-winning history featuring the driver who many experts view as the best of the best – Jim Clark.

Perhaps surprisingly, Clark's Team Lotus Cortina was durable enough to finish all its 1964 British Championship outings: he was always on the podium or an outright/class winner. Creditably, J. Clark also steered another Ford-backed Cortina Lotus to double race-weekend finishes at Sebring, supporting Ford's USA marketing push of the 'English Line', including 250km (157-mile) and 12-hour events. Not so fragile! Those US performances can be viewed on YouTube today.

Thanks to *Motor Sport* and *Autosport* magazine reference sources, I have adapted a detailed résumé of Jim Clark's nine-round progress to the British Saloon Car Championship (BSCC) and BJH's 1964 season. That year Clark used BJH 417B to win his class in every round he contested, with three outright wins, three overall second places and two overall third places. For the British GP meeting at Brands Hatch, where Clark won the headline event, Whitmore took over his Cortina Lotus for the support race, scoring another outright win.

Under the skin: the ultimate collector's Cortina

Compared with today's prestige classic-event winners, this legendary Clark Cortina Lotus was almost a showroom item. It was wonderful to inhabit 24 years after its championship heyday, but provided a startling contrast, with its low power, cabin simplicity and delicacy of controls, when I drove it alongside the brutal turbocharged RS500 Sierra.

So what went into that 1964 Lotus-Ford champ, aside from the stellar talent of 1963's World Champion, James 'Jim' Clark Junior, OBE? Prepared to the Group 2 international FIA rules for 1964, Clark's mount was still recognisable as a kissing cousin to the showroom item. By October 1964, a showroom Cortina Lotus cost £992, down from the 1962–63 launch price of £1,100 in recognition of increased Ford mass-produced content, rather than limited-production Lotus race-conscious hardware.

My automotive upbringing included a lot of rally and race cars featuring either complicated cockpit instrumentation or the bare minimalist approach where bare metallic surfaces abound, possibly plus the reassurance of a roll-over bar that became more like complex scaffolding as the decades rolled into the '90s. BJH 417B was a refreshing change that spoke of carefree

▲ The grid at South London's Crystal Palace parkland track departs in a blare of exhausts and squealing tyres. Although V8 power has prevailed off the line, David beat Goliath on this occasion. The American Ford blew a tyre and Clark's Cortina led team-mate Peter Arundell home for a victorious 1–2. *Motorsport Images*

driver safety standards – no safety belts, no roll cage, standard-size foot pedals, with none of the metallic anti-slip finish that became so fashionable. Another missing item was the fabled Ford 'Aeroflow' ventilation 'eyeballs', because the Clark Cortina pre-dated the 1965 mass-production adoption of that widely praised system, probably the most effective cockpit-cooling system outside air-conditioning.

True, there was a pair of bucket seats – apparently sourced from a period Lotus Formula Junior – and the carpets had disappeared, but the Ford presence prevailed in the flowing chrome italic Cortina script that adorned the metal dashboard. All visible front trim remained including the door cards and chrome handles for window winders and door catches. The road-car Smiths instrumentation had survived, laid out in the standard left-to-right pattern, from the oil temperature/oil pressure gauge along to an 8,000rpm tachometer (with the red sector beginning at 6,500rpm) and a matching 140mph speedometer.

Although the turn-key ignition/starter switch remained, there was a proper low-oil-pressure light in orange adjacent to it. A long, thin gear lever was fitted, topped with a wooden Lotus-embossed knob, but the star patina feature in evidence on the museum car today, favoured by Clark in 1964, is a three-spoke steering wheel with very dark brown leather rim. It does feel flimsy at first, but then that initial impression is common to many more Lotuses right through to the 21st century. Even the entry-model Elise, saviour of Lotus in 1996, featured delicate steering inputs and the most sensational feedback – a subtle chattiness that makes you feel as though you have inherited a generous measure of driving talent from the gods.

Actually, the steering on this race-pedigree Cortina was heavier than you'd expect from a showroom Lotus, because of the race-track camber and caster settings employed.

Now let's get into a little more detail courtesy of Andy Middlehurst. In 2021, Andy recalled the technical highlights of BJH 417B as I drove it 33 years previously, and we started with that steering system and associated suspension, the heart of any legendary Lotus: 'Remember the Cortina had the recirculating-ball steering box [the 1968 Escort marked the belated introduction of rack-and-pinion steering] and that was uprated to a 2.5 turns lock-to-lock ratio.' That was significantly quicker in response than the production 3.0 ratio for a showroom Cortina Lotus, but the production Lotus-tweaked Cortina already had elevated camber angles, thanks to longer bottom front suspension arms in forged steel. They were relocated slightly closer to the ground than those fitted to mass-production Cortinas.

Notable body changes included brace bars

in the boot to guard the external panels against extreme body-flex damage precipitated by the movement of the A-bracket rear-axle location. The Group 2 specification allowed a boot-mounted 20-gallon fuel tank, serviced by a Bendix fuel pump.

The suspension layout still featured a live rear axle, but instead of the usual Ford leaf springs, Lotus-engineered coil springs over vertical dampers were fitted, and the A-bracket location principles of the differential they had used on the ultra-light Lotus Seven were employed. At the front, Ford's MacPherson strut layout survived. The race cars, particularly BJH 417B, had stiffer spring rates (around 140lb/in front and 100lb/in rear) than the road cars.

The Lotus-overhauled suspension cooperated in race or road format with immensely strong steel wheels of 5.5x13in dimensions. Andy recalled: 'When you drove the car it would have worn White Spot Dunlop CR65 race rubber with 4.75-section front and 5.0 rears.' Braking? The Clark car used the oversize 9in units of the Cortina Lotus, but with Girling P14 callipers and uprated (probably Ferodo) anti-fade friction materials.

Weight, or a severe diet to reduce kerb *avoirdupois*, was central to the Lotus/Colin Chapman creed, and the ultimate competition Cortina was very well prepared in its initial form to carry as little flab as possible. Aluminium castings encircled gearbox remote-change extension, clutch and rear-mounted differential. Aluminium sheet substituted for steel in the outer skins for the bonnet, doors and boot lid. Some weight was relocated – the main battery can be found in the boot – and some scraped off, and any trace of showroom sound insulation was removed for competition Cortinas. The final result in Andy's experience was a recognised race weight of just 1,595lb (723kg), which is truly astonishing when compared to 1,820lb (826kg) for the already lightened road car, or the homologated weight of the miniscule Mini Cooper S at 1,435lb (651kg).

Back in 1964, the twin-cam engine within BJH 417B was supplied by Cosworth to yield around 138–140bhp with a rev limit of 7,500rpm, instead of a production 105bhp and 6,000rpm. Andy laughed and added: 'Today's classic racers expect at least 180 horsepower and a 9,000 rev limit.' Andy's quote was delivered in 1988, and by 2023 some 200bhp was reported, but only with race-by-race engine refreshments!

Originally, Clark would have worked a slick four-speed synchromesh Ford 'Rocket' gearbox, those close gear ratios originally chosen for the Lotus Elan and stacked tightly, with an elongated first gear that allowed over 50mph.

Behind the wheel: Jim Clark's winner

Externally, this Cortina Lotus has the familiar production clues – Wimbledon White with Olive Green Lotus stripe carrying the emotive gold italic capitals to spell out TEAM LOTUS. As expected, the car has a low-riding stance over the functional steel wheels, but race Cortinas delete what the Americans aptly dubbed 'dog-dish' shiny hubcaps. Compared to a production Cortina Lotus, it naturally sits lower, the rear arches squatting over the tyre tops, with no rubber above the road wheels visible. The front isn't so radical: you can still see most of the tyre. Chromed bumpers were retained front and rear for owner Franchitti to complete the 2019 road drive for the car to be exhibited at the Clark Museum, but it raced in period without front quarter bumpers or full-width rear bumper.

▼ A classic period photo of Jim Clark's all-conquering 1964 Cortina Lotus, again at Crystal Palace. The track was set in a fine parkland venue in South London and you could actually look down on the racers from the elevated spectator zones. Alternatively, as here, media privileges allowed you to photograph from trackside foliage! *Motorsport Images*

The early Cortina interior is an experience in itself for any baby-boomer motor racing fan who saw Clark in action. I know Andy Middlehurst did his absolute best to preserve the authentic aura of this 1964 legend. Such care extended to the dull brown leather rim of the skeletal three-spoke wheel that carries period Lotus green-and-yellow ID alongside Colin Chapman's stylised full initials ('ACBC').

Subsequently, Franchitti explained to Ed Foster of *Motor Sport* the key connection you feel with a supreme racing driver: 'This steering wheel was favoured by Jim. It's the one he liked to have in his cars. It's really flimsy, but it feels good. The steering is actually quite heavy because it has a load of caster. If you pull away from a junction with a lot of lock on, you have to give it some throttle so that it doesn't stall.'

Even for this reporter, used to driving competition cars from 1968 onwards, the lack of any safety features in the Clark Cortina was sobering. The deeply contoured, near wrap-around, fixed seats offer excellent support, but there is no roll-over bar – not even the single bar and stay I was used to in 1971. Lap seat belts, never mind a four-point harness, are absent, and the foot pedals really are standard, with production rubber pads over brake and clutch pedals, plus a small production accelerator pedal.

The Clark Cortina was already warmed up for me on that 1988 Silverstone date, so the engine start came with a twist of the key, and a cautiously progressive press on the standard-issue metal throttle pedal to ensure I didn't suffer the embarrassment of flooding the twin-cam and stalling. As soon as I had the front wheels straight, I eased out of the pits to the brisk bark of four cylinders warming to their work throughout a long first gear, snapping raucously between 3,000 and 4,000rpm, allowing encouraging acceleration.

'OUTSIDE THE ENGINE BAY, IT WAS ALL GOOD NEWS. THE CLOSE-RATIO FOUR-SPEED WAS ALWAYS SET TO DELIVER NOT JUST QUALITY GEARSHIFTS, BUT AN EXCELLENT RANGE OF RATIOS FOR THE BEST TRACK PERFORMANCE.'

As I steered gradually on to the full Grand Prix circuit, I appreciated the all-round vision afforded by the generous glass allocation, particularly compared with the fully roll-caged hatchback Sierra RS500 later that session. The Cortina, even in race trim, almost offered that sense of all-round glasshouse vision delivered by BMW's pillarless coupés of 1968–75.

The expansive acres of Silverstone were a little overwhelming in a then 24-year-old Ford. I used only 6,000 of the available rpm, as there was an obvious misfire that would see the engine lapse onto three cylinders when trying to sustain 5,000rpm or more.

Outside the engine bay, it was all good news. The close-ratio four-speed was always set to deliver not just quality gearshifts, but an excellent range of ratios for the best track performance. The handling, braking and cornering became rewarding as the treaded Dunlops generated a little heat. I particularly liked the steering, which tended toward the lighter end of the effort scale in the high-speed swerves that characterise the Northamptonshire track, but requested more effort in the third-gear speed zone of Becketts.

The outstanding memory was of the detailed feedback at the steering-wheel rim, but neither steering nor the tightly located rear axle were subjected to the ultimate strains of trying to squeeze 1.6 litres past a 7-litre Galaxie, as was Mr Clark's frequent task on twistier tracks.

Postscript *After I drove BJH 417B during Andy Middlehurst's ownership, the car also lurked within the Tom 'TWR' Walkinshaw collection and was raced regularly by Kerry Michael. As noted, three-time Indianapolis winner and Jim Clark fan Dario Franchitti became a long-term owner. The August 2022 issue of* Octane *magazine contained features revealing that both BJH 417B and 997 NUR, a pre-production Elan Coupé of Jim Clark pedigree, were for sale at undisclosed prices through London specialist Fiskens. In 2023 my unofficial research disclosed that an established historic racer and collector had become the new owner. Apparently the Cortina was now slightly more civilised, modified with occasional road use in mind, so it may well have changed hands again.*

RACING A CLASSIC CORTINA

I was fortunate that in 1994 I was allowed a chance to race a successful classic Cortina Lotus owned by Nick and Amanda Torregiani, and usually driven by Amanda. By the mid-1990s, 9,250rpm was almost normal for the constantly redeveloped twin-cam engines within the ICS Historic Racing Saloon Car Championship competitors. This 1.8-litre twin-cam was from Anderson Racing Engines, running on larger (Weber 45 DCOE) carburettors and was a pleasure to deploy on the fast (since modified) Castle Combe layout, revving raucously between 6,000–9,000rpm, supplying at least 180bhp.

I qualified midfield with an 88mph lap versus the 92mph of the quickest rival Cortina, conducted by Sean Brown. Sadly, mechanical gremlins ended my classic-racing day, with an inevitable post-practice rear-axle-repair welding job, followed by a loosened carburettor in the race. Yet that Cortina Lotus gave me the chance to see how a Lotus version felt in flat-out racing combat, so nimble it could initially run amongst far more powerful V8s, just as in period. Fabulous!

This Cortina Lotus belonging to Amanda and Nick Torregiani was a 9,000rpm race revelation, albeit fragile! I drove it at Castle Combe when Amanda was busy frightening all the front-running blokes with one of the Torregiani couple's two Mustangs. *Author*

Amanda Torregiani deployed the 180bhp Anderson-prepared twin-cam motor in the Cortina Lotus very effectively on the track. My helmet sits on the floor of the functional cabin. *Author*

A 21ST CENTURY DRIVE IN A MkI CORTINA LOTUS

An opportunity to use an excellent example of the face-lifted MkI Cortina Lotus occurred in August 2006, when Ford of Britain's press people green-lighted borrowing FGF 113C from their Heritage warehouse at Dagenham for a trip across France to Belgium's reborn city of Ypres, known to the British army over the years as 'Wipers'. Ravaged, and literally razed to rubble by artillery and continuous heavy fighting during World War I, Ypres is now a magnificent testament to Belgian abilities to rebuild architectural landmarks such as the Cloth Hall to recreate an historic site. This small city remembers the millions who died in the international conflict with a simple but moving daily Last Post service at the Menin Gate – a brief, standing ceremony that attracts large attendances, especially young people, who have come to remember the fallen.

Under the skin: a low-mileage road car

First registered on 13 August 1964 by Ford in Britain, this authenticated Cortina Lotus left company ownership and was sold to a private buyer during May 1965. The car stayed close to home, kept by four Essex-based owners before it passed back to Ford. I believe this quartet of purchasers were probably Ford managers: for sure I know that Michael R. Platt, a Ford PR employee who dealt with commercial vehicles, owned it. Those FGF 113C owners enjoyed their motorsports judging by the number of old competition scrutineering stickers around the cabin, and the retention of Yokohama 185/70 S306 tyres (production sizing was 165-13) paired with race-height suspension.

▲ In August 2006, before I departed on a classic car rally to Belgium, I took a detailed series of pictures of this 1964 Cortina Lotus. Originally Ford owned, subsequently it had four private Essex custodians, before passing back to the company. This well-sorted example is retained by Ford UK in their English Heritage collection. *Author*

➤ No fancy fat exhausts here, but this was the ultimate middleweight race car of the mid-'60s and remains a classic racing choice today. *Author*

Ford recorded the mileage as just 5,575 in August 2005! The car led a sheltered life at Heritage, as when I logged the fuel consumption in June 2006 my start mileage was 5,662. Sadly, I was forced to abandon it to Ford Heritage technicians in Belgium after we had covered another 286 miles at an average 23–24mpg: more later about that Belgian handover.

This example of the Cortina Lotus proved that the specification of these twin-cam cars not only shifted when produced at Lotus in Cheshunt, but also at Ford of Dagenham. It was a leaf-sprung example, but the lightweight battery, with a regulation motorsport master switch, remained in the boot along with the brace bars that were thought only necessary with A-bracket axles. Similarly, a keen owner could still acquire the lightweight body panels and transmission castings, if motorsport was the priority for a later Ford-built Cortina Lotus. Any serious, or factory-blessed, competitor would build from a bare shell anyway.

'My' loaned C-plate Cortina cockpit featured 'Aeroflow' ventilation, but the umbrella handbrake still protruded threateningly towards your clutch leg, and the heater/ventilation control central slide and pull-push control was large enough to serve on a period hotel air-conditioning unit.

Living within the cockpit was mostly a visual pleasure, thanks to the period all-round vision provided by generous glass areas, plus classy wood finish to the slender steering wheel and Lotus-embossed gear-lever knob. The black-and-white six-dial instrumentation offered clarity and much more information than today's cars, but sadly bad news was often relayed by the ammeter, oil-temperature, water-temperature or oil-pressure gauges.

The 140mph speedometer and 6,500rpm tachometer spoke of exciting mileage, but the plain plastic seats with no backrest adjustment were a sticky experience for occupants in continental June heat. Adding to discomfort levels, the stark seating had no chance of absorbing the jolts of strapped-down live-axle suspension over Belgian cobbles.

Behind the wheel: a noisy cruiser

The 1,558cc twin-cam fitted to this example was a blue-cam-cover unit, producing 105bhp at 5,500rpm, officially coupled to 108lb ft of torque at 4,000rpm. The four-speed gearbox was a slick-shifting Ford unit that shared ratios with the contemporary Cortina 1500 GT, so no more 50mph first gear. Overall gearing was pretty short – circa 18mph per 1,000rpm. So a motorway cruise in a comparatively lightweight Cortina two-door – one with little concession to soundproofing – battered your ears. A hearty 4,000 gargling rpm from a pair of 40mm twin-choke Weber carburettors was the 70mph norm.

Performance? Despite the obvious light motorsport modifications and usage (sprints, hill climbs), the engine didn't feel that sharp – possibly because the author then usually drove a 3.5-litre BMW. Even the lighter 1963 press Cortina Lotus only just cracked 10sec for the benchmark 0–60mph. My performance memories are coloured by a colleague's practical demonstration of his company BMW M3 CSL versus the loaned Cortina Lotus during that Belgian tour – both motorsport legends initially created to win on the track. More than 30 years and over 255bhp between the pair left me eating Munich-manufactured dust – a sharp jolt to my Lotus-Ford rose-tinted spectacles, but motorsport, like war, viciously accelerates technical development.

As for many Lotus owners over 50 years ago, my 2006 journey plans were overruled by an electrical tantrum. Despite European sunshine and no short trips, the low-capacity battery and original dynamo system gradually failed to supply enough juice to ignite the twin-cam power unit. A succession of jump-starts delivered temporary

▲ No chips that Belgian summer's day, but a good opportunity to show off the traditional green stripe and a ride height that allowed the Yokohama-shod rear wheels to snuggle in unblemished arches. *Author*

➤ The DNA of million-selling Cortinas can be seen in the classic cockpit, with the effective 'Aeroflow' ventilation and clumsy umbrella push-pull handbrake. Yet the dash-full of dials, slim wooden-rim steering wheel and Lotus emblem atop wooden gear-lever surround tell you this Cortina has serious pedigree. *Author*

⮟ Ford four-cylinder foundations and Lotus Twin Cam engineering underpinned successful European Cortina and Escort competition programmes through the 1960s. *Author*

⮝ When we assembled the participants in this Guild of Motoring Writers June 2006 classic rally to Belgium, I did not know that the Cortina Lotus would not be taking me home. After Lotus restart problems worsened, the V4 Corsair 2000E alongside did that job smoothly but lethargically. *Author*

fixes, then the boot lock failed, preventing rear battery access. Gallant Ford Heritage technician Ivan Bartholomeusz eventually snaked his way into the boot after removing the rear seat, and we were able to return to our Ypres base.

Since there was no guarantee of restarts without skilled technicians in attendance, I was swapped into an F-plate Corsair 2000 for the 200-mile trip back to UK. This 69,000-miler with four doors amounted to a stretched Cortina with restyled external panels, the V-shaped prow owing inspiration to Ford USA's later Thunderbird. It was intended to fill a Ford of Britain marketing gap between the Cortina and the larger Consul/Zephyr/Zodiac bodies, manufactured 1963–70.

For me, this Corsair was an interesting diversion, as I had only experienced its compact 2-litre V4 engine in Transit vans. It could not have been a bigger contrast to the Cortina Lotus, featuring comparatively soft suspension thanks to its settings, ride height and longer wheelbase (with 3in grafted onto the Cortina floorpan, ahead of the rear wheel arches). The engine was strictly limited to 5,600rpm, but we actually only used 3,000–4000rpm as there was good pulling power, but it tended to overheat during sustained dual-carriageway use. There were some hints of the E for Executive Fords in the wooden dash and the plusher interior than that found on basic Cortinas, but we are still talking plastic seating and splashes of chrome rather than expensive upgrades.

STEERING GRAHAM HILL'S MkII CORTINA

I spent less than two months at *CCC* magazine before they let me drive several readers' and official road test cars. A dry, overcast Monday afternoon, 30 October 1967, started a trip that would change my working and sporting life.

Ford's motorsport factory

My 1965 battle grey Mini, with obligatory massive exhaust and black-tape speed stripes, heads its bumperless and noisy way from West London to Essex. After the inevitable halts to consult a map (remember them?), I arrived in gathering late-autumn gloom at Boreham airfield, home to Ford Competitions Department. This 623-acre site opened as an airfield in 1944 and saw USAAF Martin B-26 Marauder bombers and Douglas C-47 Skytrain troop transporters flying missions to occupied Europe.

Post-war, following a period of use for emergency housing, Boreham, like so many other former World War II airfields, briefly became a motor racing circuit. From 1949 to 1952, the old perimeter roads were graced by world-class drivers such as Stirling Moss, Mike Hawthorn and José Froilán González. At the Festival of Motor Sport in August 1952, a big meeting that attracted a reported 50,000 crowd, Luigi Villoresi set the final official lap record at 103.45mph in his works 4.5-litre Ferrari 375 when winning the *Formule Libre* race. The circuit closed shortly afterwards due to financial pressures.

When I arrived that October 1967 evening as a 21-year-old newcomer to motorsport journalism, on an annual living wage of £1,040, Ford owned Boreham, having bought the site for £60,000 in 1955. The company redeveloped those outer perimeter tracks into a three-mile, four-lane, circuit. Used initially for heavy trucks, it was also the test circuit for the Transit – the original white van – to run development disc-brake mileage.

Boreham eventually became Ford's 40-year motorsport base, occupied by the competitions engineering, preparation and subsequent RS engineering and managerial staff from 1963 to 2003. In 1967, when I shamefacedly hid my grubby 850 Mini in the car park at the rear of the long hangar-like HQ, former GP driver Henry Taylor was the Competitions Manager. His deputy – and the man who hired the late, great Hannu Mikkola as a youngster – was Bill Barnett. My guides were chief rally engineer Bill Meade and Ford's sports media 'Mr Fixit' Stuart McCrudden, later better known for coordinating the Fiesta Challenge one-make races, and for running media-friendly race programmes, such as fielding disc jockeys Noel Edmonds and Mike Smith. The latter competed to a pretty serious level as Stuart managed Mike in Escort RS Turbos, including bringing home a 1986 Snetterton 24 Hours race victory, before Smith graduated to Sierra Cosworths with 1990 British Touring Car Champion Robb Gravett.

Under the skin: Graham Hill's Group 5 Cortina

I was present in 1967 to see how the company prepared its second-generation Cortinas with Lotus-Ford twin-cam power for the rigours of rallying. Specifically, the doomed 1967 British RAC Rally, a world-class international that was subsequently cancelled due to bovine foot-and-mouth disease. You can judge that rally's importance to Ford back then, as the single-storey workshops and stores reverberated to the soundtrack of the preparation of six – yes a full half-dozen – Cortinas, including one for double F1 World Champion Graham Hill. Two factory drivers – Sweden's Bengt Söderström and Welsh hero Tony Chappell – were in less-modified Group 2 machines, as Söderström was well placed in the European Championship for Group 2 cars,

'BOREHAM EVENTUALLY BECAME FORD'S 40-YEAR MOTORSPORT BASE, OCCUPIED BY THE COMPETITIONS ENGINEERING, PREPARATION AND SUBSEQUENT RS ENGINEERING AND MANAGERIAL STAFF FROM 1963 TO 2003.'

having scored an outright victory earlier in the season.

The man who represented British rallying at international levels, Roger Clark, was in one of four more radically reworked Group 5 Cortinas. Most were built from bare bodies, with just basic wiring and brake-pipe plumbing installed. They carried Perspex side and rear windows, and reworked versions of the live rear axle: Ford factory Cortinas went rallying with their own locating-link/radius-rod layouts, not the Lotus A-bracket.

Tecalemit Jackson mechanical fuel-injection featured on the 1967 Cortina sextet, with a boot full of intricate high-pressure pumps and associated plumbing around the 10-gallon safety fuel tank. The boot also held a pair of 6J Minilites within chunky loose-surface tyres, and the battery with an adjacent electrical master-cut-off switch.

The twin-cam motors were at that time built at Boreham's own engine department, led by Lancastrian Peter Ashcroft. Peter would be British Ford Competition Manager in later years, and the softly-spoken engineer was also a key human component in the racing overhaul of the German Ford Capri V6 motor, so we'll encounter him again later.

Boreham's engine-building department housed some strong talents, who later stood out long after Ford stopped constructing its own competition engines. Notably, David Wood (Metro 6R4 engine guru decades later) and Terry Hoyle, who became such a specialist engine wizard at his Maldon premises that a classy diet of Merlin V12 aviation engines (best known in Spitfire, Hurricane and Lancaster warbirds) passed through his Essex unit, alongside Ferrari's classic V12s and Audi's unique five-cylinder rally units for David Sutton-tended teams.

Back in 1967, those Lotus-Ford four-cylinder engines were subjected to 60 hours intense labour during the RAC Rally. Power wasn't the priority – just under 140bhp at 7,500rpm – but accessible low-down grunt and durability were demanded. In addition to the consistent fuel delivery provided by the fuel-injection system, an essential ingredient in allowing access to torque was the camshaft-profile choices provided by Cosworth's CPL2 design or BRM's Phase 3 alternative.

'THE MAN WHO REPRESENTED BRITISH RALLYING AT INTERNATIONAL LEVELS, ROGER CLARK, WAS IN ONE OF FOUR MORE RADICALLY REWORKED GROUP 5 CORTINAS.'

The iron blocks were 'selected' – a favourite Ford PR expression that has at times encompassed alternative castings, which wasn't so difficult in period, as Ford factories still followed in-house manufacturing policies that meant they had their own iron foundries. The factory rally cars ran 1,599cc (80.98mm bore x 77.62mm stroke) engines, but later 1,800cc was a common Escort stretch, and club-racing legend David Brodie had an expanded 2.1-litre twin-cam that was utilised in his Escort and Elan.

Cylinder-head work was extensive and – as for the BMC A-series powering Minis yesterday and today – expert hand-finishing over one and a half days yielded results beyond the mere fitment of performance parts. Sure, 1.562in oversize inlet valves were installed, and the compression ratio was raised to 11:1, but it was the manual skills employed in expertly reshaping the combustion chambers, and fine-finishing the inlet tracts, that were particularly precious in the years before computer-aided measurements and digitally programmed milling machines became commonplace.

To complement that head handiwork, all reciprocating components were precision-balanced, including Cosworth pistons, BRM connecting rods and a crankshaft of unspecified origin operating to a 7,500rpm limit. The rally engines were tested in conjunction with their four-speed, uprated Ford gearboxes (complete with 50mph Lotus first gear) for a running-in period. Once installed, the twin-cam utilised a production Cortina Lotus exhaust manifold mated to Ford Performance Centre underbody tubing to the tail pipes.

While the Cortina bodies sat in their birthday suits, electricians moved in to prepare the systems for the multiple additional tasks that rallying demands, including supporting a quartet of Lucas high-output spot and fog lamps, utilising an

▲ The 1967 factory Cortina with Lotus twin-cam power assigned to Graham Hill was loaned for my first factory rally car test, published in *Cars & Car Conversions* early in 1968. The muddy venue was on Surrey farmland, as rented by my brother, Hugh. *Jasper Spencer-Smith*

alternator (dynamos were used previously) and providing an accessible centre console to carry the switches for those auxiliary lamps and the wipers. The sheet-metal console also carried Tudor four-jet screen-washer switchgear, a flexible navigator's lamp and a two-pin auxiliary electrical socket. Incidentally, Söderström's Cortina carried Hella lighting, a supplementary voltmeter, and an uprated seat to support the burly Swede.

The interior, in 50 shades of black, was dominated by a fixed-back seat for the driver and an adjustable-rake version for the co-driver, who also had the traditional Haldex time-and-distance measuring equipment before him, featuring twin clocks. Instrumentation remained as for the contemporary Cortina Lotus/1600 GT, with the addition of a large orange warning light, triggered by low oil pressure. Other sensible interior kit included a fire extinguisher, quick-action jack (adapted from the contemporary German Ford Taunus) and a wheel brace.

Job done, after six weeks of skilled labour? No way! Next, the versatile fitters had to prepare the service vehicles.

On this occasion they were fielding rare (fewer than 950 built) Corsair estates, introduced in 1966 and constructed by E.D. Abbott at Farnham, Surrey. The Corsairs were equipped with 2-litre, 88bhp V4 engines, and oversize Minilite wheels carrying the fat and chunky treads of Goodyear Ultra Grip tyres (as I would experience on the factory rally Cortinas). Those Corsairs weren't the Ford factory service-vehicle legends remembered and often recreated fondly today, but they proved capable enough to carry a roof rack with six spare wheels and tyres, plus all the welding gear, tools and spares needed to support a substantial works team.

Behind the wheel: taming a thoroughbred

Thursday 30 November 1967, just a month since I first visited Boreham, and I was transported back past the high-security fencing by *CCC* editor Martyn Watkins, this time to pick up a Group 5 Cortina to test drive over the weekend and return the following Tuesday. I knew it was a significant assignment for me, as both Watkins and Ford PR's Harry Calton had to trust a newcomer – who had yet to complete three months as a motoring writer – with a full-blown factory Ford competition car. Just getting the temporary insurance cover for an under-25 driver had been a nightmare for both men – and now I was about to disappear into the December murk conducting triple the horsepower and ten times the value of my Mini daily driver.

It was an astonishing act of faith, but the Competitions Department staff knew how to tame young hooligans. Bill Meade stepped forward and introduced me to UVX 649E, a traditionally side-striped Group 5 Cortina hiding behind a quartet of large auxiliary lamps. It carried 'Graham Hill' script on the right-hand driver's door, a big deal, as the elegant and moustachioed GP star had won his first world title in 1962 and would win a second in 1968.

Bill Meade – the legendary technician who made so many hands-on modifications to Cortinas and Escorts – is no longer with us. However, Bill's legacy, which ran through to later classic RS showroom Fords, lives on, together with that of his equally resourceful and creative colleague, Mick Jones, who worked on competition cars not just in Britain but also South Africa. Mick did much to make the 4x4 rallycross Capris a reality in the usual (for Ford at Boreham) abbreviated development timescale, making use of parts-bin adaptability.

Valued at the time at £3,000, this Cortina edged close to showroom Porsche 911 money after six weeks of skilled Ford labour lavished on stripping and rebuilding almost every component. The first step in picking up this thoroughbred Ford was to sit in the reclining Microcell passenger seat, to see how this radically reworked Cortina should be driven free of public-road restraints. Meade took first stint behind the deeply dished leather-rim steering wheel and we both fastened our four-point safety harnesses.

Before us lay a nearly empty 3-mile perimeter track with a central 1.5-mile ex-runway that would serve to ascertain the acceleration and gear-speed statistics. The road-legal exhaust crackled to life as the whine of the injection pumps was overwhelmed by a conventional key starter. I reached out for the grab handle and ensured my feet were braced on the navigator's essential foot bar.

The perimeter-track preview of this special-brew Cortina's capabilities was initially an overwhelming welter of 7,500rpm blaring, as the backdrop of bleak Essex airfield scenery blurred while the Ford assumed angles of drift and three-figure speeds that were totally outside my limited experience. I think it is relevant to quote a sentence from my March 1968 report: 'This was an enthralling experience, watching just how easy it is to hold a Cortina Twin Cam in a 100mph controlled slide (yes, fully controlled!). Even a sudden transition to a damp surface didn't upset

GRAHAM HILL'S GROUP 5 RALLY CORTINA LOTUS

Collated December 1967

Engine

Configuration: Four-cylinder inline, twin overhead camshafts, eight valves **Capacity:** 1,599.1cc **Bore and stroke:** 80.98mm x 77.62mm **Compression ratio:** 11.25:1 **Fuel system:** Tecalemit Jackson fuel-injection **Maximum power:** 137bhp at 7,500rpm

Transmission

Type: Four-speed Ford, close-ratio gears **Final-drive ratio:** 4.7:1, with Salisbury Powr-Lok limited-slip differential

Suspension

Front: Uprated MacPherson struts **Rear:** Semi-elliptic leaf-sprung live axle, with Group 5 location links and radius arms. Adjustable rear dampers

Brakes

Front: Production-based 9.62in front discs, with Ferodo DS11 pads
Rear: Production-based 9in x1.75in drums, with VG95 linings

Weight

Kerb weight: 2,016lb/914kg (showroom, 2,025lb/919kg)

Performance

Measured by Ford at Boreham, Meade/Walton, 30 November 1967
Gear speeds at 7,500rpm: First 48mph; Second 73mph; Third 90mph; Fourth 110mph **Acceleration:** 0–30mph 2.0sec; 0–40mph 3.5sec; 0–50mph 5.8sec; 0–60mph 8.2sec; 0–70mph 11.8sec; 0–80mph 16.9sec **Standing ¼-mile:** 16.1sec at 77mph **Maximum speed:** 110mph at 7,500rpm limit **Fuel consumption:** Overall test figure, 19mpg

the car at all, it merely slid a bit more.'

Those initial track laps were particularly awe inspiring, as I knew I would soon have to demonstrate to this hard-bitten expert that I was competent to take their hand-built baby away for five days. Knowing that Bill Meade had sat alongside legendary drivers such as Jim Clark, Roger Clark and Graham Hill, and most works-contracted rally and touring-car aces, just made the idea exponentially more daunting. The task was slightly simplified for a stranger because Hill's machine had been uniquely specified with showroom Cortina Lotus SE camshaft profiles, to allow the broadest and most-accessible path to rapid-response pulling power.

For my track laps under Meade's courageous mentoring, I started from rest with about 2,000rpm as I eased the clutch home. We needed at least those revs to overcome the surprisingly high level of tarmac grip afforded by plump Goodyears and a Salisbury Powr-Lok limited-slip differential.

The Cortina proved very friendly, and under Bill's tuition I learned to apply opposite lock and progressively release power to produce a lame, lower-speed imitation of those slithering initiation laps. Meade ensured I built up some tarmac speed and familiarity with this 2,016lb (914kg) package of cutting-edge 1967 engineering. Just the next day I would slip over to my older brother's rented Surrey farmland to put in some serious time learning those loose-surface power-slide tricks enabled by a honed rear-drive layout – tricks that stood me in good stead for my 1968 and '69 autocross seasons and more. So, I have always owed Bill Meade and Uncle Henry Ford for allowing me such an opportunity.

Back at Boreham on that last day of November 1967, Bill built enough confidence in me that we could proceed to the central runway and assess the drag-strip performance. The full figures and an abbreviated specification appear in the panel opposite. I can highlight that a high differential ratio for rallying (4.7:1) allowed no more than a 110mph maximum speed. The sprint from 0–60mph occupied 8.2sec, and the standing quarter-mile 16.1sec, coupled to a terminal speed nudging 80mph. Not fantastic figures judged today, but back in that rallying period, the Cortinas tended to be powerful and friendly enough to win. However, rivals in really slippery-surface conditions – flyweight rear-engine Renault Alpine opposition on Monte Carlo's most challenging icy mountain passes are a good example – pinpointed that front-engine/rear-drive traction in a modified production saloon had inbuilt handicaps.

The Escort's lighter body and sheer power in Cosworth 16-valve forms (with as much as 110bhp advantage over this Cortina) made it a world-class competitor. Yet Ford did not have the homologated saloon-car hardware to win the world's most prestigious rally – Monte Carlo – until the '90s era of 4x4 Sierras and Escort RS Cosworths.

The Hill Cortina was softer and easier to drive than its siblings – or the 1971–75 Escorts I drove during my subsequent Ford Motorsport employment – because of those SE showroom-specification camshafts and their bias towards public-road use. This meant that the car's driveability was equally suited to covering Essex dual carriageways, crawling through London traffic or ploughing around a fog-dampened Surrey field.

In period, Ford of Britain's sports staff were spoiled by the instant and outrageous pace of Jim Clark in the 1966 RAC Rally with the MkI Cortina Lotus. The Scot ran as high as fourth in the event, and his pre-rally speed impressed even the 'seen it all, got the T-shirt' Ford insiders. The 1967 grapevine suggested that Graham Hill had to put in repeated hard graft to get the best out of a competition car, whereas Jim just did the business through a golden shot of natural talent that remains legendary. In fairness, Jim Clark

'THE CORTINA PROVED VERY FRIENDLY, AND UNDER BILL'S TUITION I LEARNED TO APPLY OPPOSITE LOCK AND PROGRESSIVELY RELEASE POWER TO PRODUCE A LAME, LOWER-SPEED IMITATION OF THOSE SLITHERING INITIATION LAPS.'

had much more slippery-surface experience in his early career than Hill, and stories abound of the Scot's ability to tame even the massive Ford Galaxie on icy Scottish roads.

So with hindsight, when I started stirring the long central gear lever of Graham Hill's Cortina Lotus and got a few more miles under my belt, it became obvious that this was actually a very straightforward drive, even for a 21-year-old apprentice.

Under the heading 'The Versatile Projectile', my 54-year-old period piece contained these road impressions: 'You can trickle it through traffic with clutch released at 1,500rpm. It is outstandingly tractable in traffic, but has real teeth on the road or rough. For overtaking on a crowded main road it's the cat's whiskers. Once you've got 2,000rpm on that dial, it is off. The great thing about it for road use is that the urge is right where you need it; that's to say from 20 to 100mph, which comes up bloody fast.'

This factory Cortina was a magnet for unwelcome constabulary interest. With an efficient quartet of auxiliary lamps up front – plus Graham Hill's sign-written name above the traditional Lotus side stripe, plus both Lotus and Tecalemit Jackson badges strewn about the back end – it was bound to attract attention. One line from my period report was: 'The fuzz even tried getting us in on a loitering charge during the foggy part of the test period!' Naturally, there was a tale to tell behind that incident…

It was very foggy on 1 December 1967, especially on the way up to brother Hugh's farmland above Epsom race course. Our photographer and company art manager was a fearsome bearded character with a voice to match the booming tones of Brian Blessed. Keeping Jasper Spencer-Smith in convoy, along with the inevitable lurking police car wondering why this flash motor was proceeding so slowly, was nerve wracking. We did get stopped, but it was just to ask why Graham Hill's name was displayed on anything but a single-seater racing car.

'FOR OVERTAKING ON A CROWDED MAIN ROAD IT'S THE CAT'S WHISKERS. ONCE YOU'VE GOT 2,000RPM ON THAT DIAL, IT IS OFF. THE GREAT THING ABOUT IT FOR ROAD USE IS THAT THE URGE IS RIGHT WHERE YOU NEED IT; THAT'S TO SAY FROM 20 TO 100MPH, WHICH COMES UP BLOODY FAST.'

My brother's leased 70-acre farm was one of the last still remaining in this posh part of upland Surrey, but the clinging fog saved us from interference by the neighbours. I found a fallow five-acre grassy field and got to work as quickly as possible. I was cocooned in a warm, dry, comfy cabin, learning in absolute ecstasy how to power the amiable Ford into full-lock oversteer slides, and how to turn it into a third-gear muddy curve of my own making, with a weak, but effective, imitation of a Scandinavian flick: scrub off excess speed without braking, then just steer into a broadside slide away from the corner direction before turning the steering back into the corner, and hopefully slithering safely through, flushed with success.

I subsequently found this does not work at lower speeds in an Audi quattro, without the friendly run-off areas I had at the farm. Photographer Norman Hodson captured that '80s incident (ironically outside the Boreham Ford base). The expensively tuned Audi reversed out of the ditch into which it had nose-dived, unharmed…

Back at the farm in December '67, my brother obviously thought I had been enjoying myself too much, and joined in the mud-slinging fun with a large, turbocharged Dodge truck. Looking like a World War II Jeep, the ex-film-set Dodge, steered by brother, once ambushed me in the murk. I was exceptionally grateful for the on-board four-jet screen-washers to restore vision after receiving a coating of turbocharged turf. To add a little spice to the photography, we pursued the unfortunate Jasper until he retreated into a clump of nettles, while we played an automotive version of 'Here We Go Round The Mulberry Bush'.

After an hour or more of dynamically educational playtime, we decided to leave the now rutted trails of our progress before the neighbours precipitated another police encounter.

STAFF DELIGHT: BRM TWIN-CAM MkII CORTINA LOTUS

When my *Motoring News* mentor Andrew Marriott left that specialist newspaper in 1970 to manage a Formula 3 March race team with Tom Walkinshaw on the driving strength, I was left with a wonderful present: a 37,000-mile MkII Cortina Twin Cam, complete with British Racing Motors (BRM) engine uprated from 109 to 124bhp.

I used the 'paint-it-plum' Cortina to go to my first *MN* track test at Silverstone. I got a bit over-excited and found that, even on wide Avon Safety tyres and Lotus-supplied alloys, it wasn't quite quick enough to avoid becoming rather heavily engaged with Northamptonshire Constabulary…

Originally, the car had been a standard MkII Cortina Lotus, but before I became the custodian of this staff machine it fled to BRM's Lincolnshire Customer Division. The misfiring 27,000-mile motor was rebuilt to BRM High Performance (road use) specification, retaining rejetted Weber 40 DCOE carburation alongside a host of replacement parts, including uprated water and oil pumps, with Boreham's oil cooler and plumbing added. A power curve was supplied and revealed 53bhp at 3,000rpm and 108bhp at 5,000rpm – very similar to the Lotus twin-cam's official 109.5bhp peak at 6,000rpm. Finally 6,500 BRM revs released 124 horses.

The rev limit was 7,000, and there really was little point in using that much, save for the acceleration tests. The big benefits of the BRM engine were the power to overcome the extra weight and boxier outline of the Cortina MkII body, plus a wider spread of useable power that was particularly useful for our staff in Central London. Outside the city, I reckoned 85–95mph was 'effortless', but I was young, and deafened by contemporary rock music and too many Minis. Raised on a diet of motorcycles, then noisily tuned/competition cars, with the Lotus Seven my firm favourite (this was the year three of us assembled an S4-1600 GT 'bathtub' – see Chapter 17), my definition of 'effortless' was definitely suspect.

I covered less annual mileage than expected in this Cortina, as road tests of many makes and conversions were a weekly routine. Also, I was unexpectedly gifted an almost brand-new Capri, a model that became pivotal in my working life. I do remember that BRM conversion was effective and enjoyable. Also, the MkII Cortina Lotus/Twin Cam proved a lot more predictable, spacious and reliable than its predecessor.

BRM twin-cam MkII Cortina Lotus performance

Figures recorded with fifth wheel, at Chobham test track, Surrey, by Marriott/Walton, summer 1970 (the figures in brackets are for same Cortina, same track, showroom trim, prior to BRM conversion).

0–50mph	7.2sec (8.0sec)
0–60mph	9.8sec (11.0sec)
0–70mph	14.6sec (15.0sec)
0–80mph	18.9sec (20.5sec)
0–90mph	27.2sec (31.2sec)
Standing ¼-mile	16.1sec at 77mph
Maximum speed	110mph at 7,500rpm limit
Fuel consumption	Overall 10,000 miles, 21.7mpg

I returned to my rented rooms in London SE15, just above Peckham Market, then one of London's tougher neighbourhoods. The extrovert Cortina could be safely stored overnight at a local garage, but washing it down on Sunday certainly drew an audience as word of its presence spread.

Monday, and the Cortina Lotus burbled through the South Circular rush hour to the Ealing, West London premises inhabited by Speed Sports Publications. After a full work day, the car then had to spear east to Essex and an overnighter with the Editor, before a sad early-morning return to its Boreham prison. The staff looked relieved to have their baby back unscathed, and Zurich Insurance were equally glad to have me off their books.

I never did get over the enjoyment of my first encounter with a factory Ford, and I would be employed by the company five years later. And Zurich Insurance were notably cool about having me back on their ledger with a privately purchased Escort Mexico.

THREE SAVAGE CORTINAS: 1967–95

The Cortina Savage was a conversion carried out by Jeff Uren and his Race Proved Performance and Racing Equipment Ltd (and later Jeff Uren Ltd), with the car's defining feature a 3.0-litre Ford V6 engine transplant. The majority of cars produced were based on MkII Cortinas, with around 1,000 examples built in total.

Under the skin: Savage modifications

I did a double take on these Uren/Race Proved Ford V6 Cortina experiences because the VYP 11G registration plates of the car appearing in my contributions to *Classic & Sportscar* (February 1986) and *Ford Heritage* (March 1994) were identical. Such high-quality restoration was credited to the car's first owner, Mike Birch, a restoration professional who bought this Savage new and rebuilt it, with some unique features in 1976, selling a decade later. A model-aircraft flier and designer, with aerobatic-championship form, Birch's working life featured many other prestigious pre- and post-war vintage and classic machines, including a fine V12 Lagonda.

My Cortina Savage experience actually went back further – late 1967 for January 1968 publication within the pages of *CCC* to be precise – as an anonymous new boy picking up a two-door demonstrator in West London and accompanying Martyn Watkins to performance-test that example (XJH 234F). We also interviewed Jeff Uren, 1959 British Saloon Car Champion in a Zephyr 6 and proprietor of Race Proved Performance and Racing Equipment Ltd. There were some very talented Race Proved employees over the years, including ace expat driver Brian Muir, promising British rally and race driver Mike Hibbert and David Price. Cheery Dave graduated to serious motorsport management, from Formula 3 through British Saloon Car Championship Rover V8s to international sportscar racing at Le Mans.

I revisited the Race Proved premises at 177 Uxbridge Road, London W7, quite regularly as the company continued to produce Savages in two- and four-door forms, with the occasional estate and regular conversions of privately owned Cortinas. Race Proved was fine fodder for a motoring writer, as its engineers implanted more powerful V6s in early Capris (Commanche), and V8s in later Capri IIs (Stampede), long after I left CCC. In total, Race Proved assembled over 1,700 converted Cortinas and Capris, the Savage Cortina easily the best seller.

Technically, the Savage V6 version of the Cortina MkII sounded simple, but involved detailed redevelopment beyond a simple conversion to the 3-litre Essex, iron-encased six cylinders. In 1967, that 3-litre engine was big by European standards, and its main attraction was flexible torque (192lb ft at 3,000rpm) rather than outright power. The Ford-quoted power of 144bhp was more realistically reported on the German DIN system at 128bhp produced at a sedate 4,750rpm. At that time, the V6 was a softie employed in the bulky Zephyr and Zodiac barges that Ford advertised as possessing 'A Mark of Distinction'.

Subsequently, in 1971, for the Capri 3000E/GT there was a proper V6 upgrade which provided more horsepower (136bhp DIN at 5,000rpm) and a reported 174lb ft of torque, again at 3,000rpm. It was this uprated motor that proved durable within VYP 11G after two of the earlier V6s expired during the car's first 25,000 miles.

Although substantially heavier than the contemporary Cortina's four-cylinder, the V-motor was compact, and installing it within the Cortina's generously proportioned engine bay was not a problem. However, the three-branch steel exhaust manifolding and ancillary oil and water plumbing, including enhanced cooling with thermostatically activated electric fan, demanded engineering ingenuity. Developing the suspension to cater for the higher speed and weight loads took a quoted 1,100 hours and 30,000 test miles. The results were most evident at the front, where the team employed sturdier side rails in the engine bay and reinforced the strut top mounts. Spring rates were increased to cope with the extra engine weight, delivering a ride height similar to the contemporary Cortina Twin Cam. Plain 5.5x13in steel wheels, covered by Goodyear G800 radials back in the day, also came from the Twin Cam.

BATTLE STATIONS

The V6 shoot-out to end them all: hand-built Cortina Savage versus mass-market Mondeo. Jeremy Walton sits in judgement. Photos by David Wigmore.

Over 1000 second-generation 1600E Cortinas got the Savage V6 treatment in the '60s. Can this legendary hand-built individual be compared with a mass-market Mondeo V6?

Ford is justly famed for bringing the affordable V8 to the American masses — and the unholy partnership of Bonny and Clyde to Bank of Middle America's 'unauthorised demand' tills. But Ford in Europe's track record is in manufacturing masses of V6 engines, long before every European manufacturer 'had' to have a V6.

From the production line, Ford of Germany made a variety of cast iron Cologne vees of four and six cylinders. These went to the US in large Capri numbers and propelled their version of Cortina (Taunus), Sierra and Granada/Scorpios, as well as in Capris, from 125-188 production bhp.

Ford of Germany continued to offer a 24V iron and alloy bent six of 2.9 litres. A unit that was successfully redeveloped by Cosworth to give the old Scorpio soldier better breath and competitive 204 bhp performance in 1995.

The Brits went their own V6 way. Their 60-degree V6 family was colloquially called Essex, its 1966 cast iron components created at the Dagenham foundry. Installations included the Transit van, Capris of all three generations, late '60s Zephyr/Zodiacs, early Granadas, plus plenty of export/specialist car sales. That meant the rugged but heavy V6 became the base beneath some South African production Cortinas and Sierras (XR6s).

Britain was not allowed a Cortina V6 until the 2.3-litre German unit was imported for the last editions of Cortina. That Cortina 3-litre V6 gap in the market was served from 1967 to 1976 by Jeff Uren at 177 Uxbridge Road, Hanwell in West London. He made some 1700 Ford-engined transplants of Cortina (Savage 3000E), Capri (Comanche as a V6, Stampede with a Boss Mustang V8). The Cortina and Essex V6 combination was by far the most popular.

From his retirement base in Devon, Jeff Uren told us: "We made over a 1100 such Cortina-based conversions."

More than 600 were the type tested here: a 1600E Cortina II model from the first couple of years.

The pedigree of each genuine Cortina Savage V6 traces back to Ford's first serious efforts to dominate what has become the British Touring Car Championship. John Willment and his Ford dealerships in West London

The rear suspension was simpler, relying on radius arms and leaf springs like those of the contemporary Cortina GT/1600E and Cortina Twin Cam, but adjustable Armstrong dampers were used. Traction was addressed with the inevitable Salisbury Powr-Lok differential of the period and uprated half-shafts. The final-drive on the 1967 and 1995 test cars turned at 3.77:1, allowing about 18mph per 1,000rpm. That meant 117mph at the standard 6,500rpm limit using fourth gear in a contemporary four-speed Ford Corsair 2000 V4/Cortina Twin Cam gearbox. A very clever option featured on the 1995 Savage was a Laycock overdrive, which owner/restorer Birch had ensured worked on all four gears, effectively providing eight forward speeds!

Other Savage features included an auxiliary eight-gallon fuel tank (taking the total capacity to 18 gallons), Contour front seats, a dished 1600E-lookalike three-spoke steering wheel with leather rim, plus a wooden gear-lever knob, footrest to the left of the clutch and a 140mph speedometer. Options included 6Jx13in Minilite road wheels (two of my trio were on shiny 1600E Rostyles) and Weathershields Sunway Slimline sliding roof.

Speaking in 1994, Jeff Uren assured me: 'We made over 1,100 such Cortina-based conversions' – and at least 600 of those were based on the Cortina 1600E. Each took up to 10 weeks to complete. I learned in 1985 from Mr Uren that Savage development was assisted by Ford: 'Ford discreetly gave us their blessing and a lot of test information that helped us brace the body correctly for the extra loads… What sort of welding would be needed to chassis rails, that kind of thing. We submitted the car – based on the legendary 1600E – to Ford for approval. That 1600E was a fantastic car for £865.'

Price? In September 1967 Race Proved charged a total £1,365, and Mike Birch paid £1,750 in 1969. Value today? First find a genuine one, then be prepared to pay over £40,000. One example I tracked was £31,000 in 2019 and subsequently sold by a specialist dealer with an asking price of £45,000. Another modern listing proved that you could just take your Cortina and have it

▲ Into the 1990s I still drove V6 Cortinas, in fact this was the second time I had driven this pedigree Savage edition, VYP 11G, which had the benefit of the 1600E's plusher fittings. I had also steered it for another magazine in the 1980s, when it was fresh from an extensive restoration. *Author's archive*

rebuilt by Race Proved to Savage specification. Just such a vehicle – originally a 1600 GT Estate – was authenticated and sold in summer 2021 by a Ferrari specialist for an undisclosed sum.

There are plenty of replicas, and some will be excellent with the benefit of more recent technology and a conscientious conversion, but a lot more will amount to secondhand MkII Cortinas with retro-fitted V6 power and (hopefully) uprated brakes and suspension.

Behind the wheel: Race Proved performance

Since I had Savage experiences over four decades, I will summarise the overall impressions and include some performance figures I had a hand in, published in January 1967. That first Savage demonstrator (XJH 234F) was not built on my preferred 1600E base. My preference for that starting point is that it gives you a very handsome cabin and a look that I think is appropriate to the key characteristic of the Savage – that it is quite unlike its name. This automotive Savage is actually a very civilised citizen, packing flexible (high-torque) power with a motor that really dislikes delivering much useful thrust over 5,000rpm.

On British public roads, this translated into a package that offered outstanding London traffic abilities, plus some real top-gear get up and go between 50–90mph. From standstill on runs along the Colnbrook bypass (this was late 1967…), with corrected speedometer, it would munch 0–60mph in 7.5sec. That launched the Savage into the prestige ballpark (the period 6.2-litre Jensen Interceptor could complete the sprint in 7.3sec), and it was significantly quicker off the mark than overtly sporting breeds such as Alfa Romeo GTV 1.6 (11.1sec) or 1275 MG Midget (14.6sec).

Today, that 1967 Savage remains an enjoyable road smoker because of its supremely accessible pulling power and modest size, although the braking and grip offered by contemporary 175 or updated 185/70-section radials would succumb to any modest modern hatchback attack. The Savage could also supply regular overdoses of controllable, but onlooker-disturbing, oversteer slides in any gear up to third, or on any surface of less than billiard-table smoothness.

Now on to the best of the Savage attributes revealed in the March 1994-published *Ford Heritage* magazine, a feature that benefitted from detailed colour pictures of VYP 11G at Brands Hatch by the late Dave Wigmore, tested alongside a contemporary 2.5 Mondeo V6. That ex-Birch Savage was a special four-door example, then owned by Nick Blunsden, a classic race and rally winner with an Elan and Escort RS1600, who added considerably to our knowledge and judgement at Brands Hatch.

Aside from the 'eight-speed' overdrive gearbox installed on this car (actually a clunky but beefier early Zodiac four-speed rather than the slick Corsair/Cortina Twin Cam unit), other featured options included vinyl roof, shiny Rostyles, halogen headlamps, painted gold coach lines and simple V6 badges on the rear flanks.

The cockpit was well-preserved 1600E, offering six-dial instrumentation, classy three-spoke leather-rim steering wheel and varnished wood dashboard and door cappings. It made the much younger Mondeo look dull, albeit the latter benefitted from a massive boost in safety and lowered motorway sound levels. However, there is another compensation with the classic Cortina outline: visibility. It's also a lot easier to place accurately on the road or track, with comparatively boxy lines and large glass areas.

A lack of power-assisted steering was a Cortina Savage issue, particularly for urban parking chores. Twiddling the steering wheel in earnest, the Cortina's ancient recirculating-ball system felt vague and slow to react. Yet the oldie was a lot more fun on the track, albeit slower in lap time, than the front-wheel-drive, five-speed Mondeo. Where the big-engine, middleweight Savage formula really scores is in top-gear acceleration, between 50–90mph for VYP 11G (which featured a final-drive ratio of 3.55:1), and it was a civil legal-limit cruiser too, 70mph demanding less than 3,000rpm. Modernity, represented by the Mondeo, featured much lower wind noise and could honestly run 120mph on an *autobahn*.

As for my Savage conclusions, the big engine in a smaller-body car has never been more deftly executed by a British private company. It was so popular that Race Proved tried the Cortina MkIII in Savage guise, with more radical Weslake V6 tune and a glass-fibre bonnet, but that was not such a satisfactory package.

CORTINA LOTUS TO CORTINA TWIN CAM

This section, adapted from my article submitted to *Octane* magazine in 2014, analyses the key differences in character between the original Cortina Lotus and MkII Cortina Twin Cam (Lotus), as produced by Ford. Effectively, the Cortina Twin Cam (the Lotus badge was dropped after a few months of production of the MkII) became the larger-body hardware donor for the all-conquering Escort Twin Cam. It also donated key features (principally suspension settings, and dash/multiple instrument layout) to its popular prestige Cortina 1600E sibling.

In an appreciating classic-car world – and for three-wheeling around racetracks – the original A-frame Cortina Lotus remains the choice for escalating values and victories. But which Cortina, powered by the Lotus-Ford 1,588cc motor, is the more worthwhile in period production format?

Essentially, the second edition of the twin-cam-powered Cortina is more suitable for reliably entertaining road trips. The plumper MkII combines strong period pace with an accommodating larger body of uncluttered-boot practicality. If what floats your performance boat is the overwhelming need for joyous handling and measurably superior performance, take the Lotus-created original – perfect for those rose-tinted moments when only a 46mph first gear and Lotus cross-country verve will do.

But there's more to it than that.

The post-July 1964 revamped MkI Cortina Lotus deserves serious public-road consideration, wrapping up major durability improvements alongside some street comforts, all within that emotive earlier body, now lacking those vulnerable and labour-intensive lightweight alloy panels. Externally, this 'face-lifted' Cortina Lotus gained the 'Aeroflow' ventilation chrome air-outlet slots on the rear (C) pillar. The dashboard also became more comprehensively instrumented and the umbrella handbrake was (theoretically) lost for a conventional centre-tunnel device.

The transmission became more durable, but less lovable to the hardcore driver, losing the Elan close ratios in favour of a Cortina 1500 GT gearbox with uprated second gear, which contained a 'mind the gap' step between the caravan-towing first-gear ratio and second of too many period Fords. Dispensing with all the light-alloy transmission castings and sorting out the rear-axle A-bracket location durability was important for road users.

The 1965 summer saw another radical transmission alteration, the replacement of the A-frame with leaf springs and an overall rear-axle commonality with the mass-production Cortina 1500 GT.

I also experienced the South London street ecstasy of the original Cortina Lotus in its purest form – on the days it ran predictably – when I landed the weekly 1967 task of exercising an original A-bracket example for a private owner away on military service. The slim wood-rim steering wheel connected to recirculating-ball hardware, but delivered near rack-and-pinion precision at the cost of demanding muscular inputs at urban speeds. An array of proper black-and-white dials was welcome, along with wood-topped remote gear-change of period excellence.

When you did find a clear patch of public tarmac, acceleration and handling supplied madcap memories galore. The bank robber's favourite 3.8 Jag was a faster motorway muncher, but for twisty terrain or race-track pace, the Jag's sell-by date loomed. That Lotus adaptation of the Cortina packed enough clout and cornering capability to humiliate all but the best of Coventry's graceful old warriors. The inevitable clusters of Mini Cooper buzz bombs were outgunned and out-fumbled, unless it was wet, or a diesel spill ambushed you on a favourite roundabout.

Less welcome on the first Cortina Lotus was a tough ride and the tricky clutch/engine rpm balance in traffic. For that sportier cabin, the umbrella handbrake was a shiny Ford oddity, poking your kneecaps from under the dash. Slippery showroom PVC front seats begged for replacement, delivering little more support than a temporary overdraft.

The 1965 successors became more Ford and less Lotus, in fact the C-plate original I drove in 2006 was assembled at Dagenham. It was a different drive to the original, boasting 'Aeroflow' ventilation and a slick Ford gearbox and clutch, rather than elongated Elan ratios. It retained the tough ride over road imperfections and much of the Lotus aura in terms of mind-reader steering and alert responses, but the debut Lotus format remains perceptibly better for tracks and smoother tarmac.

The larger, less aerodynamic and slightly slower 1967–70 MkII Ford Lotus edition had strong grip by contemporary standards, an uncluttered boot and a better ride, but wasn't so exhilarating to drive. Yet there was a better chance of getting to your destination!

CHAPTER 4

AMERICAN HEARTS

Homeland-built performance Fords

Lest this Englishman forget, you could argue this chapter, and the following chapter focusing on the GT40, are the most important in this book, as they underline that Ford Motor Company USA is the parent of all the charismatic cars described in these pages, wherever they were built. Ford has become a multi-national with a global reach from its Dearborn, Michigan HQ established over decades. The sprawling corporation is now a mammoth that has become not just an American institution but an iconic worldwide brand – its Blue Oval, surrounding elaborate Ford script, is now as familiar as the Stars and Stripes. Ford uniquely retains members of the founding family prominent in its Dearborn organisation; at the time of writing, William Clay Ford held the title of Executive Chair, Alexandra Ford English was Global Brand Merchandising Director, and Henry Ford III remained an active Board Member.

HORSES FOR COURSES

I apologise in advance for the jumbled order in which American Fords appear in this book. I ordered them by the years in which I experienced them, and that means some '60s examples were driven in the '90s and the 21st century.

Unlike most other chapters, the words here are devoted to multiple models rather than a single one. Here you will find Mustangs ancient and modern, Thunderbirds rented and thunderously raced, plus the Mustang's Falcon father, raced in the UK. Oh, and the biggest of all the American Fords I experienced – the Galaxie and Fairlane 500s from the '50s and '60s.

All of the automobiles in this chapter were mass-produced in huge numbers to be supremely affordable, as far from supercar exotica or tightly restricted homologation specials as you could get. The Mustang alone sold over 607,000 in its best year (1966). Over 2.9 million Mustangs in the original and classic 1964–73 formats fed the

Ford corporate cash pile. More than 10 million cars carrying the running wild-horse nameplate were reported to have been built as at August 2018, following a 2005 rebirth and return to the retro styling cues of the '60s. Ford's Marketing Manager in 2018, Jason Mase, summed up the Mustang's continuing role at Ford at its emotive best when he asserted that the Mustang had become: 'Kind of the soul of the company.'

All US Fords featured here were remarkable and enjoyable road or race-track rides, as I will describe later, but first I will explain how these heavy-metal instruments became such a personal obsession.

Born on 4 July, and hero-worshipping my father who worked in a shipping business that thrived on North American trade, I suppose my US bias was to be expected, and I've enjoyed every one of my many visits to the States. The Big Country and its equally large automotive products from Detroit's then unchallenged 'Big Three' (GM, Ford and Chrysler) bit me even as a schoolboy. A penchant for buying American car magazines (*Motor Trend*, *Hot Rod*, *Car and Driver*), rather than dry but factually strong British weeklies, drew patriotic derision from school friends and family. The US bug also meant my writing style was easier to adapt to both the slang of embryonic British publications and contributing columns and features for an American magazine over a 19-year stint.

Over the years, I received more generosity and awards for my media and book work from the US than any other country, including my UK base. I was 13 when my Dad died, so my mother and adult brother were saddled with my stroppy

▲ In 1995, I carried out a twin track test of Amanda and Nick Torregiani's 494bhp GT350 and 589bhp notchback Mustang racers at Silverstone. *Ford Heritage magazine*

'ALL US FORDS FEATURED HERE WERE REMARKABLE AND ENJOYABLE ROAD OR RACE-TRACK RIDES, AS I WILL DESCRIBE LATER, BUT FIRST I WILL EXPLAIN HOW THESE HEAVY-METAL INSTRUMENTS BECAME SUCH A PERSONAL OBSESSION.'

DRIVEN

Road cars

Mustang Mach 1 Test for *Motor Sport*, March 1970 issue; pitted 7-litre V8 against Broadspeed 1600 GT modified (road) Capri. *Motoring News* 15 March 1970 report, Frank Gardner in Group 2 Boss Mustang debut Brands Hatch. Boss fan, forever! **Hertz Mustang** Rental car, 302 V8 notchback, 646 ANE; driven Los Angeles–Newport Beach–Corona del Mar. **2.3 Mustang** USA–UK import, loaned privately owned road car, 132bhp/110mph, appeared in *Motoring News*. **Mustang SVO** PR trip courtesy of BF Goodrich tyres/MPH, turbocharged 4-cyl, 2.3-litre Fox-bodied Mustang at Nelson Ledges track, Ohio and Akron town. **Thunderbird V8** Rental car, Newport Beach, US; covered 888 miles of commuting to Riverside raceway and more. **1966 Mustang** Race Mustang feature for *Ford Heritage*, courtesy Andy Kirk; restored 289 V8 2+2 road GT fastback, with uprated steering and suspension by Andy Dawson. **1966 Mustang** Shared Californian/Monterey drive to drive-in dance; cherry red notchback, courtesy photographer owner Mike Valente, 2015 **Mustang 5.0 GT** Test-day drives around Reading, UK, EN65 NKF, RHD specification; impressed.

Competition cars

Falcon Sprint V8 Silverstone GP circuit track test, 1968, Martin Birrane's Group 2 example; exhilarating laps, but spun halfway to Becketts from Copse! **5.0 Mustang II** Snetterton track test, Vince Woodman's 5-litre V8 Group A Mustang in (compact) Fox body. **1965 Mustang 289** Goodwood track test, notchback reworked for UK classic competition, 5.3-litre V8 in place of 4.7, owner Steve Warrior, driver Phil Wight, preparation JTM (Jim Morgan) Racing Services; total 7 races with 6 wins as tested. **Galaxie 500** Mallory Park track test for *Classic Cars*, February 1992, updated for *Ford Heritage* in 1995; car filled crowded track, packed 600bhp/550lb ft torque to shift 3,300lb (1,497kg) swiftly – and civil with it! **Falcon Sprint V8** Track test and 1992 FIA one-hour Silverstone Historic Festival race, finished 14th overall, 4th in class, driving shared with Bob Sherring, owner Mark Dees, USA. **1966 Mustang** Snetterton, Birkett Six-Hour Relay 1992; practised on ice, didn't get a race within the Historic American Racers Association team. **NASCAR Thunderbird** Track test at Brands Hatch, 5.9-litre V8, 720bhp, 1,600kg, Roush prepared for Mark Martin. **Two race Mustangs** Silverstone track test for *Ford Heritage* feature, November 1995, Nick Torregiani notchback classic touring car racer, 589bhp and 1m 10.05s on Silverstone Club circuit. Also tested Amanda Torregiani's GT350 Mustang, 494bhp and 1m 10.9s.

teenage years, emphasised by my love affair with motorcycles. They were not overwhelmed when my career took me to US-owned Ford: as I was born and bred in the Thames Valley, MG at Abingdon would have been a natural geographical choice. I was just very lucky not to take that route, as the British Leyland era and the closure of the competitions department foreshadowed even grimmer times ahead for the legendary Abingdon site. By contrast, the dollar-rich British arm of Ford flexed sporting muscle and allowed opportunities to access some great period products and influential personnel, along with their motorsport stars. So, that schoolboy bias paid off long-term.

As mentioned previously, my first direct contact with Ford cars began when I was 21, via Graham Hill's RAC Rally-prepared Group 5 Cortina MkII (see pages 47–53). When I got the 1968–69 break of writing full-length features about Ford of Britain's Boreham competition centre, I became conscious of the US cash sources underpinning major motorsport programmes. Fulfilling competition commitments from Formula 1 to world-class rallying and myriad other motor-racing formulae required parent company cash and trust. Nobody at Ford in Europe measured substantial PR or motorsport budgets in English pounds or German Deutschmarks: dollars ruled.

Overturning the 1960s Auto Manufacturers' Association (AMA) agreement between America's 'Big Three' not to promote horsepower and performance to the public, and implementing its Total Performance programme, the rebellious Ford USA took its Blue Oval to legendary Le Mans victories, plus success in touring-car racing, NASCAR, rallying and, as an engine supplier partnering with Cosworth, Indy Car and Formula 1.

I retain a soft spot for the V8s America churned out by the million, yet not all my US Ford experiences were with a V of eight cylinders. The smaller-bodied Mustangs I steered in the '80s came with varying degrees of 2.3-litre 'Lima' four-cylinder power, as did the Andy Rouse XR4Ti, which was really a Group A adaptation of the Merkur sold in the USA, and is detailed in Chapter 11.

FALCON: 1968 TRACK TEST

My very first taste of American V8 power came in pretty dramatic form in 1968 at Silverstone with Martin Birrane's proven Group 2 Falcon Sprit racer, an exhilarating session that included this track test apprentice spinning at Copse corner in an expensive haze of tyre smoke. Some 34 years later, I was at the wheel of a race Falcon again, this time for a one-hour FIA international championship event for classics. Despite shared wheelwork with lead driver Bob Sherring, it had been hard work in a very original American-owned example of the 4.7-litre Falcon Futura Sprint. We had finished well down the field, amongst the Falcon brigade, and I wasn't feeling particularly chirpy, especially when Mr Birrane (who owned Lola in later years) walked over with a big grin on his face and reminded me that, following that Copse track-test spin, I still owed him for four flatted 1968 race slicks!

Back in the winter of 1968–69, the Birrane Falcon, in a sandy beige suit, was an established part of the British racing scene. We had included this car in the February 1969 *Cars & Car Conversions* feature, which also covered the Berpop special saloon (Chapter 1) and Roger Nathan's Imp-powered GT. On this day, monitored and tutored by Steve Neal (Steve was a leading British Saloon Car Championship Mini Cooper driver, and later father of BTCC driver Matt Neal, and founder of Team Dynamics), we left my baptism of V8 firepower until the afternoon of our Silverstone GP track session.

Under the skin: authentic American muscle

Back then, Martin Birrane was a youthful and very busy property magnate, so left his fearsome Falcon to our tender care while he attended to London business. As was common in British racing of the late '60s, the Falcon Futura Sprint V8 was a sold-on part of the original 1963 shipment to Alan Mann for use on the 1964 Monte Carlo Rally, where an example placed second and won the GT class. That version of the million-selling Falcon compact V8 coupé was specially developed and homologated within FIA regulations to compete internationally in 1964, the year that the Mustang would debut,

The Ford Falcon was intended as a compact economy car to fight GM and Chrysler competition, but simple underpinnings and docile acceptance of V8 performance parts meant it served as an effective showroom and circuit predecessor to the Mustang. *Author*

▲ In 1968 and 1992 I drove Falcon Sprint V8s at Silverstone. The earlier outing was part of my novice three-vehicle test session, so I went from the four cylinders/135bhp of Chapter 1's Berpop to eight cylinders/320bhp in a rather larger outline. In 1992, as pictured, I shared the Mark Dees-owned 4.7-litre Falcon Futura Sprint V8 with Bob Sherring in an FIA Silverstone event. *Author's archive*

using many Falcon underpinnings. Yet the Futura Sprint interpretation of the Falcon was no homologation rarity, as over 10,479 were made in 1963, and 13,830 were built with the 1964 coupé body.

As presented at Silverstone in 1968, the 1963 Birrane example had seen many drivers wrestle with the left-hand-drive layout. Most successful was John Whitmore, contracted Alan Mann Racing (AMR) driver when this Falcon had been converted to full race trim. Sir John (a hereditary baronet) set the first 100mph saloon-car lap at Silverstone GP circuit in the car before it was sold on to Peter Bloor Racing, when subsequent 1971 Monza GP winner Peter Gethin steered it. John Ewer ran it for veteran Formula Ford front runner Syd Fox, and Cobra conductor Eric Hauser owned it long enough to paint it his trademark purple, before Birrane bought the car for the 1968–69 seasons and had it more soberly presented in sand yellow with black side and bonnet (sorry, 'hood') stripes.

It didn't have much success when raced by either Fox or Birrane in the Group 5 form in which I tested it with Neal senior, but it wasn't fully developed in an era when the factories were going all out with Group 5 modifications. The specification embraced the Mustang/Falcon performance option of 4.7-litre (289cu in) V8. It didn't have the useful GT40-developed Gurney-Weslake alloy cylinder heads, fed by quadruple Weber downdraught carburettors, nor was the exhaust system to full race specification. More relevant to consistent lap times was the uneven action of the brakes, the rear still comprising drums (with grabby sintered linings) that did not balance out with the proper full-size ventilated discs up front.

However, the Birrane V8 did kick out a reported 320bhp, enough to deliver a good mid-range punch with a kerb weight of 2,688lb (1,219kg). For a novice race driver – I had only Hillman Imp autocross experience and an 850 Mini road car at that time – it had plenty of power and was a thrill to experience. Of course, there was a fabulous soundtrack, and memorable acceleration to 130mph, as my normal competition drive had been powered by less than 75 grass-circuit horsepower.

Behind the wheel: spinning classes

I began to feel a little foolish in crash hat and race overalls for an initial 30mph photo session with Max Le Grand, but there were a few opening comments that I'd like to share from the February 1969 published test: 'The Falcon started immediately on flipping the ignition and dropping a starter switch. I pushed the long-travel gear lever forward… and forward, until I thought all was well. Let the clutch in slowly at 2,500 revs and mumbled down the pit road. The clutch was very light, but I noticed tons of travel between gearshifts. It snorts along quite nicely in third with the engine running at 3,500–4,000rpm.'

Post-photography session, I was unleashed: 'Concentrating on keeping the willing eight under 6,000rpm, I try to remember the jumbled impressions of riding round with Birrane that morning, not much help as there was no passenger seat and clinging on for dear life had seemed the priority. I regroup my thoughts at the start of a flying lap into the first corner, Copse. Change down to third and try to place the Falcon tight around the curve, ease the throttle respectfully down and feel all those litres hurling it towards Maggotts, which I can take in fourth [top back in the day]. With light, sensitive steering, there's no discernible body roll and it settles into the corner well. Brake very hard and into third for Becketts, give it a fair bootful on the exit and feel the seat trying to force a way through my spine. Chapel Curve left is taken flat in fourth (I'm not going that fast anyway!), which means using all the road out onto the Hangar Straight. Now it is pulling an easy 5,000rpm, approximately 130mph.

'Bellowing along the straights nicely gives me confidence, far too much confidence… The brakes are reassuring at first, but I did seem to be entering Silverstone's quick curves faster than anticipated and exiting on modest oversteer lock. I think the times were beginning to look acceptable, and there were thumbs up from the pit wall.

'I started to push harder with only a few laps under my safety harness. Abbey is on with the throttle slightly eased in fourth: this means I'm on the bumpy side of the straight and the back end of the car is skimming the (rippled) surface as I put the power back on – rather like autocross. The bridge blurs past and the grandstands on the outside of Woodcote look damn solid. I chicken out and change to third into Woodcote, fourth for a resounding thunder down the grandstand straight. Flash under the second bridge, keeping well to the bumpy left-hand side. This is the corner (Copse) where it all happens a few laps later. I over-brake on the way in, and while the car is still unstable I put on too much power…

After a long spin on near full opposite lock, which failed to curb the provoked mass, the Falcon came to rest safely on the outside grassy verge – today a massive, smoothly surfaced run-off area to suit Formula 1. 'I undo the harness to take a look around [a private test-day allowed that luxury] and so that I can turn the ignition off. Everything looks fine, except for the thick deposits of rubber on the track. I can't get the straps done up properly again, so I fire up the engine and rumble back to the pits. When I get back everyone is lying over the pit rails, laughing. Steve Neal, my monitor, and an experienced front-line saloon car racer, observed shrewdly: "Stacks of power to the point of being embarrassing!"

'There's lots of gear movement, and I kept dropping out of third. The instrumentation is bitty, but adequate. It's very quick, but the impression of speed is deceptive when braking and positioning the car: however, you soon get accustomed to this. The brakes aren't so good; when you put them on the car judders, especially when they are really hot. They also tend to knock on and off quite badly. The Falcon's very stable at speed, the good steering and handling with a relaxed ride making it easy to drive. There's little or no roll on cornering and no dive under braking. With the power on it tends to oversteer…'

▼ There was a charismatic Australian iteration of the Falcon V8 (seen here at Bathurst) in obsolete two-door format, driven nobly by Allan Moffat/ John Fitzpatrick. For a Ford fan in 1979 Australia, it was a rugged experience at the hands of superior GM-Holden forces. Folk hero and outstanding driver Peter Brock allowed me into his home with a friendly welcome and I had some of the most memorable weeks of my motoring life down under. *Ray Berghouse*

FALCON: 1992 ONE-HOUR HISTORIC RACE

My 1992 rendezvous with a Falcon Sprint began, as my memorable automotive outings often did, with a casual Gerry Marshall introduction. Brave Bob Sherring had been a sports-car salesman and was a regular classic-car racer, famed for an enormous (by UK standards) 7-litre Galaxie 500. He also had some very nice road cars, including a scarlet Mustang amongst a small herd of classics kept alongside his clothing business in Hinckley, Leicestershire. Besides which, he and his wife Yvonne were excellent social and circuit company.

This Sherring meeting came after we had participated in a TVR Tuscan Championship round at Mallory Park. Gerry had won rather easily, while I was inordinately smug about a tricky sixth place in the two-part TVR V8 blast (the accelerator stuck open on occasion as a reaction tester), after fending off another freelance motoring scribe. Overcoming some naughty-boy reprimands from the stewards with a second overall in my earlier race, featuring the Collins Performance Engineering Sierra RS500 (described in Chapter 13), also set me in a good mood for post-race dinner with the Sherrings.

Bob nominated me as second driver in a 4.7-litre (289cu in) Falcon Sprint V8 of startling period originality, owned by soft-spoken but knowledgeable and generous-to-a-fault American lawyer Mark Dees. Here are some edited period observations of that exciting opportunity, as published in *Motor Sport* in November 1992.

'Silverstone, Northants. It's late July and the second-biggest crowd of the UK season has come to bask in the nostalgia of the third Christie's International Historic Festival. On race day, the jams stretch back to Towcester. Old, familiar faces are almost as numerous as the hundreds of Ferraris assembled in and around the central paddock. Some stray far from the purist crowd. They have come to witness the FIA European Challenge for Historical Touring Cars, a one-hour recreation of the pre-1966 contests waged between Alfa Romeo, BMW, Lancia, Austin-Morris and Ford. The Opel Olympia Rekord does not fool them with the cuddly toy in the side window. Some serious racing with factory money is still to be found in a series all too frequently marred by eligibility disputes and post-race disqualifications.

'Pacesetters became Scuderia Bavaria, with a trio of 192bhp 1800 Ti saloons. Backed by BMW, with a budget reportedly running beyond two million Deutschmarks per annum (over £700,000), the team deploys wickedly effective preparation. Although these four-doors are much better presented than their road-registered cousins that you will find in marque history books, their multiple-sidedraught-carburetted power units are matched by driver talent. The line-up included former World Champion Denny Hulme (who sadly died recently), Monte Carlo Rally winner Rauno Aaltonen and multiple European Touring Car Champion (for BMW) Dieter Quester. Similarly strong financial support comes from Alfa Romeo for Italian and German teams running beautiful GTA coupés.

'Ford were numerically strong, because the '60s marked the international homologation of the American Mustang and Falcon, an example of which we were to share, as well as the Cortina Lotus, winner of the 1965 ETCC in Sir John Whitmore's hands. Ford offered no financial support, but there were a dozen assorted Mustangs and Falcons, plus nine Twin Cam Cortinas. A total of 44 cars practised, and my initial priority was simply to be amongst those to get a race. Simple enough in a V8 at this circuit, surely? Not when history is being rewritten around you…

'At Silverstone, admirers of the Detroit school of iron V8 charm ran their palms slowly over my borrowed mount. Its extensive steel and glass-fibre body panels in white and blue brought onlookers into our garage as if a missing link in Ferrari history had materialised. They gazed at the Sports Car Club of America sticker on the back panel and puzzled over the number plate 'Balena'. The latter means 'white whale' in Italian, a strictly European view of what was regarded as a compact car back home in the '60s. Some even whispered reverentially of the car's "wonderful patina of constant use". A few spotted the alloy fabrication of the oil catch tank by Lockheed

Aircraft Corporation, which sent them muttering off to look at other apparently similar Fords, which actually had little in common with our car save the theoretical specification.

'At that worshipping point it was time to remind ourselves that what we had here was a re-bodied 1963 (load-bearing body) and 1964 (homologated outer panels) Ford, albeit one that has competed for most of its life with regional success in California. That this charismatic Ford, with its unique wheel arches, was present in England was due to the generosity of owner Mark L. Dees, a Californian Allard fanatic, Bonneville Salt Flats regular and attorney at law. That I was to co-drive was due to the loyal lobbying of number one driver Bob Sherring, former Speedwell employee in the Graham Hill era. A chance meeting the week before the race attracted some petrol and tyre money from the UK arm of the Australian vintage and veteran oil specialist Penrite, courtesy of well-known TR4 driver Evan Mackenzie.

'Dees had come over with the 4.7-litre Ford to co-drive it in the Mitsubishi Marathon. He retained his enthusiasm, despite a rocky ride into a Belgian marriage garden and an engine expiry that terminated the Falcon's participation in Czechoslovakia. The Falcon, formally titled Ford Falcon Futura Sprint V8, was returned to Britain and to the overall care of fellow historic Ford racer Nickie Torregiani, at Classic Affaire in Dorset. We shared Silverstone garage space with the Torregiani equipe, who ran a Mustang for Nickie and Rod Birley, plus a Cortina Lotus for Amanda Torregiani/Stephen Damant. We would not have got past scrutineering had it not been for the efforts of Classic Affaire's Doug Forbes and my lead driver's mechanic, Dave Huffer.

'By the Silverstone weekend I had grown accustomed to the Falcon's LHD girth. Besides, *Motor Sport* is not unfamiliar with the Falcon. In March 1964 the late Michael Twite assessed

▲◀ Contrasting cockpits in black-and-white and colour between the race Falcon I drove in 1992 (above) and a classic car show road-car example (right). The race Falcon carried a period competition seat and aircraft lap belts that were replaced before the UK technical inspection, contrasting with the bench seat and no belts in the road car. *Author*

a factory rallying example, one of 14 Falcons used by Ford for its 1964 Monte Carlo Rally assault. That practice car of 28 years ago had its 4,727cc uprated by Carroll Shelby to give 285bhp at 6,000rpm, a capacity we shared, though our version of the iron eight was assembled by ex-All American Racers employee Larry Ofria at Valley Head Services, San Fernando Valley. It yielded an estimated 325bhp with a safe limit of 7,000rpm (we actually used 6,200–6,500 most regularly).

'The Dees Falcon weighed about the same 2,800lb (1,270kg) as the factory Ford of 1964, and we initially used a 4:1 final-drive from the Marathon specification, not dissimilar to the 4.5:1 of the factory rally car. The latter, tested in 1964 with a then new *Motor Sport* magazine fifth-wheel speedometer and stopwatch, returned 0–60mph in 6.9sec, 0–100mph in 16.9sec and a standing-start quarter mile in 14.8sec.

'Following a fraught practice session, spent with the engine hammering away at maximum rpm for much of the 3.19-mile Historic GP lap, we opted for a 3.7:1 final-drive in the (very) live axle. This naturally meant we gave away a lot on acceleration, compared to that rally car of the '60s, but gave the engine an easier time for a one-hour race and improved consistent lap times.

'At a midweek practice session on Bruntingthorpe airfield, I had grave misgivings of whether we would qualify. Nickie Torregiani had brought along the Mustang in which he had set the 1991 record for historics in FIA European trim at 2m 21.9s (81.08mph), the same black V8 as he would share with Rod Birley in 1992. Driving said Mustang alongside the Falcon I saw little comparison. They may both have been based on the same floorpan and shared the common starting point of pushrod eight, four-speed BorgWarner with Hurst quick shifter, double-wishbone front suspension, and multiple-leaf and Panhard-rod-linked live axle. However, one was a soft tourer, rich in history and period furniture, while the other Ford was a ferociously quick device that turned into every corner very rapidly, showed a clean pair of heels to a Sierra Cosworth four-door beyond 137mph and braked with equal efficiency.

'BOB NOMINATED ME AS SECOND DRIVER IN A 4.7-LITRE (289CU IN) FALCON SPRINT V8 OF STARTLING PERIOD ORIGINALITY, OWNED BY SOFT-SPOKEN BUT KNOWLEDGEABLE AND GENEROUS-TO-A-FAULT AMERICAN LAWYER MARK DEES.'

'Sherring and I sat down with pen and pad to list what needed altering, just to get a run for the Falcon that weekend: we noted 10 items, including removal of a pair of nail scissors and the radio! Then we thought about lap speed and asked for a foot brace, plus a Corbeau race seat and modern harness. I felt like a vandal modifying such a genuine car, but Classic Affaire not only carried out that list but made sure that – in every instance – it could be returned to original specification, including blanking off that protruding fuel filler with its emotive chain-link cap retention. There had been no time to work on the handling, and this immediately proved to be our biggest Silverstone handicap. That we qualified 28th of 44 cars, with a lap of 2m 24.52s (79.61mph) and within 0.47sec of the Torregiani Mustang, was a most satisfying personal lap.

'Both drivers knew that such a sideways performance and continuous use of 6,700rpm in top were unlikely to bring a racing result. An overnight change of differential – we finally had four to choose from – and a last-minute tyre swap, saw Bob Sherring lined up for the 1pm Sunday start. We were way behind the trio of factory-backed BMWs that dominated practice in 1–2–3 formation, but even the fastest Falcons had been almost four seconds adrift of the four-cylinder BMWs' pace. I don't remember results like that when I was a *Motoring News* saloon-car reporter in the late '60s; then, V8 power reigned supreme.

'Still, I was happy just to be taking part. The meeting had the most relaxed air of affluence and pleasure, something that used to be a feature of Silverstone Grands Prix in its 'garden party' era. The opening laps saw the lead BMW of Quester/Aaltonen stamp away from its team-mates and the Alfas. Best Brits were Hadfield and Schryver in the A-bracket Cortina, but even that flying machine lapped more than two seconds slower than the factory BMW's amazing record lap of 2m 13.86s (85.95mph), which was fractions faster than Quester had managed in practice. Again, history was being rewritten, for BMW legend Alex von Falkenhausen picked the 1.8-litre capacity for the four-door saloon to avoid direct confrontations with the 1.6-litre Alan Mann Cortinas, knowing them to be faster.

'Our race progressed as quietly as it can in a 4.7-litre V8, which had a penchant for imitating a forest-rallying Escort driven by a deranged Finn with an alcohol-abuse problem. In fact a very sensible Finn, Aaltonen, climbed out of the winning BMW at the finish and smiled at me: "That slides a lot, even for a Ford. I don't think it should be that bad. How does it feel?" I replied ungraciously of my faithful race companion that it was "a very big taxi cab".

'I had spent much of my closing stint looking for the chequered flag, startled to discover that the 13 laps I covered felt more like 133. In the large and airy cabin, I cursed that we hadn't put a piece of masking tape on the straight-ahead steering position. In the slow infield sections (Priory, Brooklands and Luffield), neither Bob nor I could keep tabs on how many turns of opposite lock had been applied. It was desperately embarrassing if you came into that complex with any rival, especially one that stopped and handled, for the braking took an agonising age. Then you would put on an act like a performing seal, trying to balance 25cwt, 325bhp and tiny contact patches. To be fair to the Falcon, it never did spin. To be fair to Dunlop, there was still work left in those abused covers, despite their obvious (FIA-mandated) under-size dimensions on such a heavy car. It may help to understand the relative lap speeds when I tell you that the Scuderia Bavaria BMWs used a broader-section tyre for a car that must weigh 800lb (363kg) less.

'Much of my race was spent in solitary confinement scanning the SW gauges. I had the moral lift of passing another Falcon, which had qualified at the back of the grid, at regular intervals, and also had time to feel sorry for Torregiani as he surveyed our progress from the banking, the Mustang's motor having expired in his stint.

'Although we had lots of minor motor troubles in pre-event running (the rockers clouting the ornate aluminium covers), the high-compression V8 ran perfectly at the easier 6,500rpm of the taller differential (circa 120mph in a straight line). It completed 25 laps at a 77.32mph average, a race speed that covered the slickest of pit stops and driver changes, an art that we hadn't previously practised at all. We simply talked through a good exchange, which worked better than we had a right to expect.

'FINALLY, LONG AFTER OUR CELEBRATORY GLASS OF CHAMPAGNE BECAME A MEMORY, NEW RESULTS WERE ISSUED TO CLASSIFY US 14TH, FOURTH AMONGST THE V8s.'

'We had rumbled our way to 17th overall of 31 finishers, but that was before the apparently inevitable disqualifications. BMW lost its third-placed 1800 Ti (Denny Hulme and Prince Leopold von Bayern). In the end, five cars were excluded, most of them Minis.

'Finally, long after our celebratory glass of champagne became a memory, new results were issued to classify us 14th, fourth amongst the V8s. Our best lap was in the closing stages of the race, a tenth slower than I managed in practice. A more important factor had been that, with the taller diff ratio, the Falcon stayed in the same second bracket for both drivers, most of the time.'

Looking back, both Falcons, driven 24 years apart, delivered adventures galore and seem overlooked today amongst the Mustang masses. Probably best enjoyed as a road car with a traditional sharpening of chassis and brakes, and replacement of the original 164bhp street 4.2-litre (260 cu in) V8 with the 4.7-litre version homologated for competition.

Footnote *In 1993, as a comparison with this Falcon racing experience, I did drive a Scuderia Bavaria 1800 Ti/SA BMW at Paul Ricard for* Classic Cars *and* Roundel *magazines. Tutored by Rauno Aaltonen, the 1963–64 BMW came across as swift as a 20-years-younger M3, with many lightweight materials and 192bhp performance.*

MUSTANG: GETTING ACQUAINTED

Ford unveiled the seventh-generation Mustang at the 2022 Detroit International Auto Show. The runaway horse is the only Ford nameplate I drove to survive as a current car – and, actually, the only 'car' in the 2023 US Ford range on offer to the public in a sea of SUVs and electrically powered Fords to face the future. It appears there will always be a Mustang of some kind on offer, since the advent and sales success of the all-electric Mustang Mach-E.

Personally, I drove more Mustangs than any other US Ford product. I have listed as many as I have details for in the 'Driven' panel on page 60, but have been very selective about highlighting specific varieties in the following features. I decided to concentrate on those that either made the biggest impression or were outside the media mass-coverage of cult models such the 1969–70 TransAm racing legend Boss 302 or Steve McQueen-linked *Bullitt* models.

January 1970 marked my first physical contact with a Mustang V8 and then it was thanks to *Motoring News* colleague Chris 'Wittypaldi' Witty. The former child star (he was in the original film version of *The Railway Children*) had blagged an excellent red 1969 Mach 1, complete with a US-style record player that only offered Elvis Presley's Elvis' Golden Records Volume 1. More relevantly, we chuckled over its wayward style in a wet British winter, automatic transmission firmly kicked down to wastefully wheelspin away all that V8 energy on street or track. It was a very comfortable ride, but we mischievously pitted it against a well-modified Broadspeed Capri 1600 GT, with 106bhp, at the Chobham military test track in Surrey. The only picture I can retrieve today shows the pair locked in contact on the outer circles of the skid pad, front wheels cocked as the Capri tries to ease by inside, while those optional 7 litres (428ci) and 335bhp of the Mach 1 kick the Mustang wider. Not sure we came to any meaningful conclusion, save that a red Mustang was definitely a better choice for attracting the opposite sex and police curiosity.

▼ Expat Aussie wheelmen at Silverstone. Steering an iconic ex-Kar-Kraft/Bud Moore 5-litre Boss Mustang from 1969 is versatile champion racer Frank Gardner. In pursuit is the Brian Muir ex-Penske Camaro Z28, raced in the UK by the private Malcolm Gartlan equipe. *Mike Hayward Collection*

Later in 1970, I got an opportunity to learn Mustang V8 lore a little more thoroughly in its homeland. For an October dream debut trip to the USA, where I interviewed major motorsport and tuning people on their premises, including Roger Penske and Dan 'The Man' Gurney, I rented a cooking 302/5-litre V8 notchback from Hertz in a startling lemon yellow. I was supposed to have had a sportier Mustang direct from Ford PR and its local Los Angeles office, but my flight from the East Coast Penske appointment was so late that the arrangement blew out.

However, registration 646 ANE proved a friendly girl and took me on some fine adventures during nine days in and around LA. Our first trip was to head south in deepening 7 o'clock twilight on the San Diego Freeway, until I saw a sign that contained the word 'beach'. So Newport Beach became my base on the West Coast. I found the equivalent of a British B&B on a beachside road and lived the Californian beach dream between outings that included Orange County Raceway's drag strip, Dan Gurney's nearby nest for his Eagle operations, Traco (famed for competition Chevrolet V8s some of their best work was on Penske's AMC Javelins for TransAm) and EMPI, the supreme VW performance and custom house of the era. I also rubber-necked in Hollywood and visited the Briggs Cunningham Motor Museum, most of these doors flung open generously thanks to my masters at *Motor Sport* or *Motoring News*, as I was a 24-year-old nobody.

It is worth recalling some of the things that attracted me to America over 50 years ago. Fuel was less than 25p for their admittedly slightly smaller gallon, so I barely worried about the 250bhp Mustang's 14.1mpg in my eager hands. Even better, a full breakfast steak with two eggs and hash browns retailed at just over 50p! At first sight, Californian girls became a major plus point, but when you've seen 20 cloned in identical blonde/white-teeth/tanned-skin bikini uniforms, the company of individual females with alternative alluring outlines becomes a trip back to reality.

I illicitly checked the performance of this 10,000-mile Ford, but had no independent timing gear to call on, so called on my old-school

◀ The summer of 2022 saw me back at Thruxton for a Hendy Ford dealership day with breathtaking precision/stunt demonstrations. As this poster depicts, Paul Swift deploys a very early UK 5.0 GT RHD convertible, one that had seen UK Type Approval duties at Ford. *Author*

CCC magazine experience. We did some simple 0–60mph runs, the average hovering around 8sec, after I had observed 110mph was available, which tied in with the statistics for a 200bhp Fastback GT.

I returned the Mustang just prior to making an anticipated 1,000-mile trip to Monterey to see the CanAm series in late-season glory. Thanks to the generosity (that word again, but that open-hearted 'can-do' attitude characterised my experience) of *Road & Track* magazine staff, I then took on a press-loan Datsun 1600 SSS, a reliable four-door answer to the 1.6-litre Cortina 1600 in E or GT guises. The Datsun completed all my unreasonable demands and came back for more. Particularly memorable was the beautiful Highway 1 coastal road (even more so back then), and my open-mouth spectating of Chaparral's 'sucker' (fan) car running rings around the previously dominant McLarens at Laguna Seca.

I had a truly inspiring introduction to America and regular Mustang motoring a half century ago, and that enthusiasm lives on, although subsequent Mustang versions with bloated bodies and low horsepower, and compact four-cylinder variants, would test that loyalty until the car was reborn in the retro versions we see today.

MUSTANG: ROAD AND TRACK TRIO

An autumn 1995 track and local roads day at Silverstone became one of those sunlit days you never want to forget. The ingredients were three 1966 classic V8 Mustangs: a fastback 289 GT of 200bhp for the road, loaned by owner Andy Kirk, and two racers. One of the track cars was a GT350 with just under 500bhp from 4.7 litres, and the second a notchback sporting nearer 600bhp from a stretched 5.3 litres, both belonging to Nick and Amanda Torregiani, regular participants – and outright winners – in the contemporary Classic Touring Car Championship.

Fastback 289 GT road car: under the skin

I used media colleague Kirk's 289 GT as a benchmark for the radical performance changes carried out to make the Mustang a consistent winner 30 years after its debut. For the showroom and track the formula sounds simple: V8 motor up front, live-axle rear constrained by multiple links, double-wishbone front suspension, and low weight: around 2,400lb (1,089kg).

Andy's 1994 purchase was a very easy drive, with three-speed Cruise-O-Matic auto gear-change and power steering. However, 'Captain' Kirk had taken the 81,500-miler into Andy Dawson's Silverstone Dawson Auto Developments (DAD) premises (former international rally driver Dawson is Mr Abarth Fiat in the UK at the time of writing) and had some detailed work carried out that made it a much better drive. The list included dropping the ride height 2in, applying negative camber to the front suspension, and cutting the slack out of the production power steering. Andy commented: 'It still feels like you are driving on ice, with virtually no feel until you get used to it.' That low-effort steering acclimatisation was pretty swift, and not dissimilar to pre-'80s Jaguars, but not so accurate as a Coventry Cat.

As for the exterior, Ford wanted you to know you were in a Mustang-branded vehicle, and applied chrome plating enthusiastically. The interior quirks included a disc-brake insignia on the relevant one of just two floor pedals, and the handbrake was one of those cumbersome under-dash pull-push umbrella affairs. The plastic three-spoke steering wheel was overlaid with 'drilled' chrome effects. Small four-dial instrumentation (fuel, oil pressure, water temperature and amperes) supported a 140mph speedo but there was no rev counter. A useable rear seat was fitted, the fronts notably unsupportive for cornering, but excellent for hours of cruising American expressways.

▼ A memorable Silverstone road and track test with three V8 Mustangs, the Torregiani racers proudly bearing a Maltese cross as witness to their owners' home territory. The road car in the centre belonged to media man Andy 'Captain' Kirk. *Author*

▲ I was privileged to be a passenger and to attend 2000s classic car pub meets in this 1966 V8 fastback. One of the best examples I encountered – civil, handled well, and with enough punch to entertain. *Author*

Fastback 289 GT road car: behind the wheel

Driving the white-with-black-interior Mustang 30 years ago, my impressions included: 'The flexibility of the '60s lightweight V8 is such that a second-gear start is entirely practical.' I used Drive for fully automatic shifts and found changes were, 'delivered with more civility than a current Scorpio exhibited recently at a fraction of this icon's mileage.'

Almost inevitably, I compared the visibility within with that of the Ford coupé I knew best: 'The view ahead is that of a power-bulge Capri on severe steroid dosages. The front wings frame that view, which lacks the big central air intakes of the racers. Whereas the Capri had a sort of post-box slot of a rear window, the Mustang Fastback stretches forever and is much less claustrophobic than its European echo.'

I did not expect much in the way of perceived performance from this showroom Mustang, credited with 200bhp gross, probably 185bhp by European DIN standards: 'The nose rears up when the throttle is flattened and you gallop away to the impressive backbeat of those twin exhausts, but I doubt it would frighten a current hot hatchback. Even at 185bhp, the power-to-weight ratio (and torque) should be slightly superior to that of most V6 Capris, so it is totally at home in current traffic. It really is so easy to live with. I even reversed it into a parking bay with less bother that my '90s Euro saloon, thanks to feather-light steering, outstanding vision (by Capri standards) and that amiably grumbling giant of an engine.'

More compliments were paid following a drive on the demanding local Northamptonshire roads, after some speed assessments and photography alongside the racers at Silverstone: 'The ride wasn't at all floppy over B-road bumps. The Mustang was mechanically content to rumble along between 50–100mph: I chose 70 and was rewarded with a more spirited rendition of the menacing exhaust beat.' The brakes inevitably came in for some criticism over the distances needed to slow or stop, but this example had Kelsey Hayes front discs from the GT options board, and I'd say they were a priority on classic American cars of all brands that are to be driven enjoyably in Europe today.

No pampered garage queen, just such an easy everyday companion with a shot of film-star glamour that is hard to equal at any price. Loved it.

Racing GT350 and notchback race cars: under the skin

The two Torregiani classic race Mustangs were also from 1966, but with differing GT fastback and notchback bodies. Outside their spacious engine bays, running gear and chassis upgrades were pretty similar, but the specialist V8s, with Trick Flow Specialties (TFS) alloy heads, were assembled initially in Florida by Southern Style Racing Engines, and tended in the UK by Steve Warrior. The engines differed in cubic capacity and power: Nick's 5.3-litre (335 cu in) version was the more radical, with the best part of 600bhp and 8,000rpm capability, but I focused on his then wife Amanda's 4.7-litre (289 cu in) Mustang GT because it was the best comparison with Kirk's amiable road car. From an ego point of view, I was almost as quick in the lady's Ford, which led to racing her Cortina Lotus, as related in the earlier Cortina chapter. I had previously experienced Nick's Mustang for the 1992 Snetterton Six-Hour Relay, when it had 4.7 litres rather than this test day's 5.3-litre monster motor.

So, from the same 4.7 short-stroke litres as Kirk's road car, Amanda's race Mustang extracted more than twice the power of the street version – 494bhp at 7,400rpm to be precise. That was backed up by 383lb ft of torque at an elevated 5,600rpm – beyond maximum horsepower revs in the showroom item – and a bonus of some 100lb ft over the road car's accessible 2,400rpm peak pulling power. Horses for courses!

The transmissions were of truck strength, and straightforward, with Jericho-rebuilt BorgWarner four-speed manual gearboxes hitched to Coleman triple-plate clutches and the same company's 9in limited-slip differentials.

'THE TRANSMISSIONS WERE OF TRUCK STRENGTH, AND STRAIGHTFORWARD, WITH JERICHO-REBUILT BORGWARNER FOUR-SPEED MANUAL GEARBOXES HITCHED TO COLEMAN TRIPLE-PLATE CLUTCHES AND THE SAME COMPANY'S 9IN LIMITED-SLIP DIFFERENTIALS.'

Possessed of at least double the horsepower and more over the road car, you would hope the racers would have equal retardation. Dream on! Although these track Mustangs were sharply improved, vented 13in-diameter discs up front deployed heavily without the aid of power assistance, and rear axles wearing 11in vented items, the astounding acceleration and sheer velocities demanded a respectful slowing-distance allowance.

Of course, similarly thorough race preparation had been applied to the production steering and suspension principles of the Mustang. Forget power steering, and heave at two turns lock-to lock – nearly half the number of twirls required for the power-assisted road Mustang layout. Suspension was stiff, and I mean 750–900lb/in spring rating coupled to competition Koni damping, a stout 1 1/16in front anti-roll-bar, and a plethora of locating links to the back axle, sitting on four-leaf springs with no rear anti-roll bar. However, when I tried the cars, Nick's sported a negative-camber rear axle (shades of Broadspeed's track tricks) and a five-link layout, coupled to 750lb/in coil-over springs – a set-up that I subsequently raced with Nick.

Wheels and tyres to support these rubber-hungry, big-horsepower racers were regulated at 9x16in covered by Yokohama A008 RS 245/45 dry-weather tyres or Bridgestone VRs for wet conditions.

Racing GT350 and notchback race cars: behind the wheel

My opening test laps were in Mr Torregiani's high-power notchback. Settled into the usual stark, stripped-cabin racer's single (left-hand) seat, there was a sizeable 10,000rpm rev counter to contemplate, along with the usual dials for water and oil temperatures and a gauge for the leaded 99-octane fuel level (consumption around 3–5mpg). That was the fuel needed to generate 588bhp at 7,400rpm in Nick's Mustang, coupled to 459lb ft torque by 5,800rpm, statistics that told you this would be some beast to drive even on a dry Silverstone track. When I subsequently came to race it, the early-morning surface at Snetterton would see light ice… Tricky!

Back to our sunlit day, and the main memory was of the enormous straight-line speed. I quote

1966 MUSTANG 289 GT ROAD CAR AND GT350 RACE CAR

(Torregiani classic race Mustang GT350 in brackets)

Engine

Configuration: Ford cast-iron head and block 289 V8 (cast-iron block with TFS alloy cylinder heads; race camshaft, oversize valves, plus uprated reciprocating components) **Capacity:** 289cu in/4,736cc (same) **Bore and stroke:** 101.6mm x 72.9mm (same) **Compression ratio:** 9:3.1 (12.0:1) **Fuel system:** Autolite division Ford twin-choke downdraught carburettor (Holley four-choke downdraught carburettor) **Ignition system:** Dual-advance distributor and coil (MSD electronic, magnetic pick-up distributor, control module and coil) **Maximum power:** 200bhp at 4,400rpm (494bhp at 7,400rpm) **Maximum torque:** 282lb ft at 2,400rpm (383lb ft at 5,600rpm)

Transmission

Type: Three-speed Cruise-O-Matic (four-speed Jericho-BorgWarner, Coleman triple-plate clutch, limited-slip differential) **Final-drive ratio:** 2.8:1 (choice of final-drives, limited-slip differential)

Suspension

Front: MacPherson struts, double wishbones (MacPherson struts, single lower track-control arms, anti-roll bar) **Rear:** Live rear axle, replacement hubs, struts, anti-roll bar and links; four-link axle location, vestigial single-leaf springs, vertical coil springs, Bilstein gas dampers (live axle, leaf springs, two locating rods, double-action dampers)

Brakes

Front: 11in/279mm solid discs (13in/300mm ventilated discs)
Rear: 9.25in x 2.25in drums, hydraulically servo-assisted (11in/279mm ventilated discs)

Wheels and tyres

Wheels: Steel, period embossed hubcaps 5in x 14in (Minilite-style alloys, 9in x 16in) **Tyres:** Non-standard BF Goodrich 205/70 radial (Yokohama A008 dry/Bridgestone wets, 245/45-16)

Body

Type: Mass-produced, unitary steel two-door fastback GT body, no spoilers; front seat single-strap lap belts, rear 2+2 seats (same two-door GT outline, Corbeau race seat, six-point safety harness, integrated steel roll cage, on-board fire extinguisher systems, glass-fibre external panels, including doors, bonnet and boot)

Weight

Kerb weight: 1,210kg (1,200kg race weight)

Dimensions

Length: 4,613mm/182in (4,520mm/178in) **Width:** 1,732mm/68in (1,706mm/67in) **Height:** 1,308mm/51in (1,207mm/48in) **Wheelbase:** 2,743mm/108in (same) **Front track:** 1,422mm/56in (1,480mm/58in) **Rear track:** 1,422mm/56in (1,480mm/58in)

Performance

Maximum speed: 108mph (170mph)
0–60mph: 10.5sec (4.0sec)

from *Ford Heritage*: 'Boosted by three times the horsepower of the road car and a lot less weight, Nick's Mustang is awesomely quick. It bellowed on to Silverstone's two straights (we used the short Club layout) with nose-up Mustang stance at a shattering 8,000rpm and terrifying velocity.' It was harder to settle into a cornering rhythm, partially owing to the high power versus traction, and equally because we were out amongst a pestilence of single-seaters that the Mustang would see off in a straight line then obstruct in the corners. However, my lap time was surprisingly good (1m 10.05s/84.8mph) and led to other adventures.

I actually preferred Amanda's GT, as it was more balanced in terms of power versus chassis capabilities, plus I had a clearer session than that with the more powerful Mustang. The result was a 1m 10.9s lap (83.77mph) that was a genuine pleasure. Here's how that third Mustang outing in a day appeared over 25 years ago: 'As a direct comparison with Andy Kirk's road car, the most

➤ In 1992, I co-drove Nick Torregiani's Mustang at the Snetterton Birkett Six-Hour Relay. Always quick, especially in icy practice, unreliability plagued that 1992 team and I did not get to race the car in that event. *John Colley Photography*

immediate impressions are of the extensive safety gear. Plus, how severely you are trussed into an enveloping seat via a full safety harness, especially as I didn't even discover the lap-only belts on Kirk's Mustang!

'Amanda's racer starts on the twist of a master switch, the flick of two high-pressure pump toggles and the push of a modest black button. The motor blares into life on a bit of throttle, mollycoddled constantly to hold a 1,500–2,000rpm "idle". The Coleman triple-plate clutch is as surly as the gearbox at rest, but the car provides good racing starts. A Coleman LSD paints black lines with glee as 385lb ft of torque melts Yokohama's finest.

'The Ford Motorsport tachometer, dominating a quintet of dials, reports that 7,200rpm is the chipped limit for this motor. Those revs flick up in the first three gears with electric immediacy. In fact second became too short for some of Silverstone's tighter curves. So, you have to balance this bucking bronco against the throttle, and steer with one hand, then whip third home and get the whole plot straight again. If that all sounds fun, it was.

'This Mustang flattered with friendly cooperation, even when its bottom was rearing way out of line and the kerbs were all that would set you back on the path to another majestic swoop past the pits.'

'THIS MUSTANG FLATTERED WITH FRIENDLY COOPERATION, EVEN WHEN ITS BOTTOM WAS REARING WAY OUT OF LINE AND THE KERBS WERE ALL THAT WOULD SET YOU BACK ON THE PATH TO ANOTHER MAJESTIC SWOOP PAST THE PITS.'

There were some detailed niggles on the obstructive four-speed at rest, and the long movements needed for accurate shifts thereafter, and a comment about the brakes not being quite as convincing as those on the more powerful Torregiani Mustang, plus there was notably more of a hop, skip and jump routine through 100mph Woodcote, but the killer concluding comment became: 'On 94bhp less than Nick's Mustang, Amanda's Mustang was within fractions of equalling the more powerful car's lap time, delivering a lap that will not be forgotten.'

As I said previously, it was just one of those perfect track days, and the generosity of the Torregianis was outstanding. Their goodwill was behind the scenes in some other personal fast Ford outings, including Nick's Mustang in the Birkett Six-Hour Relay at Snetterton and, as mentioned previously, Amanda's Cortina Lotus when she was busy showing the boys how it was done at Castle Combe in her Mustang.

COMPACT MUSTANGS: TURBO AND V8 POWER

In 1974 a new breed of minimalist Mustang, badged Mustang II, debuted and sold strongly with appropriately minimal horsepower by previous standards: a four-cylinder base engine offered 88bhp and a 2.8-litre V6 (Cologne Capri) 105bhp. Significantly, 1974 marked an historic point on the sales front: Ford hit the top of the light-truck charts with the F-branded range, but remained in its established car-sales slot as number two to GM-Chevrolet, shifting nearly 2.2 million automobiles that year.

The following year, the company cynically popped in an emissions-influenced emasculated 5-litre/302cu in version of its V8 for the Mustang II, delivering a mere 134bhp beneath emotive white and blue paint. Such eight-cylinder Mustang II variants would lead to cars that could race under international Group A rules in the '80s.

Turbocharged four cylinders

However, there was another route to enhanced horsepower for compact Mustang IIs, and that was to turbocharge the 2.3-litre South American Lima version of the 2.0 Pinto, an idea that would mutate into the Sierra (Merkur) XR4Ti that we will meet later (Chapter 11). It was the turbocharged version of the Mustang II that much-missed and highly respected colleague Clive Richardson and I initially assessed in the UK. It was 1979 before I steered the new breed of Mustang and then it was offered to us at *Motoring News* via Frank Hedley, private owner and proprietor of the Removatop sunroof business. It pre-dated the official import model that Clive would subsequently drive and loathe for *Motor Sport*.

Mr Hedley's Mustang II wore Cobra war paint (blue with white stripes) and was privately imported at the dollar equivalent of £3,500 before shipping and UK taxes, which made the final total £4,671 complete with air-conditioning, leather trim and uprated suspension in association with alloy wheels carrying Michelin TRX tyres. It was a substantial saving on the official Ford UK pricing of around £7,000.

On a truly awful British downpour day, I was grateful to have a rather better chassis than previous US showroom products had delivered in the high-horsepower days of thunder. The baby Mustang sported vented front discs (then absent on the British 3-litre Capri), carried coil springs and four-link location for the live rear axle, and benefited from the accuracy of rack-and-pinion steering that had been absent on classic original Mustangs.

Sadly, the single-overhead-camshaft turbo four-cylinder was no match for such chassis sophistication (we're talking Ford standards here, not Lotus). The Garrett turbocharger huffed just under half a bar (6.75psi) of boost from a non-intercooled location, buried beneath a twin-choke downdraught carburettor. In the USA, the result was 140 gross horsepower, but in Europe we estimated 130–132bhp at 5,200rpm – pretty close to the contemporary 3-litre Capri. Yet the Mustang, even in compact format, weighed an astounding 200lb (91kg) more than the Capri, so the acceleration runs for the benchmark 0–60mph occupied over 9sec versus a second less for the 3.0-litre Capri II. Maximum speeds were also markedly different, at 110mph for the Mustang II, and nudging 120mph for a good 3-litre Capri. This was before the introduction of the near 130mph Capri 2.8 injection of 1981 onwards.

A thunderstorm had fully developed by the time I wheeled the Ford out onto the excellent but soaked roads local to Washington, Tyne & Wear (as opposed to DC). It says much about the chassis, grip and steering that I felt happy to explore their abilities to the limit in such conditions. We had a tachometer, so I could see that the engine unsurprisingly disliked life beyond 5,500rpm, but gave its best from 1,500rpm upwards. No boost gauge was fitted, but there was a green light on the dash to tell you all systems GO!

I concluded for *Motoring News*: 'The latest Mustang struck me as attractively blending American creature comforts with European handling. What a shame the Americans weren't allowed to keep the breath-taking acceleration they used to offer in the Mustang.' Incidentally, when Clive tried the official 120bhp Ghia-badged version for *Motor Sport*, he disliked it in the strongest possible terms. On reflection, he was right about the watered-down product officially retailed in the UK, but in the mid-'80s I did get another chance to drive a turbo four-cylinder Mustang that was a lot better. Before that, however, I had a proper racing Mustang to assess…

V8 power

By 1983, a Mustang II 5-litre V8 variant was homologated as an international Group A race car, although sadly unable to compete in the British premier league, where a 3.6-litre limit prevailed. I had a unique chance to drive a Belga cigarettes-liveried example (which still appears in Belgian classic events at the time of writing) before Europe's premier 24-hour saloon car race at Spa-Francorchamps, thanks to British race friends and the patience of Belgian race-driver brothers Philippe and Jean-Michel Martin.

The backdrop for my test was the flatlands of the former USAF Snetterton airbase and the quick race track created there post-war. The 5-litre Mustang was presented in concours condition via Dave Cook, best known, like Spa-entered driver Vince Woodman, for his talents in preparing and driving winning (but by this time obsolete) 3.0 Capris.

I had driven with Vince in the now infamous second Capri celebrity race (at Brands Hatch in 1972), tested his back-up 3-litre race Capri and knew him well as a Ford dealer and independent with a long race career already behind him. The V8 Ford was an exciting prospect, but actually never fulfilled its promise in Europe. Fortunately I didn't know that, so seven laps in a 350bhp package, ballasted to weigh 2,910lb (1,320kg), were keenly anticipated.

Looking back, I can see that a carburetted and Group A-constricted 350bhp wasn't enough in an era when 5.3 litres of Jaguar V12 from the TWR XJ-S were proving their worth in the European championship. The fabled Jags were pitted against a fleet of fantastically reliable and hard-driven BMW 635s, TWR's rival Rover V8 squad, and the dawn of the Volvo turbochargers. However, this race Mustang impressed me, even from a cautious 4,000rpm starting point to a 6,000rpm rev limit. It set 1m 22s/83mph laps for an unfamiliar driver, mainly because of the relatively low 130mph maximum speed at this track, plus fabulous four-wheel disc brakes (the first time I had tested a quartet of discs on a Ford), and an easy Getrag five-speed synchromesh gearbox, used by many others including contemporary race BMW coupés. Not so good was the numb power steering and driving a left-hand-drive car at Snetterton, where the quickest right-hander sat you close to the bank at 90–105mph.

I felt privileged to have this somewhat pressured access in front of paying sponsors and the waiting Belgian drivers. I should have known from my media reporting activities that this brave project would struggle against the serious factory opposition during that European Touring Car Championship era.

'THIS RACE MUSTANG IMPRESSED ME, EVEN FROM A CAUTIOUS 4,000RPM STARTING POINT TO A 6,000RPM REV LIMIT. IT SET 1M 22S/83MPH LAPS FOR AN UNFAMILIAR DRIVER, MAINLY BECAUSE OF THE RELATIVELY LOW 130MPH MAXIMUM SPEED AT THIS TRACK, PLUS FABULOUS FOUR-WHEEL DISC BRAKES (THE FIRST TIME I HAD TESTED A QUARTET OF DISCS ON A FORD).'

Footnote *At the Spa 24 Hours in 1983, the Mustang qualified 13th and was out of the running in an accident before nine of the 24 hours had been completed. It also retired from the Silverstone TT European championship round that year after qualifying 12th when Vince Woodman shared with Jonathan Buncombe in Woodman's traditional Esso sponsorship colours. The Mustang was back in Belga livery for the late-season Zolder round, qualifying further up the grid (seventh), but again failed to finish.*

THE MUSTANG TURBO RECIPE REPEATED

Back to the 2.3 turbo Mustang formula now, and for that we need to travel to 1985 and Ohio in the USA. Here we find BF Goodrich chief tyre-development technician and part-time Porsche racer Bob Strange in an open-hearted mood. We have completed some detailed office work around the Akron base of BF Goodrich (a brand bought by Michelin in 1988) and feel the need for some track time. Bob's personal 1984 SVO (Special Vehicle Operations) Mustang and a rewarding scenic road trip out to the tight Nelson Ledges race track are the perfect answers.

After years of redevelopment around the 2.3 turbo formula, the smaller-capacity four-cylinder was offered as an alternative to the 5.0-litre V8 in the third-generation compact Mustang built on Ford's all-purpose Fox platform. By this time Mustangs had posted over five million sales.

A 1984 *Road & Track* full test drew a comparison between the diverse Mustang power units, reporting similar performance in the 0–60mph sprint, at 7.2sec for the V8 versus 7.8sec the 2.3 turbo, and the cars were even closer over the quarter-mile at 15.8sec and 15.9sec respectively. The 1984 Mustang was biased towards comfortable cruising, using a five-speed BorgWarner (the first four ratios shared with Sierra RS Cosworth initially) with an overdrive fifth allowing an 82mph pace at just 3,000rpm.

The big everyday difference between the V8 and turbo – more relevant today than 35 years ago – was fuel concumption: the turbo returned 21 US mpg versus the 5.0-litre V8's 15.5. Unfortunately for SVO higher-tech supporters, this cut little ice with traditional Mustang V8 buyers, who shied away from the turbo in the showroom (the car had become a hard sell after earlier reliability issues). To encourage buyers, an extra $2,500 worth of extras had been thrown at SVO clients in 1984, the turbo then retailing at $16,100 (roughly £11,500).

This 31,000-mile Mustang wore road-legal track tyres – back then a big area of development in all production-regulated racing formulae, and a recipe I subsequently raced or tested regularly in 1986–88 Sierra XR4x4 and RS Cosworth Fords. This meant that this street SVO Mustang had significant bonus grip on a track that was covered in slimy autumn leaves. There were other individual touches to cheer up the handling, including Koni adjustable dampers.

The basic engine became familiar to us in the UK through the winning Rouse Sierra XR4Ti (*née* Merkur), but for the showroom in 1984 it was rated at 175bhp and 210lb ft torque. It was a lazy unit, producing peak power at just 4,400rpm, and presumably designed to wean buyers off the effortless but thirstier V8s. Well, that was the intention, but Bob had installed a supplementary microprocessor to switch boost according to octane ratings, culminating in a full 1.0bar/14psi with appropriate fuel.

Despite the presence of a Porsche-embossed key fob, the SVO Mustang started readily and steered with the usual light effort on the street or circuit. The gear-change flopped around a bit on this example, but the shift was as amiable as the 1986 RS Cosworth. On track, the limitations of the powerplant interrupted playtime: although 2,000–4,500rpm was perfectly acceptable on the road, when stretched on track the engine just ran out of breath even before the low 5,500rpm warning band, never mind the indicated 6,000rpm limit.

My conclusion was much the same as for other compact Mustangs: good chassis (even better in Bob's painstakingly calibrated example), fine cabin creature comforts, but not an engine I wanted to spend time extending.

▼ A very different Mustang for the 1980s. Bob Strange, test and development ace at BF Goodrich, lent me his personal SVO turbocharged four-cylinder. After a scenic road trip, we spent an afternoon driving rugged laps at Nelson Ledges, often the scene of packed-grid SCCA endurance races. *BF Goodrich*

THE MUSTANG RETURNS: BETTER THAN EVER

A routine media drive day in Berkshire on 21 September 2016 held a welcome surprise. There, in powerful Shadow Black right-hand-drive guise, was a 415bhp Mustang 5.0 GT, part of a more serious Ford effort to sell the American icon in Europe. The base price was £35,745, that special paint and two other boring options taking it up to £37,430 – but for me that is a seriously seductive offer when weighed up against pricey prestige badges of similar performance (0–62mph in 4.8sec and 155mph limited maximum) and with less showroom equipment.

I had to fight my way to the top of a media queue to drive it, but it was worth every cajoling minute. The cockpit was finished in modest greys with leather trim adorning the compact steering wheel and everyway power-adjustable seats. There was even a footrest, and the foot pedals were sensibly spaced for serious motoring alliance with a six-speed manual gearbox and limited-slip differential. I greedily managed to take in a route that embraced dual carriageways and rumpled lanes, and returned impressed by both the bellowing power and the replacement of the traditional live axle in favour of an independent rear-suspension layout.

I was also exhilarated by the acceleration supplied by the free-revving 32-valve V8, with peak power delivered at 6,500rpm and a healthy 391lb ft of torque by 4,250rpm. Yet this sixth-generation Mustang was about so much more than power: I became thoroughly satisfied with the way it rode and handled on 19in alloys. Retardation was now up to frontline European standards, via an equipment list that included ABS-monitored six-piston front callipers and four-piston rears. There were also some track-use tech tricks, such as Line Lock (locking the front brakes while releasing the rears to enable burn-outs), launch control and a specific track app (telemetry system) deploying accelerometer technology.

I returned after those rewarding miles, convinced this was something of a performance bargain compared with comparable sporting BMW (M3) and Audi (RS5/quattro siblings) offerings – machines that were my regular fare at the time. On paper it looked as if the Mustang could hold its head up alongside my favourite modern Aston Martin (Vantage V8) and it happened that I liked the Mustang's image, although I appreciate all that filmed

▼ A media test day brought a rewarding hour with the freshly arrived Mustang 5.0 GT. This is a 2018 model, rather than the car I drove, but the muscular lines and fundamental mechanical specification remained unchanged. *Ford*

◀ ▼ Interior contrasts between 1965 and 2016. You don't realise how sombre and safety-conscious we have become until you inhabit these red and grey contrasting Mustang cabins. *Author*

glamour and reflected bad-boy glory would not sit well with the class-conscious Aston set.

Reflecting on Mustangs, it has been a hard task to decide what to leave out of this book. For me, the sentimentally best in the Ford herd are the 1966–70 models that have some welcome development over the 1965 model-year originals. Yet that 2016 demonstrator had me hooked: living in the UK, its right-hand steering and major improvements in all dynamic aspects (especially independent rear suspension), alongside that powerful image, make it my personal choice.

THUNDERBIRD: A PERSONAL TREAT

For me, driving around coastal and inland California some 30 years after Ford's 1955 Thunderbird weighed in to fight Chevrolet's 1953 Corvette, the 'T-Bird' had suffered through many bloated mutations that obliterated the original clean-cut vision. However, by the time I chose a Thunderbird to rent in the autumn of 1985, Ford had regained some pride in the 'Bird. Smoother lines owed something to the Blue Oval's belated discovery of aerodynamics as a buzz marketing word for the ninth-generation 1983 model. Plus, V8s worth their multiple cylinders were staging a tentative comeback, recovering horsepower under fuel-injection after being castrated during the early emissions era through the '70s.

This late-1985 USA working trip, courtesy of Athene Karis of BF Goodrich Tires, featured Porsche's 962 as the star turn, plus two other visits to Riverside Raceway (long since turned into housing) to try a production-racing Chevrolet Corvette and for a Mazda media driving day (my American forays were not always Ford-themed).

Following that wet outing at Nelson Ledges in the SVO Mustang related earlier, the Henry Ford Museum in Greenfield Village was my destination for a flying pilgrimage, literally, as a light aircraft took us from Ohio to Dearborn over sections of the Great Lakes and into Michigan. This vast collection with extensive outdoor acreage, subtitled Museum of American Innovation, is thoroughly recommended for all because it is not just about automobiles. It also covers the backdrop of everyday life, from complete transplanted historic houses, to the Wright Brothers and Ford Tri-Motor aircraft, and so much more.

Looking over my diaries, I found that we actually rented a black Thunderbird for the museum visit, and booked ahead for a manure-brown V8 to be collected on my return to Newport Beach. That second example would remain with me for a neat 888 miles of double commutes to Riverside, and a run out to the Simi Valley, returning after a memorable Trivial Pursuit party, via Santa Barbara and long-time expat Brit John Rettie's beach establishment. I also enjoyed some naughtiness at a cocktail bar when I met up with a British Mazda press party, plus a bizarre drive in a borrowed and battered Volvo sedan through evening fog around Simi Valley's twistiness; survival, rather than sportiness, became my priority.

The brown T-Bird certainly had a V8. After several traditional Californian sunny days, I lifted the bonnet/hood, seeking the source of an occasional fast-food odour wafting through the ventilation system. Besides 140 anaemic ponies from 5-litres, I found a foil-wrapped chicken burrito takeaway nesting cosily in the vee!

I was grateful to that Thunderbird for its unfailing, comfortable company on those exciting outings, but it would be 10 years before I experienced another. This time back in the UK with monstrous horsepower trapped within the confines of the tight Brands Hatch track.

THUNDERBIRDING AROUND BRANDS HATCH

At Brands Hatch Indy circuit in August 1995, it had been 10 years since I drove a Thunderbird, but this war-painted monster seemed to have little to do with those low-powered road cars, and everything to do with winning in NASCAR. Compared to the previous 'T-Birds' I had driven, this example had a V8 and a badge in common, but that was pretty much it, aside from the cosmetic sheet-steel panels to enable a recognisable Thunderbird outline. Ford of Europe had persuaded its American masters to allow a factory-backed Jack Roush-prepared example of the breed out to tour selected circuits, including an outing at Zandvoort for 1990 Indianapolis 500 winner Arie Luyendyk (who would also go on to win at Indy in 1997). For the UK, the circuit was Brands Hatch and separate outings for then Mondeo pilots Andy Rouse and Paul Radisich (public) and myself (earlier track day).

'ALTHOUGH THE DETAILED PREPARATION AND PRESENTATION APPEARED AS THOROUGH OR BETTER THAN THAT FOR ANY INTERNATIONAL MOTORSPORT CATEGORY, SOME KEY TECHNOLOGY DEMANDED BY THE RULEBOOK IN 1995 HARKED BACK 30 YEARS, EVEN THEN.'

Under the skin: NASCAR thunder

The test Thunderbird was one of a round dozen prepared annually for the factory Ford equipe, but they divided to face three main tasks: superspeedways like Talladega and Daytona, with a mandated 200mph speed target from motors clipped by over 200bhp; shorter banked ovals between 0.5 and 1-mile lap lengths (the bulk of the calendar), requiring staggered suspension-geometry set-ups; and finally, rarer road courses, such as Watkins Glen, demanding more cornering agility – a tough call when preparing a machine the size of a European limousine that weighed 3,520lb (1,597kg). This #6 Ford was as prepared for regular winner Mark Martin to drive on road courses, and was retained as a back-up Roush Racing/Team Valvoline Thunderbird.

Although the detailed preparation and presentation appeared as thorough or better than that for any international motorsport category, some key technology demanded by the rulebook in 1995 harked back 30 years, even then. Transmissions were a good example, with a heavy-duty four-speed gearbox, Hurst shifter mechanism and a live back axle carrying a 9in differential.

The 5.9-litre V8 ran on old-school carburation, not fuel-injection. The iron-block engines retained pushrod valve operation, and Carillo race rods and Holley downdraught carburation had to be used, even though Ford's showroom Thunderbirds sported overhead camshafts for their 4.6-litre fuel-injected V8s. Yet the road-car engines produced 208bhp, while the track-test Ford was still rated close to 720bhp on European 102-octane petrol, rather than its usual US diet of 108–110 octane, which allowed us to understand how Roush could reliably run the engines with a sky-high 14.75:1 compression ratio.

Behind the wheel: a gentle giant commanding respect

Enough of the details: this NASCAR hero had just as much character as any Grand Prix car and a lot more physical presence, thanks to its appropriate Big Country size. I wrote the story for several English-language magazines, including *Autocar*, but the memories that follow are excerpts from previously unseen editorial, as they are taken from my text supplied for translation and publication in French magazine *Auto Hebdo*. I have chosen them because they have more emotion.

'The racing Thunderbird was unseen from my waiting position on the Brands Hatch infield, but I could hear it being warmed up three-quarters of a mile away. That mighty V8 boom soared contemptuously over the wasp-like buzzing of the skinny Formula Fords around the shorter 1.2-mile Indy track layout that we would tackle. That *Days of Thunder* representative quivered into our pitlane garage a little later, bulging limousine large, filling the two-car garage with a brutal bongo-drum roll as the ankle-searing side exhausts eased the war-zone small-arms soundtrack.

'Roush T-Bird guardian Ken Fackender activated the release mechanism for the truck-sized steering wheel. It dawned on me that I was to be admitted to a very exclusive club, steering one of the world's biggest single-seaters. As there are no doors, I Fosbury-flopped through the open driver's window aperture (no glass, just netting on NASCARS), feet first. Leverage to wiggle within came via the roof and roll cage. I have never felt safer, and reassurance was immediate on squatting into that single high-sided race seat.

▲ A publicity poster for the Roush-prepared Mark Martin Thunderbird's appearance at Bands Hatch in August 1995. *Author*

'You sit low within the Ford, so that it engulfs you. There is a faded aroma of honest sweat meeting garage lubrication bay. The scuffed seat offers a view of multiple instruments and their cheeky yellow needles. The roll-cage tubes – massive in central sections, encircled by smooth plastics – form the load-bearing chassis structure. The stout cage includes transverse door-aperture protection and extends to embrace every corner of the cabin. Similarly, sturdy steel tubing provides a forward cradle for the engine and running gear, but the vulnerable front and outer rear panelling is in skimpy Kevlar, formed to represent a Thunderbird silhouette.

'Externally, there were some crude touches, such as metal plates that form the two-section rear wing. They supply an 'air shovel' to cut aerodynamic lift at speeds European touring-car racers cannot reach. More sophisticated were neatly recessed roof flaps (two, set at differing angles), which flip up to cut speed if the Ford starts to spin out of control.

'The driver is trussed up in full Simpson aircraft harness, and starting required a trio of switches to be flicked. The plump Holley cascaded a waterfall of volatile mixture into the Ford Motorsport-branded alloy cylinder heads. TRW inlet valves of marine diameters took up the story as the motor snorted, ignited by high-voltage Ford MSD ignition energy to stir Bosch spark plugs.

It spat twice and muttered something guttural about English strangers. I persisted.

'War was declared in Kent. I pushed against the hefty spring-loading of the manly accelerator. Ultra-high-compression pistons from Wiseco take some prompting, but the 90° iron-block eight now shook like a dog at the 2,000rpm "tickover". I slid the Jericho-rebuilt four-speed BorgWarner gear-change lever through elongated avenues to net first gear. All a bit like talking to a farmer: surly at first, awkward to shift, but totally reliable. Determined not to stall in front of the inevitable onlookers lured by that V8 siren song, I used half the 8,000rpm limit set for my outing. Bassett 9x 15in alloy wheels supported bulging Goodyear Eagle-branded slicks, but they yelped just as much as the driver when the clutch juddered home to set this flamboyant show on track.

'The view ahead was partitioned by three vertical strakes in the windscreen sight lines; rearward, there was a swirling panorama of impressions via a 2ft elongated mirror that serves all backward glances, as there are no external rear-view mirrors. I felt part of the monster, but only the section that surrounded me in scaffolding: there was a lot more automobile stretching beyond my privileged safety cell. The jaunty yellow tachometer needle indicated 3,000rpm with the eight now firing properly, and then flicked to 7,000 revs with rewarding gusto. The regular NASCAR professional's power band was 4,500 to 8,400rpm. The Hurst clacked through ratios cumbersomely and I was integrated into the world's noisiest video game, but instead of virtual reality, I witnessed absolute reality.

'An abbreviated blast of throttle up to Druids hairpin and efficient Wilwood 12in diameter discs slaughtered the speed with no hint of the bulk to be slowed. I had to take a second bite at the slow power-assisted steering in tighter corners, but that was partially because the high-sided seat obstructed flailing elbows. The T-Bird wasn't above modest kerb hopping and it was the only competition car of many I have steered round Brands Hatch that seemed to make the run-off areas run off in fright! The stability on the elongated wheelbase was outstanding and the strong initial understeer on turn-in was much the same as any period European BMW, followed by the inevitable power slides that over 700 horsepower can summon so easily. So long as you showed a bit of respect, this proved to be an amiable member of the gentle-giant family.

'Past the pits on a high at the completion of each lap, the experience augmented by the reverberating 7,600rpm wall of sound beating back at you from the pitlane brickwork. On a flying lap that meant you were now braced for a fabulous ride on the Kentish version of a Cresta toboggan run, plunging more than 1.5 tons over the crest into Paddock Hill. On the approach to Paddock, the revs slumped alongside the 126mph velocity and I got a fourth-gear rollercoaster ride down that formidable curve. I peered across the imposing real estate acres of Thunderbird bonnet at the long, occasionally off-camber, unravelling right-hander. The apex was a far-off mystery, but I must have been doing something right as we emerged on the exit kerbing with the midday sun still burning through the window-aperture netting beside me.'

There were several surprises, more so with 27 years' hindsight. The Big Bird didn't feel ungainly within the tight confines of Brands. Sure, it felt fatter than anything I had driven there previously, or since, yet Roush Racing experience had filtered through to this NASCAR Ford from earlier generations of road-racing Roush Mustangs. Despite the ferocious exterior and the harsh hours of racing it was born to contest, it wasn't such a brutal experience as expected. The rolling start/continuous high-speed elongated gear ratios and the long wheelbase took the violent edge out of both the acceleration and the cornering characteristics that I had expected from earlier drives in 500bhp+ Mustangs.

Footnote *Mark Martin (no relation to the Brit of the same name in the Capri chapter) had an extraordinarily long race-driving career, which spanned victories beginning in 1977 in an enormous variety of teams and race series to semi-retirement in stages from 2005 to 2014. He took 40 NASCAR Cup wins. He was most successful at Roush Racing during his 1988–2005 Ford appearances. Yet his complete racing career included finishing second in the NASCAR Sprint Cup five times and third four times. 'Probably the best driver never to have won the NASCAR title outright' in the opinion of contemporary observers.*

GALAXIE 500: GIVING IT LARGE... LITERALLY!

The Bob Sherring connection (see previous Falcon race stories) yielded much social laughter, serious classic racing, and in 1992 and 1995 the chance for me to track test the largest saloon car I ever conducted. In view of the Ford Galaxie 500's length (over 17ft/5m) and considerable girth (nearly 7ft/2m), Bob naturally selected a smaller British circuit, Mallory Park, for our assessment and his local entertainment. Other headline oversize statistics that seem to belong in Texas rather than a successful British historic racer included the 7-litre/429cu in V8 that generated 600bhp and 550lb ft of genial torque to propel 3,300lb (1,497kg). In the 1963–64 NASCAR stock car formula, for which Ford commissioned Holman Moody to rework these two-door Galaxie 500 variants of its range-leading machines, 170mph was expected for hour after oval-track hour. For comparison, the 6.3-litre/304 gross horsepower Ford UK demonstrator tested for *Autosport* by John Bolster in the 10 January 1964 issue of the magazine was geared for 110mph and covered 0–60mph in some 10sec at a reported 10–12mpg…

Bob discovered the car, a 1964 road model, in California in 1989 and imported it to the UK, where Jim Morgan's JTM operation crafted it into a racer. When I drove the big white Ford hardtop, Bob and a cheerful mechanic were tending it in-house. By that time it was an established top-class winner in the Historic Racing Saloons Championship, and Bob – armed with a variety of large-capacity Ford V8s – had annexed the annual class title since 1986 with one exception. Yet, the most prestigious victory came in France, as outright winner of the Grand Prix Historique in 1991.

Under the skin: heavy-duty engineering

Bob ran the big beast over many seasons, so the specification and personnel altered regularly, the biggest changes coming when racing under FIA international homologation rules rather than in British domestic events. Spike Winter built at least three versions of the 7-litre V8 and when I tried the car in 1992 it was in 600bhp mode. When the Galaxie was fielded in later FIA-regulated races, it became, said Bob in 1995, 'a little more historically accurate with a little less power', specifying as an example that the maximum torque had dropped to 485lb ft rather than the earlier 550lb ft.

However, the basics remained of that giant V8 up front, heavy-duty four-speed gearbox, and beefy live rear axle carrying a Detroit Locker LSD. When I drove the car, it had a Ford Sport Top Loader 'box, operated by a Hurst shifter paired with a single-plate clutch, which would be replaced four times a season. A subsequent and obvious update was a move to a triple-plate Alcon 7in clutch that would endure a season of 7-litre startline thrusts.

Naturally, the suspension, wheels and tyres saw plenty of development over those seasons. I steered it on quad Spax dampers at the front, with a pair of the same brand at the multi-link leaf-sprung back axle, restrained by a Panhard rod. That meaty front layout rested on 1,800lb/in springs with a stout anti-roll bar, while six spring leaves per side were required to restrain that beefy back axle. Tyres were by BF Goodrich, with 245 and 255 section resting on 9x16in Revolution alloys. Similar muscularity was demanded of the brakes, of course: I had 13in AP Racing discs with Alcon callipers at the front, but the rears were finned drums. To my surprise, those back drums survived the 1992–95 seasons. By 1995, a Tilton twin master-cylinder layout and balance bar were in use.

▼ In all its substantial glory, Bob Sherring's well-presented and successful classic racing Galaxie 500 makes its presence felt at Mallory Park in 1995. *Maurice Rowe*

▲ Mr Motorvator. The 7-litre V8 delivered some 600bhp and 550lb ft of easy-access torque. *Maurice Rowe*

▼ Bob Sherring (left) and the author discuss the massive task ahead, threading Ford's US full-size coupé through the Minis and more within Mallory Park's tight confines. *Maurice Rowe*

Behind the wheel: all aboard!

Surveying the race Galaxie 500 at my 1992 Mallory session, the strongest impression was made by the scale of this racer. Sure, it had lightweight glass-fibre panels for bonnet, boot and bumpers(!) and Perspex thin side 'glass', but it also demanded a sturdy six-point roll cage and full Lifeline fire extinguisher system, and that 7,003cc motor displaced a considerable amount of kerb weight in all-iron block and cylinder-head configuration. Once I was installed, the Galaxie cockpit impressions were fairly naturally of generous space around a single Sparco race seat in Kevlar, and a Luke harness. I remarked of the ambience from behind a three-spoke Mota-Lita steering wheel: 'Big enough to hold a reasonably social wedding reception, but the atmosphere within is of functional racing saloon, rather than period juke box. The car was physically easy to drive at track speeds.' Lacking the power steering and automatic gearbox of Bob's first race Galaxie, it was tougher to manoeuvre around the paddock.

'It started on the first push of a central black button. Equipped with two side exhausts, the V8 made its voice heard as we rumbled towards the track, but the motor and its dinner-plate-dimensioned clutch were easy enough to handle. What proved tricky was the gear-change: "Just feel it into second from third on the down-change," advised Bob.

'That 7-litre V8 was intent on delivering 600 horsepower almost as soon as the foot-long accelerator pedal was depressed, and exhibited the sort of brawny power between 2,000 and 6,000rpm that the NASCAR originals must have deployed when collecting 30 stock-car victories in 1964.

'The iron V8 allowed majestic acceleration and the feeling that I was captain of an exceptionally spritely aircraft carrier – one that had somehow become beached amongst a flotilla of public-test-session minnows. I could even look down on a delectable Lola T70 as we sailed into the everlasting starboard turn normally described as Gerard's.

'The big surprise wasn't the power; you expect that of such a vast V8 in competition

◀ Left-hand-drive room with a Galactic wide-screen view. Mallory Park's short track allowed enough straight and an elongated corner to enjoy the Big Country Ford's surprisingly fleet progress. *Maurice Rowe*

▼ Coming and going in the thunderous Ford was extremely enjoyable and gave the author the challenge of threading an aircraft carrier through noisy tugboats. It handled well, and the accessible V8 power was deeply satisfying. *Maurice Rowe*

specification, but the track manners of this substantial chunk of automotive Americana. It was very much better behaved than I had any right to expect on the way out of Mallory's tight hairpin. That was forcefully demonstrated by the way the beast avoided two spinning cars that had slewed to an unexpected halt on the (unsighted) exit to a track section that barely has width enough to accommodate American automotive real estate in powerful motion. Viewing a screen full of Triumph TR6 and Escort Mexico, I dared to pull hard at the wheel, accelerate harder and escape to the inside of the mêlée.

'I was a little wary of the brakes, for a mixed disc/drum system didn't sound the complete cure for the well-publicised dramas of slowing large American vehicles from high velocity. In fact the large AP front discs were so efficient that quite a lot of overtaking in crowded conditions was completed on that approach to Mallory's hairpin. Mind you, the spectacle of the Galaxie bearing down on skeletal British racers tended to improve track behaviour no end!'

Looking back to the '90s, I had experience of similarly powerful turbocharged Cosworth RS Fords – albeit in rather lighter and smaller bodies – but the Galaxie still lives bright in my memory as a totally larger-than-life character worth cherishing.

Thanks for the trust in sharing your enormous boys' toys, Bob!

CHAPTER 5

GT40

Le Mans-winning thoroughbred

The Ford GT40 story has expanded way beyond quadruple successive (1966–69) Le Mans victories, plus numerous other long-distance sports car wins. From initial early reports in the specialist motoring press to the stuff of bestselling books and on to Hollywood movies, via the sensational Ford-versus-Ferrari headlines, the GT40 has captivated generations over the decades. Today, over 50 years after the mid-engine V8 Ford (unexpectedly) last won Le Mans, hundreds of 'GT40s' attend events, although fewer than 140 originals were assembled between 1964 and 1967 at the Ford Advanced Vehicles (FAV) production facility in Slough, England, and by

◀ The legendary 1966 Le Mans 24 Hours moment that underpins the Ford GT40 racing legend. Not the dead-heat that corporate suits wanted, but a crushing 1–2–3 that saw the mission to defeat Ferrari at the French endurance classic achieved. *Ford*

Ford sub-contractors such as Kar-Kraft in the US.

Numerous copies and authorised continuations now exist, just as for the contemporary Anglo-American Cobras. The difference is that the '60s GT40 legend also prompted the parent Ford USA company to back the construction of specialist follow-ups. These included 4,035 supercharged V8 GTs (produced between 2004 and 2006) without an assigned factory race purpose. A second-generation Ford GT was later created, these materialising as hardcore race Le Mans LM GTE Pro class winners. They ran V6 'EcoBoost'-branded turbo motors, wrapped in purpose-built bodywork that served the race cars and road adaptations. Such racers were specialist-built from 2016 to 2019, securing an essential class victory at Le Mans in 2016 to remind us of Ford's first defining outright 1–2–3 defeat of Ferrari and the rest at Le Mans in 1966, 50 years earlier. Some 1,350 road cousins of these second-generation GTs were scheduled at the time of writing, with more than 1,000 delivered to carefully vetted first owners.

▲ Always a special outline, this GT40 is chassis 1008, owned for its entire life by Ford UK. This photograph was taken during the car's green era (its original colour), but it has appeared in many other liveries at different events during its lifetime. *Jeff Bloxham*

THE GT40 STORY: A LEGEND IS BORN

For hardcore motorsport followers and the American parent company, the GT40 (40in high at the 1964 coupe's roof) was part of a vast commitment to motorsport and spin-off performance cars by Henry Ford II. Summer 1962 saw the Blue Oval break the 'Big Three' Detroit manufacturers' (General Motors, Ford and Chrysler) Automobile Manufacturers' Association (AMA) agreement. This formality barred participation in motorsport and any emphasis on outright performance aspects (such as horsepower) of showroom vehicles in advertising. Essentially, this was a move to pre-empt the introduction of any US legislation to force the manufacturers' hands, following various campaigns aimed at highlighting the risks of powerful cars and high-speed driving on US roads.

Ford used the slogan 'Total Performance' for its

➤ I track-tested this competitive GT40, prepared by Chris Long and privately owned during the earlier '80s, at Silverstone. *Motorsport Images*

breakaway from the agreement and reconnection with motorsport. Initially, that meant hitting home-grown speedfests such as quarter-mile drag racing and NASCAR stock-car racing, the latter one of the biggest audience draws in American motorsport. A major aside to that thinking became the goal of winning America's most internationally famous motor race, the Indianapolis 500. This was enabled via an agreement with Lotus for a single-seater wrapped around a Ford V8 engine, and the winning objective was achieved in 1965 (with Jim Clark driving) to break decades of engine dominance from the Offenhauser company.

Ford US management then turned its attention to the events that would mean most to a global audience, particularly those that could boost the commercial prospects of Ford in Europe. This took the marque into Formula 1, rallying and endurance racing to varying degrees, and Ford-badged cars and/or engines would achieve spectacular success in all three categories. Two of the premier prestige single events were French: the Le Mans 24 Hours and the Monte Carlo Rally. Success in the Monte Carlo Rally would prove elusive until the advent of the 4x4 Escort RS Cosworth, François Delecour winning in 1994 against the best opposition from Toyota and Subaru. However, the Blue Oval did take two World Rally Championships with the rear-drive Escort, in the hands of Björn Waldegård in 1979 and Ari Vatanen in 1981.

Meanwhile, back in 1963, Ford executives intensified their examination of what it would take to win at Le Mans. As Ferrari had won four years in a row, corporate logic decreed that buying the legendary Italian company would be an expensively neat short-cut to victory, and would bring some halo-effect showroom performers to sell. Famously, that Ford-Ferrari diplomacy (or lack of it) did not work out. Yet that 1963 Ford look at what it would take to win provided some bullet points for victory at Le Mans. It was reckoned it would take not only a mid-engined car but also a 200mph maximum speed and 120mph race average speed, including all pit stops.

The laboriously costly business of building a Ford-branded challenger began. This much-told tale involved a fraught beginning with the cars plagued by unreliability, with background pressure from the Ford parent company to achieve rapid results.

The prototype drew heavily on UK-based Lola personnel and engineering, not least the Lola Mk6, designed by Eric Broadley. As presented to the press to make the front cover of *Autosport* dated 3 April 1964, the squat two-door coupé looked identifiable as a GT40, although period wire wheels did look more British sports car than 200mph Le Mans challenger, and the official name was 'Ford GT'. Throughout 1964, the car would be developed significantly from the prototype incarnation, but would still be known as the Ford GT (widely referred to retrospectively as the GT40 MkI).

The full story of the GT40 has been extensively told elsewhere, but, essentially, the GT40s would struggle during 1964 and 1965, due to insufficient development time and the pressure of a 'barrel load of monkeys' wriggling on corporate and race-team shoulders as the expected success failed to materialise.

The prototype and 1964 race cars were developed and built by Ford Advanced Vehicles (FAV) in Slough, under the management of John Wyer, but for 1965 responsibility for the US operation and overall factory race programmes transferred to Carroll Shelby in the US. FAV meanwhile would continue to build customer cars and would later develop the GT40 MkIII road car, powered by a 4.7-litre (289cu in) V8, of which only seven were built.

The 1965 season saw development of what became known as the GT40 MkII, packing even more American iron in the form of a 7-litre V8. Originally the plan was to race this 7-litre car in 1966, but the appeal of the MkII's Le Mans potential overwhelmed endurance-race development caution, and a pair of the MkIIs debuted at Le Mans in 1965. They were backed by multiple 4.7-litre GT40s run by assorted teams, but all that achieved was a clean sweep of front-running GT40 retirements.

In preparation for Le Mans 1966, the Americans resumed development of the 7-litre MkIIs. Significantly, a great deal of practical development mileage was run prior to the race. The hard graft paid off: the 1966 opening races in America – the Daytona 24 Hours and Sebring 12 Hours – both returned 1–2–3 results. Another factory MkII got an outing at the fabulously fast Spa-Francorchamps road circuit prior to Le Mans, that Alan Mann machine finishing second.

Now, it was time for an historic Le Mans that became the stuff of Ford dreams. Although ten GT40s retired before that infamous Ford attempt to stage a tied finish, a trio of MkII 7-litre cars finished in 1–2–3 formation, the finer points of the finishing order proving controversial, as has been well documented.

The US operation now focused on development of a MkIV, built by the Ford-owned Kar-Kraft company. The MkIV featured a completely new, and significantly lighter, aluminium honeycomb chassis and heavily revised bodywork. The 7-litre V8s had been extensively revised too, carrying a pair of Holley multi-choke carburettors and iron cylinder heads. The compensation was

◀ Celebrating 50 years since the 1966 GT40 1–2–3 outright victory at Le Mans, the totally new Ford GT took 2016 LM GTE Pro class victory at Le Mans, against stiff opposition including Ferrari, Porsche and Chevrolet. *Author*

▲ A rare sight and with an extremely selective customer base, including celebrity owners such as Jay Leno, the roadgoing 21st century Ford GT was every bit as tricky to manufacture as the GT40, and a lot more sophisticated. *Author*

extra grunt, with 530bhp the top quote.

The lead American GT40 MkIV crew of Dan Gurney/A.J. Foyt won Le Mans in 1967, finally achieving Ford's ambition to win using an all-American car with an all-American crew.

With a rule change for 1968 dictating a maximum engine size of 5 litres, the MkIV GT40s were effectively made redundant, but this did not stop John Wyer's Gulf-backed team, JW Automotive, run from the ex-FAV premises, from continuing the GT40 challenge with revised MkI cars, with great success, including winning the French classic in 1968 and 1969, remarkably taking the same chassis (1075) to victory in both races.

Aside from Le Mans, the GT40 won many other international and national events, cementing the car's legendary status and ensuring its place as the most valuable fast Ford today.

STEERING FORD'S FABULOUSLY FAMOUS GT: IT STARTED WITH A FRENCH DOWNPOUR...

For me, the regular experience of driving two Ford of Britain-owned GT40s became a thrilling reality 30 years after the heyday of the legendary factory and JW Automotive/Gulf entries, and those legends are recalled after a few more-recent GT40 driving experiences.

Friday 13 October 1995 brought not bad luck but the best of French fortunes. A small press party flew into Nîmes and coached onwards to the Goodyear test track at Mireval, used by northern European automotive manufacturers such as BMW for winter testing. Ford had laid on a spectacularly varied driving day for the assembled UK media, but it was wet, verging on a continuous downpour that worsened during the day, besmirching southern France's sunny reputation with a proper British bad-weather day.

We drove a selection of contemporary Ford showroom products such as the Orion (front-drive Escort plus a boot) and Scorpio (final rework of the honourable Granada name). That top-of-the range Scorpio was not a bad drive, but its front and rear stylists – apparently a flock of them according to contemporary insider comment – obviously had no meaningful conversations before knitting a fish-eyed faux Mercedes face with a plump rump. Then it was time for some fun: we were allowed to roam in various competition Fords and some crazy two-seat Westfields (lookalike variations on a Lotus Seven theme) with Sierra Cosworth turbo power.

Did I mention how wet it was? Good, because even world-class navigator and gifted driver Gunnar Palm, Hannu Mikkola's co-pilot on many earlier Escort rally wins, including the London–Mexico World Cup, spun us around in the celebrated 1.8-litre crossflow FEV 1H Escort of the kind used on that marathon rally. While we shared much laughter about this, back at the temporary Ford test-track HQ the PR people's faces were turning to stone as conditions deteriorated and more gyrations ensued.

'YOU CAN IMAGINE THIS METALLIC SILVER FOX MACHINE WAS, AND REMAINS, DEAR TO FORD IN BRITAIN: COMPANY PERSONNEL WEREN'T KEEN FOR IT TO GO OUT PADDLING IN THE HANDS OF DRIVERS WITH ASSORTED TALENTS.'

There were no worries about the Andy Rouse Mondeo that 1990 British Touring Car Champion Robb Gravett steered through the deluge and that many of us enjoyed as passengers. More problematic was that Ford had also brought along its ultra-rare (one of seven) MkIII road version of the GT40. Registered DWC 8G in June 1969, this machine has been with Ford UK since its delivery from FAV at Slough, although it did escape to the private domain of Ford PR supremo Walter Hayes for long periods, even shipped to the USA during Walter's elevated Ford Vice Presidential status in the '80s. You can imagine this metallic Silver Fox machine was, and remains, dear to Ford in Britain: company personnel were not keen for it to go out paddling in the hands of drivers with assorted talents.

The dilemma was solved in a casual tea-break moment that changed my 1995–96 driving life. 'We're looking for a pair of safe hands to drive our GT40: would you mind stepping outside, away from the hack pack?' said my former race-reporter colleague Graham Jones. The Canadian, known for his authoritative North American race reports and spells at my former CCC employers, was deadly serious.

Tea and biscuits forgotten, along with my sporting CV, to which the expression 'safe hands' had been a stranger, I gabbled my reply to a dream offer: 'I'd be honoured, lead me to it…'

Under the skin: road-car cousin of a Le Mans winner

A decade earlier, I had track tested a racing GT40 – a privately prepared and very competitive example for a national British classic racing series

– but this MkIII was utterly different from the race/semi-race examples and numerous replicas that abound today. Even compared with the classic MkI and MkII GT40s that are so often imitated, the MkIII is visually different, with a longer tail hiding still limited (and heated!) luggage accommodation, and rerouted (lowered) exhaust system. The front end is filled with quad headlamps, but it was when I climbed within that the biggest furniture changes materialised.

This public-road semi-civilised version of the purpose-built race GT40s was comparatively easy to access, for the broad side pontoon/sill structure no longer carried the gear lever. Instead, a conventional road-car central position was employed, meaning that left- and right-hand-drive examples would not need different gear-change layouts. Just seven of 20 planned MkIIIs were constructed, and two of those were prototypes. This chassis (1107) was numerically the last to leave Slough and was one of four originally retained by the company, so truly a rarity.

This final Ford Advanced Vehicles, Slough-built iteration of the GT40 was also comparatively luxurious, with adjustable full-leather seats rather than the iconic ring-studded items of earlier GT40s, plus an attempt at sound deadening, including a foam-filled roof and a safety-conscious polyurethane wrap to the alloy fuel tanks (rather than the bag tanks of the race cars). Naturally, some public-road legalities were also added, including a weedy horn, central and untrustworthy handbrake, plus flashing indicators. So, the cockpit was stronger on familiar road-car controls and quieter than any other genuine GT40, but there was absolutely no improvement in rearward vision. This was a primary factor in my cautious exploration of the soaking French circuit, which I insisted on tackling alone before passengers joined me for the rest of my driving day.

Behind the wheel: a civilised character

A blue leather-rim steering wheel, with three flat spokes and the FORD 'G.T.' logo set within red, white and blue backing, greeted me in the comfortable cockpit. With clear black-and-white markings, the dials remained at least as comprehensive as any race GT40, the 200mph speedometer a bonus, although you probably would not want to look at a speedo anyway in a race situation. The tachometer was properly in the line of sight, with a red line at 6,500rpm. However, many of the planned road-car instruments illustrated in Ford literature had disappeared. For instance, the separate gauges for the left- and right-hand fuel tanks were absent on this Ford-owned example by 1995, replaced by an oil-temperature gauge to the right of the rev counter and central 'all-contents' fuel gauge to the left of the water temperature dial. I didn't find the cigar lighter either!

There was a mundane ignition key with no

◄ The Ford of Britain-owned MkIII GT40 road car travelled to the USA along with Walter Hayes when he was a Vice President of the American parent concern. *Richard Hayes*

FACTORY MkIII GT40 ROAD CAR

Data published in June 1967, courtesy of Car & Driver

Engine

Configuration: V8 **Capacity:** 4.7-litres (289cu in) **Fuel system:** Single four-choke Holley **Maximum power:** 306bhp at 6,000rpm **Maximum torque:** 329lb ft at 4,200rpm

Transmission

Type: Five-speed ZF, synchromesh gears **Final-drive ratio:** 4.22:1

Suspension

Front: Unequal-length wishbones, coil springs and anti-roll bar **Rear:** Triangulated lower arms, single upper transverse link, trailing arms, coil springs and anti-roll bar

Brakes

Front: Girling 11.5in ventilated discs **Rear:** Girling 11.2in solid discs

Weight

Kerb weight: 2,340lb (1,061kg)

Performance

Gear speeds at 6,200rpm: First 48mph; second 80mph; third 109mph; fourth 124mph; fifth 140mph (6,000rpm). **Acceleration:** 0–30mph 1.6sec; 0–60mph 5.0sec; 0–100mph 12.7sec **Standing ¼ mile:** 13.8sec at 105mph **Maximum speed:** 140mph **Fuel consumption:** 10mpg

fancy pushbutton-activated start. Contemporary Ford road-car rocker switches, with the usual faded symbols, commanded the lighting and the radiator fan. A confusing single stalk to activate indicators, headlamp flash, dip-switch and horn lay to the right of the lower dashboard rocker switches; I remember a similar single stalk on the steering column of many Mercedes, but the idea lost something in translation from classy Stuttgart to cost-cutting Ford. Another gripe was that screen demisting was borderline in the damp conditions, although the eyeball ventilators did deliver a lusty airflow. Sorry to say, these ergonomic shortcomings showed that the GT40 was so obviously a hasty, cut-price conversion for road use that Ford lost any chance of joining the supercar ranks dominated by the Italians – and membership that the Blue Oval had earned with its race results.

But let's return to all the good aspects. A single four-choke Holley carburettor was happy to cooperate in supplying a reasonable but not overpowering V8 beat. The docile V8 offered an official 306bhp at birth, as a 4.7-litre (289cu in) unit, although it was also listed with a 5-litre (302cu in) engine with the same power and 329lb ft of torque. For me, it didn't really matter which engine capacity was installed as – heresy I know – this GT40 felt rather slower than the independent performance figures implied from US road-test examples, with figures for Shelby American-supplied 4.7-litre versions yielding an average of 5sec for 0–60mph. To me, this MkIII felt best as a long-legged, high-velocity device, gaining speed unspectacularly (by GT40 standards), but getting the job done where speed limits were absent.

The long-travel and heavy clutch was no immediate impediment, but the convoluted gear-change mechanism that had been necessary in evolving to a centrally located lever was a knottier problem. I learned to change with the measured patience of the double declutch, but the synchromesh action of the five-speed ZF remained sympathetic, despite a variety of drivers over the years, and a history of them trying to beat the sequential change. Appointed contemporary guardian of the Ford-owned GT40s, Bryan Wingfield, had removed the sequential system on that MkIII, and I always operated both of Ford's own GTs without skipping gears. That habit was prompted by previous experience with ZF five-speed units, which I judged to be in a smooth and reliable class of their own, until Getrag transmissions became common in the BMW world.

Fortunately, forward vision was excellent over the sloping bonnet and contoured wings, even from the semi-reclined driving position necessary with that celebrated 40in roofline. The single wiper coped with speed-boat spray kicked up whenever that race-spec Mondeo passed us on the track, which occurred many, many times…

I was slightly fazed to see that this GT legend remained on wire wheels, which I tend to associate with classic British sports cars, although

▲ Ford of Britain's MkIII was docile enough for any duty, from course car at Goodwood Revival to driving senior executives around the gardens at Beaulieu! *Jeff Bloxham*

these 15in-diameter 'wires' were actually by Borrani. No matter, the 6.5in front and 8in rear rim widths supported treaded tyres reliably on the softer road suspension settings. At a modest wet-weather pace, with a spread of passengers from 17-stone male photographer to brave featherweight females, my 1½ hours at the wheel proved rewarding.

From a dynamic viewpoint, the near 30-year-old design was incredibly capable and sure-footed in these conditions – both in the front end cutting a swathe through standing water and the consistent traction afforded by fatter rear rubber. I left the French track with a different respect for the legend that was Ford's first GT, for it had stolidly carried me through some stressful track and passenger duties with total faithfulness and charisma that captured the admiration of every occupant.

I didn't know it then, but the silver dream machine had delivered an entry ticket to a very exclusive driving world that I would enjoy through 1996 and still relish 26 years later…

Footnote *In 2021 I learned that this public-road-friendly road MkIII (1107) had been thoroughly refurbished and returned to the silver shade I knew, having worn various other colour schemes over the years. The steel tub had corroded significantly, so the opportunity had been taken to restore it internally and externally. It still lives with the Ford Heritage collection.*

MILLBROOK MAGIC IN A SEMI-RACE GT40

The annual Society of Motor Manufacturers and Traders (SMMT) day is an event at which the UK car manufacturers and importers field their latest hardware for media and VIPs to drive. The event is staged at the Millbrook test track in Bedfordshire, at the time of this drive owned by GM, with multiple facilities from simulated smooth and rough road surfaces, to a stop-start city course and a snaking hill-circuit layout of outstanding twists and crests that has been used on some enterprising British special-stage rallies. Oh, and there's a speed bowl where the magazines I tested for ran to 150mph routinely, but more of that later.

The back-story

In late April 1996, I got a call from Ford of Britain's PR department: 'Could you do us a small favour tomorrow?'

'Depends what you want,' I replied warily.

'We're stuck for a driver for our GT40 in the lunch break, when all the exhibitors show off their classic cars in a ride-and-drive exercise. We're up to something different: we want to run our semi-race GT40 [chassis 1008] – the Green Machine in Mario Andretti #1 livery – around that two-mile speed bowl. It's a very special adventure for youngsters who are terminally ill, but well enough to experience something different via the

➤ ⮟ Despite a varied history and many drivers at the wheel over five decades, chassis 1008 remained a stable steer whether at lower infield speeds or unleashed beyond 120mph at the Millbrook two-mile bowl. *Jeff Bloxham*

Make a Wish charity,' explained the patient PR.

Only half believing this dream opportunity to drive the car and also do some good for a worthwhile cause, I answered: 'You mean while Vauxhall run their duotone Cresta with castrated American fins, and Rover a '50s MG, you are going to blast them all with that half-race GT40? Cheeky beggars, count me in.'

Possibly five minutes later, Ford GT40 guardian and preparation specialist Bryan Wingfield phoned with cold-start, ZF sequential gearbox and maximum rpm instructions. That vital briefing took a bit more than five minutes.

Behind the wheel: wishes fulfilled

The morning of the SMMT day – which also included the chance to take some of the demonstration vehicles out on public roads – I reported for this delightful duty. In fact I did not have to go through the usual cold-start routine of master switch on, select fuel tank, turn ignition on, five stabs at the downdraught carburettors for rich mixture, then push black-button starter. Paul Wilson from the Ford press garage – whom I knew from racing production Sierra RS Cosworths – had already whipped over from his demo fleet and warmed the GT40 up, ready to go. Yes, luxury for me, but it also meant we could maximise the number of Make a Wish rides in the time available.

In the event, they did allow me an exploratory lap, and the passenger sessions continued long after the official lunch break. My first passenger was very much an adult, Ford PR Product manager Stuart Dyble, best known for his later time at Jaguar in Formula 1, and for delivering not one but two babies on frantic road trips that did not quite make hospital on time!

Our conversation covered neither of these topics: 'Dyble?' I queried, 'Sure, I've heard that name at Ford before.' 'Probably my Dad, Rod. He was the senior engineer at Ford Advanced Vehicles in the '70s,' explained Stuart as I held the tough clutch against a minimum 2,000rpm departure speed. Between bursts of that conversation, I navigated my way through the right-hand gearshift pattern to 4,000rpm, about the point where the carburettors and I stopped spluttering.

The reverse slot in the cast-alloy gate is carefully blanked off from careless hands with a collar. All five ratios started life with synchromesh, but that's often absent when pulling back from second to first in a dead-stop hurry. Although the V8 in this car during that period had been publicly dismissed as nothing more than a tired old 'bitza' that just had all the correct bits to mate with a ZF gearbox, it made all the right rorty noises. After the bad-mannered coughing and spitting spell from 2,000 to 3,500rpm, I could stop easing down on the throttle and attack with vigour.

This rescued and hard-worked GT40 went hard enough through the gears both to silence experienced occupants and to reach 100mph in 10sec. Millbrook allowed 120–130mph as a sustained pace in fifth, when the car felt stable enough to turn to a passenger and await a thumbs-up, confirming that they were equally happy to proceed at these exhilarating speeds. I never had a thumbs-down from my 15 passengers. Some were as seasoned as Carol Horsman (daughter of JW Automotive's John) and Stuart Dyble, and there were those with less experience of high-speed passenger rides, such as 15-year-old Jonathan Brook from the Make a Wish queue. Just broad grins and outright laughter at our transportation into an entrancing alternative universe of blurred and Armco-ringed wonder, to the thunder of an eight-cylinder soundtrack.

My shaky notes were made after two Millbrook sessions ran one of the tanks totally dry (just flip a switch for the other tank). The afternoon runs included VIP GM Millbrook managers who had allowed such sessions to steer their way through official obstructions, and Jeremy 'TV' Clarkson (we had both worked for *Performance Car* before he went stellar). However, the star for me was my final Make a Wish teenager. Cheerfully installed in the GT40 cockpit, despite obvious handicaps, I heard him say above the V8 in getaway mode, 'Now this is what I call a proper car!'

I could not hear any more of his comments, just see the grin that accompanied a vigorous thumbs-up for another lap to follow the 125–128mph openers. We hauled towards a demonstrator Rolls-Royce displaying the factory's 1800 TU registration plates, as he dropped down from the banked track's outer (no speed restriction) lane. We whooshed by, the yellow upper track-definition and Armco barrier bolts whipping by the speed-hungry Ford's snout. The GT40 felt rock solid.

◀ The GT40 cabin is emotive with those ringed seats, 200mph speedometer and a wooden-topped right-hand gear lever to select five ZF gears. The car bears the noble patina of decades of use, yet is a 'one-owner' (British Blue Oval company) example. *Jeff Bloxham*

All too soon it was time to return to spluttering infield pace as we turned back into our improvised pits. Temporarily, we listened only to idling motor and low-key transmission metallic cacophonies as we hiccupped past the latest in showroom glamour and a few of yesteryear's classics. My co-pilot said: 'Told you this was a proper car. Even Rolls-Royces have to get out of the way!'

The pasted-on grin and outbreaks of laughter from my passenger, as his minders extracted him from the car back to a harsher life than I can imagine, just made my day.

▼ Now that's a proper power unit! Lots of downdraught Weber twin-choke carburettors, snake-pit of an exhaust system – it's all there! *Jeff Bloxham*

THAT SEMI-RACE GT40 AT THE LE MANS 30-YEAR CELEBRATION

Although there were many memorable drives for me between 1995 and 1997 in the pair of GT40s owned by Ford of Britain – from Beaulieu stately-home gardens to a two-mile speed bowl – there could never be any doubt about the star turn. The 30th anniversary of Ford's historic 1–2–3 finish at Le Mans in 1966 was held on 13–17 June 1996. I had the stunning privilege of driving the pre-event high-speed parade laps in the 1965 semi-race chassis 1008, which was, and still is, retained by Ford as a MkI demonstrator, serving over the years in many guises and many paint schemes.

During the various times I experienced its fierce charms, chassis 1008 was liveried in the original Linden Green with a #1 door decal. Previously, 1008 – never road registered – had also appeared in black and silver with the #2 in honour of the winning 1966 Le Mans car. This chassis also appeared in Gulf colours with #6 to tour British dealerships, reflecting the unexpected final 1969 GT40 Le Mans win for the JW Automotive team.

Under the skin: my pedigree chum

Externally, chassis 1008 had been a chameleon of mood changes in both paint and panels to match occasions, but within it felt pretty original to me. We did have seats belts for the Le Mans sortie, but there are pictures of it without in previous incarnations. The prime impression centred on the grand entrance through doors extending into roof cutaways, and straddling the wide side sill that houses the right-hand ZF five-speed gear-change, the stubby lever topped by a classy polished wooden knob.

The gate surrounding the gear lever was peppered with white-on-red commands, and there was a lock-out mechanism to enforce sequential shifting. You did have to pressure the lever with some sideways force across the gate, but that came naturally enough because the lock-outs virtually tell you what you can and cannot do. Overall, I thought the change quality was excellent, but the synchromesh had been abused by many hands.

The cockpit was a rewind overload in 50 shades of black. It featured many dials, individual fixed-backrest seats with brass ventilation ringlets, and a sturdy three-spoke, leather-rim steering wheel that carried the Ford GT motif centrally in early black-and-white pictures of the car, but had lost that identity in favour of a large alloy disc when I drove it. There were many more dials, but as I did not usually warm it up – they did not trust me that much! – I only kept an eye on oil temperature. Most notably this was necessary when we did repeated two-lap banked track runs at Millbrook test-track's two-mile bowl, interspersed by stops to exchange passengers.

I never did drive 1008 with the far-left-mounted 200mph speedometer working with any credibility. I never had time to look at it anyway, making sure no oil-pressure lights blinked on, and that water temperature stayed below 100°C. I was fiercely commanded to stay safely between 3,000 and 6,500rpm. Tickover? I expect it could have ticked over, but it seemed to splutter and fart so much that I would idle it at over 1,500rpm and not let the ferocious clutch in much below 2,500.

The Le Mans adventure began with an elongated road trip from Thames Valley to Le Mans, via the Dover–Calais ferry, in an unlikely but efficient contemporary custard-yellow Escort cabriolet. This car moved frugally with 1.6-litre CVH front-wheel-drive power, but was also fast and calypso-colourful enough to be trapped by the

'I NEVER DID DRIVE 1008 WITH THE FAR-LEFT-MOUNTED 200MPH SPEEDOMETER WORKING WITH ANY CREDIBILITY. I NEVER HAD TIME TO LOOK AT IT ANYWAY, MAKING SURE NO OIL-PRESSURE LIGHTS BLINKED ON, AND THAT WATER TEMPERATURE STAYED BELOW 100°C.'

inevitable French police speed guns. Fines were on the spot, so the credit card was battered before we ever passed through the circuit gates.

It was worth the hassle. Not only did we stay at a château with a swimming pool in proper French summer heat, but there was 1008 to steer in a parade that included celebrity GT40 owners such as Noel Edmonds. Bryan Wingfield had organised our appearance and a mighty post-parade lunch with the familiar efficiency of experience.

Behind the wheel: a personal highlight

My job that hot Saturday race morning was to conduct 'our' GT40 with parade sponsor and Sun Alliance Director John Knightly as passenger. This would be a three-lap celebration around the full Le Mans circuit, albeit one savaged by high-speed chicanes since that legendary '66 result.

The car was not in the best of moods sitting around in 30°C temperatures with the engine running, but the fabric seating with its period ringed ventilation holes was an appreciated bonus. I do remember it was quite obstructive that day in terms of getting underway smoothly, demanding almost 4,000rpm, which meant it was also difficult to communicate with VIP passengers as eight Weber throats gargled, spat and finally cleared themselves as we gained respectable velocity up to the Dunlop Bridge. We had a pace car, but I had been cunning enough to find us space towards the rear of our emotive GT40 traffic queue, so we could motivate the green Ford with sufficient rpm and speed to entertain the occupants.

The opening lap was pretty pedestrian by race-circuit standards, plenty of sprints to 100mph and firm braking to keep vehicle spacing respectable. By the second tour I could relax a little and enjoy split-second peripheral views of blurring scenery, and imagine what it must have been like to race these fabulous mid-engined cars along an unfettered Mulsanne straight – especially braking from 200mph to negotiate the tight right at the conclusion of the most revered stretch of public highway in the history of sports-car racing.

I must confess I did lose concentration with all this fabled motor-racing tarmac and the clear view over the GT40's prow. So much so, that at one point I crunched the usually suave ZF five-speed gears with clumsy coordination of clutch, shift and throttle. Black mark, but the Ford forgave me, along with the quietly courageous passenger: Mr Knightly had military service to his credit along with 37 years at Sun Alliance.

Actually, because of the chicanes in more recent times, we got a lot more cornering action than back in the day, and we both got some fabulous bursts of speed and charisma from the one-owner Ford. My diary says we got up to 150mph during catch-up spells in the parade. Certainly, the other participants had all determined they were not going to waste this unique opportunity to exercise such automotive icons around one of the world's most famous race tracks.

Personally, the Le Mans GT40 weekend was the high point in my driving life. Just the view out of the raked front screen with the vision of that fabled track unravelling before my very eyes, plus an iron-block V8 stomping out the beefy soundtrack, marked this outing as perhaps the finest Ford experience of all.

▲ GT40 chassis 1008 at the Goodwood Revival in 2007, masquerading as the 1966 Le Mans winner, with appropriate bodywork, when it was driven by Jacky Ickx – Le Mans winner in a GT40 in 1969 – to lead a parade of Ford Cosworth DFV-powered cars. *Motorsport Images*

FORD'S REAL GT40S VERSUS REPLICAS

It is 12 February 1996, and Chobham's damp ex-military test track awaits four very different GT40s. We feel lucky not just because we have 1,300 horsepower from four V8s to assess, but grateful that the winter snows have just melted away. For the April 1996 issue of *CCC*, we borrowed a pair of the real-deal Ford-owned examples to face two replicas – redeveloped versions that had the benefit of hindsight and affordability – created by Poole-based GT Developments (GTD). It was a measure of the respect in which the GTD reincarnations were held by Ford and sub-contractor Bryan Wingfield that this confrontation was allowed to happen at all, but it did, and neither photographer Dave Wigmore nor I were going to turn down such a tasty tale.

Under the skin: variations on the theme

Both Ford originals were familiar to me by then and I have already introduced them in this book. The racier one was the unregistered chassis 1008 in Linden Green and the other was the silver MkIII (chassis 1107 registered DWC 8G) that had been created and kept for road use. Facing them was a striking pair of GTDs: 804 AJB in red, owned by Alan Barlow to road-biased specification, and K10 ARW, property of Tony Gordine and featuring scoops and the full yellow and black-striped finish of a race car, packing a 320bhp wallop.

Appearances can be deceptive, and Gordine's track look-alike was actually the best developed for road use, with plenty of sound-proofing, a luxury leather-clad cabin, and a single quad-choke carburettor sitting atop a 5-litre V8 carrying a soft camshaft profile for accessible pulling power. It went convincingly, not so raucous as the quad-downdraught-Weber-fed V8s of the others, but probably the most able road warrior, even if it was the heaviest, at more than 2,250lb (1,021kg).

What about the original Ford-owned MkIII, surely that was road capable? Certainly, very flexible with the now-tired 305bhp V8, but vision with the reprofiled rear deck was poor, even compared with the earlier MkII rear screens. Also, the road version suffered with the conversion to a central gear-change, and the shift quality was only on a par with the skilfully converted Renault V6 gearbox attached to the GTD replicas.

Behind the wheel: originality shines through

However, one area where the Ford originals both scored heavily compared with the GTDs came in the form of the handling on a slippery test track. That '60s rigid monocoque chassis (the replicas were built around assorted square-tube space frames) and significantly harder spring rates, allied to anti-dive front suspension geometry, paid off notably in back-to-back drives. GTD owner Gordine came out for a ride in DWC 8G and admitted after observing an 80–120mph helping of high-speed banking and tight corners: 'I'd never have got through those

MKI GT40: PERIOD FACTORY-QUOTED SPECIFICATIONS

Engine
Configuration: V8 **Capacity:** 4.7 litres (289cu in) **Maximum power:** 380bhp at 7,500rpm **Maximum torque:** 330lb ft at 5,500rpm

Transmission
Type: Five-speed ZF, close-ratio gears

Weight
Race weight: 2,016lb/914kg (showroom, 2,025lb/919kg)

Performance
Gear speeds: First 58mph; second 90mph; third 120mph; fourth 133mph; fifth 160mph **Acceleration:** 0–30mph 2.0sec; 0–40mph 3.5sec; 0–60mph 5.3sec; 0–80mph 7.7sec; 0–100mph 11.8sec; 0–120mph 15.9sec **Maximum speed:** 160mph at 7,500rpm limit **Fuel consumption:** 7–10mpg

GT40 REAL v REPLICA

GROUP TEST

We didn't think it was possible. Snow made it impossible. Then it happened... two of the legendary Ford GT40s faced their GTD counterparts in road and competition trim. CCC assembled 32 cylinders and over 1300bhp just for you... and that lucky sod, Jeremy Walton, who drove

Pictures: Dave Wigmore

Spawned alongside the Beatles and **CCC**, the Ford GT40 celebrates the 30th anniversary of its first win at Le Mans this year. Unlike the Beatles, the original line-up is still in good voice. The replica business treats GT40s as absolutely premier league: only the AC Cobra attracts so many clones. We wanted to know if the best of the publicly-available replica GT40s – specifically the GTD – could share tarmac with the real thing, and Uncle Henry indulged our curiosity. Under the care of Mr GT40, Bryan Wingfield, and this former Ford engineer's DRL Engineering concern, a brace of road- and race-spec GT40s faced their young pretenders. Both GT40s have been in continuous Ford ownership from their birth. The Mk3 quad-lamp road car, which was registered as DWC 8G in June, 1969 – has covered a genuine 18,000 miles, utilising its original brake pads! The second, a Mk1 from August, 1965, has led a more chequered life as a race-winner look-alike for publicity. It did not race, but it carries Mario Andretti's name in honour of the American Le Mans win. It also has the clout of almost 400bhp from a rebuilt Ford V8 that carries full race breathing apparatus (quad Webers, hot cams, big valve, iron heads) allied to a production bottom-end.

A genuine no-shit value for a genuine GT40: "... spans £175,000 to a million quid. The best racer known would fetch the upper figure: a rough but undoubted original would cost under £200,000, with some seriously expensive restoration to do," said our informed trader.

GT Developments' (GTD) machines cost a fraction of that, but then so can a house. Although you can buy a body/chassis unit from £4150, both the GTDs tested cost in excess of £35,000.

Our GTDs came courtesy of leading members of the 80-strong GTD 40 Car Club. Chairman Alan Barlow brought his Brian Pepper-built Porsche Guards Red machine (804 AJB) to represent Concours and Sprint competition within a Mk1 outline while club magazine editor, Tony Gordine, brought his superb, self-built, K10 ARW in Lotus-like yellow and green to Mk2 spec. ▶

14 CCC

CCC 15

▲ I was slightly surprised Ford lent us their precious single-owner GT40 and MkIII for a track session versus their counterpart 1996 GTD replicas. They did, and there were some surprising conclusions from the replicas' private owners. *Dave Wigmore*

at that speed in mine. This is far stiffer. You can feel the basic difference in construction coming through. What a car! It's my first time in a real GT40 and I love it!'

Alan Barlow's GTD was alluring in show-finish red, and also beautifully detailed from cabin to engine bay. However, it provided a trickier drive in wintry conditions. Wafts of petrol fumes are always worrying in a glass-fibre-panelled car and the steering feedback was mixed. The effort required at the larger-diameter steering wheel rim was lighter than for an original – despite the presence of wider 265mm rear and 215mm front BF Goodrich tyres – but the connection between driver and car felt disconcertingly loose, versus the comparatively taut originals.

For me, the favourite remained 1008 – the original green hulk. Weighing in at the lightest – around a metric tonne – and energised by a narrow-rev-band version of the Ford 4.7-litre V8, it was then reckoned to sport around 400bhp. The other goodies came at all four corners, with stiffer track-tuned spring rates and dampers allied to Dunlop race tyres. Although it had never raced, it sent out all the messages of a competition car: taut, accelerative and so aggressively rorty that we actually got banned from further track action in Surrey's stockbroker belt almost as rapidly as the car lapped!

Chassis 1008 was no road car, and mechanically a trifle awkward to get moving, as the tough clutch needed to be fed in whilst taming an engine tuned to operate cleanly between 3,000 and 6,000rpm. Yes, it could go another 500rpm at the top end, but back then it carried a frangible cast-iron crankshaft in what was a budget-conscious 'bitza' engine.

It had been an educational day. It reinforced how replicas can give so much more pleasure than is possible with the originals with their collectability and massive values, and I felt doubly privileged to repeatedly taste the charismatic pull of a genuine GT40.

CHAPTER 6

SUPERVANS

V8s in Transit

On 23 March 1965, Ford started the white-van revolution with a name that was to become generic for a flock of imitators – clones that still chase Transit's definitive work/recreation image and multi-million sales today. Personally, my Transit drives are at the radical end of the scale. They covered Ford's V8 Supervan series of mighty performers built between 1968 and 2023 while their more mundane Transit stablemates were busy in the transportation and towing roles that this versatile vehicle was created to tackle.

EVOLUTION OF A COMMERCIAL CLASSIC OVER NEARLY 60 YEARS

Back in the late '60s and '70s, I experienced the Transit for trips such as towing a rallycross car from London north to the Croft circuit in the depths of winter. The Transit performed perfectly, and competition car as imperfectly as possible. I enjoyed the Transit's commanding driving position and the well-thought-out adaptation of car controls, features that remained commendable as the workaday Ford progressed through generations of updates.

I tested Transits as the base for V6 motorhome conversions and race-car towing vehicles, and sat terrified within an armoured Group 4 security-van version as a mock shotgun raid was enacted to show off its defensive capabilities in 1974. In 2003, I moved my worldly possessions from Berkshire to Hampshire in a rented high-roof diesel.

In summer 2022, a neighbour called to ask if I would help rescue his seriously modified pick-up truck with his GT-customised 2021 Transit. The stricken 1949 'Light Duty' truck wore 'Ford' badging, but had sported a variety of V8 power transplants prior to the failing motive power requiring my call-out. I found it now deployed a

▲ Up she goes! This crowd-pleasing party trick made the original Supervan the most spectacular of all in so many public displays. *Motorsport Images*

BMW 330d diesel engine mated to a five-speed GM-ZF automatic transmission, all stranded on a blazing-hot day, outside an active military camp.

On such a 30°C afternoon, I appreciated the powerful Transit air-conditioning and controls that were light and logical to use, even after a near 20-year gap in my Transit experiences. The diesel powerplant was sharp to respond to the accelerator (it featured remapped management) and I also appreciated power steering to free it from its tight parking-area confines. Blue-needle dials of considerably better clarity than the earlier-era Fords were welcome too. Other plus points extended to a slick six-speed manual gearbox and a fine driver's-eye forward view, matched to comfort levels that compared with those of earlier Range Rovers.

It was not an easy task for the Transit to tow the inert 1.5-tonne pick-up, especially with its automatic transmission, and moving along at sub-20mph pace over country lanes and main-road junctions. The owner did a great driving job, but we did have a couple of tow-wire and cable-connection failures. I do not want to repeat

▼ In 2003 there was a Transit mission to accomplish with my friend, photographer Peter J. Osborne. A rented high-roof diesel variant gallantly lugged the remnants of my earlier life across southern Britain at improbable speeds. The lurchers were a hindrance when it came to unloading 30 years' worth of automotive books, and more. *Deborah Parrett*

➤ In 2022 I acted as a callout guy to a neighbour's distressed classic Ford pick-up truck (then deploying a BMW turbo-diesel), using his GT version of a 2021 Transit. Aside from the dress-up items, there was much hardcore progress for me to absorb since my 1960s race-car towing days. *Author*

looking at a flat-backed Transit's partially visible registration plate on zoom close-up anytime soon, especially when working hard at the non-power-assisted steering and brakes of a powerless pick-up.

At the time of writing, the van that became a legend was on course to lead the British commercial market for a 57th straight year, with over eight million examples sold in multiple derivatives. The market had changed, with ever-fancier versions, some quite racy with a bow to three generations of V8 Supervans described here. Alternative recreational Transits are designed to meet the 21st-century challenge of rivals such as VW's Transporter, a vehicle with just as much pedigree and with a successful high-profit transformation for the camper van/recreational market. In recent years, Ford's Transit name has been applied not just to core workhorses, but also to the bestselling Custom versions and a light van adorned with Transit Connect badges.

However, there was nothing glamorous about the Transit's original design brief. Aiming to replace both Ford of Germany's 1953–65 Taunus Transit and Ford of Britain's 1957–65 Thames/400E vans, it was targeted at hard graft. What became notable in automotive history was the multinational interactions the parent Ford Corporation imposed on their European satellites. From an American perspective, teaming British and German personnel in the workplace was legendarily 'don't-mention-the war' tricky. Ford at Dearborn pulled it off, engineering and marketing the first pan-European Ford years before the formation of Ford of Europe (1967) and a decade before Britain joined the European Common Market (January 1973).

Back in 1963, the joint van project was coded Redcap and by the 1965 launch year was scheduled to be publicly dubbed V-series, which fitted Ford's commercial vehicle names and the use of German and British vee-layout engines of totally independent design and manufacture. Although V4 motors were the production mainstay, higher performance was readily available at extra cost in the UK by installing the 3-litre Essex V6, as sold in the Zodiac MkIV and earlier 3.0-litre Capri. Specialists often carried out this V6 implant as a conversion, but so did Ford for ambulance or police requirements. Ford's Competitions Department also utilised V6s for its heavily-laden rally service vans.

Much pre-production testing was conducted in Britain using the then recently acquired Boreham

➤ The 2021 Transit GT featured a mass of features nobody would have thought of including at its debut over 50 years previously. *Author*

airfield facility, which also became home to the Ford Competitions Department. However, some of the sustained higher-speed testing was carried out on Essex public roads, with police monitoring activities in the era before Britain received a blanket 70mph maximum speed limit in December 1965.

Just weeks before its debut, Ford of Britain Chairman William Batty reviewed progress. The vehicle he evaluated was to LHD specification, badged as Transit in line with that market's German predecessors. Batty went batty for the Transit appellation, and such was his influence that it became the multinational name. The initial styling buck came from the USA. It broke new European ground by dispensing with the flat front seen in many contemporary vans, introducing a short snout to house compact V4 petrol engines. This had the added advantage of allowing more sound-proofing for the engine and transmission than was possible with the flat-front predecessors. That American design buck was transported to the UK for final styling touches, emphasising the headlamps and grille. Initially they went for a more truck-like appearance, but would revert to a more car-like outline in the '70s and an aero-friendly streamlined nose for 1986.

The first million Transits had been manufactured by 1976 in either Britain or Belgium. A second million were recorded by 1985, by which time many more production sources were available, from Turkey to Australia, with Hyundai also building the van in South Korea under its own HD-branding. The Transit had become a truly international phenomenon.

THE FIRST OF THE FAST AND FURIOUS SUPERVANS

Easter Monday 1971 marked the official debut of Ford's first publicity-hungry V8 Transit. Seen at many British circuits in front of thousands of spectators attending bigger international race meetings, it was the genius idea of John Dale, whose day job at the company was Sales Promotion, Truck Merchandising. The beast was a magnet for contemporary film cameras, and enormously successful, but thanks to my former colleague Clive Richardson, I learned it was not the first V8 Transit to pound the publicity trail.

Barry 'Leapy' Lee had wrestled with a much

◄ One of a series of photographs taken at Silverstone of the first official Ford Supervan on a 1971 media day when some, such as my former colleague Clive Richardson (pictured), got to ride and then drive the white beast. *Motorsport Images*

▲ Plenty of commercial decals adorned the first-edition Supervan, which was driven by a number of professional racing drivers and a few media men. *Peter J. Osborne*

more primitive American Galaxie V8-engined mongrel in faded red before the Lydden Hill TV cameras. Barry even took the 7-litre monster round the quarter-mile oval at Walthamstow as his Escort '351' hot-rod racing career bloomed. That must have been some feat, as the giant 7-litre motor lay alongside the driver; production suspension was employed alongside standard Transit front and rear axles.

I watched the first official Supervan V8 at Thruxton, where Ford saloon-car legend Chris Craft entertained the 1971 Formula 2 crowd with 'how high will it go?' wheel lifts through the chicane. Normally demonstrated by Terry Croker, it attracted enormous spectator and media interest, even from legendary Denis 'DSJ' Jenkinson of *Motor Sport*. Mass-media and specialist publications would feature that original beastly Transit, which was followed by two more such machines of increasing Cosworth V8 sophistication for the '80s and '90s – Supervan 2 and Supervan 3 – that I drove in official forms.

By the '90s, rival manufacturers such as Renault had cottoned on to the idea and wheeled out headline-grabbing mutants based on commercial-vehicle outlines, leaning heavily on Grand Prix V10 and carbon-fibre technology. A Renault built by Matra mustered up to 800bhp within a cartoon interpretation of the Espace people carrier, one that catered for brave passengers, always with ear protectors! All a far French cry from 1971's Ford of Britain Supervan.

Under the skin: delivering something different

Created around a production steel body with flared arches to accommodate 15.5in x 15in Firestone tyres and oversize Revolution alloy wheels, Supervan was the workshop-brewed creation of Terry Drury, a Ford-favoured Essex-based man of practical skills who was previously behind a record-breaking outing for Transits at Monza.

Earlier and subsequent endurance speed runs for Ford had demanded Drury use all his ingenuity to adapt a production vehicle. Supervan demanded serious modifications to capture as much public awe and admiration as possible. Externally, those bulbous wheel-arch extensions, and the stance of a rearing bronco with off-road spaciousness between front tyres and prominent arches, were spectacular, even at a standstill. Your ears were battered by seriously oversize twin exhausts, ready to singe anyone foolish enough to step too close to the double rear doors.

Open those standard rear doors and you witnessed disgraceful disregard for payload but outstanding attention to delivery times. Boldly rising through the steel flooring were a Ford-based 5-litre (302cu in) V8 and attendant ZF five-speed synchromesh gearbox. The clutch was a sintered, triple-plate Borg & Beck unit, with drive delivered to the fat rear tyres via Rotoflex couplings and drive-shafts fed through magnesium hub carriers.

The Ford V8 was retrained to Dan Gurney's Eagle specification, with Gurney-Weslake cylinder heads, high-lift camshaft, and oversize porting and valves, as usually installed within a GT40. The Transit became mid-engined; the eight cylinders wrapped in exhaust plumbing, and sporting quadruple Weber 48 IDA carburettors – all too ready to dump high-octane fuel in a heavy petroleum downpour, laced with air from an otherwise bare, but multi-tube-braced cargo area.

A tube spaceframe carried not just the engine but also a Ford D800 truck radiator and adjacent fuel tanks. So the body was relieved of stress at most points, and some handy Formula 1 hardware was devoted to that peculiar nose-up stance. The front end combined GT40 unequal-length upper wishbones above obsolete Cooper F1 lower wishbones, with the usual period single-seater inclined coil spring/damper units. Steering was via GT40 rack-and-pinion componentry. Such a tall vehicle on such wildly offset, fat wheels and tyres demanded careful consideration of where the rear roll-centre might hide. So, more freestyle ingenuity went into Drury's rear top location links, working in association with Cooper F1 wishbones.

Brakes? Yes, it did have some – Lockheed ventilated discs with four-piston callipers. Some evidence of a stronger servo action, according to contemporary reports, may have eased applying such racing-technology stoppers.

The driver/passenger cabin bore some strong hints of Transit furniture, including the slippery vinyl bench seat for victims – sorry, passengers. Standard, widely spaced pedals were provided, along with a steering wheel with an upright stance, then uncomfortably familiar to Mini drivers. The triple-spoke wheel was a contemporary deeply dished and drilled leather-rim item, as used in factory competition Escorts. A large rev counter sat where the production speedometer had been, with the showroom fascia top and instrument cowling preserved.

Distinctly non-standard were a Corbeau race seat and the right-hand ex-GT40 ZF five-speed gear-change, awkwardly squeezed into far less door-to-seat space than for the GT Ford.

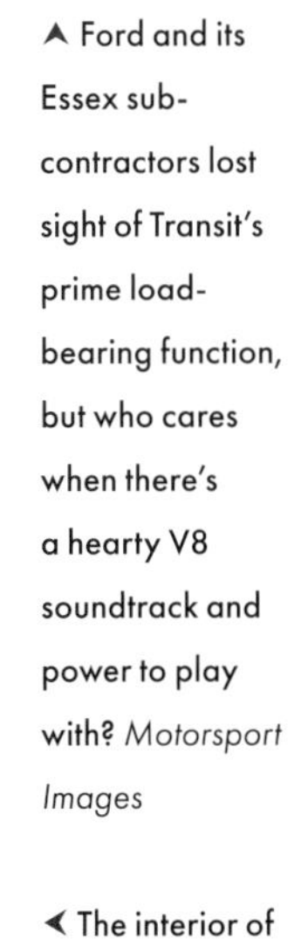

⮝ Ford and its Essex sub-contractors lost sight of Transit's prime load-bearing function, but who cares when there's a hearty V8 soundtrack and power to play with? *Motorsport Images*

⮜ The interior of Supervan was a marvellous mishmash of 1971 track and warehouse furniture. The bench seat explains why it was so tricky for passengers to stay upright, whilst the driver was cosseted by a Corbeau race item. Note the right-hand gear-change for a five-speed ZF transaxle. *Motorsport Images*

COMMERCIAL VEHICLES!

Courageous Clive tussles with Ford's frightener and examines a haulage hybrid

SUPERVAN SAMPLED

SHOULD ANYBODY be looking for a delivery van driver then I could be the man for the job. Qualifications? Demolishing a farm building with an Austin J4 milk van at a full 10 mph in reverse (at the not so sweet age of 16), and more recently, lapping the Silverstone Grand Prix circuit at nearly 86 mph in the fastest Ford Transit van on earth. The first being an experience I certainly don't wish to repeat, and the second one that I definitely do.

The idea of John Dale, Co-ordinator Sales Promotion, Truck Merchandising for Ford, Supervan is acting as the big plug for Ford's sales theme 'Transit the Supervan' for their rather less demon load carriers.

Designed and built by Terry Drury Racing, Rainham, Essex, Supervan is a follow-up to the 7-litre Galaxie engined Transit Barry Lee drove last year for publicity film purposes at Lydden Hill and a public demo on the Walthamstow short oval. That was a pretty lethal device with the V8 alongside the driver, standard suspension and axles and wide steel wheels. Supervan the Second is a much more sophisticated machine with mid-chassis mounted 5-litre GT40 engine and what amounts to F1 suspension and brakes.

To be precise, the 18 cwt Transit shell is powered by 4999 ccs of wet-sumped Ford-Gurney Eagle V8 — big-valve heads, Sullivan camshaft and four twin-choke 48 IDA Webers give it 435 bhp which it's claimed will push the ungainly beast over the standing quarter-mile in 14.9 sec, only 1.8 sec slower than a Racing GT40. Potential top speed in the fifth of its ZF gears is calculated at 150 mph.

The shell is standard apart from flared arches to accommodate the huge 15.50 x 15 CanAm Firestones ('with rigid Indianapolis treads') on special, alloy Revolution wheels.

With the shell ends Supervan's resemblance to any of those Transits which fly past you at 85 mph on the motorways. The engine sits in the middle of the load space, carried with the rest of the mechanical bits, D800 truck radiator, fuel tanks and running gear in a multitubular space frame, fixed to the shell at six points.

Terry Drury worked long and hard on the suspension to get Supervan to handle anything like right. At the front, unequal length wishbones come from a GT40 at the top and F1 Cooper bottom, the coil spring/damper units are inclined and the front uprights are borrowed from an embarrassingly non-Ford speedy luxury carriage which I won't name in case of reprisals! Wheels are pointed in the right direction by a GT40 rack.

Chris Craft lifts the nose and scatters the spray at Brands. Later he spun it.

To get the roll centre right no rear bottom links are used, top ones being of Drury manufacture and the wishbones F1 Cooper again.

Transmission is a sore point with John Dale in that his small budget has just been considerably drained by the antics of a procession of human apes, who swung on Supervan's gearlever during a tour of dealers' showrooms, irreparably mangling the innards of the ex-Le Mans winning GT40, ZF all-synchro, five-speed gearbox. Hence a very big bill for a new one. The rest of the transmission consists of a Borg and Beck three-plate sintered diaphragm clutch, Rotaflex couplings and fixed length driveshafts supported in magnesium hub-carriers.

Terry Drury's BDA Escort and Lola T210 driver Terry Croker is the number one Supervan demo pilot round the circuits.

My own chance for fame and fortune as a works Transit driver came when, with several other journalists, I was invited to tame the monster at Silverstone.

Kitted up with waterproof underpants I arrived at the 'Stone just in time to be

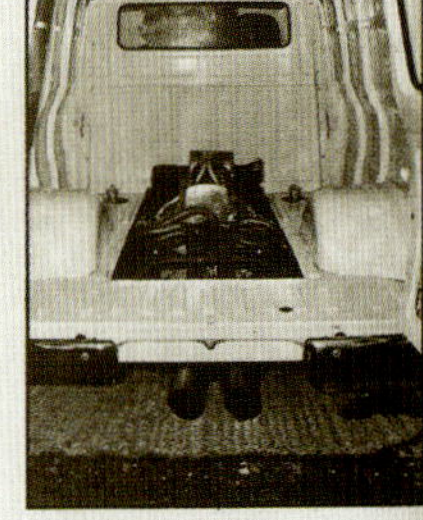

And today's load is 4999cc of Ford/Gurney Eagle V8 and numerous yards of piping.

40

▲ The original Supervan, with GT40 V8 muscle, excited much media comment – as intended – but this is one of the more detailed accounts, by my late colleague Clive Richardson. It appeared in *Cars & Car Conversions* magazine during 1971. *Author's archive*

Behind the wheel: a high-speed work-out

From the spectator areas at Brands Hatch or Thruxton, the 1971 Supervan was an awe-inspiring sight, lurching from slow-corner acrobatics to thundering dashes along any semblance of a straight. However, our view was just a spectacular teaser. Here's how my trusted colleague during my time at three magazines – and friend over decades – Clive Richardson experienced Supervan at a Silverstone day held for a trio of motoring writers. Naturally, this proper Yorkshireman recorded the fastest time of the three, less than 2sec adrift of Drury's professional demo driver (not Chris Craft).

'From the first Copse corner into Maggotts, I struggled with the right-hand gear-change to get appropriate cogs. The lever sits in the narrow gap between seat and door, with first to the left and down, moving logically across the gate through second and third to fourth, which is to the right and forwards. Fifth, straight back from fourth, wasn't really required for Silverstone, gearing being fairly high.

'I can't remember driving a vehicle that needed so much effort to steer: muscles rather than skill was the key to driving it fast, to forcibly tame the beast. If I'd relaxed for a second at holding lock on its understeering course, there'd have been no getting it back. As it was, under control the understeer took the fat tyres on a perfect line. A wide exit on the concrete verge at Copse in third, up to fourth and flat through Maggotts with the V8 banshees doing their stuff in the back, Supervan's nose high in the air.

'Heeling and toeing across the brake and clutch pedals was an impossibility, so quick footwork was the order of the day, banging it down through the gearbox to second. Timidly booting it round Becketts [the tightest corner on the circuit] with the back end giving a quick flick. Then into Chapel, understeering almost onto the grass, and hit fourth for a 110–120mph burn down Hangar Straight to dramatic Stowe. Here, the bump by the marshals' post had me on two wheels, pitching and tossing across the circuit in third gear. A snatch in fourth, then it was time to brake and change down a cog to clip the apex at Club, before being taken forcibly across the circuit [by the sheer speed and momentum of Supervan's modest 2,016lb/914kg].

'Back across the tarmac again to line up for the fast Abbey curve, taken in fourth, then flat-out down the straight towards Woodcote [the ultra-quick final corner]. At 120mph in a vehicle that seemed about 10-foot high, I was ducking involuntarily under the road bridge. Woodcote was hellish tight in the monster, so it was down to third again, the offside front wheel waving in the air.'

So our Clive completed his intrepid reporter's ride with an automotive bruiser. Sadly, hearing these recollections would be the last working connection Clive and I had, as my colleague and friend from 1968 onwards died in tragic circumstances in June 2022.

The original Supervan was obviously some experience, and my personal admiration goes to Clive for sharing an illuminating and exhilarating perspective on the Supervan most still regard as *the* original.

SUPERVAN 2: A 1980s REVIVAL

A decade on from Clive Richardson's Silverstone heroics, I had an appointment on 13 June 1985 to drive the second iteration of the Supervan – a totally new vehicle – at Donington. Rex Greenslade, Ford PR, former *Motor* journalist and regular BSCC racer, would mentor me for the day. By then, *Motor* magazine had run proper fifth-wheel tests on Supervan 2 and reported that the 3.9-litre Cosworth DFL V8's 590bhp hurled the vehicle from 0–60mph in 3.2sec, and covered 0–120mph in less than 10sec. I had already been told the truly special Transit would be geared for 155mph at 9,500rpm. So, I knew this iteration of super Transit would be a special experience, and so it proved.

Under the skin: sports-racer in van clothing

First, a little history. Created by John Neale and John Thompson within Auto Racing Technology, with the help of renowned designer Tony Southgate, the foundations under the trick glass-fibre/carbon-fibre/Kevlar superstructure came from the cancelled Ford C100 Group C sports-racer. That meant that TC Prototypes in Northamptonshire converted the C100's aluminium-honeycomb and carbon-fibre composite chassis to fit beneath the squashed Transit outline, all concerned noting that the C100's track and wheelbase were actually surprisingly close to that of the second-edition production Transit!

By this point, Ford had sold two million Transits, but had become conscious that aerodynamic-drag factors were a sales weapon, even in the commercial-van market. Audi's large 100 saloon of 1982 had made the public more aware of the fuel-consumption benefits of a low drag factor, with a heavily publicised – and proven – 0.30Cd for the Audi. Whilst front-drive Escorts and other showroom Fords simply had to demonstrate such sleekness, particularly to the sceptical German public, even this show Transit was debuted with a claimed 0.33Cd. Much of that smoothness was owed to taking strategic slices from the overall body height compared with a showroom van (which was rated at 0.37Cd). Later, that bold low-drag claim was crucified by the addition of an effective and generous rear wing that boosted Supervan's Cd up to 0.47,

▼ Supervan 2 was a bit more sophisticated than the original, using much Group C sports-car racing technology and an enlarged 3.9-litre Cosworth V8. *Motorsport Images*

➤ Designer Tony Southgate (centre) discusses key features of Supervan 2 in the paddock with Rex Greenslade (right) looking on. *Motorsport Images*

➤ Pure pedigree racer layout within the 'load bay' of the 1985 Supevan 2, with its endurance 3,995cc DFL version of the Cosworth V8. *Motorsport Images*

but had the notable advantage of restraining this Transit V8 from tail-happy aerobatics. Also assisting in making this edition of Supervan more manageable was the C100 hardware, including wishbone front suspension with concentric coil spring/damper units and various anti-roll-bar options. The rear featured a combination of lower wishbones and upper triangulated location links cooperating with inboard spring/damper units, these prompted via pull-rods, again with anti-roll-bar thickness options.

Officially weighing in at 2,200lb (998kg) – a little more than the original Supervan – this second-generation version's ability to stop had progressed mightily. Fitted with 13in-diameter AP ventilated discs and associated race callipers and pads all round, there was cockpit adjustment for front-to-rear brake bias, with a handbrake thought superfluous.

The 16in-diameter BBS wheels, neatly incorporated beneath the bulging body, were 11.5in wide up front and 14.5in for the back. They wore Goodyear Eagle slicks for our dry-weather Donington circuit test, carried out on the shorter track, without the Melbourne Loop. It offered some fun times through the chicane and surrounding tarmac, which swoops, crests and curves just as attractively as Brands Hatch or Laguna Seca.

I was guided around the Transit that summer of 1985 by members of the David Price Racing (DPR) team, people I had known since my late-'70s freelance PR days with Unipart. At that time, DPR ran Formula 3 Marchs with Triumph Dolomite 16-valve power for Nigel Mansell and Brett Riley.

As with its predecessor, the engine was hidden away, mid-mounted in the cargo bay. A pure racing unit, the Cosworth DFL was a well-known stretch of the original 3-litre Cosworth DFV, with a capacity of 3,995cc on a bore and stroke of 90mm x

77.7mm. The fabled four-valve cylinder-head layout and double overhead camshafts were a given, and the compression ratio was a comparatively mild 11:1. The V8's alloy-encased weight was a skinny 352lb (160kg), much less than the production-based iron-block GT40 unit within the earlier Supervan. The official power output was 590bhp at 9,250rpm, coupled to a rather obvious 7,000rpm torque peak of 340lb ft.

Behind the wheel: heavyweight Group C performance

Because of the substantial C100 sports prototype foundations and underlying chassis, you stepped up to the cabin, which was significantly more civilised than you would expect of a thoroughbred racer. A Corbeau touring-car race seat (with a second for a passenger) and fully restraining six-point safety harness sat you behind the awkward upright steering column and an RS200-style red leather-rim steering wheel.

Beneath the clip-secured windscreen, a squared-off snout carried a matt-black grille and production Blue Oval badge, all promoting the Transit identity above strictly race, full-depth wrap-around skirts and air dam. Although the five instruments were surrounded by a production fascia, the mechanical tachometer's tell-tale pinned at 9,600rpm was a prompt reminder that this Transit hauled speed, not cargo.

Of the four minor dials, those for oil and fuel pressure, plus a steady water temperature read-out of 90°C, were reassuring, but the oil temperature shot by the end of the scale at 160°C, owing to the surrounding enclosed engine bay according to DPR minders Mark Julyan and Michael Sholl.

I had the benefit of a warm Supervan to start, but the full procedure from cold could occupy all of 30 minutes, requiring an external generator and a fan blowing warm air around the powertrain to preheat the gearbox/transaxle unit. Surrounded by so many flat panels, the enclosed V8 demanded earplugs for my eight- and five-lap sessions around the 1.9-mile abbreviated version of Donington.

Departing the long pitlane, I let the clutch out fully once we had reached 4,500rpm in the long (70mph) first gear. The DFL pulled cleanly at 4,500, but 5,500 cleared gargling inlet throats, ready for the run to maximum torque at a tingling 7,000rpm.

▲ Into the 1980s and Supervan 2's cockpit abandoned Transit pretensions and became a high-rise racing cell for two – a rapid-response office, featuring a raucous 9,600rpm soundtrack. *Motorsport Images*

Tingling? Certainly, for there were no good vibrations from this soundtrack, which penetrated not just the flappy flyweight Transit panels but ran right up the steering column to punish your gloved hands with an accompanying drubbing from the agitated flooring. However, seeing a former journalistic colleague performing at a quicker pace was sufficient encouragement to use the full 9,600rpm, and to penetrate beyond heavy steering efforts and a pronounced desire to plough straight on in slower corners.

It took me most of my allotted session to descend from laps in the 1m 28s bracket to 1m 22.4s (85.5mph), but it was worth it. Just short of 90mph was supplied in second gear. The acceleration was simply outstanding, as 0–120mph took less than 10sec at a time when most road cars struggled to return 0–60mph below that 10sec barrier. That boot in the back and sheer speed, plus the shattering sounds of eight cylinders wailing beyond 7,000rpm, were major thrills.

The AP brakes were fabulous, for you had to haul it back regularly from 120–150mph bursts, remembering that the cornering pace was far from sports prototype, more tin-top touring car. Those 13in dinner plates worked through the massive Goodyear slicks to cope flawlessly with the erratic demands of a newcomer at the steering wheel.

➤ A much more focused drive than the original, Supervan 2 had become an elevated and enlarged version of a Group C race car. Thrilling for a stranger, but not as charismatic as the original Supervan. *Motorsport Images*

Motor measured those independent fifth-wheel acceleration times via a passenger reporter, Chris Craft whirring the pedals and flicking through mechanically precise Hewland five-speed gearshifts. So I became a cliché, the first journalist to drive this device, and some of my contemporary comments as a contributor to the cult *CCC* magazine sound strange now.

I described the spectating experience when watching the Transit mutant as 'travelling at Group C straight-line speeds, bursting into bends with all the grace of a randy hippo'. Not sure I had first-hand experience of a severely over-excited hippopotamus, but the cornering technique certainly did provide serious excitement, as Supervan 2 could hop and skip into slower curves if you tried to enter beyond the feasible speed of such a big mass. I only hit fifth gear along the main Starkey's straight, where it would pull over 8,500rpm before the brakes were summoned for the second-gear twitches through the end-of-lap Park Chicane.

'I DESCRIBED THE SPECTATING EXPERIENCE WHEN WATCHING THE TRANSIT MUTANT AS "TRAVELLING AT GROUP C STRAIGHT-LINE SPEEDS, BURSTING INTO BENDS WITH ALL THE GRACE OF A RANDY HIPPO".'

Just as for Clive Richardson's encounter with the first Supervan, physically restraining the Transit within the bounds of the tarmac under full V8 power was a challenge, particularly from the slowest lap-opening Redgate right-hander to the exhilarating plunge through the markedly downhill right-left into Craner Curves, plummeting into the hard right at the Old Hairpin. The fourth-gear run via Starkey's Bridge was remarkable only for the pleasure of impressive acceleration, but the tight right through McLean's was a physical challenge, followed by a blast of uphill energy to the double-right of Coppice, where one could often ignore the driving-school-suggested apex, seeking a clean line to launch a near 600-horsepower beast onto that final Starkey's straight.

It was a moving experience conducting the second of the Supervans, but I probably was not in the best place to appreciate how fortunate I had been. That is because I had recently comprehensively failed an audition to become a BBC TV *Top Gear* presenter, so was even grumpier than usual. We now all know how important that was to the media career and financial status of a certain other Jeremy!

SUPERVAN 3: MORE OF EVERYTHING

Silverstone on 10 May 1996 was the occasion of my last outing with the penultimate edition of Supervan. Naturally dubbed Supervan 3, it had some common features with Supervan 2, as it again rested on a modified C100 sports-racing chassis and had a pedigree Cosworth V8 race engine installed, albeit this time from the later Grand Prix HB series, and offered significantly more power within a van with lower kerb weight.

Under the skin: a more-focused package

From a Ford perspective, Supervan 3 was required to mark the Transit's 30th anniversary, and to celebrate manufacture of more than three million examples. Most of those workhorses would have been motivated by 80 or 100bhp diesel motors, whereas this track-only show car had been detuned from Formula 1's 730bhp at 13,500rpm to a more moderate 650bhp, with an 11,500rpm limit for my outing. The 3.5-litre unit delivered by Cosworth under Ford branding for the 1993 Benetton F1 season was another generation on from the DFV/DFL series, featuring a 12:1 compression ratio and a 73° (aluminium) block angle rather than the traditional 90° for a V8. Compared to the DFL unit I tried in Supervan 2, the HB was lighter, more compact, and utilised more lightweight materials.

The engine was located in the traditional Supervan position, lying low in what would have been the cargo bay, breathing high-level cold air supplied via roof-mounted tubing to the air box. The fuel-injected race eight was mated to a Benetton six-speed gearbox via a sintered metallic AP Racing clutch, with final-drive via a Quaife limited-slip differential.

Supervan 3 was based on the previous Supervan's C100 foundations, built in a four-week hurry by GT40 custodian Bryan Wingfield of DRL Engineering. It made the tight schedule for the 1995 Earls Court motor show, appearing complete

▼ I was able to tell the inside story of Supervan 3 on ITV Meridian South as well as in magazines. *Alamy*

from the outside, but actually minus most of its mechanical hardware.

From a driver's point of view it was more agile than its ancestors, weighing 1,962lb (890kg) and built on a 110in (2,794mm) wheelbase to seven-eighths scale on the sleeker September 1994 showroom silhouette. The cabin featured two race seats, with a near-central driving seat, a friendlier steering rake angle than the previous incarnation, and radical composite exterior. The stripped-out body displayed swoopy Ford Electronics branding and a massive 'cow catcher' front spoiler, complemented by a full-width wing at the back.

Below a body with overall length of 181in (4,597mm), width of 74in (1,880mm) and height of 66in (1,676mm), Goodyear Eagle tyres were carried on substantial cast-magnesium OZ 13in-diameter F1-specification wheels with 10 and 12in rim widths. The suspension featured conventional double wishbones with inclined coil-over dampers. Anti-roll bars were fitted at both front and rear, but we had the back one disconnected for the test to cater for my level of experience in such a challenging automotive mongrel. The braking system had moved on since the C100 era, with Italian supplier Brembo providing 279mm (11in) ceramic discs and pads, plus four-piston callipers.

Behind the wheel: Grand Prix racer meets urban workhorse

As you would hope of a third evolution, this Supervan was a much more resolved driving experience. I was mildly surprised at the attention to detail in the cabin, as I knew neither budget nor project time had been generous. The Sparco seats and red six-point Sabelt safety harness delivered functional comforts for both driver and passenger; the pilot was set forward of the co-pilot, free to flail arms and elbows without contact.

There was less drama on-track, as this Transit super-edition was much better mannered in action, especially when conducted by professional Warren Hughes, then a reserve driver for the Mondeo BTCC race team, but better known for his single-seater prowess. From a personal viewpoint, I knew I was in good hands before my turn at the wheel, as I track tested new road cars regularly through the '90s with Warren's brother Mark, now one of the finest international Grand Prix correspondents, then a colleague at *Motoring News* and *Motor Sport.* Mark needed our test-day lunches of ham, egg and chips to survive on the traditionally skeletal wages of that management period.

The track for this test was the simplest Stowe short circuit, formed over 1.08 miles of the Silverstone infield. It features one long back straight, and the rest left-right wiggles down to a 30mph hairpin loop back to the start/finish.

Supervan 3's astounding acceleration and retardation qualities were well demonstrated by Warren, although we did have the indignity of one push-start. Then it was my turn to be installed in the Sabelt harness, grasping the top of the steering wheel and dubiously studying the right-hand gear-change and the linked rods disappearing alongside the race-car pedals sprouting up from the floor. I was right to be cautious about the transmission, but not the gears: the problem for a stranger was succinctly summarised by Bryan Wingfield in a pre-flight briefing: 'Remember, it's only a 20-second clutch-engagement life. If you've got to turn it round, leave a really big turning circle, enough for a race transporter!'

I leaned against the belts to drop three switches with extension pieces to assist the harnessed driver. I cracked the throttle open slightly, and signalled through the plastic side 'glass' that we needed the external energy of the Formula 1 starter. Out of sight, our commands are met: a distant whirr, sounds of mechanical components straining as they churn against that 12:1 compression. Cough, splutter and 32 valves prodded by four camshafts get the upper hand.

The stark cab shook, the load area and comprehensive front-to-cargo-bay roll cage

'SUPERVAN 3's ASTOUNDING ACCELERATION AND RETARDATION QUALITIES WERE WELL DEMONSTRATED BY WARREN, ALTHOUGH WE DID HAVE THE INDIGNITY OF ONE PUSH-START.'

reverberated sympathetically at the 4,000rpm 'tickover.' Thumbs up from me to signal disconnection of the external battery and starter by the hidden crew behind us. More throttle, and 6,000 of the 12,000rpm available according to the tachometer scale blare consistently. Dip the heavy clutch, and pull back against the spring-loaded gate of the simple H-pattern gearbox. The revs drop to 5,000 as I let the clutch in, grateful to have forward motion before all of Mr Wingfield's 20sec allowance is swallowed.

The clutch went fully home before the 50mph capability of first gear disappeared. Grand Prix regulars were unimpressed by the acceleration of the comparatively plump Transit mutation, but for me a calculated 747bhp-per-ton power-to-weight ratio with 2.5sec 0–60mph ability, coupled to 0–100mph in 7sec, were simply addictively exhilarating. Supervan 3 smashed through the first four gears in a click-clack rush, one that delivered 100mph with immediacy. This was strongly emphasised on such a small circuit because my next thoughts had to be about slowing and swerving round the next curve.

As with Supervan 2, the brakes were consistently and efficiently admirable. The harder I trod on them, the better they got, to the distress of the front spoiler snuffling tarmac, although there was some weaving courtesy of the components serving our urgent retardation requests. Incidentally, as a passenger I had to hang on to the big dashboard steady-bar desperately under our heaviest braking, otherwise even a belted body desired only freedom, via the windscreen!

The shorter 400m start/finish straight allowed me to slap through the gears at 11,500rpm to reach 124mph, followed by that impressive slow-down into a 30mph corner. The gear-change and its solid joints allowed a precise change, so easy that even this unfamiliar operator could dispense with the clutch on the straightforward 1–2 and 3–4 shifts, but I did dip the clutch a fraction to slip across the gate from 2–3 and 3–4.

So, go and slow – impressive. How about slipping around some racy corners?

Mmm, if there is an Achilles Heel to building a Supervan, it is the conflict between the desired showroom appearance of a commercial van, and the sleek, low body of no carrying capacity whatsoever ideally suited to track demands. As with its predecessors, slowing sufficiently, especially for sharper corners, was a challenge. That was because I was lured by that adrenaline-

▲ Supervan 3 continued the V8 track Transit tradition of drawing the spectators (here at Goodwood Festival of Speed) and was also a much better drive, as I discovered in a session at a short-circuit Silverstone. *Alamy*

▲ Accelerating, braking or cornering, the third edition of Supervan, with 600bhp thrust, was an impressive sight and a challenging drive. *Alamy*

pumping acceleration – and such magnificent brakes – into over-estimating cornering capability. Supervan 3 was a significant improvement over its forerunners, but still had some party tricks.

I learned very quickly that they had disconnected the rear anti-roll bar because that eased a tendency to switch rapidly between gripping and tilting – to the point where I was riding a two-wheeling stunt machine. As it was, the slower corners had the inside rear wheel hopping and the limited-slip differential snatching at every instance of intermittent power delivery.

Once I had wrestled it back into a semblance of composure, the full power of that obsolete Grand Prix V8 tweaked the back tyres in any gear from first to fifth. Warren Hughes reported: 'For the debut at Brands Hatch it was very wet and slippery. It would spin its wheels in fourth and fifth along the top straight.' I reported for the Anglo-French translator on *Auto Hebdo* in May 1996 that during my dry test, full throttle in third gear 'snapped the whole thing sideways for a lurid instant. I was glad of such rapidly responsive race-car steering to bring it all back in line.'

'I LEARNED VERY QUICKLY THAT THEY HAD DISCONNECTED THE REAR ANT-ROLL BAR BECAUSE THAT EASED A TENDENCY TO SWITCH RAPIDLY BETWEEN GRIPPING AND TILTING – TO THE POINT WHERE I WAS RIDING A TWO-WHEELING STUNT MACHINE.'

Supervan 3 was the perverted cross between Grand Prix power, sports-prototype racing car and urban mule. For me, the 1993 Formula 1 Cosworth-Ford V8 was the headline act. The pedigree powerplant guided this stranger cleanly from 5,000–6,000rpm, rushing onwards with majestic conviction. The 7,000–11,500rpm soundtrack was composed of premium mechanical automotive soul, wailing resoundingly within those bizarre Transit load-bay quarters to permeate the cabin and spread the silly grin all across my helmet-encased face.

The final words go to Warren Hughes, the professional who drove most Supervan 3 miles: 'You've got Formula 1 power and brakes – but not the full effect because of the weight and lack of distance-generated heat. All from a van driver's position. Not the easiest drive I've ever had – but hell, it's different and Big Fun!'

SUPERVAN 4: ELECTRIFYING

We had to wait a little over 25 years, until 2022, for a fourth Supervan spectacle.

No V8s this time: instead this iteration boasted pure quad-electric-motor propulsion equating to a neatly rounded 2,000PS (1,972bhp)! The public and TV audiences saw the road-registered (SV04 BEV) carbon-fibre and spaceframe concoction at Goodwood Festival of Speed in 2022, when it carried a simple 'Electric Supervan' logo amongst many more messages. We subsequently had a look beneath the surface of the electrified commercial, produced to promote the 2023 E-Transit production van, which reportedly supplied Supervan 4's floor.

Ford Performance project monitors and sub-contractors STARD Motorsport revealed some of the technology that double Le Mans winner Romain Dumas forcefully guided that debut day, via an old-school conventional steering wheel. Key to the project was a 50KwH battery and the all-wheel-drive configuration of the electric motors on front and rear axles.

Instead of the quick-shift, multi-speed transmissions of the V8 Supervan trio, the Electric Supervan had just two gears. The first provided blazing acceleration, aided by cockpit-programmable launch control, which delivered 0–100kph (62mph) in a quoted minimum of 2sec.

The secondary ratio could deliver up to 300kph (186mph). More usefully, there were three drive modes, covering that drag-strip ability, track use and what looked to be passable road manners.

To reinforce the Ford SYNC Connect and electrification programmes, sophisticated satnav and a fleet managers' guide to recharging needs and power points could also be configured. The workaday utilities were summoned via two screens, one a tablet layout as provided on the contemporary Ford road-car range.

And so the Supervan story continues for the 21st century.

▼ Wearing an appropriate 'personal' UK registration plate, Supervan4 was an awesomely powerful electric 4x4 iteration of the 55-year-old concept of Ford's track-biased Supervans, but this time it had road-going ability and comprehensively programmable electronic systems. *Ford*

CHAPTER 7

ESCORT

REAR-WHEEL DRIVE

Bestseller and motorsport legend

For decades I wrote in depth about the first two rear-drive Escorts (1968–80), driving on the road and racing Ford's replacement for the Anglia extensively between 1968 and 2013. The emphasis here is on stories untold in my earlier Ford work. The list of these conventional, small but extraordinary fast Fords, that I was in the right place at the right time to experience, extends to 53 examples (see pages 120–121).

APPRECIATING CLASSICS

These simple, tough Escorts still give plenty of competitors, collectors and those who just love their motor cars generous helpings of pleasure. Notice, I did not say cheap. Back when I drove most of these cars, yes, they were affordable, disposable, quick cars – Fords that could see off their similarly priced competitors and more prestigious marques, attracting a dedicated following. There are still some 2020s owners and racers who avoid the over-hyped collector/auction market and enjoy their Escorts affordably today, but some authentic rear-drive performance Escorts fetch prices that owners of used Rolls-Royces and many Mercedes and BMWs can only imagine.

At the time of writing, a glance at the catalogue for a specialist auction will tell you that £30,000–£75,000 is common for many RS Escorts. For example, an RS1800 Custom from 1975, with authenticated history, was offered at a Dorset auction with an estimate of £70,000–£75,000; it achieved the higher estimate, which meant that the buyer paid £80,414 including buyer's premium.

Some 48 years earlier, the MkII RS1800 had debuted in January 1975, but was not on (rare) public sale until June 1975, at £2,825.15p, the Custom trim taking that up to £2,990.12p. When my February 1976 test of the RS1800, RS2000 and Mexico was published in *Motor Sport*, the RS1800 was listed at £3,049 excluding inertia-reel seat belts

(£27) and delivery charges (£35), and you needed £117 more if you wanted it to sit on alloys!

As with the RS1600 and any other RS/Twin Cam with competition potential, an unmolested RS1800 road car is a rare find today, particularly as my research indicates that just 109 showroom examples were originally converted from the second-generation RS Mexico. Subsequent ex-Ford sources suggest that just 50 were made at the Aveley pilot plant – not Ford's Advanced Vehicle Operations (AVO) at South Ockendon – that made its last Escort on 24 January 1975. The same sources add 15 more assembled at Boreham, and those would have VIP media and show vehicles to support the initial motorsport homologation.

In 2021, a 1970 Escort Twin Cam evocation (restoration replica) sold at auction for £41,450. It was in exceptional, shiny, restored condition, sat on black-tinted alloys under rally-style extended arches, and was fitted with period bucket seats, and a battery in the boot. It was exceptionally well executed, in the spirit of the period, but with extra touches rather than an exact clone of what you might have found in period at a Ford dealer.

▲ The author wheeling the Broadspeed flyweight version of the RS1600 – John Fitzpatrick's 'Giant Killer' – around the Silverstone GP circuit. With around 250bhp, the car was capable of running with, or beating, the best of America's V8s on the UK's twisty tracks. *Motorsport Images*

▼ A rare beast worth over £70,000 in 2022, the road version of the RS1800 was made in lower numbers than any RS 'production' Ford, but did the homologation job for international Group 4 motorsports. *Ford/Dave Wigmore*

➤ The original Escort on any kind of loose surface was – and remains – a potent tool for epic sideways travel. Here, the equally original 1973 European Rallycross champion and subsequent Ford Motorsport management member, John 'JT' Taylor, demonstrates the oversteer art at a rare dust-laden Lydden Hill TV rallycross. *Peter J. Osborne*

I do not want to even think about what you would need to pay for an original road Twin Cam (much rarer than modified/rally examples), such as I drove back in the day, which lacked any extras. That basic showroom Escort Twin Cam press car remains my fantasy favourite today, rivalled only by the practical 'beak-nose' RS2000, preferable for everyday excitement.

So we can now put these performance Escorts firmly at the more expensive end of the post-'70s classic-car market, but which are the favoured models? I would include the 1968–70 Escort Twin Cam, as well as the official RS production cars: RS1600, Escort Mexico and first- and second-generation RS2000s. Excluding competition Escorts and the 'beak-nose' RS2000, I believe that the original MkI (1968–74) is generally the more valued, and my preference.

'THAT BASIC SHOWROOM ESCORT TWIN CAM PRESS CAR REMAINS MY FANTASY FAVOURITE TODAY, RIVALLED ONLY BY THE PRACTICAL "BEAK-NOSE" RS2000, PREFERABLE FOR EVERYDAY EXCITEMENT.'

How it all began...

First, we should just outline the basics of these over 50-year-old automotive heroes to inform younger readers, refresh memories and possibly underline some uncomfortable truths about vehicles reborn from mass-market million sellers. So far as mainstream Escort production is concerned, Ford likes to round things up for the sales of Esorts bearing that 1968–2000 European use of the nameplate to more than 4.1 million, but that figure encompasses technically very different motor cars – including rear-wheel-drive, front-wheel-drive and four-wheel-drive evolutions, and a wide variety of engines and power. Here, we cover just the front-engine, rear-wheel-drive first and second generations.

As late as 2014, Ford revived the Escort badge for a Chinese-market entry-level Focus derivative. Back in the '50s, the earliest use of the Escort badge in Britain was in the period 1955–61 for the lower-cost alternative version of the square-rigged 100E predecessor to the 1959 105E Anglia. That original Escort shared an estate body with the 100E Squire, but had less trim and equipment. The Escort proved the better seller, with over 30,000 units shifted during a production run extended a couple of years beyond the Squire.

The Escort MkI, 1968–75

We start with the basics that made the higher-performance Escorts possible. The first-edition Escort saloon arrived in 1968 in two-door form, but in a matter of months a van and three-door estate were available. By 1969, the four-door layout that the Anglia never offered was in the showroom for saloons, along with van and estate models featuring up-and-over tailgates.

At their debut, mass-production Escorts came with inline four-cylinder engines of 1,098cc or 1,298cc, all short-stroke designs that would rev amiably to power peaks of 53–75bhp over 5,000rpm. In reality, the 75bhp 1300 GT was the only true performance Escort initially available. Prices in 1968, rounded off for clarity, ranged from £635 for an 1100 de Luxe to £765 for the 1300 GT.

The first 115bhp Lotus-Ford Twin Cams were promoted effectively through motorsport (particularly winning at TV rallycross) and the high-profile mass-media launch. The initial batches of these cars were hand-assembled at Boreham, or taken off the line at Halewood to be transformed into a Twin Cam (TC) specification that involved a lot more than just a powertrain transplant.

In reality you could not buy an Escort TC easily in 1968, unless such a sale was advantageous to Ford publicity: both *Autocar* and *Autosport* had a Boreham-built Twin Cam (XOO 352F), listed at £1,123 in 1968. Even the double-page-spread advertisements for 'Escort Twin Cam model' had some weasel wording: 'Now you probably will not be able to sleep tonight' plus a small-print declaration, 'Especially when you hear that most of the early models are heading for the export market.' The slanted black-and-white picture featured a car at Boreham carrying the well-travelled XTW 368F registration – a plate that had appeared on the Moroccan press launch and even survived on a blue 1970 rallycross Escort for privateer Ron Douglas. I had the privilege of driving a TC with the same registration at Boreham for a May 1968-published CCC article, and the editor and I also had the Tony Chappell ITV *World of Sport* Croft championship-winning rallycross Escort to try that damp day.

Even some specialist magazines had to wait until the summer of 1969 to assess production examples. For CCC, in April 1969, I had another

TWIN CAM IMPRESSIONS

Back in 1969, I was allowed a production(ish) Twin Cam for a test week. This TC was a Boreham-built test and development machine that had covered 18,000 miles.

I ran a set of timed figures at the Chobham test track that averaged out at 0–60mph in 9.2sec. The highest standing-start acceleration we could measure on the then military facility was 0–80mph, and that occupied under 15sec. With a rolling start we measured 70–90mph in 11.5sec, and the more usual top (fourth) gear flexibility benchmark of 50–70mph in 11.5sec. We got within 1mph of the official 115mph maximum at the 6,500rpm (distributor) cut-out. During the test week – including the performance-test sessions – I got 23.7mpg on 98-octane fuels.

That test Twin Cam wriggled into my soul and has never really been displaced. Here's what I thought for April 1969 *CCC* publication when I was 22 years old: 'The handling, brakes and road holding are well capable of coping with the performance, so long as you remember that it is a very quick machine, so it is probably travelling faster in any situation than more staid machinery. So, if you take time to acclimatise and don't stuff your foot into the works until you've a measure of the extremely responsive engine in a pretty light shell (15.5 hundredweight: some 787kg or 1,736lb), then everything is fine. I do think Ford could improve the TC's resistance to side winds and probably most owners would add a sound-proofing kit. Apart from those two points, I think it is worth the money they ask for it as a completely tractable and speedy way of getting round the UK.'

▼ The original Escort high performer with Lotus-Ford Twin Cam power at £1,171 in 1968 still draws serious attention at classic-car shows. During 2022, bids beyond £50,000 for rare, unmolested originals were established – and replicas commanded over £35,000. *Author*

DRIVEN

Road cars

1300 GT Road test published CCC, June 1968; £851, 71bhp, 0–60mph in 13sec, 93mph; sweet four-speed gearbox. **1300 GT Broadspeed** Published February 1969, CCC; 1968, LOX 981F, 0–60mph in 9.8sec, 107mph, 7,000rpm Stage II engine. **1.3 Allard supercharged** 1968, Escort Twin Cam acceleration, 22.5mpg; pinking handicapped max mph. **Escort Twin Cams (two examples)** First road and track impressions at Boreham in May 1969, then road test for CCC; 0–60mph in 9.8sec, 112mph, 23.7mpg; my favourite MkI. **1600 GT Lumo Piranha** Road test for *Motor Sport*, 1969; TXE 940G, plus Peter Gammon's similar 1.6 GT recipe; 0–60mph circa 10sec, 12in steel wheels, 26mpg. **3.0 Escort V6 (three examples)** Road tested for *Motor Sport/ Motoring News*; Crayford Eliminator, Willment Race Proved and SuperSpeed conversions, all based on 138bhp early Zodiac V6; handling better than expected, fabulous mid-range acceleration. **3.5 Escort V8** *Motoring News* road test/*Motor Sport* comment, 1970; Richard Martin Hurst's 214bhp Weslake-Rover conversion; great West Road drag racer! **Escort RS1600 (two examples)** Long-term test, April 1970, *Motor Sport/Motoring News*; GNO 420H full test for *Motoring News*, 0–60mph in 9.3sec, 109mph, 21.2mpg (2021: undergoing restoration at UK specialist!); I also used a LHD RS1600 for commuting to Essex from Ealing. **Sport/1300/1300 GT (three examples)** Janspeed Stage2 road car, 1971, BYD 528J, 0–60mph in 12.9sec, 24.1mpg, 97mph; Ford loan 1300 Sport, OPU 685K, 1972 road test, pre-race-season, 0–60mph in 13.8sec, 25mpg, 92mph; same car in race spec, 0–60mph in 11sec, 98mph at 7,000rpm; official uprated press test 1300 GT, KHK 580J, +8bhp bonus, 0–60mph in 12.7sec (vs standard 13.2sec), 26.1mpg, 98mph (vs standard 93mph). **RS/Mexico** Owned by me 1972–73, PEV 805K; former brochure 'RS1600' photography car, metallic-marmalade paint, extended wheel arches, definitely a 1.6-litre Mexico for my 11 months; sold at 29,800 miles, then destroyed during Essex club rallies. **Mexicos (three examples)** Company cars, 1973–74, 6,000 miles in Clubman-spec YOO 633L, and 9,200 in yellow NHK 272M, plus official road test in LVX 904J, 0–60mph in 11.9sec, 27.9mpg, 103mph. **RS2000 MkI** Beautiful silver special-build, flared arches, mine at 4,160 miles, TWC 679M, December 1974; sold early 1975 to Gerry Marshall, when I left Ford. **1300 GT estate** Ex-Ford management, 1974–78, SPU 330K. £860, silver, then resprayed blue; served 34,268 family/rally-service miles. **Escort Mexico** Jesse Crosse restoration, JKP 238N, June 2013; better than new, since sold on several times. **RS1800/ RS2000/Mexico** Full road tests, 1975, for February 1976 *Motor Sport* + Brooklands Books *RS Escorts 1968–80*; rare RS1800, JJN 981N, tested twice, cost £3,049, 0–60mph in 8.6sec, detuned 115bhp; RS2000 'beak-nose', LHJ 931P, best value (£2,857), 0–60mph in 9.8sec, 110bhp, little slower than RS1800, and road-use friendly; MkII Mexico, LHJ 938P, with 95bhp, 0–60mph in 11.1sec, slow seller. **1.6 Escort Harrier** Diamond White rally reporter's car, Jan–March 1981, 2,000 miles for *Motoring News*

Boreham-built 'road' 115bhp TC to performance test. We were obviously pushed for space that month, as the piece appeared in my 'Walton's Wanderings' column with no picture (see 'Twin Cam impressions' on page 119).

The MkI Escort was sold from 1968 to 1975, and some 1,297,308 examples were built at Halewood, outside Liverpool in the UK. The British made it their bestseller and adored the car in period – and today – as an unmatched motorsport competitor in GT/Sport and RS derivatives.

Also manufactured in Germany, the Escort MkI was much less popular there than in the UK. Some 848,388 were assembled there, but Germans bought less than 240,000 of those, and most Cologne-constructed Escorts were exported. The Escort sector of the German market favoured the Opel Kadett, Fiat 128 and Renault 12. The German market simply did not accept Ford as being on par with then General Motors-owned Opel.

Despite heroic and often very successful efforts from their Köln-Niehl (Cologne) base, harnessing the halo effect of motorsport and manufacturing the very first officially badged RS types, it was tough to defeat that Blue Oval negative image. Ford even took on the best in home-team prestige German opposition from BMW and Mercedes, deploying successive sporting models from the Escort RS to Sierra RS Cosworth, via the Capri, including the phenomenally powerful Zakspeed-Cosworth turbos (cartoon caricatures of the Escort and Capri). However, nothing seemed to make a lasting difference to German-market sales or Ford's image.

covering UK internationals; £2,393, 90mph a struggle on 83bhp, but excellent handling and equipment for period.

Competition cars

Club rally Escorts (two examples) Courtesy Sevenoaks & District Motor Club, 1971; Bill Shewan's winning Escort Twin Cam and '1800 Escort DL' economically and efficiently created around an F-plate first-generation Escort by Ron Eyers, DAA 782F; part of a six-car off-road day, maximum car budgets set at £1,000! **Escort Twin Cam** Published *Ford Heritage* June 1995; Safari Rally 1971, LVX 945J, ex-Ford, courtesy David Jenkins, road only, 1.8-litre, 140bhp. **Race Escort Twin Cam/ BD** Reborn 'Run Baby Run' (Dave Brodie), June 2022; Club-race legend, presented static only, Silverstone; mostly original rolling chassis, Richardson fuel-injection 16-valve BD motor. **Willment-converted 1600 GT** Player's No6 'Celebrity' autocross, May 1968. **Group 2** ITV Croft rallycross winner with Tony Chappell in May 1968; Boreham 'XTW 368F', ex-Söderström engine, 153bhp, 0–60mph in 7.9sec, 99mph at 7,500rpm. **RS1600/ Mexico (five examples)** Boreham airfield circuit, 1970 media debuts; Mexicos (3), 86–109bhp (KWC 434J most power), 2x RS1600 at 120–140bhp; all uprated engines by Brian Hart. **Mexicos (four examples)** All 1971–72, race GNO 463H 'Purple Passion' at Brands Hatch, and Lime Green loan car; track test of Gerry Marshall 1971 championship winner published in *Motoring News*, plus Andy Rouse's 1972 champion. **Group 2 Broadspeed RS1600** Silverstone *Motor Sport* and *Motoring News* track tests, October 1971; 250bhp, 0–60mph in 4.8sec, 0–100mph in 12sec, 138mph at 9,000rpm; my best Escort track drive; regular driver John Fitzpatrick ran the sensational acceleration figures. **Group 1/Production 1300 Sport** Full Castrol/Britax seasons, 1972, OPU 685K, second overall Castrol over 35 races, two rolls. **Celebrity Mexicos (two examples)** Oulton Park instructor day and Harewood Hill pre-RAC Rally media day, 1973, second fastest time of the day by author. **Ford Factory RS1600s (three examples)** 1973–74, return Roger Clark's LVX 942J Scotland–Essex. GVX 883N, Timo Mäkinen 1974 RAC winner, and OOO 96M (seventh 1974), Chobham media track test day; OOO 96M again for 1974 Beaulieu media test day, autocross. **Ford factory RS2000** Roger Clark Tour of Britain winner, PVX 445M, shared for 1974 Beaulieu media test day, autocross. **Ex-Ford RS1600 Dave Watkins restoration** Published November 1996; REV 119K, Monte Carlo/BP colours from 1972, fourth and first on Scottish Rally and fourth on East African Safari. **RS2000 turbo** Published July 1997 CCC; UFW 45V, 400bhp eight-valve; Ken Foster's third-placed reader's car, Castle Combe. **Ford tarmac-rally RS1800** *Motor Sport* track test, 1975, Boreham; KHK 982N, 245bhp at 8,000rpm, 0–60mph in 6.5sec, 0–100mph in 18sec. **Ex-Press Fleet RS2000** Club Rally season, 1976, and Tour of Britain 1977; LHJ 928P. **RS2000/500 hybrid** 1992, VPK 99S, astounding 560bhp road/racing-champion Escort by Paul Bailey.

The Escort MkII, 1975–80

Right from the outset, Escorts were judged too cramped and technically primitive by the German market, and those criticisms influenced the more capacious and civilised Escort MkII of 1975–80. Yet the 'Brenda'-coded successor was still technically faithful to the original rear-wheel-drive layout, with simple front MacPherson struts and a leaf-sprung live-axle rear layout. The body featured squared-off lines, with enhanced glass area, but was less aero-friendly, heavier and had slightly lower torsional strength.

Entering production in Britain and Germany from the opening week of December 1974, the second-edition Escort remained on sale from 1975 until summer 1980 (August in UK), at which point Ford of Europe's belated switch to front-wheel-drive Escorts took place. Production figures for the Escort MkII did not quite match the original output, with nearly 1.7 million were made. The plants at Halewood in the UK and Saarlouis in Germany, produced similar numbers in the 840,000s, Germany ahead by 7,984 cars. However, eight countries contributed sites that assembled Escorts, including Israel, Ireland, The Netherlands, Australia and New Zealand, plus South Africa's well-established Ford bases.

From a performance/competition Ford viewpoint, those last three Commonwealth countries were front-runners, particularly Australia, where some 50 unsold showroom Capri RS3100s were shipped, and a race RS3100 was imported for Allan Moffat. South Africa contributed an ultimate low-production example of the Capri (Perana V8)

1968–2013 COMPETITION CONNECTION

The (in)famous XTW 368F rallycross/advertising star car was a bit overwhelming in 1968 when I encountered it at Boreham.

As set up for rallycross, it was all about acceleration to just short of 100mph at 7,500rpm – the limit we were allowed as visitors that day driving around the Boreham facility and on local roads. As it was wet, and the Escort was stripped out for its recent TV rallycross role, it was even more of a sideways animal than most, but the quick rack-and-pinion steering and competition seats allowed a semblance of control, even if you were an apprentice in an international five-star competition car.

On a wet day at Boreham we weren't going to get a set of really representative figures, but a standing quarter-mile in 15.2sec appeared after four runs. By then, the hard-worked warrior was running into a period limit of 7,500rpm at 98mph in top on the 5.5:1 differential, as used in the televised winning Tony Chappell rallycross campaign.

The outstanding memories were of an engine that flew to maximum rpm with that high differential ratio, an exhilarating soundtrack, and handling that compared to that of a four-seater Lotus Seven – a true compliment. It was an honour to drive as a 22-year-old – and all of the Boreham-built road and competition Twin Cams I experienced left indelible and emotive memories: only the thought of bankruptcy deflects me from seeking to own one today.

That rallycross TC of 1968 was connected to the Dave Watkins YVW 591F of 2013 concours quality, but how? Pardon the period slang, but here is what we said in 1968 of XTW 368F's contemporary competition Lotus-Ford engine: 'The power unit is the Twin Cam go-box out of Bengt Söderström's Swedish Rally car, hung beneath two monster Webers to gargle the go-juice until it had turned it into 153 decently hairy horses.'

Apparently, this engine had run in a Boreham-built Cortina Lotus for Söderström to win the 1967 Swedish Rally, and was also subsequently installed in the Escort TC, YVW 591F, that Söderström steered to third place on the 1968 Finnish 1,000 Lakes. The registration plate also appeared on other factory sorties, but in 2013 it was firmly attached to one of Dave Watkins's star reborn factory Fords.

and New Zealand has some enthusiastic owners of ex-factory Escorts and Capris. Today, Australia has a thriving RS Owners' Club, having reportedly assembled over 2,000 RS2000 'beak-nose' Escorts in two-door, 2-litre form, plus their unique four-door version.

Announced in January 1975 and sold from 4 March that year, Europe's MkII Escorts cost from £1,100 (almost 1968 Twin Cam money – that's inflation for you!) as a two-door in 1,098cc format, plus another £340 for the extra pair of doors in 1.1-litre trim. Top of the mass-produced Escort range was the extra-equipment 1.6-litre Ghia four-door, at £2,125, while Estate models were initially quoted at £1,701 for a 1300, or just beneath the £2,000 barrier in 1300 GL trim. I mention the estates because they carried over the entire rear load section from the MkI, as did Escort vans. No wonder Ford was so profitable!

Gradually, German influence on Ford's European hardware increased – a good thing for customer quality, but the Brits continued to conceive and enthuse over competition-biased high-performance Fords. That rebellious UK sporting spirit lasted into the '90s, with the road and rally Escort RS Cosworth, plus the race-only 2-litre Mondeo touring-car programme.

The British influence

Back in the 1968–74 era, the British had an important marketing and engineering hand in the Escort's creation, testing and all the high-performance goodies, courtesy of Ford of Britain's financial and manufacturing links with Lotus and Cosworth.

If you have been reading this book in chapter order, you will suspect that the key to high-performance RS Escorts lay in the Cortina Lotus, but more specifically in the later Cortina Twin Cam. However, for most road cars, and some motorsport categories, the MkI and MkII Escorts featured much less sophisticated powertrains than the Lotus-Ford twin-cam or 16-valve Cosworth. Here, my thoughts centre particularly on the pushrod engine in the 1300 GT/Sport, Escort Mexico (pushrod 1600 GT), plus both MkI and MkII RS2000s, which both carried transplanted Cortina/Capri single-overhead-camshaft 2-litre Pinto power.

FIRST ENCOUNTERS: BAPTISM OF FIRE IN BROADSPEED'S 1300 GT

For me, Escort enthusiasm wasn't ignited at the initial launch. I was far too junior to attend the January 1968 Moroccan media debut, but we did have three showroom Escorts in for UK testing at *CCC*. Fortunately, I wasn't present for that magazine's photography session on Chobham Common (there was still a struggle to complete the M3 motorway, it was that long ago), as the art editor and volunteers managed some minor paint scraping between the previously immaculate 1100, 1300 and 1300 GT examples supplied. When *CCC* finally published those tests (June 1968), we had experienced the pre-production and rallycross Group 2 Twin Cam at Ford Competitions Centre: that was published as a twin test in May 1968, featuring the notorious multi-vehicle-assigned number plate XTW 368F.

CCC also had a bunch of modified Escorts for evaluation, from a 1300 upgrade featuring a 1600 GT engine implant to Allard's supercharged 1300 GT. The latter accelerated almost like a Twin Cam, but pinked threateningly on a 9:1 compression ratio, so any attempt to accelerate beyond 90mph, never mind match the factory TC's 110mph pace, returned increasing misfires, then some exciting bangs.

I had some Escort experience when I reported in at Brands Hatch on the chilly morning of 7 March 1968, but nothing prepared me for the quality passenger ride in store. For I was to be the 'sack of potatoes' in a radically modified Group 5 Escort deviant.

While my editor hobnobbed with established motoring writers, Ford PR people took me over to a shiny plum-painted Escort with a silver roof panel. Purporting to be a 1300 GT, it had modestly flared front wheel arches and rear wheels disappearing into pulled but not obviously 'bubble' arches. A cluster of Broadspeed personnel surrounded the bumperless lines.

I suspect there were a lot of surreptitious grins all round, as PR ace Harry Calton confided in Ralph Broad and contracted race driver Chris Craft. Harry asserted that: 'Jeremy here has never sat in a proper racing saloon car, could Chris possibly give him an idea of what it's all about?'

'Oh, indeed, certainly. Would Mr Walton care to try this crash hat, our VIP passenger seat, and strap himself in?' said Mr Broad with an angelic grin. Nudging Chris – formerly Ford at Boreham's fastest ever service-van driver – Mr Broad predicted: 'I'm sure Mr Craft can educate him.' More evil grins passed between crew and driver as the hapless apprentice was secured rather too thoroughly in the safety harness, and advised just where to brace his feet.

Under the skin: a Mini beater

Looking back, it astonishes me that potent saloon cars with intense engine and running-gear modifications (this Escort had twice the normal power in a severely lightened body) hadn't at that time reached into the era of comprehensive roll cages. There wasn't even a hoop in the Alan Mann or Broadspeed Escorts of 1968. Yes, they had the safety fuel tanks and harness, and drivers wore fireproof overalls, helmets and usually a balaclava of some kind. But the advantages of roll-over protection and extra chassis stiffness, with a handling bonus, had yet to cross the Atlantic from America's NASCAR and TransAm categories.

My outing would have been in one of the two Escorts 1300 GTs that Broadspeed regularly

▼ The Escort burst onto ITV screens as a British rallycross debutant at Croft, with this hand-built Lotus Twin Cam conversion. The car, with equally hand-crafted number plate, delivered a debut win for Ford and a championship title for Tony Chappell. It was present at Ford of Boreham for my road and track test sessions in 1968. *Ford*

WHO WERE ALAN MANN RACING AND BROADSPEED?

Alan Mann Racing

Throughout the early racing tales of Ford in Europe, the names Broadspeed and Alan Mann Racing (AMR) recur. So who were these Ford sub-contractors?

Alan Mann, a racer from the Anglia days, was a team owner based in Byfleet, Surrey who surrounded himself with outstanding specialists. His operation became a vital part of Ford's global Total Performance assault. During the period 1964–70, AMR employed the talents of John Whitmore, Jacky Ickx, Bosse Ljungfeldt, Graham Hill, Frank Gardner, Jackie Stewart and Bruce McLaren amongst others.

AMR were headliners in Europe and America, first campaigning Falcon V8s on the Monte Carlo Rally and Cobras in the GT World Championship, then winning the 1966 European Touring Car Championship with a Cortina Lotus steered by Whitmore and running its own operation at Le Mans with GT40s, developing a lightweight 4.7-litre version of that car.

Yet AMR did so much more, as I discovered when I visited the premises for a colour magazine feature in the winter of 1968–69. Not content with winning the British Saloon Car Championship twice with Frank Gardner (Falcon Sprint and Ford Escort Twin Cam), AMR was busy tending Ford Zodiac-based 'Chitty Chitty Bang Bang' creations for the major Hollywood movie when I called in.

The fabulous custom metal-flake gold-and-red colours that adorned some of AMR's most successful racers (particularly the Escort, but also seen on Falcons, Mustangs and Cortinas) were later traded for 1969 Ford TransAm corporate colours. Frank Gardner was assigned one of the legendary Boss 302 Mustangs for a 1970 British campaign and just blitzed the opposition with outright wins, but the UK class points system demoted the AMR Boss Mustang and Gardner to second in the series.

At the end of that 1970 season, Ford Competitions Manager Stuart Turner dumped Mann and his operation. Again, Mann proved his business acumen by buying Fairoaks aerodrome in Surrey where he created an extremely successful aviation business. Into the 21st century, he revived AMR, running classic Fords at major meetings. Today, son Henry runs this operation and still campaigns classic Cortinas and Mustangs in their trademark gold-and-red livery, the V8 often seen with race legend Steve Soper driving.

Broadspeed

Broadspeed was most influential in Ford's British programmes, and its Anglias and Escorts, plus unique Capri RS2600 derivatives, were regulars for customers in European events and championships. Just how effective those customer Escorts were can be judged from the fact that the Italian Jolly Club – usually 110 per cent patriotic with its competition hardware – opted to buy a pair of Broadspeed Escorts, one of which resided in Britain in 2021. I had much to do with Ralph Broad, and loved his company and being with him off-track as well as at the races.

Ralph delivered front-line fast Ford saloons for longer than Alan Mann Racing. Working from the family garage and engineering premises in Bristol Street, Birmingham – and Broadspeed competition cars frequently carried the logo of major Ford dealer Bristol Street Motors – Ralph was naturally first allied with Midlands-based BMC and the Mini Cooper. He had an inherent feel for the black-magic art of gas-flowing, porting and uprating cylinder heads, which is at the heart of every old-school engine, especially the Siamese-port A-series head used in Minis. However, when BMC had to decide where its best competition options lay, Broadspeed was last in, first out, versus the Cooper Car Company.

Thus, in 1966, Broad and his team headed off to Ford, his staff including some multi-talented technicians such as subsequent triple British Saloon Car Champion Andy Rouse, Carl 'Tivvi' Shenton and Clive Parker (both later renowned for their craftsmanship in the USA), Alan King, Andy 'TWR' Morrison and Graham Goode. Graham prepared his own immaculate race Fords, from Anglias to Sierra Cosworth RS500s, and he still runs the successful Graham Goode Racing (GGR) high-performance preparation and sales business in Leicester.

The Anglia race programme that Broad took on for 1966 fortuitously benefited from a change in the regulations that allowed more radical (Group 5) modifications. Broadspeed and Mini Cooper graduate John Fitzpatrick became British Champions outright using the Anglia in that first season, and right on the pace in 1967, with some giant-killing European forays too.

The Escort was just made for Broadspeed to display the depth of knowledge it had accrued with the Anglia. Broadspeed expanded and moved to new Southam premises outside Banbury, becoming the leading Escort

➤ I had the honour of driving the 1968 British Saloon Car Championship-winning Alan Mann Escort at Oulton Park when it was owned by Yorkshire's Ken Shipley. Although immaculate, it was not competitively track ready. *Dave Wigmore*

race-car providers using Twin Cam and Cosworth 16-valve power. The company also dealt with the Capri, and offered well-crafted road upgrades under the brand Bullitt, as well as a racer ith radically re-engineered suspension. The Capri racing move didn't work for Broadspeed, although the company did also venture into production racing with Dave Matthews. Behind the scenes, Ralph and his team worked on one of three factory prototypes for the lightweight BMW CSL, which ran once at the Salzburgring for John Fitzpatrick, placing third in a 1972 European Championship round. There were also links with General Motors, and a number of road-car turbocharger upgrades, including the Essex V6 as found in the Capri, which was also supplied to TVR.

Sadly, Ralph finally went back to a descendant of BMC – troubled British Leyland – to deliver the charismatic but ultimately flawed Jaguar XJC V12 for use in the European Touring Car Championship. It promised much in 1976–77, but never won a race. Leyland boasts of beating BMW *et al*, and consequent raised expectations, weighed heavily on Ralph, who also suffered the death of his daughter in a tragic road accident.

By late 1977 he had packed up for Portugal, subsequently running a successful wood-burner installation business, although the Broadspeed brand was sold to former Mini factory driver and team principal of the UK Dealer Opel Team, John Handley. Ralph died in September 2010 at the age of 84 – a good age after a heavy smoking habit and an obviously highly stressful life.

fielded, and the only one available at the time, as they simply didn't have two 1300 GTs race-ready by the time of this pre-season test day. Both racers were road-registered (XOO 341F for Craft and 342F for John Fitzpatrick) as they had passed through the regular Ford system, in exactly the same way as contemporary – and subsequent – factory rally cars.

Yet nothing was missing in the quest for track pace and the goal of usurping the dominant factory-funded Mini Cooper S teams. The 1.3-litre pushrod engines took some redeveloping to deliver reliability, but even for this pre-season outing it was obvious that Broadspeed had found the extra urge to give them an edge over the A-series motors and integral transmission layout of the rival Minis.

Fully sorted on Tecalemit fuel injection, and incorporating some tricks learned from the Holbay 1-litre Anglias, these conventional four-cylinder engines, with Ford's later crossflow induction/exhaust layout, were persuaded to yield a regular 145bhp at 9,200rpm. That was fed into a five-speed gearbox with Hewland race internals, delivering to the stretched and redeveloped coil-spring live axle that Broad had run for the Anglia. Power losses were less than those of a Mini with its integral transmission, and there were extra horses in hand over the contemporary Cooper 1,293cc unit anyway.

Handling was generally excellent on a dry track, with plenty of Dunlop grip on 13in Minilites. The team's Anglia experience, plus use of rack-and-pinion steering and a wider track than its predecessor, ensured the Escort was a better chassis than the Anglia, although a little bulkier in terms of weight and frontal aerodynamics.

From the passenger seat: observing and learning

I had thought I could drive before I sat in on Chris Craft's early Escort abilities. Brands Hatch was built for agile race cars. John Fitzpatrick proved capable of winning outright in the 1971 RS1600 against the best of American V8 opposition. John Fitz, with former Formula 1 racer Trevor Taylor, very nearly beat the factory BMW 2002 turbo with the 1300 GT in a 1969 six-hour race. These were memorable Brands Escort feats.

▲ The Escort, even with 1300cc on Silverstone tarmac, could be spectacular. The late and great Chris Craft demonstrates the 145bhp Broadspeed 1300 GT's tricycling prowess. *Author's archive*

Craft wasn't as consistent as team-mate John in the 1300 GT during 1968–69, but he was ultra-quick once he'd woken up from trademark tardy starts in those early days. Chris graduated to drive far more powerful machinery up to Le Mans and Formula 1 level, and that inherent lap speed was pretty demoralising for this novice in 1968. Come to think of it, Chris was still in another driver class when we tested the Hammonds Sauce Capri together at Silverstone, prior to Chris lending me that Capri II for a race at Donington's 1977 British Championship round.

Back to 1968 Brands, and the 1300 GT has clanked and whined its bare metallic cabin out onto the shorter club circuit. Already warmed up, I get the full 9,000rpm with soprano wailings from a stressed 1,298cc – treatment that lasts for a couple of breathless laps, ensuring that I sit rigidly on my side of the cabin under extreme cornering forces.

Determined to learn something in four minutes of rank fear turning to a speed freak's paradise, I watched the gear lever blur as it was whirled through the non-synchromesh ratios. I recall that the media information promoted the peak horsepower as the important value, but the ramifications of just 95lb ft of torque by 7,200rpm become most obvious in this virtuoso performance. The driver shifted between 7,000 and 9,000rpm to get the most out of that narrow rev-band race motor. That made sense for the small-capacity Escort running a full-race five-speed gearbox, while the early Escort race Twin Cams could live with a four-speed thanks to their superior mid-range pulling power.

Then I studied the foot pedals. Craft was accelerating hard and exchanging gears for more speed where my judgement anticipated heavy braking and a snappy down-change as URGENT. As we survived the first plunge at full-speed-ahead into the plummeting challenge that is Paddock Hill Bend, I learned to trust my pilot, and a sense of exhilaration set in.

I started to learn a little more. Perhaps the most important lesson was how late a professional applies the brakes, or dismisses them as irrelevant in circumstances that normal mortals feel demand firm middle-pedal applications NOW. This observation proved particularly true of subsequent passenger outings with Gerry Birrell and Tom Walkinshaw.

Ford's factory rally aces Hannu Mikkola, Roger Clark and Timo Mäkinen remained on another talent planet, particularly if you were privileged enough to see them in action over forest roads, perhaps with fun elements such as freezing fog or multiple surface changes, and cornering cocktails from off-camber hairpins to 'Is it flat?' 110mph kinks.

Back at Brands Hatch 1968, my four-minute lesson taught me just how much I had to learn…

Footnote *The 1300 GT Broadspeed Escorts only won class victories six times in 1968 owing to initial unreliability. John Fitzpatrick was their most successful British Saloon Car Championship driver, sixth overall and runner up to John Rhodes (Mini Cooper S) in the up-to-1,300cc class. Ford weren't too downhearted, as the fabulous Escort and Cortina Twin Cams from Alan Mann Racing took outright and class success, steered by laconic expat Aussie Frank Gardner.*

By contrast, Chris Craft was regularly in the premium points in the 1969 British Saloon Car Championship and finished second overall and a convincing class winner. It was a good result, but brought no cigar, as the class-divided championship points system resulted in Alec Poole's 1-litre Mini taking overall honours.

FIRST-GENERATION: 1300 VARIETIES

I have multiple Escort 1.3-litre examples logged in my 'Driven' section, via diaries or recorded in published tests. From the earliest 1968 models, to owning a rare 1300 GT estate from 1974–78, they all shared seductively high-revving and smooth four-cylinder engines, and the four gears shifted suavely too, even when asked to change without the clutch at racing rpm.

Tuned Escort 1300/1300 GT

The fastest, but least dependable, was an Allard-supercharged 1300 that mirrored – or beat – Escort Twin Cam acceleration figures to 60mph, but pinked like mad under full throttle because the high production compression ratio had been retained. That also meant any attempt to exceed 100mph could be injurious to the specialist motor's health.

The tuned examples included a Janspeed 1300 uprated beyond GT specification and modified in two stages. This had the hardest of lives, including being handbrake-turned in posh Belgravia Square, and performance-tested with a fifth-wheel attachment in awful Silverstone weather. Co-driver Alan 'The Bear' Henry, the late and great Grand Prix journalist and author, noted the 0–80mph figure in fog as '0-to-Accident'!

One of the quickest and best-equipped of these cars was the 1971 official road test 1300 GT, which appeared at a similar time to the Capri 3000E, during a period in which Ford switched to more conservative DIN statistics. Both would be excellent buys today, but finding a 1300 GT Escort would be tough, and 3-litre Capris now sell at Porsche prices.

1972 Escort Sport

The longest period I spent with a 1300 Sport was during the 1972 racing season. Frankly, it wasn't a very good Escort 1300 on the road, as it had the big-steel-wheel RS look (albeit with 5in rims rather than RS 5.5in versions) and the front-arch flares of a genuine RS Escort. Running 13in wheels instead of 1100/1300 Escort 12in rollers raised the gearing and noticeably sapped acceleration. It was prone to wheel-waving antics on the taller wheels and tyres too, but when Janspeed reworked the suspension as a low-riding racer, and Dunlop loaned us some Formula 3 low-profile slicks, it became a different animal. It still waved the front wheels about, but could corner at least a gear up on what you thought possible. During the season we had exhaust and camshaft upgrades, so the hand-finished and laboriously assembled engine could deliver around 85bhp (72bhp was the showroom DIN start point), and that allowed 0–60mph in 11sec over an average of four measured runs with a fifth wheel.

That 37-race 1972 season, which I started as a novice faced with two more races to complete without adverse reports in order to secure my racing licence, was the busiest I have had in my privileged life. The Escort – always turned out to the highest professional standards – was entered in 34 rounds of the Castrol and Britax Production Saloon Car Championships. Both these series were unusual, as the four classes were based on car retail price – not engine size – and full racing tyres were permitted, rather than the trick road/race rubber of subsequent production-race seasons.

The main events for us were 21 championship rounds of the Castrol series, with massively supportive Castrol backing: not so much financially, but featuring the car and driver in print and film media. This was gold-dust to any ambitious driver, and such publicity opened up opportunities I would never have experienced without such backing. Appearing in regular motoring-magazine race reports helped, plus some in-depth colour features, and not all in magazines that I worked for, which was generous of my commercial opposition.

We finished overall runner-up to the legendary Tony Lanfranchi and his Moskvich class-dominator in the Castrol series – and the Escort/Walton combination became £601–£800

'IT STILL WAVED THE FRONT WHEELS ABOUT, BUT COULD CORNER AT LEAST A GEAR UP ON WHAT YOU THOUGHT POSSIBLE.'

▲ ➤ For 1972 I drove a Janspeed 1.3-litre Escort Sport in production racing. The fifth wheel behind the Escort was used to measure performance for many cars mentioned in this book. The Castrol advertisement notes that we won the class and finished runner-up to outright champion Tony Lanfranchi. *Author's archive*

class champions. The parallel 12-round Britax Championship began with a debut Brands Hatch class win, but thereafter I made too many mistakes to figure prominently in the points. We came away from that season scrappily, as I had joined Ford by September, and was only allowed to complete the season by negotiation with senior management, as they had employed me on a clear 'no racing' basis. However, we went out with that Castrol class title, seven division wins and three lap records (Brands GP, Mallory and equalled at Oulton).

We also had some terrific circuit experiences, including a GP circuit finale at Brands, Oulton in full longer-circuit glory, and northern trips to Croft and bumpy Rufforth, which saw the Escorts in 'will they roll?' aerobatic contests.

The other three races I drove before my Ford race ban were: 24 hours of the warp-speed Belgian Spa-Francorchamps long track, a FordSport Day Capri 3000E celebrity race at Brands, plus a race at Brands in August driving both David Brodie's production 3-litre Capri and Janspeed's Escort. At Spa, we finished seventh overall and class winners in a BMW 3.0 CS. In the Brands celebrity race, paired with Vince Woodman, I finished ninth and Vince fifth overall. At the August Brands meeting, the smaller Ford set a lap record on a recovery drive from the back of the grid after practising out of session (no ignition key – a screwdriver did the job). 'Brode's' Capri snatched a good grid start position, but I made an arse of myself in the race, spinning in front of the pack. I worked the quick Capri back to eighth and spun it away again to finish 'well down' in *Autosport*'s accurate words.

That 1300 Sport was the key to my writer/racer early-'70s life: it appeared in *Motor* and *Hot Car* magazine features, besides my progress reports in *Motoring News*, but there was no sentimental send off. The red Ford was sold off swiftly, and the Sport's original OPU 685K registration is no longer listed by Britain's DVLA.

1300 GT estate

Most useful of the 1300 GTs was the ex-Ford management lease-scheme estate, which did all my family stuff, lugged some competition hardware – including an RS2000/Pinto motor – and lived the harsh life of a workhorse over its 34,000 miles with us. One major fault was the period Ford silver metallic paint peeling off at motorway speeds. After rubbing it back to bare metal with a mate's Black & Decker, and prior to a Ford sub-contractor's respray in a rather royal blue, it had a brief 'slick steel' identity. Sadly, returning rust poked through (hastened by being driven in the bare metal), so it was sold after a little over 40,000 miles.

The 1.3-litre Escorts all served me well, but are often overlooked today, overshadowed by the massive prices and pedigree of the RS derivatives.

FIRST-GENERATION: ROAD MEXICOS

Although I was fortunate enough to drive full-blooded factory Ford/Broadspeed Escort Twin Cams and RS1600s, most of my RS-pedigree Ford mileage and time was recorded in more mundane, but rather more practical, everyday Mexicos. This experience spanned 1970–2013, thanks to the generosity of Jesse Crosse, my former editor at *Performance Car*.

Best-known of these cars was PEV 805K, which had been a high-profile subject for Ford Photographic at South Ockendon. The lens-men and processors of film in the pre-digital age were based literally around the block from the Advanced Vehicle Operations (AVO) factory.

That particular Mexico appeared in a studio shoot to show examples of the Special Build options, complete with Minilite wheels, a quartet of Cibié spot- and fog-lights, plus the 'bubble' wheel-arch extensions that so many early Escorts now carry – to the point where road RS1600s and Twin Cams without the body modifications are distinctly rarer than examples with the extended-wheel-arch body. Mine had worn both RS1600 and Mexico badges for photography, but for nearly a year of ownership and 20,000 of my miles, there was no doubt that it was all-Mexico powertrain, slightly slower and thirstier (25mpg overall) than standard, as it dragged along oversize RS-prototype four-spoke wheels of 7in width, fitted with 185/70 Goodyears. I bought it for £840 in

▲ The Special Build Escort I bought for under £1,000 in 1972 served me as a staff car at Ford initially, but had appeared with RS1600 badges for company photography. It was later destroyed in a local club rally. *Ford*

◄ Before Ford mass-produced the Escort Mexico with 1600 GT ex-Cortina/Capri power, there were many unofficial Escorts running with that 1.6 litre engine conversion from specialists. This is a 1968 Willment-prepared 1600 GT for Players No6 'celebrities' to drive in the cigarette company's autocross series. Here I am sliding around the Hillingdon, West London track that *Cars & Car Conversions* hired for its Autocross Festival. *Author's archive*

▲ The finest Escort Mexico road car I drove was Jesse Crosse's extensive restoration, which went beyond the bounds of anything that could be done in series production. *Author*

late 1972 with less than 10,000 miles recorded. I sold it privately for £900 the following year. More recently, I received pictures of it literally flying in Essex regional rallies, where it was completely destroyed to the best of my knowledge.

As logged in the 'Driven' notes, I had a few more Mexicos as company cars, and always found them comparatively economical (27–29mpg) in showroom-horsepower spec, despite all of them being conducted with spirit, as my fuel was supplied free of charge. Only that first example mentioned previously was unreliable, as it was really a show car – indeed I drove it to the Paris motor show on Walter Hayes's command to demonstrate to other Ford delegates what those AVO Brits could build.

My Mexico memories were refreshed by JKP 238N, an original example extensively restored in every detail by Jesse Crosse. Incidentally, if you are suspicious that the registration is too recent for a 1970–74 Mexico, it was first registered a year later than most, as many RS products (particularly Capri RS3100) were a tough sell during and after the 1973–74 fuel crisis.

Jesse is an example to many professionals carrying out forensic rebuilds, especially in his diligent attention to research to ensure historical accuracy. Crosse restorations range from aged MGs, through Jaguar E-types to his present pride and joy, a Sierra Cosworth RS that you will read more about in a later chapter. Legend has it that Jesse even built a scale replica of the Supermarine Spitfire fighter with operable gunfire, but that's a little beyond the scope of this book!

I was able to access that Mexico at a June 2013 meeting of my favourite original Escort and Capri car club (Ford AVO Owners' Club) at Hatton Country World, in the Midlands, UK. The day was quite emotional, meeting a number of the factory rally and road cars, some still in the process of being restored, but the Crosse Mexico was a revelation, showcasing the standards that can be attained over 30 years after production ceased.

Most striking was the paint, fit and finish of the exterior. Yes, it was always said that Ford AVO factory-builds were superior in quality to standard production cars, due to both the pristine factory, with overhead line, and the snail's pace of production (20 on a good day), compared to the usual mainstream line aiming at over 1,000 Fords every 24 hours. Yet the AVO Escorts – sadly they never built any other models in quantity at that gem of a factory – could only be mildly pampered; paint might be renovated where obviously below showroom standard, and similarly with door fits, lock operation, interior trim and minor defects. Yet the individual attention of a dedicated owner, without profit and time constraints, allows a far higher standard to be achieved, and that was my impression of this car, inside and out.

In terms of performance, the Crosse car was even further removed from the 1970 showroom item, as it featured a carefully and laboriously rebuilt 1.7-litre engine, replacing the usual 1.6 crossflow and offering 40 extra horsepower over the production 86bhp. That bonus power reached the road efficiently, thanks to a limited-slip differential and precisely sorted production suspension.

Over local twisty minor and major roads, the Crosse Mexico proved to be the Escort you always promised yourself, a clear demonstration of what a versatile and pleasurable vehicle it could be. Even within that special AVO Essex factory, I don't believe it would have been possible to build such a fine sporting Escort. Yet, as an example of what a dedicated private owner can achieve, it is unbeaten in my experience.

As I write this, I know JKP 238N has now been resold at least twice, and will now be worth comfortably more than the £30,000 benchmark that good-quality Mexicos are achieving at the time of writing.

FIRST-GENERATION: RS2000

The first RS2000 initially appeared in my Ford working life during the eventful summer of 1973. Tragically, 23 June saw Ford RS development and race driver Gerry Birrell killed in practice for a Formula 2 race at Rouen, France. Gerry's development driving work on the RS2000 lived on, particularly the uprated suspension settings, which delivered better comfort and admirable failsafe handling in an inherently nose-heavy vehicle. Our memories were of a charming, mischievous and often laughing colleague – one of such talent that he was tipped as the next Jackie Stewart.

Initially created, engineered and developed on the Ford AVO site at Aveley, South Ockendon, at which I had media publicity responsibilities, the first RS2000 went far beyond just inserting an ex-Cortina 2-litre Pinto motor and booking a few advertisements. This third offering in the RS Escort trio, joining the Mexico and RS1600, was overhauled in many detail respects, and formed the basis for the even better second-generation version with 'beak-nose' aerodynamics.

Largely forgotten now is the fact that the RS2000 was launched to the media, and initially sold, in left-hand-drive form, because a £2 million German-market order – particularly valuable to offset the high operating costs and low production numbers of the Ford AVO facility – had to be fulfilled before UK right-hand-drive examples could be released. The RS2000 subsequently became available with UK-specific steering-wheel placement in dealerships from 11 October 1973, at £1,441.82.

Customers received a far more practical daily driver/occasional sporting steer than the RS1600, the 2-litre Pinto offering a much stronger mid-range pull than the Cosworth-Ford 16-valve could deliver in showroom trim. Useful changes were made to a Pinto unit that normally reported 98bhp. A cast-aluminium sump and bellhousing were engineered, co-operating with a snappier (raised pivot point) gearshift. An electric cooling fan was installed, so Ford credited the engine with another two horses, the 100bhp at 5,700rpm matched to a strong 107lb ft of torque at only 2,750rpm. That meant overall gearing could be raised to allow nearly 19mph per 1,000rpm, so a motorway 70mph was served up by 3,750rpm, rather than the usual performance Escort demands for 4,000rpm or more.

This Escort RS debut was one of the most significant for Ford of Britain's small-scale factory, and for me. I enjoyed a particularly smart silver example as a staff car for a few months – sadly sold on when I left the company – and I was permitted to play with uprated examples at media days, promoting the sale of RS performance parts. Yet, the most influential RS2000 sales-support exercise was to follow, publicising the participation of two Boreham-built entries on the 1974 Tour of Britain.

Personally, that July meant a change of scenery and Ford product. On 10 July I presented David Brodie's vision of the Capri II as an all-black racer, with wild and wide detachable panels, to then new Styling Vice-President, Jack Telnack, at a secure Dunton studio. Next, an 11 July trip to Birmingham for scrutineering of two rather special competition RS2000s, crewed by two bar-room legends from very different motorsport disciplines: Roger Clark, multiple international rally winner and long-term Ford-contracted ace, and Gerry Marshall, perhaps the best-known British saloon-car racer at the time. Choosing Gerry was perhaps surprising because he was contracted to Vauxhall, then owned by Ford's deadliest multi-national rival, General Motors.

Clark and Marshall's long-suffering co-drivers comprised the quietly spoken intelligence of Jim

'THIS ESCORT RS DEBUT WAS ONE OF THE MOST SIGNIFICANT FOR FORD OF BRITAIN'S SMALL-SCALE FACTORY, AND FOR ME. I ENJOYED A PARTICULARLY SMART SILVER EXAMPLE AS A STAFF CAR FOR A FEW MONTHS – SADLY SOLD ON WHEN I LEFT THE COMPANY.'

Porter, Clark's regular navigator, plus Paul White organising Marshall. White was another excellent example of the co-driving talent the UK then regularly delivered, and was later even better known at world championship level with Talbot Sunbeam and Henri Toivonen, winning the 1980 RAC Rally and more.

Ford homologation personnel had worked their magic, and the engineers transformed the single-carburettor 100bhp showroom RS2000 using a dual Solex 40/42 carburettor, 138bhp package. The FIA regulations permitted a close-ratio, four-speed gearbox and a 4:1 limited-slip differential for maximum acceleration and traction, the latter assisted by a new and subsequently popular retro-fit, featuring a single-leaf spring layout for the rear axle. Bilstein dampers were then a default sports fitting, but both wheels and tyres were quite modest 5.5 x 13in steel rollers, wearing Dunlop SP4 road tyres throughout the race and comparatively unchallenging rally stages.

The homologation papers weren't ready for Birmingham inspection, but there was a telex to wave at the officials, and the two hastily homologated blue-and-white Fords were allowed their romp. And boy, did they rampage across the UK! Friday saw them at compact Mallory Park, twisty Cadwell Park (a miniature Nürburgring), with some rough stuff in between, before Norfolk night-racing at Snetterton. For me, Snetterton was the highlight that sealed event victory. Clark, knowing that his closest rival was team-mate Marshall, who had raced and won at Snetterton regularly, decided on a lights-out policy until the rally man headed the field. That track was one of the fastest you could find in the UK, but from my trackside viewpoint Clark edged ahead through one of the very quickest sections (from the Bomb Hole to Coram). It was an ultimately assertive move, designed to dishearten any opposition; I think Marshall was surprised, but he played the role of wingman with precision and honour throughout an event that saw them finish 1–2 in their debutant RS2000s.

For me, the event coverage, with Ian Keresy, the PA/Reuters sports correspondent, was pretty physical. We hurtled into a second night, and an evening rough-road stage, having monitored progress at Oulton Park and Castle Combe. The next day was a contrast, as I travelled with international co-driver John Davenport's wife Alexa and her baby, Franca, to cover the Welsh Epynt rally stages. Here, the story was the 'crash-bang-wallop' departure of then BBC Radio 1 disc jockey Noel Edmonds in another RS2000. Edmonds had conducted the car with admirable, but suddenly excessive, brio over those famously deceptive tarmac twists and crests. He was uninjured, but the story in the national media tended to overshadow the Blue Oval's 1–2 win.

Sunday 14 July's afternoon-into-evening celebrations at the Birmingham Post House went on through a midnight BBQ. Monday was a day of personal reckoning, sleeping on a Hertfordshire forecourt and phoning in reports when finally awake. It all got done, including sponsor-related editorial for Dunlop, but Monday evening at home in Essex saw a visit from a doctor (yes, a 28-year-old could get a house call!) and I spent Tuesday necking a bottle of Septrin pills.

FIRST-GENERATION RS1600

On 9 April 1970, *Motoring News* published my impressions and performance figures for 'a fortnight hurtling around the countryside and Silverstone' in GNO 420H, a press-fleet example of the rarest RS1600 of all, an unmolested production vehicle. The bare facts are listed in the 'Driven' section, but here's a little background with the benefit of hindsight.

First of all, the public-road performance stats for the car's 120bhp showroom trim were no better, and in some cases worse, than those for the 115bhp Escort Twin Cam, and my impression was that the TC had the stronger mid-range. We actually recorded a better top speed for the Twin Cam (114mph against 110mph for the RS1600). For me, as the Escort RS/TC hardware and the Cosworth and Lotus-Ford four-cylinder engines were so similar, I would select the TC for non-competitive use, especially as parts supply is so strong for the

◄ By 1970 the substantially unchanged Escort Twin Cam running gear and body was ready for Cosworth-Ford 16-valve power. Here the author enjoys the informal RS1600 launch at Boreham's then high-speed perimeter track. *Ford*

Lotus-branded motor today. The big advantage of the 16-valve Cosworth motor is the extra power and rpm that can be generated over the eight-valve Lotus-Ford unit.

The car I tested wasn't intended for me or that weekly newspaper, but was on long-term loan to legendary *Motor Sport* editor William 'Bill' Boddy. He had always been generous about pushing performance or track cars my way. In a later working life at Ford, Boddy's daughter, Felicity, occasionally facilitated the efficient production of printed bulletins for both FordSport and Ford Motorsport.

I did enjoy the RS1600 back in the day, and it was used hard attending Thruxton races with three occupants. I hated leaving it parked outside my maisonette, so would often take it out just for the pleasure of driving and the soundtrack at higher engine revs. The best miles for me came on solo night raids from West London to the wooded back lanes of the Thames Valley, where it was exhilarating company.

BROADSPEED ESCORT RS1600 'GIANT KILLER' TRACK TEST

Probably the biggest honour in my Escort-driving life came on 6–7 October 1971, when I had the opportunity to steer the Broadspeed Group 2 class-dominating RS1600 around the Silverstone Grand Prix track. It was the car that, even 50 years later, regular lead driver John Fitzpatrick dubbed 'Giant Killer', and with good reason: besides exerting a stranglehold on the sub-2-litre UK championship category, this fabulous Castrol-liveried RS could, on two wheels and with John fired up beyond his usual calm demeanour, beat the American V8s to outright victories.

Sadly, just 15 days after I drove it, John's continuing outright battle with one of those V8s (Frank Gardner's Chevrolet Camaro) ended in the most enormous Brands Hatch long-circuit crash, and the gallant Escort was smashed to pieces. The incident was overshadowed on the day by the fiery death of Jo Siffert, driving for BRM.

Fitz graphically recalled his massive Escort accident, as he tried for David and Goliath glory, in his 2016 book *Fitz: My Life at the Wheel*: 'I pulled alongside [Frank Gardner's Camaro] and out-braked him. We went through Stirlings side by side, with me ahead slightly, but not completely. As usual, Frank left very little space, as I'd have done. Exiting the corner, his front left wheel brushed against my right rear. His tyre exploded and he veered left and pushed me sideways. I went off on the left of the track and somersaulted

➤ Key personnel that made the baby Ford a giant killer: John Fitzpatrick (orange overalls) and Ralph Broad in the Silverstone pitlane for our unique track test. *Motorsport Images*

along the earth bank, finishing up on my wheels. As the car came to a halt, facing the right way, it was just in time to see Frank slamming into the concrete pedestrian bridge. Gerry [Birrell] waved as he went by into the lead [in his Capri RS2600]. Both cars were complete write-offs.'

That Broadspeed/Fitzpatrick RS1600 – originally Ford-registered as MEV 34J – exists today in a reborn guise that clones its glorious 1971 season. Derrin James, who in 2017 had sold the respected Brian James race-trailer business created and sustained by his father and brother, became the car's owner in 2014.

The Escort, in a glorious Ermine White by Jaguar Specialists XK Engineering, and with authentic Castrol livery, now resided about five miles away from where it lived at Broadspeed's Southam, Warwickshire premises 50 years ago.

The shimmering restoration was displayed – and reunited with John Fitzpatrick – at the NEC Classic Car Show in November 2021, the same month that courteous Derrin allowed me to view it at his premises, and a day after the painstakingly rebuilt iron-block 1.8 BDA was started for the first time. That was seven years after Derrin purchased the car, much of that time spent on the restoration, initially under the care of former Broadspeed employee Roger King. Roger had acquired another 1971 race-specification Escort – EVX 266H – which he had originally built as a young Broadspeed employee. Derrin reports that the legendary Fitzpatrick racer that he bought really came as a partially reworked body and some spares, amounting to the 'identity' of the car and the chassis plate.

I should have remembered that the car I drove in 1971 was actually on its second body shell anyway: the original was destroyed at an earlier Thruxton 1971 round that I attended as a reporter. That original body shell found its way to Ford Advanced Vehicle Operations, where there is a Ford photographic sequence of it undergoing repair, unearthed by Ford AVO Club registrar Mark Heath. I think this may be the reason why there are pictures of a standard-bodied Escort registered MEV 34J out on the 1977 Tour of Mull and other rallies, resulting in dual identity for the car.

Under the skin: reworked to the limit

Today, we talk of rebuilds and restorations lasting years, and these 50-year-olds have to be rebuilt with current safety and electronics standards

in mind, if they are to see competition action. I think this usability/originality dilemma is most apparent in historic aircraft, particularly warplanes: if you want to fly them today, numerous updates are required to enable the issue of an Airworthiness Certificate. For race saloons half-a-century old, the differences are most apparent in the cockpit. Today, the roll cage must have cross-bracing and door bars to improve driver protection. Also, the seats will be lighter and stronger than those used in 1971, and six-point safety belts must be fitted.

Back in the '70s, an organisation like Broadspeed, with fast access to a bare body and the best components, and craftsmen to make ultimate use of them, reported 'a minimum of 12 weeks hard labour' to produce this RS1600. Yet it was insightful to note that the all-conquering 1300 GTs with 150bhp, and subsequent Escorts with 16-valve BDA power, used most of the same chassis hardware.

Back then, Broadspeed quoted £6,500 for the Group 2 1300 GT or a basic RS1600, plus another £500 if you wanted to clone the Fitzpatrick flier. In today's money, that £7,000 for the Fitzpatrick Escort would represent £100,000, and I would be astonished if this car changed hands for much less than £200,000 in 2023 or beyond. The cost for a season's racing in Group 2 for 1971 was quoted at £24,000 by Broadspeed, some £350,000 today.

Really, the key to this Escort racer's heroic speed was that in period, every detail was reworked to the limit of the regulations and durability. The small Ford was then driven with a combination of calm intelligence and pace that brought consistent results. However, that competitive streak in Fitz, as with every driver worth his name, could turn to aggressive speed

▲ Reborn! Derrin James's recreation of the Broadspeed-Fitzpatrick 1971 Group 2 racer packed period detail, 50+ years after the original's heyday and huge accident. *Author*

▲ The mighty 1.8-litre Broadspeed-Cosworth 16-valve motor was fed by Lucas fuel injection and lubricated by a full race dry-sump system sponsored by Castrol. I was able to use 2,000rpm for paddock paddling, or the proper 9,000rpm on Ralph Broad's advice to 'give it a good hiding'! *Motorsport Images*

when thwarted by the big bum of an American V8 waddling through corners rather slower than the flyweight (1,825lb/828kg) Escort, with its revamped and considerably more able chassis.

The low weight and 250 horses from 1.8 litres were admirable, and so was the often-maligned performance of the Joe Lucas (mechanical) fuel injection. In 1971, a perfectly set-up engine allowed us to chuff through the paddock at 2,000rpm, or obey Mr Broad's pleas to 'give it a good hiding. Use the full 9,000rpm – don't be shy!' The ZF dog-leg, five-speed gearbox and race clutch stayed sturdy throughout the day, which included moderately uphill starts for acceleration runs on Silverstone Club straight.

Yet the key to this Escort's raceability lay in that braced body – albeit with a simple roll-over hoop and cross-bar cage – and in the replacement of every performance-enhancing ingredient. Yes, it had oversize Minilite wheels and Dunlop's first-rate tyres (10in fronts/11.5in rears), but the support for those wheels and tyres was equally important, Ralph using a leaf-sprung rear-axle layout, restrained by rods and links galore. Efficient long-travel rear Bilsteins were fitted, stretching to high-level mounting points above the main boot area, with its beautifully fabricated aluminium oil and fuel tanks.

Up front, the Bilstein struts were supported by auxiliary front and rear links, but Ralph wasn't keen on fierce anti-roll bars, so the live-axle/strut layout delivered a Rolls-Royce ride by racing standards, and traction was outstanding for a front-engine/rear-drive machine. However, John Fitzpatrick warned me after warm-up that the Brands Hatch set-up installed on the car 'means maximum engine rpm are too easily reached here at Silverstone, while the anti-roll bar makes the tail a mite twitchy.'

Balancing out the tendency towards nose-heavy weight bias, the boot contained not just the mandatory safety fuel tank, but also dry-sump oil reservoir, all associated pumps and plumbing, plus a modest race-weight battery. You could also see the elongated rear dampers reaching up to their location points, where on the road car a rear parcel shelf might have butted into the boot, behind a back seat.

➤ A boot with no family luggage, but full of plumbing, premium petrol and dry-sump oil tank. Fuel consumption for track and acceleration tests was 8.9mpg. *Motorsport Images*

Behind the wheel: the ultimate Escort drive

I was ensconced in a Southam hotel for a pre-test overnight stay, with Ralph Broad and sales manager Nick May. They briefed me without threats: masterful, considering I was still a novice with only four races completed, and the Fitzpatrick/Escort still had that Brands season finale to contest. I greedily gobbled a 16oz steak during their detailed briefing, rare fare on a Teesdale Publishing salary! That was fine, but sleeping went pretty badly in the excitement/fear of what could go wrong during my unique access to a Ford that had become a legend in less than a season.

The morning of 7 October 1971 dawned bright and crisp, but my staff Ford (running subtle Broadspeed suspension upgrades) only let in occasional sunbeams. A local tractor, whined my diary 'covered Capri in shit on the way to the track' on this driving life-changer of a day. I learned about 48 years later that I was lucky I hadn't been dunked totally in poo at the track, as Broadspeed technicians Clive Parker and future BTCC champion Andy Rouse would cheerfully have drowned me for this opportunity!

It wasn't an exclusive booking, so there were plenty of faster 2-litre sports-racers whizzing around the old long-circuit Silverstone that bright Thursday, with the mighty Woodcote right-hander unfettered by chicanery. From memory, Fitz warmed up the Escort with some laps that showed it was still on its proper pace, albeit already in Brands Hatch specification. John's fastest lap from July 1971 was 1m 38.7s, an average of 106.7mph, and just nine-tenths slower than Gerry Birrell's Boreham-tended RS2600 Capri.

Back to October 1971, when I enjoyed 12 unforgettable circuit laps, the best in the 1m 46s bracket, 'enough for mid-grid' in Ralph's diplomatic words – sixth based on the May 1971 GKN support event for Group 2, or eighth for the better-supported 1971 Grand Prix support event. More realistically, much the same times as Broadspeed's blessed class-warrior Dave Matthews recorded regularly in the 1300 GT, with 100 fewer horsepower!

Back at our unique test day, this novice driver benefitted from a swift 2.5-turns-lock-to-lock steering rack and Formula 1 ventilated disc brakes. The Fitzpatrick Ford was passed fit by Broad's personnel for another task: standing-start routines, including safely attaching the fifth wheel to a rear towing eye. They ensured Walton dispensed with borrowed race helmet (thanks Andrew Marriott) and crept into the passenger seat with a period bank of stopwatches and the electrically driven speedometer head. Now we could get representative figures (see panel on page 139) from the Ford's regular conductor. The Escort ran faultlessly throughout, deeply impressive and just what you would hope to discover from an internationally successful, professional race team.

▲ Some bits of cabin furniture had to be retained under Group 2 race rules, but the gridded foot pedals, fabricated footrest, race seat and three-spoke steering wheel speak of function over showroom appeal. *Motorsport Images*

My first impressions were of a neat cockpit, mainly finished in black, from the production instrument cowling to the bare-metal floors. The deeply dished steering wheel featured four 'lightweight' holes in each of three shiny spokes. The foot pedals, plus the essential transmission-tunnel-located footrest, were all bare, gridded metal. Black-and-white analogue Smiths dials tapped into the soul of the Escort, rather than being electronically prompted. Thus, having

➤ Front or rear, this racing Escort was almost delicate to look at static, but ferociously effective in action. *Motorsport Images*

fastened the four-point Britax harness, I noted that oil- and water-temperature gauges were combined, and that the rev-counter was the period's fashionable chronometric type, with a mechanical drive that delivered that characteristic white needle flickering, and a red 'tell-tale' pointer that recorded the maximum rpm attained, usually only reset by the team's personnel after each trip, NOT the driver!

I was told oil pressure would settle around 80psi, water at 80–90°C. Ralph Broad's cheerful parting comments were: 'Use nothing less than the full 9,000 revs – give it a good hiding.'

Now that was an act of faith, especially as the plan before the subsequent Brands smash was for the very special Escort to head to New Zealand following its already agreed sale! That never happened because, as mentioned previously, the fabled Escort was destroyed just 15 days later.

There were plenty of switch and button choices, but I just had to activate ignition, high-pressure fuel pumps and press the push-button starter (it was 2007 before I had that push-to-start on my road car!), then lightly press the accelerator pedal. To engage the dog-leg first gear, the collar was lifted, then the lever moved over and to the left, with the remaining four forward gearbox ratios simply shifted in the conventional pattern.

No more advice possible, as the clatter of 2,500rpm filled my metal quarters. I asked for another 1,000rpm, eased the clutch up rapidly and the Escort moved away. Discordant clatter turned to enraged blare, as second, third and fourth slipped home as we headed towards 120mph, all with barely any need to dip the clutch. A tentative pull back into fifth, just because I could and needed to find out if the changes were all that easy. They were.

I worked back down to third for the slowest corner on the old GP track, Becketts. The Escort was forgiving as I learned how it worked. A stab of full throttle and immediately there was a need for fourth and fifth to travel the main Hangar Straight. For the opening laps, this was the only chance to really take stock of what was occurring, particularly to check rpm and oil/water readouts: from memory this was one competition Escort that didn't feature the usual orange or red dash-light to warn of low oil pressure, so required regular checks on those dials.

➤ After over half a century, a memory that doesn't fade: a sunlit Silverstone track and a package of sublime saloon-car racing joy. *Motorsport Images*

A few laps conquered, and I began to get the measure of the stunning standards of braking, adhesion and acceleration. I would drive Brian Muir's 5.7-litre Camaro, courtesy of Malcolm Gartlan, late that year, and we measured the Chevrolet V8's performance in damp Silverstone Hangar Straight conditions. The acceleration rates to 100mph were within the same-second bracket as the Escort, but the big V8 could exceed 160mph, so would naturally draw away from the Escort on any long straight.

Even for a stranger settling into a ride totally outside his experience, the little Ford, with a solid four-cylinder soundtrack pinned to 9,000rpm, delivered 138mph so quickly that I had to ease off the loud pedal at least twice a lap to keep within the rev limit.

The swift section from Copse to Becketts allowed the excellent feedback from steering and brakes, and the levels of grip, to be explored well within the RS1600's capabilities. A hearty 9,000rpm and 100mph on the third-gear exit of Becketts released enough rear-axle energy to demand eased – rather than opposite – steering lock. Fourth was slotted across the gate from third as I steered the Escort into the following left, then I could savour nearly 120mph, ready to play rocket ships as fifth hauled us towards Stowe School and 140mph. Initially, I took the ultra-quick Stowe and Club corners in fourth, but fifth was possible for me after half-a-dozen laps. The true high-speed test of this live-axle racer proved to be Abbey's bumpy, swift left, and even in such a well-behaved and professionally set-up race car I had to pick a path that didn't jar the RS1600 off-line.

The legendary and elongated Woodcote right-hander may have been a 160mph blur for Grand Prix aces of the period, but for me, behind the wheel of the two-door Ford, approaching at an eased 9,000rpm/138mph, it was overwhelming. Fortunately, I opted for a dab of the brakes and fourth, rather than allowing Broadspeed the dubious pleasure of collecting all the component parts from the grandstands…

BROADSPEED ESCORT RS1600

Figures recorded with fifth wheel, at Silverstone long circuit, by Fitzpatrick/Walton, October 1971. Figures taken from two-way averages with 9,000rpm limit.

Acceleration

0–50mph: 3.8sec **0–60mph:** 4.8sec **0–70mph:** 6.4sec **0–100mph:** 12.0sec **0–110mph:** 14.5sec **0–120mph:** 17.7sec **Standing ¼-mile:** 13.4sec at 107mph (to calibrate timing monitors, we ran a few ¼-miles with Fitzpatrick's factory-issue RS2600 road Capri: my diary says that took 15.4sec)

Gear speeds

(At 9,000rpm, with 4.66:1 differential ratio) **First:** 60mph **Second:** 76mph **Third:** 100mph **Fourth:** 118mph **Fifth:** 138mph (Max speed up to 155mph with long-circuit final-drive)

Fuel consumption

On track: 8–9mpg

My first foray into books, in October 1971, showed my bias as it featured the Broadspeed/Fitzpatrick RS1600 on the cover, alongside the factory rallycross Capri 4x4 and a dragster. Jackie Stewart was relegated to the back cover in his World Championship Tyrrell! I was an editor, and contributed alongside established contemporaries William 'WB' Boddy, Denis 'DSJ' Jenkinson, Andrew Marriott and Mike Kettlewell. *Author's archive*

ROUGH-ROAD PERKS: WORKS RALLY RS1600

Perhaps the biggest perk of working for Ford was the opportunity to drive during the period 1972–74 what were at the time exceedingly special competition cars. Half a century on, these cars have become fabulously valuable icons that were steered by legendary competitors, survivors of an age when factory drivers weren't strangers to high-octane pranks and alcoholic drinks. I'm not suggesting that was a good thing, because there could be severe downsides, as a few of these gods of forest rallying could turn nasty during a long session at the bar. I doubt whether such men and machines will come our way again, so it is worth recounting some of those experiences with the benefit of hindsight.

For me, the epitome of the dominant Escort rallying legend in the UK was wrapped in the white-and-blue body of an Esso-sponsored first-edition machine, LVX 942J. I knew that Ford as a reporter when Roger Clark won the 1972 RAC Rally with it to claim the first all-British win since the event's adoption of its rigorous Scandinavian forest-rally format in 1960. Incidentally, that 1960 game-changing event saw subsequent BMC and Ford competitions manager Stuart Turner as the winning co-driver in a Saab driven by Erik 'on-the-roof' Carlsson.

From a Ford competitions viewpoint, Clark's RS1600 was technically interesting in the fact that it survived and conquered that 2,000-mile night-and-day menu of muddy forest trails and icy northern hillsides with a then unique powerplant. It was the only factory Escort to finish, never mind score a historic victory, with an unproven alloy cylinder block incorporated in a 220bhp, 2-litre, 16-valve Ford-Cosworth BDA engine.

'FOR ME, THE EPITOME OF THE DOMINANT ESCORT RALLYING LEGEND IN THE UK WAS WRAPPED IN THE WHITE-AND-BLUE BODY OF AN ESSO-SPONSORED FIRST-EDITION MACHINE, LVX 942J.'

That RAC was memorable for me in any case, as it was my first experience of an international rally, and previously I had no experience of phoning-in reports from every service area and relaying dramatic occurrences using public telephone boxes, or by knocking on local residents' doors (in the pre-mobile-phone days). Fortunately, I had back-up in a number of media colleagues. Some got to use fragile and rare RS1600 road cars, but I spent most of my time in a comfy Range Rover, accompanied by Ray Hutton (subsequently editor of *Autocar*) and the man who always guided me in rally lore, Michael Greasley, later editor of *Motoring News*. So I got to all the right places, right on time, to see the key Finns in the Ford team retire, and to grab a few words with Mr Clark, his service crew, and the glorious Mrs 'Goo' Clark.

It was the following year – 1973 – when I got my unworthy paws on this heroic Escort, during a season in which Roger Clark dominated the British championship with the oil-company-backed Ford, now nicknamed 'Esso Blue'. It was blazing June and Ford's UK competitions personnel and hardware had a date with the annual International Scottish Rally. Tough, because some of the special stages could be uncomfortably stony, but the social side – particularly at the Aviemore ski resort prize-giving venue – was wonderfully relaxed.

For me, it started on Friday 1 June, from the Ford base at Boreham, where I picked up a Consul GT (RPU 396K) that had been used in that year's Commander's Cup record run. The Granada-type Consul was quite a special car, for it doubled up as a fast liaison conveyance between service points and a delivery vehicle for some specific half-shaft spares for Finn Hannu Mikkola, Ford's serial world-class winner (from 1000 Lakes to London Mexico Marathon) and subsequently the leading driver behind the Audi quattro's World Rally Championship success.

First, a local pub lunch with Hannu, his co-driver John Davenport and chief mechanic Mick Jones, plus Pam Goater, Boreham guardian of records and all secretarial needs. We formed a bizarre convoy departing Essex for Scotland, as my Consul GT followed Jones in Clark's 'Esso Blue' Escort. I think Davenport got the job of ensuring Mikkola's recently acquired left-hand-drive Mercedes road-smoker travelled to Glasgow. We made it by 9pm, and it was a quiet hotel evening by '70s motorsport standards, as Clark's co-driver/team planner, Jim Porter, had politely, but firmly, commanded that I go out to the airport the next morning to pick up a spare differential for Clark's rally car.

The Scottish Rally started for me at 7am on Sunday 3 June, driving some support staff out to *parc fermé* and the start at Blythswood Square, Glasgow, on a perfect Scottish summer morning. Competitions manager Peter Ashcroft had travelled up in the Capri that stars in Chapter 8 (XWC 713L), but I drove him and Stuart Turner's co-driver protégé Tony Mason (better known later for BBC *Top Gear* appearances and rally commentary) for most of the next 24 hours in the Consul. I was thankful for the enhanced-horsepower V6 in that big Ford as we kept to a tight rally schedule.

I was woken from a brief snooze in the spacious Ford near Dumfries, as Mikkola rattled by, a rear wheel busy detaching itself thanks to a half-shaft failure! That was a literal wake-up call – near panic as we tried to locate the nearest service crew. We succeeded, and subsequently saw some wonderful rally action against Scottish backdrops that included Loch Lomond and Loch Ness – Scotland at its absolutely beautiful best.

The social side squeezed in between reporting duties included racing both go-karts (competitors versus media) and a ski-slope assault armed with tea trays, the fun finishing at 6am on Thursday 7 June. Just two hours later, we staff were on parade for the prize-giving.

A stunning drive by my favourite flying Finn, Mikkola, earned him the biggest bottle of whisky I have ever seen, Hannu second to Roger Clark after a masterful recovery drive from virtually last place, with Escorts in the top 10 places! I was lucky that newspapers would still print reasonably meaty results and feature pieces on the rally Escorts, as some of the Scandinavian names really irritated the Press Association copy-takers. National-media sub-editors were naturally suspicious when I recited so many Fords in the results.

▲ The stuff of British rally legend. For me, the ultimate Boreham rally Escort in Twin Cam and subsequent alloy-block RS1600 guises was this Roger Clark/Jim Porter machine, which I was allowed to drive back to Essex from its winning run on the 1973 Scottish Rally. *McKlein*

After prize-giving, the day was far from over. I was astonished to be trusted with LVX 942J, to be driven back to Essex. This time it was just two works Escorts in convoy, my brief to stick with Tony Pond and co-driver Frances Cobb in XPU 216L, as far as Chester. Tony, then a newcomer to factory RS1600 power, had finished a promising seventh overall. Sadly, there were too many British drivers chasing Ford works drives, and the talented rally and race star established himself outside the Blue Oval for better financial rewards.

I completed the long haul south solo. The iconic Escort was a raucous dream to drive for a motorsport addict, and sensational acceleration to 90mph felt like that of a V8 Cobra. Even as a well-thrashed winner from those Scottish torture trails and standing tall on long-travel suspension, LVX loved corners, and the all-disc braking was way more effective than that of contemporary road cars, as it only had to retard a metric tonne. Most of the motorways were completed at a resounding and sustained 6,500rpm-plus in fifth, so it wasn't surprising that I got stopped for speeding (Mr Pond was smarter, bless!).

Yet that M6 interview with the men in blue would not be the last time I talked to police that night as my progress on London's South Circular road was also followed. When I had to stop on the pavement to make a hasty wire repair to the drooping exhaust system, I got fined for that too, setting a new personal record for legal and illegal excitement.

MORE MEDIA/CELEBRITY DAY CRASH-BASH ACTION!

A standout day at Oulton Park on 14 September 1973 saw a collection of us Ford rally-jacketed PR folk host a miscellany of early 'celebrity' racers. They were armed with the batch of Escort Mexicos that had been set aside for just such events via MCD (Motor Circuit Developments), circuit operators then based at Brands Hatch and managed by John and Angela Webb. The 'celebrity' moniker can mean anything today – from a blogger to TV soap actors. On this occasion FordSport Club pulled together some sports people, including marathon runner Ron Hill, heavyweight boxer Brian London (in a purple suit) and international show-jumping champions David Broome and Ann Moore.

Bearing in mind this was just a test day, there were plenty of incidents, including one inverted Mexico. However, damage was low by rally standards, so I got the job of driving that nibbled Ford back!

The show-jumpers were in a class of their own around the superb Oulton Park long-circuit. I was passenger/instructor with Ann Moore. She and Broome had the kind of deft steering inputs that come from extracting the best from horsepower controlled by very sensitive steering-by-mouth and intermittent brake-by-rein systems. The weekend race result proved the point: show-jumpers 1–2, Broome ahead.

The following 1974 season saw me allowed further playtime – sorry, serious attempts to influence cynical media personnel regarding the virtues of factory Fords – via more rally weapons of proven winning pedigree. An attractive long weekend over 11–14 October saw me complete a Friday set-up for a figure-of-eight grassy track within Beaulieu's spacious grounds – good fun in my now mud-spattered staff Mexico.

However, Monday's Guild of Motoring Writers' media day saw some much more serious weaponry. Ford fielded Roger Clark, that year's winning Tour of Britain driver, and his RS2000 (PVX 445M). There was also a full-factory RS1600 rally car (OOO 96M), which Roger would use regularly in the UK during 1974, having taken it to second place on the 1973 RAC Rally, the filling in an RS1600 1–2–3 sandwich result that year. Roger wasn't a great fan of media days, so he did his best on the grassy circuit and left the Walton assistant his helmet to complete rewarding televised slithers in OOO. Roger was in a particularly good mood at our lunch, as he had blagged a drive in a Rolls-Royce, and afterwards wandered over to the Rootes people to try his hand in their automotive prank: a reverse-steering Hillman Hunter that went right when you steered left and vice versa. Yes, the works Ford driver set the quickest time over a tight course!

I had two favourite Fords from those October 1973 work days. The first was a pre-production Capri RS3100 (probably NHK 250M) that I used to commute to Beaulieu, then to meet Jackie Stewart's Formula Finesse and publicity picture needs. We travelled to Mallory Park on the Sunday, and the London motor show the day after Beaulieu.

My second favourite was the Clark Tour-winning RS2000. With just 138bhp and an excellent chassis, it was fun on the short grass-and-mud course, and easier than the full-blown works RS1600 to extract the performance for an unfamiliar driver. However, the factory BDA motor and five-speed ZF 'box in the RS1600 were just a glorious step into another performance dimension.

▼ By 1973 I was a Ford PR employee, my duties including presenting that year's RAC Rally-winning Timo Mäkinen/Henry Liddon Escort RS1600 to selected journalists for a session on the stony forest sections of the Chobham test track. Amongst those ready to steer, three sadly now deceased, are Sue Baker, Roger Bell and Martin Holmes. *Ford*

A couple of months later came my final 1974 media outing with the factory Fords. On 18 December we hired what was then a military test track at Chobham in Surrey. A great photographic and filming venue, the tarmac sections were often used for car chase/crash scenes in TV dramas, although I did once experience it for a lush Ford Granada TV commercial insert.

On this final media day, the Ford name and full-bore factory rally cars were sufficient to attract a good cross-section of national and specialist print journalists. Particular draws were the RS1600s in which Timo Mäkinen/Henry Liddon had won the 1974 RAC Rally (GVX 883N) and Roger Clark/Tony Mason had finished seventh (OOO 96M). Most of the journalists didn't need me to conduct them around, but I did get some quiet mileage in these rally machines.

I drove the Mäkinen car a few more times on public roads as we loaned it out to journalists. On one occasion I had to check that it really had five available ZF gears, as one outstanding local Liverpool reporter had spent his loan period using only the 2–3–4–5 quartet, ignoring the dog-leg first. That's how flexible power delivery had become!

'RUN BABY RUN': 1970 BRODIE ESCORT MkI RE-RUN, 2022

On 8 June 2022, an opportunity arose at Silverstone to revisit British motor-racing legends David Brodie and his pinstriped black-and-gold first-edition Escort, known to many British spectators simply as 'Run Baby Run'. This Brodie/Escort combination won regularly in British club racing in the late '60s and early '70s. I had last examined this eye-catching Escort in detail for the issue of *Motoring News* dated 22 January 1970. On that occasion, David had collected me for my Brodie initiation from Hounslow tube station in a vast 7.4-litre Buick Riviera, proving that it could be driven as sideways as a rally Escort on the wet run out to his Thames-side lair.

Back then, famously flamboyant Brodie's attacking driving style featured in many Ford

▼ A club racing legend at Brands Hatch – a mecca for spectators and a driver favourite. David Brodie wheels the 'Run Baby Run' Escort, with 2.1 litres of Twin Cam, ahead of some quick 1970 opposition, including a rare outing for the Rover 3500 of Roy Pierpoint, holding down fourth. *Jeff Bloxham*

▲ The Brodie 'Run Baby Run' pinstripe Escort was reborn in massively updated form and unveiled by the man himself at Silverstone in the summer of 2022. *Author*

versus GM battles, with GM represented by another larger-than-life character, Gerry Marshall. Gerry raced successful redevelopments of the Vauxhall Viva saloon and Firenza coupé. The GM-Vauxhall effort evolved from the early cars with oversize slant four-cylinder motors, built by the lone creative genius that was Bill Blydenstein, into dominant club racers with GM V8 power and a professional Dealer Team Vauxhall outfit.

Meanwhile, 'The Brode' and his heroic Escort antics became involved in premier-league Group 2 racing, but this was terminated by the biggest touring-car accident seen at that time, during the 1973 Silverstone GP support event. That 130mph incident infamously also involved Dave Matthews's Broadspeed Capri and – tragically – lapped 1-litre Mini driver Gavin Booth. Brodie suffered multiple fractures and took months to recover, Matthews lost the sight of one eye, and Booth eventually died. As Brodie lived five miles from my Henley home, I visited during his convalescence, towards the end of which he progressed to building a swimming pool!

For techies, the Racing Services 2.1-litre stretch of the 1.6-litre production Lotus-Ford Twin Cam engine in Brodie's Escort was equally acclaimed, with 170bhp a healthy output for '70s club racing. When I visited the Brodie set-up in 1970, the engine was away being rebuilt at Charles Lucas Engineering, having suffered a cracked block.

For my 2022 reacquaintance, 'Run Baby Run' carried some serious updates, including a Richardson Racing, fuel-injected, 16-valve motor from the Cosworth-Ford BD series. With the benefit of the intervening 50+ years, there was also a much tidier and safer cockpit, beefier roll cage and sturdy two-seat furniture.

Also present on that 2022 Silverstone summer day, within the exclusive BRDC (British Racing Drivers' Club) premises, was the original Brodie Roadie Escort: 12 were planned, but only a single J-registered example was sold, and survives in private hands as an RS1600 today.

Other Brodie Escorts were sold in period; David and I 'tested' a black estate in London with an oversize 1,800cc pushrod motor running Weber carburation, which was sold to a private customer. Courtesy of Steve and Jill Bennett on Facebook, I also discovered that Jill's dad (Bob) bought an ex-race Brodie Twin Cam (YMD 931H) and went rallying with it – including proper full-length 1970 and 1971 RAC Rallies.

For me, it was emotional seeing the black-and-gold race Escort shimmering at Silverstone in 2022, for I vividly remember it racing in events that also featured Roger Williamson and his championship-winning Anglia (see Chapter 2), or occasionally televised battles with Gerry Marshall and whatever weapon DTV had to hand. Yet my memories paled into insignificance compared to the impact on Brodie when he found the fabled Escort's rolling chassis after 43 years padlocked within Wiltshire storage!

The period 2.1 Twin Cam motor and 'box had long been sold. For 2022, the radically revamped 16-valve BD-series race unit sat in a fundamentally uprated racer. Much of the rest of the car had survived, albeit today it has been updated in so many areas that it's really the visual aura of days gone by that remains, for it was always beautifully presented.

Footnote *For the January 1972 edition of* Motor Sport, *I conducted another Brodie-driven 2-litre Twin Cam (this time of 1,974cc) at Silverstone. The engine was installed in a very special former hill-climb Lotus Elan, concocted by Jeff Goodliff of British Vita Racing Team (BVRT) with a small spaceframe to carry Chevron single-seater components at the front. The rebuilt BVRT engine, by Racing Services, developed 178bhp at 6,800rpm. For more about this excellent similarly pinstriped Elan, see Chapter 17.*

A DIFFERENT KIND OF RALLY WARRIOR: MÄKINEN'S 1975–76 RS1800 TARMAC CHARGER

I left Ford of Britain early in 1975 amidst some bitterness, as AVO had closed and the fuel crisis had sent me into mainstream corporate PR at Ford's UK HQ in Warley. I wished to return to the Thames Valley and a writing career. Motoring books were now on my agenda after several tasters contributing and editing, and attempting the Capri book that eventually materialised in the '80s. I wasn't permitted to write about Ford for two years after leaving the company, thus my first titles, written in the late '70s, were ironically centred on Ford sporting rivals, such as working with Gerry Marshall on his *Only Here For the Beer* or my *Unbeatable BMW* tome that eventually made my American book and magazine work possible in the '90s.

An interim move saw me return to *Motor Sport* and *Motoring News* as features editor for both publications. It was a career mistake, but I did get some very engaging drives from my former employers. Finally I bit the bullet to go freelance on my own account late in 1977.

At the tail end of 1975, for publication in the January 1976 edition of *Motor Sport*, I was allowed out at the Boreham test track in the tarmac-specification RS1800, registered KHK 982N, built for Timo Mäkinen to attack the 1975 Tour de France. This was a very different kind of event from the earlier RS1600 Escort's normal diet of forests and ice, but a truly engaging diversion that would allow Ford UK to get a clearer view of the preparations and specifications Boreham Motorsports Centre needed for smoother and drier tarmac events. When I drove this Escort, Ford was fighting against the best from Lancia-Fiat (the mid-engine Stratos-Ferrari V6 had won two straight world titles) and the Alpine-Renault that so suited the tarmac-biased French international rallies.

Here's a lightly edited and hindsight-updated rewind, recounting what I thought of that RS1800 as a 'tarmac racer' 45 years ago – an opportunity

▼ Built in 1975, this 'yellow peril' RS1800 was deployed on the 1975 Tour de France for Timo Mäkinen/Henry Liddon plus the tarmac-centric 1975 Sanremo and 1976 Monte Carlo rallies. It was steered by Simo Lampinen for the 1976 RAC, and used in 1977 UK home internationals by Andy Dawson. *Motorsport Images*

▲ The 2-litre, 16-valve, Brian Hart-built motor delivered 245bhp and 9,200rpm, with an ability to run without fuss on public roads. *Motorsport Images*

that made a very interesting comparison with the famous forest-specification RS1600s and the 1971 Fitzpatrick outright racer of similar power.

'In 1975 Fiat, Alfa Romeo and Opel were joined by Ford Motor Co in the quest for honour and some useful European rallying prestige. Ford found itself in very unfamiliar country at first, and there were some mistakes made while feeling out the temperature of European rallying water.

'On this year's Tour de France Automobile, an event dependent on tarmac performance (especially at race tracks), Ford made its point when the Zakspeed-arched Escort, driven by Timo Mäkinen, went into a brief overall lead, but failed to finish. Later in 1975 Roger Clark took another Escort RS1800, a car not quite so highly modified for tarmac use, and defeated the previously unconquered private Porsches on the closed public roads that provide the meat of the Manx International Rally.

'The Tour de France RS1800 suspension is designed to tolerate the odd spot of loose gravel and rough surfaces, but those glass-fibre wheel-arches (as developed by Zakspeed for its Cologne-based Escort racing team) and redesigned rear dampers allow wheel-rim widths up to 11in. To clear the racing wheels and tyres it has been necessary to mount the body of the shock absorber on the back parcel shelf, with a long rod leading from the damper down to the rear axle. The factory Capri RS3100 provided single-leaf springs, which were installed with a Panhard rod and four long axle-location links (a parallel pair per side) for our test. On the Tour de France, a Watt's linkage was used instead of the Panhard rod, tightening up the back end's inevitable tendency to slide under the application of over 240bhp.

'At the front, the MacPherson strut has the benefit of a locating compression strut, but this is reserved for very smooth going at present, so it was absent for the test. We did have a

➤ With a much neater dashboard than the first-generation factory Escorts, using mostly production dials, the RS1800 was also an easier drive. That is until you were the professional charged with beating rivals such as the mid-engined Lancia Stratos. *Motorsport Images*

⅝in-diameter anti-roll bar, complementing a spring rate that is slightly softer than that of the rear. The high-ratio steering rack is coupled to new steering arms, the entire layout working round 4½° of caster, 2° of negative camber and a parallel adjustment of track, so there is no toe-in, or out.

'The showroom RS1800 tail spoiler is carried on the boot, but its counterpart at the front is part of the German racing equipment, though 2in of vertical depth have been removed for rallying clearances. Incorporated in the spoiler are ducts for the 10½in-diameter front ventilated disc brakes, with callipers that wouldn't be out of place in Formula 1: at the rear, simpler 9½in-diameter solid discs are installed at either end of the purpose-built axle. Braking balance, front to rear, is adjustable via a bias bar on the foot-control pedal box.

'Since its introduction in 1970, the four-valves-per-cylinder Cosworth development of the Ford 1600 GT engine has grown to a production 1.8 litres with an aluminium cylinder block. In this application, Harlow engine builder Brian Hart produces 1,977cc from a bore of 90mm and the production stroke measuring 77.62mm. The Cosworth racing cylinder head operates on a compression of 11.5:1 and breathes in through paired Weber 48mm-choke sidedraught carburettors. Maximum power is about 245bhp, with excellent torque delivery through the 4,000–8,000rpm band. Complete with our fifth-wheel measuring device, this Escort recorded a dash from rest to 60mph in just over 6.5sec and was comfortably under the 20sec barrier sprinting from standstill to 100mph.

'Those are impressive figures for a 2-litre saloon car weighing 19cwt (albeit notably slower than the race-weight Broadspeed RS1600), but that wasn't the outstanding impression left after our acquaintance. No, my memory will record how consistent development of the same basic theme has left Ford with the most docile 200bhp+ 2-litre I have ever experienced. Literally, you could send an aged relative out to the shops in it, though I think there would be complaints about the heavy operation of the triple-plate clutch in Surbiton High Street. No complaint about the gearbox though; the latest five-speed ZF is obviously tough and very easy to operate after a series of Boreham shift modifications that are rather unexpectedly necessary on what is an extremely expensive gearbox.

'Because of the low (5.3:1) final drive, top speed was only a little over 100mph at our rpm limit of 9,200rpm. In turn, this meant that the car's handling was unstressed through our laps on Boreham's Silverstone GP-style test track. In the slower corners we'd quite often be reaching fourth as soon as the accelerator was depressed. We were surprised how much body roll there was under these conditions, but it was nice to be able to read all the information we needed quickly from the production RS1800 instrument layout, instead of the older rally RS1600 mass of aircraft dials in every corner of the cockpit. The only change to that showroom instrument layout is the installation of a chronometric tachometer in place of the standard electronic device, while the centre console also carries an ammeter.

'Once we were able to find some looser going, the car's revised handling could be fully appreciated at lower speeds. Compared with previous works Fords, the car has very sharp reactions indeed, responding instantly to throttle or steering correction, without the strong initial understeer that used to be such a feature of the forest Ford Escorts. The same applies to the tarmac Escort's overall character – it's lighter, swifter and altogether a very attractive dual-purpose machine.'

TIMO MÄKINEN'S 1975–76 TARMAC RS1800

Acceleration

0–30mph: 2.5sec **0–40mph:** 3.4sec **0–50mph:** 4.4sec
0–60mph: 6.1sec **0–70mph:** 8.0sec **0–80mph:** 10.2sec
0–90mph: 13.5sec **0–100mph:** 16.6sec **Standing ¼-mile:** 14.5sec

Maximum speed

At 8,500rpm limiter: 108mph

Fuel consumption

On special stages: 10–12.5mpg

A MIXED MkII BAG – AND THE CHARM OF RS2000 RALLYING

▼ Two of three MkII Escorts tested at a damp Chobham proving ground for *Motor Sport*: RS1800 (left) and RS2000. *Motorsport Images*

More MkII Escort competition miles arrived in three very different formats. These were a 'celebrity' Escort 1600 Sport at Brands Hatch, a track test of former *Motor* journalist Gordon Bruce's second-generation RS Mexico, and a financially curtailed season of RS2000 rallying in the ex-press fleet LHJ 928P that finally (briefly) became mine.

In hindsight I can see that of these three missions, the single-overhead-camshaft Mexico was straightforward – a neat, reliable and fun package, only troubled by other competitors barging into its panelwork, but then it was always thus.

In contrast, my own outing in the 1.6 Escort Sport, #13, was all that the unlucky race number forecast. For this was a grudge match between *Motoring News* and *Autosport* journalists, with some other quick men of CCC and *Motor Sport* heritage thrown in. I hadn't long returned from Ford employment to the owners of *Motoring News* and *Motor Sport*, so I appeared for them. Practice and race were dominated by Chris Witty while I had a bit-part bundling into then

Autosport editor Ian Phillips and nursing the bruised Ford home sixth.

Strangely, I loved the 1976 rallying interlude. It wasn't the results success it should have been, but the quality company of co-driver Roger Jones, and developing the RS2000 gradually over three club events and that season's Tour of Britain, was satisfying. The initial trio of club events saw the car in basic regulation safety trim (roll cage, fire-extinguisher system and Bilstein uprated suspension). That initial preparation was by 'Demon' Dick Prior, Tony Pond's personal mechanic at Uxbridge – and Tony gave me some invaluable pre-event coaching at the Chobham rough-road course. Sadly, nobody could cure my hot-headed, race-orientated mentality.

We could do some learning with that basic equipment – and I needed a hands-on education. I remember too many departures into the solid scenery, from bushes to bridge parapets. In the latter case – following a bent-wing slither, destined for an Otterburn military training-ground stream – I was ready to quit. Brave co-driver Roger was restrained – no swearing – but firm: 'Get back on, that's not a proper accident!' Despite more adventures, and a shortage of wheel nuts (!), we finished 13th overall on that season's Hadrian Centurion, which had a quality entry. There was a good night out in Newcastle into the bargain...

That Hadrian result encouraged Reed's Rallyesport, a Ford dealer in Torquay, Devon that had taken over responsibility for preparing the car and supplying manpower for our event appearances. So the RS2000 and crew were entered for the 1976 BRSCC-organised Tour of Britain, a proper UK-based international, with the formula of smoother rally stages and circuit races retained. The RS2000 had really progressed, with a quick steering rack (2.5 turns lock-to-lock) and a close-ratio gearbox, now joined by a 160bhp Group 1 motor from former Broadspeed employee Tony Jones. That unit had all the homologated tricks on it, including downdraught Webers, tubular exhaust manifolding and a sharper camshaft profile.

▲ Instruction from Chris Sclater (left) was one of the encounters that made me want to go rallying. *Motorsport Images*

◀ An infield area at Donington Park was the venue for this Ford Rally School exercise using a shoal of 1.6-litre flat-front MkII Escorts. *Motorsport Images*

It was still perfectly amiable in traffic, yet able to drag it out with the slower 3-litre Capris around Silverstone.

I wasted that competition opportunity, especially since I was donated the front struts from Roger Clark's RS when he was forced to retire with clutch failure on the first day. That transformed our car's mistaken suspension set-up and it went from underperforming circuit competitor to stage whizz. A brief spell in the top ten ended with another trip into the bushes and a discovery that some foliage was hiding hostile tree trunks.

Effectively, that financially ended our season, as my overdraft and the Ford dealership's support expired. However, dealer proprietor 'Jumbus' Aly Khan generously allowed his staff to rebuild the RS over several months: I took over possession and LHJ 928P was sold on to Dorothy 'Dot' Warner to settle my debts. The car was then successfully rallied locally in the area around my old home town of Henley-on-Thames.

I would not have missed that season, especially the Kielder-Otterburn class result, which I researched more thoroughly in late 2021 and found very satisfying in such a competitive entry. Others who shared my rally experience probably regretted their involvement.

◀ ▼ In 1976 I took time out from racing to follow my rallying enthusiasm with a Group 1 uprated RS2000, seen here on a Dursley Club event stage, and testing at Chobham on tarmac as pre-Tour of Britain preparation. *Author's archive and Motorsport Images*

RACE MkII RS2000 PACKING RS500 HEART

This astonishing road-registered Escort RS2000, bulging with RS500 racing muscle, was built by Paul Bailey and raced by him, starting in 1990. A former karting exponent as well as an ingenious engineer, Paul's creativity and race-craft were bound to be rewarded and he became a multiple club-racing champion in the UK with his 'RS2000-500', his successes showcasing his Bailey Motorsport business.

Within a year of Paul's car coming on my radar as a casual spectator at an RS Owners Club (RSOC) track day, informed by now-established classic-rally entrant and competitor Steve Rockingham, I found myself racing against it, armed with a professionally prepared Sierra RS500 Cosworth from Collins Performance Engineering. I marvelled regularly at my rival's prowess, sometimes through clenched teeth after suffering another runner-up race result. All this culminated in a multi-page cover feature for *CCC* in 1992.

Back before we witnessed the Bailey RS2000-500 sweeping around all-comers at RSOC track days, Martin Sharp – a *CCC* colleague best-known for his outstanding technical appraisals of World Rally Championship hardware – had uncovered the car's early story. In the September 1989 issue of *CCC*, Sharp-eyed Martin revealed that Paul had converted this effective Sierra Cosworth crusher purely for road use from the remnants of a torched £400 MkII RS2000 hulk with the registration VKO 99S. Chucking around £17,000 at his project, Paul had tweaked and implanted the Cosworth turbo motor and just enjoyed spanking Cossies.

Paul did progress to trying a few 1990 rounds of the national series for Fast Ford Saloons, then sponsored by Vecta's complex but effective anti-theft systems. For 1991, he decided to contest the full series, winning the championship outright that year and again in 1992, the latter against particularly fierce RS500 opposition.

I did most of that 1991 series in the Collins Performance Engineering RS500, as a result of an unexpected mid-winter deal, but my technical material and driving impressions of the Bailey RS2000-500 came from a 1992 day at Snetterton with Paul and the car, which by this time had consumed £30,000. Paul commented then: 'Since that first *CCC* article the car has changed a lot in detail, but the basic idea is still the same. To make the quickest RS2000 go quicker than the Cosworths. All at less cost, but to have more fun than they do, building off-the-shelf.'

I certainly had fun racing against this concoction in 1991, and track-testing it myself for 1992 magazines. A full description follows, so let's clamber in…

◀ The astonishing Paul Bailey-engineered and raced RS2000, with over 500 Cosworth turbocharged horsepower, featured on the cover of *Cars & Car Conversions* magazine in September 1992, with my account of driving it featuring inside. *Author's archive*

'SINCE THAT FIRST CCC ARTICLE THE CAR HAS CHANGED A LOT IN DETAIL, BUT THE BASIC IDEA IS STILL THE SAME. TO MAKE THE QUICKEST RS2000 GO QUICKER THAN THE COSWORTHS. ALL AT LESS COST, BUT TO HAVE MORE FUN THAN THEY DO, BUILDING OFF-THE-SHELF.'

MkII RS2000 WITH RS500 HEART (1992)

Engine

Type: Bored-out Cosworth YBD **Configuration:** 4-cylinder, 16-valve with thick-wall RS500 block, oversize valves, and solid valve-lifters for 8,200rpm limit; Bailey-Kent-profiled camshafts, forged connecting rods, forged and machined pistons, steel crankshaft **Capacity:** 2,050cc **Bore and stroke:** 91.05mm x 77mm **Compression ratio:** 7:1 **Induction/ ignition:** Stag-Bailey reprogrammed Weber-Marelli system with eight injectors and 3-bar MAP sensor; Garrett AiResearch T4 hybrid turbocharger, RS500 air-to-air intercooler, with secondary RS500 unit; gearbox monitored and boost managed electronically to give lower boost in first and second gears, plus wet-weather ignition setting **Maximum boost:** 2.5bar (35.6psi) **Maximum power:** 560bhp at 7,500rpm with 2.2bar (31.3psi) boost **Specific power per litre:** 273.2bhp **Power-to-weight ratio:** 543bhp per ton

Transmission

Type: Four-speed, Quaife internals and reinforced casing, unique propshaft and Quaife half-shafts. **Clutch:** Twin-plate **Rear axle:** Atlas, with ZF limited-slip differential **Final-drive ratio:** 3.44:1

Suspension

Front: Fabricated MacPherson struts with triangulated lower arms, coaxial coil springs, Koni damper inserts, and 25.4mm (1in) front anti-roll-bar **Rear:** Live Atlas-Ford axle with five-link location and Panhard rod. Coil-over spring layout on Koni telescopic shock-absorbers. No rear anti-roll-bar

Steering

Type: Ford Motorsport high-ratio rack-and-pinion, 2.4 turns lock-to-lock

Brakes

Front: Tarox discs, 330mm (13in) vented, four-piston callipers, Tarox ceramic pads **Rear:** RS Sierra Cosworth, 285mm (11.2in) vented, four-piston callipers, Tarox ceramic pads **Brake bias control:** Cockpit-adjustable brake bias, with five pre-sets

Wheels and tyres

Wheels: Bailey-modified Speedline centres on BBS rims, front 9in x 16in, rear 11in x 17in **Tyres:** Tested on Goodyears, but race category demanded Yokohama A008R, front 245/45 ZR16, rear 315/35 ZR17

Body

RS2000 steel body, radically braced internally with extended Safety Devices roll cage in 18-gauge steel picking up front and rear suspension loads with X-braced door bars. Glass-fibre external panels, extended 'X-pack' wheel-arches, full-depth front air dam with rubber splitter, RS2000 outline for rear spoiler

Weight

Race weight: 1991: 1,050kg (2,315lb) **For 1992 test:** 1,008kg (2,222lb)

Dimensions

Length: 4,143mm (163in) **Width:** 1,639mm (64.5in) **Height:** 1,270mm (50in) **Wheelbase:** 2,438mm (96in) **Front track:** 1,395mm (55in) **Rear track:** 1,421mm (56in)

Performance (estimated)

Maximum speed: 170mph **0–60mph:** 3.5sec **0–100mph** Under 10sec **Fuel consumption:** 3mpg on full-race boost

Behind the wheel: colourful functionality

On 29 June 1992, I paid £70 for a half-day testing at Snetterton in exalted company, as the British Touring Car Championship contenders were out in force, including Janspeed exercising its Nissan Primera for Kieth Odor.

I walked around my special ride for the morning, Mr Bailey's Escort now insured for £25,000 for our track test. I took in a riot of red, including the RS2000's exterior, rainbow rocker cover and Ford oval, plus red safety harness and solo seat insert. Installed behind the Cosworth RS steering wheel, I surveyed the bare-metal cabin, absorbing sheet alloys for instrument and switchgear sub-panels. A home-created but functional multi-tube roll cage was a comfort for what turned out to be one of my wildest Ford rides.

With a high power-to-weight ratio of 543bhp per ton, a live axle, wheels and tyres of considerable

▲ Paul Bailey still owned his RS2000-500 during 2021. The fabulous Escort was preserved for demonstrations rather than racing. *Courtesy of Paul Bailey*

girth, and unassisted, high-geared steering, Snetterton's swift 1.9-mile layout tested every aspect of this Ford's dynamics, from first-gear rush to top-gear fear, exploring the potential 170mph maximum. It was hard work, especially as Paul had a particularly tough Silverstone race the prior weekend, so the pads and discs delivered plenty of long-pedal action.

The exhilaration, particularly from the massive turbocharged acceleration, came in adrenaline surges, countered by equally violent chemical changes in my body as the brakes tried to slash that velocity. Most challenging were the longer and faster Snetterton swerves that set the flaming red RS skittering over ripples and bumps, and the brief first-gear interludes where I wished I'd taken weight-training seriously to assist at the steering-wheel coalface.

Footnote *Paul Bailey was always fascinated with aviation, and characteristically got involved hands-on. In 1999 he formed Bailey Aviation and became particularly involved in paramotors (powered para-gliders), applying his fabrication skills in this area just as he had through Bailey Motorsport, when he created specialist components such as dry-sump oil tanks.*

In 2021, Paul explained what happened next: 'The old RS2000 is very much still alive, I don't race it anymore, but I do a few track days. It has been modernised a bit with a new engine and management system and a sequential gearbox. It's still as mad as ever and always puts a grin on my face. Bailey Motorsport kind of faded out, as I didn't have the interest and enthusiasm for car stuff I once had. I was in full-swing developing engines for paramotors. More recently, I have started working in the film and TV business (again), but this time as a stunt performer, mainly doing car stunts. This led to me building some special-purpose vehicles that I now rent out to productions. Kind of a hobby turning into a business for the third time…'

'THE EXHILARATION, PARTICULARLY FROM THE MASSIVE TURBOCHARGED ACCELERATION, CAME IN ADRENALINE SURGES, COUNTERED BY EQUALLY VIOLENT CHEMICAL CHANGES IN MY BODY AS THE BRAKES TRIED TO SLASH THAT VELOCITY.'

CHAPTER 8

CAPRI

Promises kept and lost through three generations

The effective European English-language strapline from the 1969 official advertising debut of the first-generation Capri, the most popular of the breed, became 'The car you always promised yourself'. That advertising line, plus a question mark, became the default for slothful media headline writers until long after the affordable coupé's elongated death. Sales continued in Britain into 1987, but production ceased during December 1986, after three generations and 1.9 million examples.

The Capri finally, and ironically, died in Cologne. As with Audi's original Ur-quattro, it had been the stubborn Brits who kept on buying Ford's European interpretation of the USA's Mustang bestseller well after continental Europe's appetite for the Capri waned. Yet, some 35 years after production ceased, that 'car you always promised yourself' slogan was still being spouted by lazier UK classic-car media.

FORD'S SPORTING TWO-DOOR COUPÉ ARRIVES

To recap, the first British Ford Capri was an earlier two-door coupé adapted from the Consul Classic saloon and formally named Ford Consul Capri after the Mediterranean island. It sold slowly following its 1961 announcement, and was dropped in summer 1964. Ford sold just short of 20,000 examples – initially only for export – powered by three very different four-cylinder engines. Most desirable was the 1500 GT, redeveloped by Cosworth, and about a tenth of all sales (circa 2,000) were manufactured in this 78bhp trim, the engine becoming famous as the tough five-bearing power unit for the Cortina 1500 GT.

The Capris highlighted here were assembled from November 1968 in Britain and Germany, and made their public debut at the January 1969 Brussels motor show. Coinciding with the official sales and advertising debut, a totally unavailable 3-litre V6 Capri, with all-wheel-drive, was splattered across British TV screens a month later.

The black, white and blue Ford gained valuable exposure as millions tuned in to watch Saturday afternoon televised rallycross.

For mass production, Capris carried rakish long-nose outlines containing simple rear-drive hardware. The in-line four and V4 or V6 powertrains, plus some running gear, featured in other mass-market Fords. The Cortina shared many items with the Capri, and thus pricing could be aggressively competitive, with only a few specialist press highlighting how many common components lay beneath the striking Capri styling.

To seduce the naïve, Ford of Britain listed a basic 1300 at less than £900 including all UK taxes. Initially, Capris only escalated beyond £1,000 in 1600 GT or 2-litre V4 forms. The fabled and now highly valued 3-litre V6 wasn't sold until September 1969. That 115mph 3000 GT model was priced at £1,584, offering pace, but very little rear seat and luggage space. However, it challenged many more expensive prestige marques, both on the road and on the production-racing tarmac, especially when the V6 was mildly uprated from 128bhp to 138bhp, and then 140bhp. That allowed

▲ 'My' staff Capri 1600 GT XLR, complete with mandatory period matt-black bonnet and metallic gold paintwork. It took a year to personalise, via Broadspeed stiffened and lowered suspension, competition bonnet pins(!), plus Revolution alloys and fresh tyres. *Motorsport Images*

▼ Ford Competitions personnel took TV rallycross seriously in the late 1960s and early 1970s, debuting the Capri in 1969 on the small screen, here at Cadwell Park. The 4x4 Ferguson-Ford created Capri trio powered all over the opposition for championship glory and priceless publicity. Here in 1971, Roger Clark, with a very special hand-built 3.1-litre fuel-injected engine, leads Rod Chapman and brother Stan Clark in 3-litre V6s. *Peter J. Osborne*

➤ Ford created and backed many 'celebrity' races for the Escort, Cortina and Granada-type Consul GT, as well as the Capri. I participated in this 1971 event for the 3000E black special edition 3-litre, sharing the almost tidy #16 with Vince Woodman. *Jeff Bloxham*

⮟ Probably my favourite UK Capri: the 1971 Broadspeed Steve McQueen *Bullitt!* movie-inspired 3-litre. The changes took the MkI Capri way beyond showroom capability and added serious driving pleasures. Genuine examples were always rare, more expensively so today. *Author's archive*

the showroom Capri 3-litre models to become genuine 122mph machines, with 3000E or GXL models that smoked from 0–60mph in less than 8.5sec.

As for the all-American Mustang low-cost coupé formula that it echoed, the 1969–73 Capri was a rapid seller. Some 380,000 shifted by September 1970, and over a million had been produced by August 1973. That included over 50,000 shipped to the USA from the Cologne production line, in a cliché 'Coals to Newcastle' marketing move.

However, behind the shiny stats, some harsh realities intruded, including a quality gap in manufacturing between the UK and Germany. Capri sales slowed dramatically in Germany, where there was much stronger showroom opposition, including Opel's pretty and effective Manta coupés, which divided that niche two-door market and managed to inch ahead in the German sales race.

An update was needed. The Capri II was quite a radical departure, although that conventional and affordable rear-drive ironmongery remained. It incorporated a hatchback (as opposed to the boot lid on the MkI) and – with a plusher Ghia version – seemed a softer iteration altogether, for which leading British executives claimed the creative credit. That didn't sit well in Europe, as the Capri's image was now diluted, especially as Ford didn't deploy the Capri II outline in its premier-league competition-car programmes. The company stuck with the proven low weight and outline of the first-edition Capri for competition.

That also meant that the high-performance Capris appearing in showrooms and high-profile media exposure remained focused on the first-edition RS2600, which boasted Ford's first European fuel-injection motor. The RS2600 was UK-engineered and featured uprated sports equipment, but was assembled in Germany, and Cologne was home to the factory racing team too. Finally, an RS3100 belatedly appeared alongside the Capri II. This Halewood-manufactured road car was a rarity, assembled by the hundred and created to accommodate a Cosworth 415bhp/24-valve motor for the fuel-crisis-ravaged 1974 Group 2 racing season, with the racing team again based out of Germany.

Meanwhile, Capri II sales slumped. The peak manufacturing year for the first Capri was 1970,

The Capri II was a significant step away from the original Mustang-focused format, featuring a hatchback and initially a much softer, more practical image. This 2.9-litre show car is not quite what it seems, packing a 24-valve Scorpio Granada Cosworth V6. I drove a similar 24-valve Capri that finished in the top three of a Readers' Car Day for *Cars & Car Conversions* magazine. *Author*

with 238,913 examples produced (almost matched in 1973, the final year), versus a Capri II best of 183,706 in 1974. Inevitably, Ford's cost-conscious accountants saw the opportunity to concentrate all production in one factory, so from October 1976 Capri II manufacture moved to Cologne, with the car no longer built in Britain.

March 1978 saw the launch of a third evolution of the Capri, but note that this last Capri didn't carry a logical Capri III badge. This time the facelift was a lot lighter, reflecting the car's declining status in Ford's European model range. The hatchback was retained, but the lines were more aero-conscious, the front end carrying a slatted grille and lip spoiler, and the rear also sprouting a modest spoiler for the 'S' trim level. However, the best was yet to come from beneath the power-bulge bonnet.

In March 1981, the Capri 'III' benefitted from the addition of the ex-Granada 2.8i fuel-injection V6, a development of the German 2.0/2.3/2.6-litre 'Hummer' six-cylinder engine. Arguably, this was the best mass-production example of all European Capris, and a fine performer, boasting vented front disc brakes and suspension conscientiously reworked by some of the engineers who formed

Highlight of a trip to Cologne during the winter of 1971–72 was to drive the road RS2600, this one apparently in basic (lightest) homologation form (no rear bumper). Yet it was far from stark, offering a sunroof and a comfortable interior. *Motorsport Images*

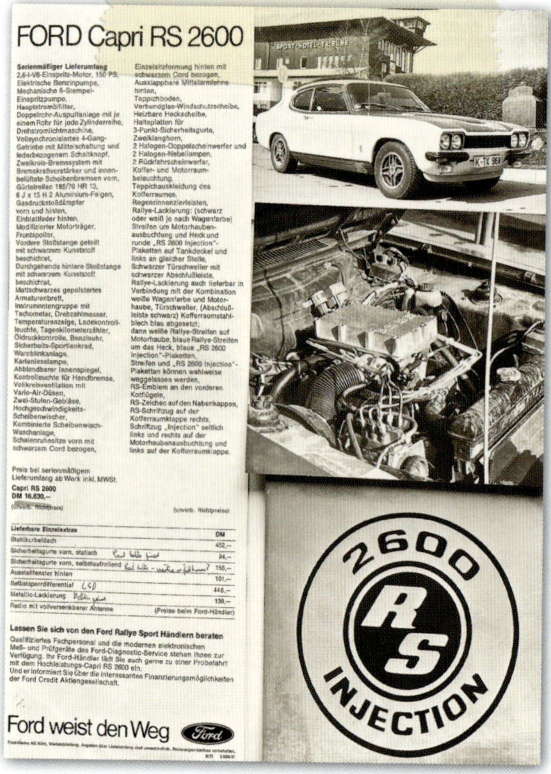

FORD Capri RS 2600

Serienmäßiger Lieferumfang
2,6-l-V6-Einspritz-Motor, 150 PS,
Elektrische Benzinpumpe,
Mechanische 6-Stempel-Einspritzpumpe,
Hauptstromölfilter,
Doppelrohr-Auspuffanlage mit je einem Rohr für jede Zylinderreihe,
Drehstromlichtmaschine,
Vollsynchronisiertes 4-Gang-Getriebe mit Mittelschaltung und lederbezogenem Schaltknopf,
Zweikreis-Bremssystem mit Bremskraftverstärker und innenbelüftete Scheibenbremsen vorn,
Gürtelreifen 185/70 HR 13,
6 J x 13 H 2 Aluminium-Felgen,
Gasdruckstoßdämpfer vorn und hinten,
Einblattfeder hinten,
Modifizierter Motorträger,
Frontspoiler,
Vordere Stoßstange geteilt mit schwarzem Kunststoff beschichtet,
Durchgehende hintere Stoßstange mit schwarzem Kunststoff beschichtet,
Mattschwarzes gepolstertes Armaturenbrett,
Instrumentengruppe mit Tachometer, Drehzahlmesser, Temperaturanzeige, Ladekontrollleuchte, Tageskilometerzähler, Öldruckkontrolle, Benzinuhr,
Sicherheits-Sportlenkrad,
Warnblinkanlage,
Kartenleselampe,
Abblendbarer Innenspiegel,
Kontrolleuchte für Handbremse,
Vollkreisventilation mit Vario-Air-Düsen,
Zwei-Stufen-Gebläse,
Hochgeschwindigkeits-Scheibenwischer,
Kombinierte Scheibenwisch-Waschanlage,
Schalenruhesitze vorn mit schwarzem Cord bezogen,
Einzelsitzformung hinten mit schwarzem Cord bezogen,
Ausklappbare Mittelarmlehne hinten,
Teppichboden,
Verbundglas-Windschutzscheibe,
Heizbare Heckscheibe,
Halteplatten für 3-Punkt-Sicherheitsgurte,
Zweiklanghorn,
2 Halogen-Doppelscheinwerfer und 2 Halogen-Nebellampen,
2 Rückfahrscheinwerfer,
Koffer- und Motorraumbeleuchtung,
Teppichauskleidung des Kofferraumes,
Regenrinnenzierleisten,
Rallye-Lackierung: (schwarz oder weiß je nach Wagenfarbe) Streifen um Motorhaubenausbuchtung und Heck und runde „RS 2600 Injection"-Plaketten auf Tankdeckel und links an gleicher Stelle,
Schwarzer Türschweller mit schwarzer Abschlußleiste,
Rallye-Lackierung auch lieferbar in Verbindung mit der Kombination weiße Wagenfarbe und Motorhaube, Türschweller, (Abschlußleiste schwarz) Kofferraumstahlblech blau abgesetzt;
dann weiße Rallye-Streifen auf Motorhaube, blaue Rallye-Streifen um das Heck, blaue „RS 2600 Injection"-Plaketten,
Streifen und „RS 2600 Injection"-Plaketten können wahlweise weggelassen werden,
RS-Emblem an den vorderen Kotflügeln,
RS-Zeichen auf den Nabenkappen,
RS-Schriftzug auf der Kofferraumklappe rechts,
Schriftzug „Injection" seitlich links und rechts auf der Motorhaubenausbuchtung und links auf der Kofferraumklappe.

Preis bei serienmäßigem Lieferumfang ab Werk inkl. MWSt.
Capri RS 2600
DM 16.830,–
(unverb. Richtpreis)

Lieferbare Einzelextras	DM
Stahlkurbeldach	452,–
Sicherheitsgurte vorn, statisch	94,–
Sicherheitsgurte vorn, selbstaufrollend	155,–
Ausstellfenster hinten	101,–
Selbstsperrdifferential	446,–
Metallic-Lackierung	130,–
Radio mit vollversenkbarer Antenne	(Preise beim Ford-Händler)

(unverb. Richtpreise)

Lassen Sie sich von den Ford Rallye Sport Händlern beraten
Qualifiziertes Fachpersonal und die modernen elektronischen Meß- und Prüfgeräte des Ford-Diagnostic-Service stehen Ihnen zur Verfügung. Ihr Ford-Händler lädt Sie auch gerne zu einer Probefahrt mit dem Hochleistungs-Capri RS 2600 ein.
Und er informiert Sie über die interessanten Finanzierungsmöglichkeiten der Ford Credit Aktiengesellschaft.

Ford weist den Weg

▲ This montage of Ford RS2600 sales material shows that the company in Cologne did a good job in promoting and selling the left-hand-drive Ford way beyond initial 1,000-off homologation numbers. The subsequent British RS3100 was not so thoroughly promoted through a 1973–74 fuel crisis, with fewer than 260 produced. *Clive Richardson*

▲ I took the RS2600 to the Nürburgring for an excellent run, as the shorter *Südschleife* old circuit was open and almost deserted. It's not always the sunny days that are most fun! *Motorsport Images*

➤ The RS2600's V6 was Ford's first European fuel-injected petrol power unit. The 2.6 litres were rated at 150bhp, but it was the smooth power delivery that impressed me. I thought the German 'Hummer' nickname appropriate. *Motorsport Images*

Ford Special Vehicle Engineering (SVE) a decade later, those engineers then based at the Advanced Vehicle Operations (AVO) site. The 2.8 injection engine was rated at an official 160bhp, and evolved from a four-speed 2.8i through to the final five-speed Brooklands Green run-out Capri 280 of 1986. Prices escalated from a debut £8,000 to a final £12,000. Production figures for the final generation of Capri peaked at 85,420 in 1979. Output plummeted thereafter, bottoming out around the 10,000 mark for the final 1985–86 manufacturing years in UK right-hand-drive specification.

There were many low-run production Capris, some factory-blessed, some not. The most extrovert, with pedigree, was Zakspeed's turbocharged version of the German-market 2.8 V6, trading on its *Deutsche Tourenwagen Meisterschaft* (DTM) championship success with four-cylinder turbo Capris of 600bhp potential.

Also eye-catching were the 205bhp British white-out Tickford Capris, built from 1983 to 1986, but their extensive redevelopment – including the addition of disc rear brakes – and hand-built status pushed prices from £15,000 to beyond £17,000 by 1986. Late in the production run, 200bhp Turbo Technics turbocharger kits were made available for retro-fitting to 2.8i Capris with Ford's blessing. Some 200 such turbocharged Capris were delivered via Ford showroom sales, and another 300 were retro-fitted, according to contemporary Turbo Technics data.

▲ Before the third incarnation of V6 Capri was listed, with triple Weber carburettors and extended X-pack body options, Bill Meade organised a personal day at Brands Hatch for the author with this Capri II, to demonstrate the performance options Ford was developing. Besides the engine and body modifications, the car featured wider RS alloy wheels and tidily uprated suspension. *Ford*

MY CAPRI ROAD TO FORD

My path to Ford employment – and therefore to the astonishing opportunity to race XWC 713L, perhaps the most valuable 3-litre Capri of all – was strewn with more than 40 examples of the long-snout breed (see 'Driven' section on page 160). In the period 1969–72, I raced 3-litre models in celebrity or production/Group 1 events and covered thousands of miles in my *Motoring News/ Motor Sport* company Capris. The first was a 1600 GT XLR, registered ALW 746H, the second a mash-up of German RS2600 GT/RS parts, originally built for PR supremo Walter Hayes and registered LHK 573J for use in Britain.

My first Capri became modified with £70 Broadspeed 2in-lowered and cambered suspension, and Revolution LM25 alloy wheels shod with Goodyear G800s of 185 HRx13 Capri 3-litre size, costing £122 for wheels and tyres – about what you'd pay for one performance tyre in 2023.

Finishing 1600 GT personal touches included functional bonnet pins from GM Dealer Sport-contracted Coburn Improvements, and a pair of Marchal driving lamps. Those Marchals were installed after a threatened libel action when I criticised the Ford-Wipac showroom items in *Motor Sport*. It cost editor Bill Boddy and I a factory tour around Wipac's Buckinghamshire factory and a printed apology to slip the noose on that casual comment!

'My' Capri was driven ruthlessly by *Motoring News* staff members and for my winter rallycross TV reportage, plus many race-track and media assignments. It only ever failed once, after 17,000 of the 25,000 miles I covered. An Autolite coil failed on the A1 dual-carriageway. A walk over a roadside grass bank revealed a duck farm, and a barn with a box-full of spare Lucas coils for a Rover 2000 that a pessimistic farmer kept handy. Even I managed to fit one: the Capri fired up and performed so perfectly that I eventually bought a new Lucas sports coil 4,000 miles later.

My second Capri brought me ever closer to Ford Public Relations employment. For £1,300 I bought the Sand Yellow RS2600 GT mongrel created for my future ultimate boss, Walter Hayes. This

DRIVEN

Road cars

1600 GT CCC; modified by Peter Gammon. **1600 GT XLR** First staff car, *Motoring News*, ALW 746H; later lowered via Broadspeed suspension, Revolution alloy wheels. **1600 GT Broadspeed** *Motor Sport* test, CVW 327G, March 1970 issue; extra 20bhp plus chassis/aero upgrades for £160; 0–60mph in 10.4sec, standing ¼-mile 17.4sec/74mph, max 107mph, 22.9mpg. **3.0 Broadspeed** *Motoring News* test, BNX 743J, 25 February 1971 issue; 3000E base, cost £2,400; 185bhp; 0–60mph in 7.2sec, max 126mph, 16.6mpg. **3000 GT (four examples)** Standard and Weslake 170bhp race-proved; disliked early showroom 128bhp versions, liked 170bhp Weslake dealer demonstrator tested for *Motor Sport*, CLR 300H; 0–60mph in 7.9sec. **3000E 'Facelift'** *Motoring News* test, PEV 697K, 1971; best showroom MkI. **RS2600** *Motoring News* test, 10 February 1972 issue, K-VZ 117; Cologne test/commute, Nürburgring *Südschleife*; fab! **'RS2600 GT'** Second staff car, *Motoring News*, LHK 573J (right-hand drive). **Capri V8s** SuperSpeed 5.0-litre 'Boss' for *Motoring News*, reported 140mph/0–60mph in 6sec; Arrow Rover 3.6-litre V8 for CCC, misfired; *Motoring News*, 23 April 1970, account based on Andrew Marriott's encounters with developer Basil Green and a Jackie Stewart track outing. **3000E** David Mills *FordSport* outing, Leeds. **3000 GT** XWC 718L, Boreham test-day commuter **RS3100 (three examples)** XWC 718L again, now to RS3100 spec, followed RAC Rally route; NHK 250M, Jackie Stewart retirement PR; TDH 224M, Donington, published *Ford Heritage*, September 1994; via Ford AVO Club's Capri ace registrar Len Pierce and RS Owners' Club. **3.0 Capri II Ghia automatic** Nice road cruiser for Walter Hayes. **3.0 Capri II GT** *Motoring News* test, PNO 575R; safer, useful, less fun. **3.0S Capri 'III'** MJN 706V, Scottish Rally coverage; clutch cable failed in Glasgow but Ford dealer round the corner; nae problem! **2.0 GL Capri 'III'** Second staff car, *Motoring News*, XVE 902T; abused, reliable. **3.0 X Pack** MJN 710V, star car, triple carbs, wide body. **2.8i injection (five examples)** Mine was KJD 7X, plus Nürburgring debut/John Miles (VVW 2W); featured in national TV commercial 'Goes Like Lightning'; tests for *Car* (five-speed) and *Performance Car* (B122 XVX). **2.8 injection turbo** Turbo Technics 200bhp, outpaced suspension! **2.8 AM Tickford (two examples)** Quality uprating, turbo 205bhp, aero body kit and relocated rear suspension with rear disc brakes. **3.0 Oselli 230S** Exciting non-turbo, tuned to 230bhp (NBW 345W); 7,000rpm, 130mph, 0–60mph in about 6sec. **2.8i 'Brooklands' (two examples)** Multiple outings in Britain's official last Capri (D194 UVW); also tried D193 UVW when on 1,800 miles, as later run by *Classic & Sports Car*'s Mike McCarthy, from whom it was stolen and recovered. **3.2 WM Capri III** Ex-*Autocar* test car, reworked by John Miles in multiple suspension specs; 155bhp at 5,000rpm, 22.3mpg driven hard; 10/10. **Capri 280 Turbo Technics** *Ford Heritage* feature, June 1995, owned by Trevor Potter (E223 MMM); 112,800 miles of which 72,000 were as 200bhp turbo. **Capri V6s (two examples)** *Ford Heritage* feature, November 1995, Norfolk Capri Spares; drove restored 3000E (YRO 4K) and 3000 GXL (NEB 300M) with 3.1-litre V6 transplant.

Competition cars

3000E 'Black' Celebrity race, Brands Hatch 1971, ninth (RWC 388K). **3000 GT** David Brodie's production racer, Brands Hatch 1972; two spins, recovered to eighth overall; 'well back' said *Autosport*. **3000 GXL Group 1** 1973 Spa 24 Hours, shared with Nigel Clarkson, 13th overall, 10th in class (XWC 713L). **3.0 Capri II** Mike 'DJ' Smith's Radio 1/Anchor Records car, Snetterton, third in practice amongst Capris, top 10 in race, fourth amongst Capris of first and second generation. **3.0 Capri II** Ford Performance, Brands Hatch, triple carbs. **3.0 Capri II GT 'Group 1½'** Chris Craft's 1977 British Saloon Car Championship car, Donington Park, 215bhp, 10th overall **3.0 Capri II 'Group 1+'** Ford UK-built for 1974 Spa 24 Hours, Hermetite livery, 3.2-litre, triple Webers, 280bhp; also driven at Brands Hatch for July 1996 article. **3.0 Capri 'III' 'Group 1½'** Vince Woodman/CC Racing, test for *Motoring News*, wet Donington, 255bhp; damply impressive. **2.9 Capri 'III' Cosworth** Tim Morris 24-valve V6, Castle Combe, top three CCC 'Readers' Converted Cars' competition. **3.0 Capri 'III'** Classic saloon race, Brands Hatch, May 2000; my last race, second in practice, fifth overall in race.

carried the 2.6-litre V6 in carburettor form, with some complementary tuning (higher-lift camshaft and more) to provide a good impression of the pre-production 150bhp fuel-injected German RS2600. I assume all this work – the interior carried superb Scheel RS2600 cord-inlay seating and resolutely British right-hand-drive steering – was to satisfy Mr Hayes. He was regarded as a god of sporting and performance image within Ford UK and USA, sporting a pedigree that included contracting Colin Chapman and Cosworth, which brought corporate Grand Prix glory with Lotus and Ford-branded DFV power. Walter was a master of manipulating cash from corporate coffers, and he also supported the infamous early Cortina Lotus and the creation of the Boreham Competitions Department (see Chapter 3), along with many other influential Ford performance and motorsport projects.

That yellow Ford, with vinyl roof and *faux* Lamborghini rear slats, was a youthful dream that I had to wait for. I picked it up on 24 May 1972, and it was stolen five days later! That loss occurred despite parking it jammed between two other vehicles outside the office and with the under-dash fuel cut-off switch activated. The experience of stolen hot Fords was to be repeated into the '90s, including inevitably with the Sierra Cosworth.

The difference with the Capri crime was that I recovered the car myself. Walking into the central London office a few days later, I spotted the distinctive paint and vinyl roof up on a pavement, just two blocks from where it had been pinched. Naturally, the radio was gone and that precious uprated interior had been vandalised. It proved exceptionally hard to replace the limited-production interior trim, even for Ford insiders.

I had that Capri back in June, but my working and racing life was moving apace. I started the long process of interviews at Ford for the job of Motorsport PR, based at Boreham and Advanced Vehicle Operations, South Ockendon, with on-event continental European intervals. Simultaneously, I had two test sessions with BMW Concessionaires production-racing 2002s and experimental/pre-production CS coupés. British BMW competitions boss John Markey decided to take an almighty punt on a man who had inverted Janspeed's 1.3 Escort Sport in the week before these BMW test sessions. He offered me a drive in a production/Group 1-specification 3.0 CS for the Spa 24 Hours – a Le Mans for tin-tops. John paired me with ace sports-car driver Peter Hanson and ensured I stayed in that BMW despite some heavy political pressure from regular team members who had other ideas.

⮝ Formula 1 team owner Frank Williams drove with passion. Here he chases the author's Janspeed Escort in 1972 after a number of Silverstone Club Circuit spins in the David Brodie/Norman Reeves 3.0 Capri. *Paul M.P. Thompson*

From a Capri viewpoint, my returned road RS2600 GT became my transport and practice car around the old and very quick eight-mile Spa-Francorchamps road circuit. It performed faultlessly, and I fell in love with its smooth engine and predictable, but exciting, handling all over again. The BMW came good too, scoring a class win (seventh overall). You can see how quick that old track was from our 24-hour average of 104.9mph, including all tyre and fuel stops. For perspective, Jo Siffert had recently lapped in a Porsche 917 at 162.087mph, a world record for a road circuit.

Back in Britain, Ford had decided to employ me, but I didn't start until September, owing to my wife's impending hospitalisation and Ealing–Essex house move. I nearly didn't make the financially seductive transition to Ford. One reason for not immediately snatching up the Ford offer was that unique Capri mongrel. My publishing employer owned it and wouldn't permit its release, so I finally left it for Murray Taylor at *Motoring News* on 15 September This was doubly painful as my initial Ford contract didn't include a staff car.

FACTORY 3-LITRE CAPRI MkI (XWC 713L) ON ROAD AND TRACK

The public debut of a trio of (Boreham) factory-built Capri 3000 GTs occurred on the first Avon Tour of Britain in July 1973. This was a very British event, themed around a confrontation between race and rally drivers over established circuits and mild loose-surface stages. The latter couldn't be premier-rally-tough, as the event was open only to recognised production cars. These were meant to carry just basic safety modifications, as BRSCC organiser Peter Browning sought to restrict manufacturers and importers, who were keen to gain maximum media coverage.

That first Avon Tour – and successors run under various sponsorship until 1976 – received massive publicity. Ford of Britain went for it in a big way, with modestly uprated Capris, rather than its usual radically modified rallying Escorts. The Blue Oval fielded diverse crews. Roger Clark, Britain's most successful rally driver, steered a unique plum-coloured Capri, XWC 712L, alongside Tony Mason. For absolute maximum publicity, Prince Michael of Kent accompanied experienced sports-car racer Nigel Clarkson in XWC 713L. Aiming for the best results on tarmac, leading Capri touring-car hand Dave Matthews got the nod in XWC 718L, navigated by Ford Competitions Department organisational ace Charles Reynolds.

In my PR role for Ford, I followed the Capris through scrutineering at Melksham (Avon HQ). The only drama for the factory Fords was that their trick Dunlop tyres, aimed specifically at coping with this event's varied surfaces, were rejected during that technical inspection. Next day, wet Wales called, and a visit to pockmarked Llandow circuit. The production Michelins adopted for the Capris allowed Clark to be reasonably competitive with fellow Capri conductor Gordon Spice (entered through Wisharts dealership, steering an ex-Ford 3-litre), until Clark's front wheel worked loose in protest. We Ford followers checked into a Birmingham hotel at around 10pm and my diary noted 'no food'.

Next day was sunny, for a circuit diet of Oulton Park and Silverstone, but our factory Capri effort suffered a major blow when Clark's car succumbed to repeated electrical gremlins at Oulton. A mandatory cut-out-switch circuit burnt out, and Roger resorted to completing some laps on the ignition key, but 712L expired out on track, and that was our major hope for a result gone.

Via innovative night racing, the survivors finished with a high-profile public performance at a sunny Sunday Brands Hatch. James Hunt nursed a privately entered Chevrolet Camaro to a slim

➤ Royal start of a long road and competition trail for factory Capri XWC 713L. Prince Michael of Kent and capable co-driver Nigel Clarkson hit the dirt on the 1973 Avon Tour of Britain. *Motorsport Images*

win over Spice's Capri, showing unusual restraint alongside the powerful brain of co-driver Robert Fearnall. Hunt's victory was achieved despite the Chevy V8 suffering a number of afflictions, including sagging oil pressure.

The best-placed factory Capri was the car of Matthews/Reynolds in fifth: Nigel Clarkson in Prince Michael's machine took 16th, just behind the winning Ladies' Award crew of Rosemary Smith/Pauline Gullick in yet another Capri.

My 24-hour bit part in the history of XWC 713L

None of us knew it, but that Royal performance on the Tour was just the beginning of XWC 713L's long and illustrious career. At the time of writing, that dark blue Ford with a vinyl roof has been reborn in Ford corporate colours after more than 45 years away from motorsport, and almost 47 years after I appeared in 713L's motorsport life.

I was introduced to Prince Michael during April 1973 when he was serving with a British Tank Regiment at Tidworth, Wiltshire, where I had my best Ford PR result, securing a national exclusive for Fleet Street legend Patrick Mennem featuring tank-testing interludes. These scored a page in the *Daily Mirror* on a national election day because we put contracted lady racer/rally driver Gillian Fortescue-Thomas into said tanks. Those eye-catching pictures pre-dated Margaret Thatcher's similar outing a decade later, again excellent media fodder. Gillian did her usual conscientious driving job under the Prince's tuition and I grabbed maximum corporate Brownie points, all gratefully received as recurrent bouts of asthma interrupted too many of my 1973–74 Ford UK working days.

I joined Ford with the managerial edict that I wasn't to carry on motor racing. A substantial number of Ford engineers did participate, particularly from Advanced Vehicle Operations (AVO), creators of production RS Escorts and designers of RS Capris. AVO became Stuart Turner's kingdom when he left the Competitions Department at Boreham, and he didn't like Ford staff racing, especially against customers. As I was to report to Stuart, as well as my mainstream boss at the UK HQ, Harry Calton, I accepted my race ban as the price of increased remuneration – and the best automotive business education a 27-year-old could receive.

▲ Under-estimated, but very important to some elements of Capri RS3100 homologation, the 1972–73 3000 GXL became a benchmark amongst non-specialist mass-production 3-litre Capris. The quad headlamps were useful, but at least 10 more horsepower were liberated from the V6, and the four-speed gearbox delivered without the second-to-third gear-ratio chasms of the earlier 3-litres. *Ford*

Yet, 'rules are made to be broken' says a well-worn cliché, and so it proved, despite Stuart Turner's objections. By 25 June 1973, just a couple of days after my witty and talented in-house hero Gerry Birrell tragically died at Rouen in a Formula 2 Chevron, a senior British Ford director confirmed that they wished me to race again.

I would accompany Nigel Clarkson to Belgium and contest the Spa 24 Hours in XWC 713L. Nigel had hired the ex-Tour Capri and two other employees from Boreham – manager Martyn Watkins, formerly my editor at *CCC*, and senior competitions technician Terry Samuels. My diary says that Mr Clarkson paid a bargain £500 for that factory deal.

I got that European Touring Car Championship 24-hour race run – the only formal competition I had in my 1972–75 Ford years – presumably because Prince Michael had been informed of the regular fatalities at this race, a day and night at Spa deemed not a place for British Royalty. Ford had somebody expendable on the payroll in the form of your author, and I had Spa experience from 1972 on the legendary long circuit, a layout used until 1979.

I first met XWC 713L, then a showroom dark blue and black vinyl-roofed V6 Capri GT with GXL options, prior to the Avon Tour. Ford Competitions Manager Peter Ashcroft strolled up from Essex to monitor the dominant performance of the factory

▲ The start of the 1973 Spa 24 Hours was chaotic at the back. 'My' Capri (XWC 713L) spent most of the startline straight clipping the grass, then dodging errant Alfa Romeos during the first few corners. *Ford*

Escorts in the Scottish Rally. Peter arrived in what was still an unmodified road car, which would compete from July onwards behind the quad-headlamp 3000 GXL facelift features, reverting to 3000 GT designation for the 1974 British saloon-car season, a configuration replicated for classic events in recent years.

I reckon my Spa-ride XWC 713L got the 165bhp mildly uprated V6 and heavy-duty suspension, 13in x 5.5in steel wheels and the discarded Avon Tour trick Dunlop tyres, plus mandatory safety items like the single-transverse-tube roll-over bar and fire extinguisher, all installed during May/June 1973. The other two 3-litre Capris for the Avon Tour were reported as being hurriedly finished on the eve of the event.

For our Spa outing, the Capri was again prepared at Boreham, but it was far from a purpose-built competition saloon. I certainly drove plenty of other production or privately owned Capris that were more powerful, better braked and sensibly lowered. There was none of today's low-rider stuff with the wheels and tyres peeping out from bulging arches. The overall weight was slightly up on a showroom car, due to the safety and night-lighting items.

Boreham quoted 160–165bhp, and that seemed correct, as it would nudge past 125mph on Spa's downhill contours, versus an independently measured 124–125mph for press-test RS3100s with an official 148bhp. At this time, the 3-litre mass-production Capris were quoted at 140bhp, the extra 20–25 horses on 713L – and siblings – extracted via a spikier camshaft profile and full-length dual exhaust systems.

Scrutineering in a public park opposite the spa baths of Spa – a 10-mile ride away from the track start at Francorchamps – held some bad omens. First, I had to fix a broken throttle cable on the staff Escort Mexico on the Belgian leg of our ferry/road trip. Next, the factory Capri was about an hour late, due to a wobbly-wheel malady, the front steel wheels sporting slack wheel nuts. If there is one thing a driver doesn't want to see before racing on Europe's quickest road-racing track…

Yet Friday practice and qualifying went off deceptively drama-free. I was tasked to see if the Dunlops would be troubled by the consistent 100–120mph Spa swerves. No bother, but it was obvious this Capri would be no match for the well-prepared and talented team of Dutch Opel Commodore conductors: my 4m 58s time was only good enough for the last third of the 60+ grid.

The 3-litre Ford performed just as faultlessly as 'my' 1972 BMW, but the tyres stripped treads as soon as we dropped below a tardy five-minute lap time in daylight temperatures. There were three driver fatalities – the worst toll on many visits to this race as a reporter or racer.

Probably best just to quote from my diary entries over 22–23 July: 'Preliminaries seemed to fly by until 17 minutes before the afternoon off, then plenty to think about on my own on the grid. Start was dreadful, took to the grass while clowns ran into each other at the start of a 24-hour race! Did first stint of exactly one hour. Exhausted by fight with two Alfa 2000 GTVs, one went off in front of me and scraped along the Armco at super-quick Burnenville.'

Subsequently, I was able to see my journalistic mates waving from the café porch on the inside of Burnenville, where the Capri would have been hitting 120mph, about 40mph slower than the leading Group 2 BMW CSLs and RS2600 Capris.

We settled into two-hour stints, some of which I enjoyed, some just survival exercises in heavy rain from 2am onwards. The worst memory was the multiple pile-up in which race leader Hans-Peter Joisten, driving an Alpina BMW 3.0 CSL, lost his life. Two more drivers were killed in other incidents. After more than 12½ hours of race driving, I was glad to hand the Capri over to Nigel for the last half hour of the event as a 'thank you' for enabling the deal and busting my Ford ban. Peter Ashcroft sent a very kind post-event letter that I have kept.

All in all, that Spa 24 Hours was a pretty harsh experience, even if it was just about worth it to say I'd raced a factory Ford! But it didn't put me off 24-hour races: I love the night sections, the feeling of achievement after a tasty stint, and the laps within sight of the finish. I did four more day-and-night outings, all at Snetterton, in three Fords (assorted Sierras) and one Fiat Abarth, all courtesy of Jesse Crosse and *Performance Car* magazine.

Back to active sporting service in the 21st century

Boreham sold XWC 713L to Stuart Rolt, who continued to race it with points-scoring results until the 1976 TT at Silverstone. That race, a round of the European Touring Car Championship, allowed some more generous tweaks, so a deeper front spoiler, wider wheels and uprated V6 (with a claimed 240bhp) featured. Shared with Jody Carr, the battle-scarred Capri finished a fine eighth overall, still in corporate blue and white, but minus all but ShellSport trade support stickers.

The car then went through the hands of two private owners before Graham Taylor bought it in 2010, although four more years passed before he could extract it from its long-term storage quarters in the Midlands. XWC 713L reappeared in public on 19 July 2018, when Taylor put it into a Silverstone auction with an estimate of £35,000–£45,000. It sold for rather less than that to a trade bidder from whom current owner Mark Martin quickly extricated it: 'I offered him another two grand for it within 24 hours and it was mine. I'd always liked Capris, I remember riding in my dad's, so it was that rather than its history that attracted me.'

▲ Early-evening pit stop and the number plate is already bug-spattered, whilst the ex-Tour of Britain tyres have told us what lap time we must run to avoid tread damage. Just a Boreham technician (Terry Samuels) and a manager (Martyn Watkins), plus volunteers, looked after our day-and-night progress. *Ford*

◄ Little-changed from the public-road specification XWC 713L wore on the 1973 Tour of Britain, the 3-litre Capri still wore a black vinyl roof and steel wheels to finish 13th overall in the Spa 24 Hours for the author and Nigel Clarkson. *Ford*

XWC 713L IN OTHER DRIVERS' HANDS

Thanks to author Jon Saltinstall and his acclaimed book *Niki Lauda: His Competition History*, I now know that XWC 713L raced at the Österreichring on 14 October 1973. Called Capri Vergleich ('Capri Comparison'), this was a celebrity event featuring prominent Austrian drivers Niki Lauda, Dieter Quester, Helmuth Koinigg and Kurt Rieder. They participated in two 10-lap heats to qualify for an overall result, and swapped cars between heats, with Niki steering XWC 713L in one of them. After finishing third in the first heat, he came through to win the second heat and so claim overall victory.

Incidentally, these Capris have been wrongly identified in many published sources as fuel-injected Group 1 RS2600s. In fact they remained in UK-specification RS3100 pre-homologation road form and were right-hand drive, unlike the RS2600. XWC 713L was changed externally following the Avon Tour of Britian and Spa 24 Hours, running RS four-spoke (road) wheels and tyres, plus RS3100 spoilers front and rear.

The Österreichring event also previewed the fact that Lauda, until then a BMW man in touring cars, had been recruited by Ford to race a 24-valve Group 2 Capri RS3100 in the 1974 European Touring Car Championship.

Walkinshaw takes the wheel

When Ford Competitions and Advanced Vehicles Operations tragically lost the skilled driving services of Gerry Birrell in June 1973, Stuart Turner's radar had locked onto another feisty young Scot. Tom Walkinshaw, 27 years old, was invited to fill the yawning chasm in RS road car driving development left by Gerry's demise, plus taking on a British race programme featuring a Group 1+ conversion of that XWC 713L Capri. It all sounds straightforward, but Ford's European motorsport programmes – as for rivals such as BMW – had to improvise around the 1973–74 fuel crisis, resulting in the curtailment of many motorsport activities.

However, Ford of Britain did see merit in the replacement Group 1+ (closer to production) race rules for the 1974 premier British series, and decided to revamp XWC 713L as a contestant for class honours (2,500–4,000cc) in the multi-class series. Unfortunately for Ford, big-banger American V8s were still permitted, and the Chevrolet Camaro was the way to go for ambitious privateers. Outright Capri wins were scarce, although there was one magnificent overall victory for Tom and the Capri on Scottish home ground.

XWC 713L and I met up again at Boreham 7 March 1974, when a factory tour for the Cambridge University Automobile Club saw it stand out from a forest of rally Escorts. The V6 was then in final preparation for the British Saloon Car Championship (BSCC), to appear in corporate white and blue, with ShellSport as the title sponsor, and Tom Walkinshaw driving the quickly redeveloped package for a debut at Mallory Park just days later. I was assigned to follow the series and support Tom's efforts with press releases and interviews.

XWC 713L became a much more honed package. When it race-debuted in its revised form in March 1974, the interior had been stripped of all but mandatory showroom trim and furniture, and the suspension had been retuned to the much higher spring rates that could be employed in smooth circuit racing, damped by Bilstein. There was now a proper race fuel tank low in the boot, as the standard tank would influence race-circuit handling when full (giving top-heavy/tilting feedback) and the new tank reduced the chances of fuel spillage.

The 3-litre V6 now sported an official 175bhp and was the work of Racing Services at Twickenham, run by ex-Willment proprietors Ken Brittain and Spike Winter, plus eight employees, including former Lotus F1 driver John Miles, who was later to redevelop several Capris and work on the aerodynamics of the Sierra Cosworth. Incidentally, the Racing Services craftsmen also came up with the stretched Lotus-Ford 2.0- and 2.1-litre Twin Cam motors used in Formula 2 and David Brodie's legendary 'Run Baby Run' club-racing Escort.

For Tom's 1974 season, the Capri's brakes remained production discs at the front, and drums at the rear, albeit with anti-fade Ferodo DS11 and VG95-branded friction materials. I often dismissed the production Capri disc/drum brakes as slackers in the affordable 3-litre Capri recipe for speed. A passenger ride with Tom around the Boreham perimeter circuit in the rapidly improving Capri proved he could unleash impressive late braking, partly because of the Capri's sub-1,000kg homologated race weight, but mostly owing to his skill.

The Capri's highlights in the Castrol-backed series were obviously the one outright win in Scotland (at Ingliston) and six class victories. The Ford season took

▲ Tom Walkinshaw, late in his Ford development/race driver contract, with a Capri II at Boreham. Some aero and body developments are evident on the Capri, but they were not sufficient to make the car a front-runner. *Ford*

months to get rolling, but class victories at both quick Thruxton and twisty Brands Hatch spoke of successful redevelopment.

Personally, the memories that linger nearly 50 years later are of that slow start to the Capri season. Mallory Park on 10 March started our schedule, notable for me because a Capri that I had manipulated from Ford for my 1972 Escort Sport rival Ivan Dutton won. That didn't happen again, as Peter Hanson (my senior BMW driving partner in the 1972 Spa 24 Hours) blitzed the next-highest total of class wins to Tom in his Opel Commodore. If Peter didn't win, Tony Lanfranchi was competitive in a BMW dealership (Nick van der Steen) 3-litre BMW CSi coupé, and he was especially impressive at the 17 March Brands Hatch Race of Champions round (Tony was the racing school's chief instructor at Brands) when he won on the GP long track with a last-minute showroom engine transplant.

I went to interview Tom about his opening impressions of the Capri season, and I attended a Boreham meeting on the following Monday. My diary noted, 'everyone touchy about Capri performance, or lack of it!' Things didn't improve at the 7 April Silverstone round, as the Capri trundled in seventh; 'depressing' said my diary.

A week later, I enjoyed watching Brian Redman thundering round Oulton Park in the Formula 5000 Lola-Chevrolet V8, but the Capri became firmly lodged in third gear for 'our' race. Just three days later, on Easter Monday, Tom's gearbox-shifting dramas recurred, with the lever breaking. Hanson's Opel Commodore carried on winning for General Motors, so that went down badly with Ford managers and money men.

We had to wait until more than half the British Championship race calendar passed before Spring Bank Holiday Monday at Thruxton on 27 May yielded our first win. Still, when it came, I noted that it did so 'easily'. I now think that my observation ignored the background Boreham development moves that powered the Capri into victory lane on a regular basis for the rest of the season. The supporting race at the British Grand Prix on 20 July underlined this form, with a fine third overall and class victory before the biggest audience of the year.

At Ingliston on 18 August, Tom motored hardest on his home tarmac and thumped in a pair of outright victories on the short showground circuit, one for the Capri and the other for a Ford Group 1+ redevelopment of the Escort RS2000, owned and more often driven by Alan Foster. The Walkinshaw Capri practised a scant 0.2sec down on the best Camaro and went on to win by 5.4sec after a rough, panel-bruising race with

◄ The 1974 British Saloon Car Championship brought class victory for Tom Walkinshaw and XWC 713L against tough BMW and GM-Opel opposition. This historic Capri was reborn in the 2020s and continues to appear in Ford corporate colours at classic events. *Motorsport Images*

jump-starting Hanson and Richard Lloyd's Chevrolet. The Escort RS2000 win was even more remarkable, as the model never beat the class-dominant Triumph Dolomite Sprints again.

Down south at Brands Hatch for the 'late summer holiday' weekend, Tom carried on his classy results. Appropriate, as that was one of many maximum-publicity Ford Sport days. Cheshire's Oulton Park in early September showers was kinder to us than the early-season disappointment at the same venue, Tom snatching fourth overall, plus the class win.

Snetterton in early October was as climatically bad-tempered as these old airbases can get on rainy days. The resurfaced track didn't help the Capri cause – or the many spinning Camaros, one sporting a 7.6-litre V8! Stuart Graham showed how to win (by over 20sec) with the pristine Brut Fabergé Z28 Camaro. The 2,500–4,000cc class battle was an uneven ride for the Opel and Capri, Hanson suffering low fuel and unable to capitalise when Tom's Capri picked up a puncture on the closing lap, so Lanfranchi resumed his initial lead to win for BMW.

The Brands Hatch finale was even more dramatic, as was often the case for the title-deciding Motor Show 200. Late October showers made the long circuit particularly slippery and it took a cool head to finish the race, never mind secure the outright title. A chaotic race – stopped after 18 of 20 scheduled laps – saw neither outright championship contender finish in this crash-fest. Ultra-quick Barrie Williams (Mazda) spun Tom aside in the excitement – both restarted, and Tom managed third overall, comfortably ahead of the class opposition and putting in laps 5sec down on the best from the quickest 5.7-litre Camaro V8.

Walkinshaw finished fourth in the championship with 63 points, behind Champion Bernard Unett (Hillman Avenger 1500 GT) with 69, Andy Rouse (Triumph Dolomite Sprint) on 67 and Stuart Graham (Chevrolet Camaro Z28) on 64. Over 12 rounds, Tom took six class wins, at Thruxton, Brands Hatch (three times), Ingliston and Oulton Park, and that one outright victory at Ingliston. The competition in the class was so close that Peter Hanson took five class wins. If the Capri had been reliable earlier, a top-three championship result would have been possible.

October 1974 was the last I saw of XWC 713L for years, although I took one of its siblings (XWC 718L, now dressed as an RS3100) around the RAC Rally a month later. It has been published that motorcycle racer Barry Sheene drove XWC 713L at a Brands Hatch celebrity event in 1974, but I haven't found any confirmation of that. Sheene was certainly keen to swerve into car racing, and saloons were the most feasible way to do that, but he didn't drive regularly until 1981, with a TWR Mazda RX-7, and then two seasons with Toyota Team GB in 1985–86.

From December 2018, this pedigree Capri resided with Mark Martin's regular Foscombe classic race team in Gloucestershire, an outfit overseen by the calm and occasionally caustic tones of Dave Lampitt. Dave has over five decades of dealing with the detailed preparation and action-packed dramas of motor racing. I first met laconic Lampitt in 1971, when he was tending the Group 2 Escort Twin Cam of Gerry Edmonds Racing, driven by Lawrie Hickman in the British Saloon Car Championship.

When I visited Foscombe Racing during June 2020, Covid-19 dominated the headlines, but as for other garage-based businesses, the Foscombe Racing team worked on in their trusted workmate bubbles. 'Garage' is a poor description of the immaculate modern unit within a traditional Cotswold stone-built estate that housed Mark Martin's then 10-strong race-car collection. The Capri now wore gleaming corporate white and blue, with ShellSport stripes, bearing multiple trade support decals from Champion, Bilstein, Dunlop and others. Some obvious bits were missing on this first encounter, including the V6 engine and associated exhaust systems, front and rear bumpers, plus the wiring loom, which had to cater for rather more electrical demands and electronic systems than were available in 1973. The unique wiring loom was the work of John Mawby.

The bare matt-black metal flooring and interior acknowledged the time-warp element, with Ford's interpretation of Formica wood for the dash panel. Aldridge Trimming supplied most authentic interior touches, highlighted by chrome window-winder handles, full-trim appearance for the door cards, 'Aeroflow' ventilation balls, and a brace of period push-buttons for the two-speed wipers (Ford generosity knew no bounds!) and lights, plus a heater ventilation panel carrying two sliders.

The dashboard retained showroom six-dial instruments. The dials within were analogue, but not as we knew it in 1973, Jim! Instead of the previous Tour of Britain/Spa production-instrumentation vagueness, there were proper black-background, white-needle numeric dials for oil and fuel pressures, plus water and oil temperatures. Additionally, we got a large Ford red-needle period 140mph speedometer and an 8,000rpm mechanical tachometer. The period Smiths rev counter was of pure competition format, with a red tell-tale secondary needle to indicate any indiscretions with the engine's 7,000rpm testing limit.

Naturally, safety standards had been considerably uprated over the years. The Custom Cages-supplied roll-cage kit came with substantial door braces on both sides. The carbon-fibre Tillet race seat gave some clues as to the progress made. Foot pedals had a reminder of yesterday, with rubber inserts for the accelerator, but bare metal for the brake and clutch pedals.

The hardware to propel the 1,002kg race-weight Capri came from the late Neil Brown in Lincolnshire, leading supplier of V6 motors conforming to British regulations into the '80s heyday of the 3-litre Capri 'III'. In the car's 2021–22 format, Foscombe had alternative engine settings to meet classic-racing regulations in a variety of British and overseas series. Power was first quoted at 225bhp at 7,000rpm in May 2021, but inlet-manifold and running-experience tweaks subsequently released a reported 250bhp by August of that season, along with an extended rpm limit and improved mid-range pulling power. The 225bhp unit ran beautifully in the June pre-season test I witnessed, a credit to its creators, and rather more refined and precisely managed than we could have dreamt of in 1973–74.

As in 1974, a Salisbury multi-plate limited-slip differential transmits the power to the rear wheels, working with a hand-built and hardened version of the Ford four-speed gearbox, now with the final homologated ratio set fitted. Back in 1974, Walkinshaw – and other Capri racers into the '80s – had these four-speed units assembled by a moonlighting Ford Halewood employee with skills that provided amazing durability under race duress. Incidentally, production rear-axle

'THE 225BHP UNIT RAN BEAUTIFULLY IN THE JUNE PRE-SEASON TEST I WITNESSED, A CREDIT TO ITS CREATORS, AND RATHER MORE REFINED AND PRECISELY MANAGED THAN WE COULD HAVE DREAMT OF IN 1973–74.'

BUSY CAPRI LIVES...

The 3-litre Capri used by Dave Matthews on the 1973 Avon Tour of Britain (XWC 718L) was subsequently photographed by Ford as an official example of the showroom RS3100. I also drove it on 8 June 1973 to unofficially attend a Boreham rival road-car comparison, staged between the promising Triumph Dolomite Sprint (16 valves/single camshaft), and the MkI RS2000 and RS1600 Escorts.

The objective was to see if Ford had an answer to the Dolomite Sprint, perhaps by dropping the 16-valve RS1600 engine into an Escort of plusher Ghia specification. We were aware of the 1973 Vauxhall Firenza 'Droop Snoot' but that pretty and effective machine, with ZF five-speed (Vauxhall's first), isn't in my diary for that test. Stuart Turner arranged a separate swap with Vauxhall, and we did get a Firenza 2300 CHP (Coupé High Performance) in for a week in 1974. I blagged it for 24 hours and was mildly disappointed by the floppy gear-change and 2.3-litre slant-four that remained slow-revving. However, I felt Ford RS models benefitted from that Vauxhall knowledge, as the 1975–80 RS2000, with beaky or flat nose, proved a smash hit by limited-production RS standards, with more than 10,000 sold, versus 204 strike- and fuel-crisis-hit special Firenzas.

I subsequently used that XWC 718L Capri again, in full RS3100 regalia, as Charles Reynolds and I covered a 2,000-mile road route for the fraught (fuel crisis looming) 1974 RAC Rally, rationed to three gallons for each Scottish forecourt fuel stop, which was therefore something we had to do frequently.

At various points we also had TV soap stars as passengers. Stuart Turner and the mainstream Ford PR machine always covered every publicity angle, but subjecting these people to the back seats of a Capri chasing a rough-and-tumble rally? Not the ideal motorsport experience, you'd have thought...

Jilly Cooper, then a *Sunday Times* columnist, later a successful novelist, had an urban ride with us, and she was happy, especially enlivened by the Boreham Boys' banter.

'THE OBJECTIVE WAS TO SEE IF FORD HAD AN ANSWER TO THE DOLOMITE SPRINT, PERHAPS BY DROPPING THE 16-VALVE RS1600 ENGINE INTO AN ESCORT OF PLUSHER GHIA SPECIFICATION.'

casings, differential gears and half-shafts proved very tough. A senior Ford engineering executive cynically told me in the '70s: 'If our cost accountants had found out how good they were, they would have said they were over-engineered and made us use something cheaper in production!'

The suspension followed the legendary Yorkshire-based CC Racing Developments (CC Racing was run by Pete Clark and Dave Cook) later work on Capris II and 'III': Bilstein front struts and Koni rears with the single-leaf layout. Spring rates were the same as back in the day for the last race 3-litre Capris, even though their homologated race-weights were slightly different, the final race-cars being slightly heavier.

Although treaded race tyres remained, compounds have advanced, and the camber and toe angles that Foscombe implemented are a major factor in providing a more-balanced, grippier drive than in the past. Another area where the team were still experimenting was brake-friction material, but the disc/drum system remained, as ever.

Externally, the 2020 paint and panel restoration by Normandale of Daventry had left 713L in fine fettle. Coats of blue and white were supported by carefully researched replica decals, plus '3000 V6' side badges for this chameleon restored to 1974 Walkinshaw 3000 GT livery. Rewind, and it had appeared as a dark-blue, vinyl-roof GXL/GT twice, before becoming an RS3100 for Lauda and Quester, still in blue with blacktop vinyl!

The main monocoque steel hull remained original, along with all front and rear three-quarter panels. 'Only the boot floor needed welding – that was in a right mess,' reported Lampitt. The originality of the primary panels in such an historic racing saloon is testament to every driver who steered it in battle. Any external panels that did needed replacing were sourced from selected suppliers via John Hills at the long-established Capri Club.

Driving XWC 713L in 2021

After a few test sessions, and a repair following a wing-bending incident for the owner at a 2021 Silverstone classic-race practice session, I finally had the chance to drive the old warrior again, on 1 September 2021. We had tried the previous Saturday at sunny Goodwood, only for an axle seal to fail and munch the wheel bearings. Now at Castle Combe, the gleaming Capri was unloaded from Foscombe Racing's regular truck by Dave Lampitt and Mark Martin, attracting a lot of envious attention at the pandemic-postponed Guild of Motoring Writers' Track Day.

The 1.85-mile track is familiar to me, right back to pre-chicane days when we used to conduct magazine group tests to assess the quickest hot hatchbacks. Back then, the 16-valve Toyotas ran close to the 1m 20s times of a classic-racing Cortina or Elan nowadays. I also held the 1991 lap record for road-tyre-equipped Vecta Challenge Fords in a Collins Performance Engineering Sierra RS500 Cosworth (see Chapter 13). No records this time: we needed to treat the 50 assorted track occupants with respect.

It had been just over 47 years since I last drove this Capri in Belgium, so it was an emotional experience. It was also a very educational one, in terms of how motorsport specialists can move apparently the same car onto a different plane with today's know-how.

The cold-start procedure required skilled carburettor manipulation for a successful result, but thereafter the fully warmed-up V6 motor was happy to run at 75–80°C on this cool September day. The 3-litre became more amenable to delivering prompt pulling power than many classics that have seen too much storage and too few miles.

The clutch snapped from in to out, but we didn't stall as we paddled out towards the track, passing marshals who – to judge by the waves and thumbs-up gestures – obviously relished the sight of this resurrected old warrior snorting by their start/finish post. The traditional three-spoke steering wheel, with leather rim, was truck-heavy at first-gear speeds as we heaved our way through the U-turn that led out into the track release area. A short run through that holding area, and first was exchanged for second with 4,000rpm synchronised simplicity, but it was a longer row across the gate to third.

We emerged onto the fastest sector of the circuit, so the yellow warning line and a look over the shoulder for any high-speed traffic both

⮝ Mark Martin (left) has owned and driven a stable of classic racing cars from Formula 1 to Lotus Cortina. XWC 713L was an accidental, sentimental purchase and now every aspect has been refurbished or rebuilt. The author put on his best red shoes for the reunion in 2021. *Peter J. Osborne*

⮜ The last time I sat in this Capri was in 1974, when Tom Walkinshaw drove me around Boreham. Now the car is a honed, functional classic racer, incorporating current safety standards, lightweight seat and extensive electronics. *Peter J. Osborne*

AN ORIGINAL COLOGNE CAPRI RACER REUNION

Courtesy of current owner Doug Titford, I was twice able to renew my acquaintance (at a 2017 Donington track day and the 2019 Goodwood Festival of Speed) with what is always referred to as the 'Kent Capri', owing to the cigarette-sponsorship colours it carried back in 1972 for Frami Racing in Holland – gold-and-white livery that was also retained that season when Gerry Birrell drove this Group 2 racing Capri to front-running status in the UK and continental Europe. What makes this car so special is that Doug has retained original construction features to this day, despite the international authorities imposing rules that would have changed features such as roll-cage diameter before they would allow it to race again. Thus, this Kent Capri remains preserved, down to the original left-hand-drive VDO instrumentation, with 250kph speedometer and 8,000rpm rev counter, and a plethora of minor dials and switchgear inserted in the original sheet-metal dash panels alongside showroom 'Aeroflow' ventilation balls.

This pedigree coupé was built in Cologne over the winter of 1971–72, alongside Ford works Group 2 machines. Sharing those factory work-bays was a

▲ These two images document a memorable day at Donington in 2017. Doug Titford allowed this amazingly original 1972 Cologne-built 'Kent Capri' out for a track session for Kara Birrell, daughter of Gerry, to drive. She was amiably mentored by experienced works Capri campaigner Jochen Mass. *Author*

▼ I visited Ford competitions in Cologne during the 1971–72 off season. Here's the squad being prepared and repainted in corporate blue-and-white for 1972, with at least two of the 2.9-litre racers destined for overseas teams Gartlan/Wiggins Teape in the UK and Frami Racing/Kent in The Netherlands. *Motorsport Images*

UK-based Capri of similar specification to the Kent car, for Brian Muir to drive on behalf of its owner, Malcolm Gartlan, in equally famous Wiggins Teape Paperchase colours.

The Donington track day was pretty extraordinary, as Doug had arranged for Gerry Birrell's daughter, Kara, to drive this very special Ford, with high-quality tuition from former works Capri and Formula 1 McLaren pilot Jochen Mass. Kara's mother, Margaret, describes her daughter as, 'a bit of a devil, real chip off the old block, but on a horse, rather than with car horsepower'. The Donington day showed that Kara has a cool brain too, settling into safe laps in a race car born 45 years earlier. Just the sound of the V6 booming along the pits straight into Redgate was worth the trip.

1972 GROUP 2 CAPRI

Contemporary 3.0 UK V6 Capri in brackets

Engine

Configuration: V6 with Weslake-Ford alloy heads (iron block and heads) **Capacity:** 2,995cc (2,994cc) **Bore and stroke:** 96mm x 69mm (93.7mm x 74.2mm) **Compression ratio:** 11:1 (9:1) **Maximum power:** 285–290bhp at 7,500rpm (138bhp at 5,000rpm)

Transmission

Type: Five-speed ZF, close-ratio gears (four-speed Ford) **Final-drive:** Choice of final drive and limited-slip differential (3.09 final drive, no mass-production limited-slip differential)

Suspension

Front: MacPherson struts, replacement hubs, anti-roll bar and links (MacPherson struts, single lower track-control arms, anti-roll bar) **Rear:** Live rear axle, with four-link axle location, vestigial single-leaf springs, vertical coil springs, Bilstein gas dampers (live axle, leaf springs, two location rods, double-action dampers)

Brakes

Front: ATE or AP Racing 300mm/12in discs (245mm/9.6in discs, vacuum-servo-assisted) **Rear:** ATE or AP Racing 270mm/10.6in discs (229mm/9in discs, vacuum-servo-assisted)

Wheels and tyres

Wheels: BBS alloy, centre castings, split rims, front 10in x 15in, rear 12in x 15in (Steel 5Jx13in, 185x13in radial) **Tyres:** Dunlop wet, dry and intermediate race rubber

Body

Two-door, single-seat, race-prepared, integral roll cage, glass-fibre external panels, including doors, bonnet, boot and wheel-arches, alloy front spoiler (mass-produced all-steel two-door body, no spoilers, four seats)

Weight

Race weight: 980kg max (1,090kg kerb weight)

Performance

Maximum speed: 165mph (122mph) **0–60mph:** 4.6sec (8.4sec) **0–124mph:** 14.6sec (0–100mph: 26.3sec)

seemed sensible. At these lower rpm, top gear was barely engaged before 5,000rpm appeared as we reached the blind crest that hides the tighter right of Quarry corner – a trap for the unwary, and destroyer of many a pristine paint job, to say nothing of the panels beneath. The rest of the lap was just concerned with adjusting to the other occupants of a mixed track day, and ensuring we had over 70psi oil pressure, and that the water did not stray above 90°C – although there was little danger of that on this official opening to autumn.

I had only used 5,200 of the available 7,000rpm (actually the limiter started curbing the Neil Brown unit's enthusiasm for more revs at 6,800–6,900rpm), so it was time to see how the wizards of competition preparation had transformed a traditional and affordable performer after nearly half a century. Let's encapsulate that experience in one flying lap.

You expect a start/finish line to be basically straight, but at Castle Combe you cross the line in top gear, already aiming at the open right-hander encouragingly titled Folly. The V6 sounds pretty good within the bare cabin, and afterwards there were a number of warm remarks about the six-cylinder's emotive soundtrack from spectators who could pick it out amongst the majority of production cars present.

Folly's tarmac ripples make the live-axle Ford, with no aerodynamic assistance (spoilers came after the MkI, and were always pretty modest by contemporary BMW CSL standards), begin to raise its snout. Now skittering modestly over the gentle downhill slope, the big brake-and-turn test at Quarry awaits. As we approach the now uphill contours to Quarry, 6,200rpm is reported on the flicking mechanical tachometer, and we are in the 125mph zone.

Charged with somebody else's increasingly valuable property, I start braking before the crest. With that surmounted, and with speed dropping rapidly, I reach for third gear and aim at an apex that remains hidden, lurking somewhere in the vicinity of the south coast.

A short sprint to the Esses fierce kerbing amounts to what is effectively the first chicane, inserted to slow the original perimeter track's top-gear approach to the fast right of Tower. Entry to the Esses demands careful selection of second

▲ Dave Lampitt of Foscombe Racing tickles XCW 713L's Neil Brown-rebuilt V6 into life. *Peter J. Osborne*

gear, for you could get lost navigating the spacious gear-change gate, finding yourself perilously close to the alley leading to reverse. Safely in second gear, and we get the first hint that this is a traditional rear-drive Ford, the steering lightening under full power as we arc to the left: what would have been slithery oversteer in 1973 is restrained in 2021 by enhanced grip. All that is required is to ease off steering lock and enjoy a swift exit and prompt gear-change across to third at little more than 5,000rpm.

The brief excitement is followed by a swift change in direction into a gentle right at Old Paddock, the Capri now sprinting through 6,000rpm in third, and I briefly change into fourth. Heavy braking is needed again into the sharper Tower right, and I find the pedal is long in the old-fashioned way, but retardation stays consistent because the current pads and linings do not fade away. The penalty is more vibration than I want, or the owner likes.

Another abbreviated third-gear sprint to the second chicane (Bobbies) and I haul the lever back into second, my size-seven feet struggling to bridge the gap between brake and accelerator. It's actually safer to forget heel-and-toe fanciness and execute a conventional downshift, allowing time for a hearty blip on the throttle, which allows second gear to slip into place a lot more readily.

Exiting Bobbies, the eager V6 vociferously hits almost 7,000rpm, and the Capri's elongated nostrils mark the car's path with dips, as third, then fourth, slip home. The exit road flicks by on the left, and we are faced with the exhilarating surge to the daunting haul right, dubbed Camp. That challenge can be a fourth-gear turn to the low kerb, but whatever ratio is employed you're looking to ease the Ford in a natural power-versus-grip arc as far down the exit kerbing as possible, before booming across the start/finish line.

It was an enormous privilege to drive this pedigree Capri again, as such devices are now so expensive to prepare and maintain, never mind buy, nearly half a century after this 3-litre provided such honourable factory service. I was sad not to be quicker in the laps allowed, but with the priority to relate how it felt, rather than boost my 75-year-old ego, I went home impressed with the 2021 evolution of what was a solid value-for-money performer back in the day.

➤ Back in the Capri groove for the author at Castle Combe in 2021, over 47 years after driving the very same car in the Spa 24 Hours. *Peter J. Osborne*

FINAL DAYS AT FORD, AS CAPRI II ARRIVES

As you may have noted in the 'Driven' section, I completed only two notable public-road outings with the official Capri II. That was primarily because I left Ford employment just as that second edition came on showroom stream, however I did cover more track miles than expected in this second-generation hatchback. As for any of the Ford performance-orientated personnel, my work in 1973–74 was focused on the RS3100 homologation special (based on the MkI Capri) for road and track, particularly the race redevelopment to incorporate wings and the 3.4-litre/24-valve Cosworth-Ford V6.

Before I left Ford, I drove a very pleasant Capri II, fully specified up for Walter Hayes, with automatic transmission and new-to-Capri power steering, plus a very plush light-grey interior to match the metallic-grey exterior. Mated to the strong mid-range pulling power of the Essex 3-litre V6, and comparatively compliant suspension settings, this specification made for a very relaxed Capri ride indeed. This was one of the last company cars I drove – for reasons I have forgotten – while employed at Ford.

▲ Yumping an RS3100 – spectacular, but you don't want to be paying the landing bills! I suspect this photograph was taken during a publicity shoot in Italy, demonstrating performance products. *Ford*

3.0S CAPRI II ROAD CAR

A 3.0S Capri II test car presented itself for a 1976 media day at Chobham test track a year after my departure from Ford. A few weeks later I had the same metallic-brown machine (PNO 575R) in for a full road test at *Motoring News*. On the military test-track tarmac, or working its way through the slimiest of winter streets on my daily M4 commute to central London, the second-edition updates were particularly obvious in this 3-litre, four-speed-manual machine.

The car featured well-weighted power steering that was a bit light by contrast with the unassisted MkI cars, but you soon adapted, and it made the second-generation model a lot easier to slot through mundane city motoring and into parking spaces. The handling wasn't nearly as snappy into the tail slides that wounded so many earlier 3-litre Capris against solid obstacles, and generally gave a rather more civilised drive. OK, the Capri II wasn't as exciting as the originals, but sometimes too much excitement can lead one astray.

The test £3,280 car was also a more stable motorway and track drive, thanks to a 2.5in stretch in rear track, Bilstein damping, and restrained front and rear spoilers. Although the front discs were also modestly uprated over the original inadequate 3-litre V6 items (from 9.6in to 9.7in), those still-small discs also had another 200lb (91kg) to slow, the weight factor also evident in straight-line performance, as the V6 remained at 138bhp. So, showroom performance dropped enough for the 0–60mph difference to be measurable (I recorded a 116mph maximum and 0–60mph in 9.1sec for the 3.0S Capri II) when compared to the earlier 3000 GXL/GT versions.

DONINGTON STAND-IN WITH A RACE 3.0 CAPRI II

My 31st birthday fell on 4 July 1977, but the top automotive present came the next day at Silverstone with a very special test session, one at which I needed to perform at my top level in order to progress to an even more exhilarating weekend. These birthday-week goodies came from my hero Chris Craft, who decided he had another pressing engagement, so couldn't attend the newly reopened Donington circuit for its first evening meeting, at which he had been scheduled to appear with his front-running British Saloon Car Championship 3.0 Capri II.

Already a winner at an earlier Thruxton British Championship round, and in the elongated Spa 500Kms, this was a premier-league Ford in period, with good backing from Hammonds Sauce (thus the brown shades of Capri and motor home) and works support. Ford Motorsport ensured all the right hardware was available to favoured drivers – Hewland close-ratio gearbox, 210–220bhp Racing Services V6, sorted Bilstein/Koni race suspension, vented front disc brakes and retention of the ultra-low race weights that prevailed from Ford homologation of the originals.

That Ford support system and homologation lasted into the mid-'80s, which is why the Capri remained a popular race car, just so long as it had the 3-litre Essex engine and associated tweaks. Sadly the 2.8i unit just never had the right ingredients for extracting enough power for any real prospect of success in touring car racing. Also, the opposition were massively armed and alert to Ford techniques. In Europe, we had Rover V8s and Jaguar V12s from TWR, plus BMW Motorsport's M3s along with 5- and 6-series models, never mind the works turbo offerings from Volvo.

For Ford, the arrival of aero-conscious and turbocharged Sierras (Merkur turbo and RS Cosworths) consigned the Capri to racing history. That is, until classic motor racing – particularly at Goodwood, the Nürburgring and Silverstone – brought back the crowds and the private-owner interest, so that Capris, especially the final 3-litre cars, formed a significant proportion of the entries for most prestige events.

Now back to July 1977. Testing the Craft Capri at Silverstone, we used the old GP track on a really hot day, hotter still in the cabin of that amiable machine, with Chris constantly urging more speed at every pit stop. I managed a 1m 53.8s, quite a lot slower than I would have liked, and significantly down on Craft's pace, but still enough to race it respectably at that unique Donington event – which was the only Friday-evening meeting I know of for the British Saloon Car Championship.

By accepting the invitation to take part in this Donington race, I seriously stretched my *Motoring News* editor's patience. My contracted journalistic mission for 1976–77 was to follow and report on the progress of British Leyland's much-hyped campaign with its Broadspeed-developed Jaguar XJ-C – the coupé version of the V12-powered XJ12 saloon – in the European Touring Car Championship. The next bout in that fight against BMW (still using the obsolescent CSL Batmobiles) was taking place that same weekend, so immediately post-Donington I had to drive to the Nürburgring for the highest-profile encounter of the season.

The Donington race went better than expected from my testing pace, as the aces all set off with great determination to Redgate and played bumper cars. That mêlée allowed me

'THE DONINGTON RACE WENT BETTER THAN EXPECTED FROM MY TESTING PACE, AS THE ACES ALL SET OFF WITH GREAT DETERMINATION TO REDGATE AND PLAYED BUMPER CARS. THAT MÊLÉE ALLOWED ME TO CREEP THROUGH AND HOLD AN INDECENTLY HIGH OVERALL POSITION BEFORE THE GORDON SPICE/TOM WALKINSHAW HARD MEN ARRIVED.'

to creep through and hold an indecently high overall position before the Gordon Spice/Tom Walkinshaw hard men arrived to shatter my dream of matching the achievements of fellow journalist Tony Dron in his Broadspeed Triumph Dolomite Sprint. My best lap times were 1m 27.0s in practice and 1m 28.8s in the race, and I finally finished 10th overall, last of the fit Capris. Yet, as a consequence I did pick up a Fiat X1/9 drive in the autumn 750 Motor Club six-hour handicap, an outing that proved a winner for our Gordon Allen Crankshaft team, so not all bad.

I was so knackered keeping to a 6,500rpm limit and hauling the heavy steering through initial understeer (I should have used 7,000rpm, as the faster you went in slow corners, the easier it got to handle) that the prospect of the Nürburgring trek paled. But we did it, my patient and now departed first wife Tricia and I.

It wasn't a Ford at our disposal, but a 1.6-litre Lancia Beta Coupé, as we left Henley-on-Thames at 6.30am on the Saturday, scraped onto a mid-morning Dover–Calais ferry and drove solidly through France and Belgium, reaching the 'Ring just after 4pm. The next day we witnessed a great six-hour race that brought Jag's best result that season – second overall.

Now the report had to be readied for the printers when I returned on Monday, and *Motoring News* in those days featured lengthy word counts that would do credit to most celebrity biographies today. After working deep into the night followed by another 6.30am alarm call, and with my report still not finished, I installed long-suffering Tricia in the travel-stained Lancia's driving seat for a foggy motorway stint across Belgium. After some scary encounters in the fog (including facing a few drivers backing down slip roads to meet us) we just caught another midday ferry and made it back to the *Motoring News* offices in central London. Even then, it wasn't all over, as my copy was really rough [*no change there – Ed*] and it took a lot of sub-editing to prepare the tale for press.

Finally, Tricia having read *The Eagle Has Landed* as she waited outside the offices, we drove back to Henley via Ealing, remembering to pick up our two kids… Glamorous life, that of the motoring journalist, eh?

PROMOTING THE MOST POWERFUL FACTORY RACE CAPRI

The fastest Group 2 race Capris got their fair share of European Ford budget. In February 1974, I took journalists from the Continent and the UK up to Cosworth to view progress of the now 415bhp Group 2 race redevelopment of the Ford Essex V6. Under Ford's GAA engine-type designation lay 24-valve heads and recast cylinder blocks for the RS3100 to battle BMW. Those very special Ford factory blocks were stronger, but not quite beefy enough for the power and revs Cosworth uncovered, as the iron could split under 8,800rpm race duress. That led to a classic comment from one Ford factory driver at Salzburgring: asked why he had retired, he proffered, 'There are now two engines in my Capri!'

Ford-Cosworth did get it right, and the engine was also used in Formula 5000, but factory Capris with that engine were rare – four constructed, some reworks of the 1973 RS2600. Group 2 RS3100s appeared in more German Championship rounds, the South African Springbok series, and there was a factory-supported car for Allan Moffat in Australia during 1976–77.

Hard-bargaining Moffat extracted his Capri from Ford during a 1975 visit to Europe, and it contributed to an Australian Sports Sedan championship title in 1976, with Pan Am Airways sponsorship. That Capri is now superbly preserved in New Zealand, in private ownership.

▼ My 1974 Ford winter PR assignment was to promote the most powerful Ford factory V6 through the depths of the 1973–74 fuel crisis. This is a retro Cosworth 24-valve conversion of what had been a Broadspeed 12-valve Weslake racer, seen at Castle Combe in 2010. *Author's archive*

DAVID BRODIE'S UNIQUE BLACK CAPRI II

Goodwood, 7 August 1974, and time to publicly unveil the boldest black Capri I've ever seen – a sensational-looking machine that was subsequently found worthy of detailed examination by the Ford Design Centre at Dunton, Essex. Since the second week in July, I had been readying official Ford PR text and media invitation lists for the public unveiling of a unique racing take on the Capri II, so it was one of the bigger publicity deals I worked on at the UK company.

This radical project was conceived by David Brodie and executed by Charles Beattie. Technically it was more advanced than the production-based, steel-bodied Capris and Escorts that Ford deployed in period motorsport, as it was destined for a Super Saloon category that enjoyed more freedom of regulations than was permitted in FIA-governed international competition.

▼ The Capri that promised so much, but could not deliver on track. Ford failed to supply competitive power units for this private project from David Brodie. However, some of the styling features were picked up at Ford. *Author*

These 'Superloon' races were dominated by Ford's deadliest commercial rivals, GM-Vauxhall, through the effective yet budget-conscious engineering work of Bill Blydenstein and the flamboyant driving of Gerry Marshall. Their intriguing new opponent was going to be Brodie's wide-track Ford, sporting wildly extended seven-piece detachable body panels over a tubular spaceframe chassis.

Those Capri II cartoon glass-fibre panels and the flyweight chassis allowed an exceptionally low race weight (1,680lb/762kg) allied to many single-seater features, including independent rear suspension based on the ISO-Marlboro Formula 1 car fielded by Frank Williams's then embryonic organisation. Wheels and tyres had an F1 look too, 13in diameter with widths of 11in front and 17in rear. Vented 10in and 10.5in disc brakes were thought generous in 1974, and other hardware, such as the ZF five-speed gearbox and the same company's limited-slip differential encased in a custom aluminium casing, were also impressive on the spec sheet.

The Brodie Capri II initially appeared with a Ford-Weslake V6 (as deployed in 1973 RS2600 Capris) producing an alleged 330bhp, but there was something errant in its componentry and build quality, for it disliked supplying competitive and durable power even with such low race weight. Inevitably, the incoming 3.4-litre GAA Ford Cosworth V6 for the 1974–75 factory Capris was required, but even that 450bhp offering, encased in a large, heavy iron block, wasn't going to deliver the power-to-weight ratios of rival club racing cars, never mind the Blydenstein-Marshall Vauxhall programme, which had produced the mighty 'Baby Bertha' – a Firenza 'Droop Snoot' coupé with 5-litre Holden Repco V8 power – in another cartoon outline of a production car.

Results for the Brodie Capri II were poor, including too many non-finishes. The car was probably more important to Ford for the inspired black body appearance.

In 2022, when Brodie was preparing to unveil both his 1970 reborn club-racing legend 'Run Baby Run' Escort Twin Cam (see pages 143–144) and the 'Brodie Roadie' Escort, he gave me his memories of his Capri II Super Saloon: 'That Capri was a beautiful car. Sure had a lot of hard work in it, like plating all the torsion suspension, adapting a lot of F1 stuff. It handled brilliantly. Ford gave me the engines but they just weren't strong enough to handle Marshall's V8 thing. I should have stuck a small-block V8 in it. I ran it for 1974–75, and shared some of the driving with Jonathan Buncombe, who did a great job with it – but it just didn't have the muscle. That first Weslake wouldn't pull you out of the paddock!'

WORKS CAPRI IIs FOR THE SPA 24 HOURS

Leading Capri privateer Holman Blackburn carried his Capri enthusiasm right from the first 3-litre cars through to his motivational role in the creation of Capri IIs built at Boreham. On 24 July 1974, I watched those cars being finished off, in black ShellSport colours for Tom Walkinshaw and co-driver John Fitzpatrick, and Hermetite livery for Blackburn and Mike Crabtree, prior to their outing in the Spa 24 Hours. I also wrote a low-key press release to support their first public appearance.

When this pair of Capri IIs debuted at Spa, neither finished, though Walkinshaw/Fitzpatrick led before retirement. However, the cars showed promise, even with a comparatively mild 255bhp from a production 3-litre engine. Holman bought two of these cars – the last Boreham-built Capri racers – for the 1975 season, with excellent sponsorship from Hermetite, and a racing link that led to employing Tom 'TWR' Walkinshaw's track talents. For the 1975 European campaign, Tom was paired again with Fitz, and Blackburn remained with Crabtree. They used an enlarged Essex V6 (3.2 litres) and reported 280bhp, or 300bhp after another (unreliable) stretch to 3.4 litres. There were a couple of wins, but also a crash that demolished the lead Walkinshaw/Fitzpatrick example, which was subsequently reconstructed.

I drove the Hermetite car briefly a decade later, courtesy of Brands Hatch, with a long lunch break allowing us to use the GP track for pictures. The images went to what was then *Ford Heritage* magazine, reborn as *Classic Ford* in 1997. It was a stop-start photo session, and what I remember was that there was already controversy over that Capri II's authenticity, following an FIA inspector's report.

Holman was living the expat life in Spain in 2022 when he explained some of the key facts that I didn't previously know about these significant final Boreham-built Capris: 'The two Boreham Capri IIs were bought by Hermetite products at the end of 1974 for £5,500 the pair: I wrote the cheque. They were M-registered with prefix SNO (the black Shell car) and SPU, both built at Halewood. Both were obviously painted in Hermetite colours for 1975. The previously Shell-sponsored 1974 car had a big off at Monza with Tom at the wheel. Peter Ashcroft, who had given me a Capri to run as a Group 1 car, pressured me to put the car he had provided into Coupe de l'Avenir trim to replace the Shell car for the next few races in Europe. I agreed to that, so by the end of '75 we had three Hermetite Coupe de l'Avenir Capri IIs. The only noticeable difference was Tom's seat, which was a Scheel and bloody uncomfortable: it was switched from car to car as needed.

'At the end of '75, we had an invite to Kyalami in South Africa, but the Coupe de l'Avenir spec was 150bhp down on a full 1975 Group 2 car. So we up-specified the brakes to 1976 Group 2 regulations, as we were building the UFO jeans 1976 Group 2 Capri II at the time. Brakes were always abysmal on the Coupe de l'Avenir Capris. We tested the car at Goodwood, and packed up to go and run in Group 2 at Kyalami. Sadly, Hermetite boss Harry Werrell was overruled by his chairman due to the period sporting embargo on South Africa, so that Capri II didn't compete.

'At the end of 1975 one of the three Capris went to Tom, who replaced the wings and brakes, turning it into a Group 1 car. A second went to Jeremy Nightingale, and we kept one: all became Group 1 cars. After I bought Jeremy's car for Chris Keen, the car was short of much of its original mechanicals, but we got it rebuilt in Capri II Hermetite livery, and got the historic passport based on the last specification it had run in period. The FIA hadn't recognised Coupe de l'Avenir rules, and these cars had raced in international events. I did spend a long time checking the chassis number of the Nightingale/Keen car at Essex records office, which verified its FoMoCo origins, and I have copies of those documents.'

In recent years, Holman Blackburn has attended Goodwood historic events, meeting both a replica Hermetite-liveried Capri II racer and the reshelled 1974–75 black 'Shell' original. In 2016, he raced again at Spa-Francorchamps in the classic event with the Shell Capri II in glossy black.

In period, Blackburn's talent for getting sponsorship stretched to Hermetite's acquisition of one of BMW Motorsport's 1976 Group 5 CSL coupés for a season in the World Championship for Makes. Walkinshaw and Fitzpatrick achieved some terrific results in that 3.5-litre flier, topped by a fabulous last-gasp victory in the Silverstone Six Hours, even as Tom fought for Capri wins in the fiercely contested UK domestic series.

The May 2007 issue of *Retro Fords* gave me a short, sharp shock. On the cover was a Hermetite-liveried Capri, but a second glance showed MkIII quad headlamps. I bought the issue (that always hurts as a freelance motoring journalist) and discovered this was the beautifully presented recreation work of Dave Weir.

RACING THE DJ'S CAPRI II

After starting his career in radio, Mike Smith worked with Noel Edmonds in the 1980s, and the two appeared pretty regularly on television, so race-car sponsorship for either of these high-profile DJs turned TV presenters was never a problem. Yet I confess that I had another earlier link to Mike Smith, because his dad, Reg, was in charge of Ford-owned vehicles in the UK, usually disposed of in weekly auctions that I attended with a very savvy member of the Ford management. Most of the Fords I bought for competition or personal use were via those auctions.

Mike was contesting the under-21 award in a 1975 British production-racing series, still run on a price basis, in the class for showroom cost between £1,799–£2,299. The series was sponsored by Smith's then employers at BBC Radio 1. Mike's very quick (claimed 180bhp) Capri II ran in a brown colour scheme backed by Anchor Records. I got the call to race the car while he was busy, and ended up at Snetterton on 19 October 1975 in an autumn championship round with some very good names out to play. These included Gerry Marshall (Triumph Dolomite Sprint), Jon Dooley (Alfa Romeo) and Mike's rival for the under-21 award, Marc Smith (no relation), plus my old Escort nemesis Ivan Dutton (now in a Capri) and Melvyn Hodgson in a very high-spec Capri.

Practice went well, the Smith Ford in strong-enough form to take its temporary driver to third in our price category, albeit spinning *en route*. Outright practice and race honours were firmly the property of a 5.7-litre Chevrolet Camaro V8 trio in the over-£2,300 category.

On-event TLC from Myk Cable and antiques restorer Nick Osborne included fitting an extra wing mirror. This allowed me a view of some really entertaining racing manoeuvres as I got a good start alongside Dutton. Fortunately for me, Dooley went off rallycrossing, leaving me to watch Dutton and Marc Smith in their Capris pulling away as I battled with Hodgson's equally effective run-out first-edition Capri. We finished third amongst the battling Capris, but it was not a brilliant overall result as the Camaros were outright leaders, and Marshall brilliantly took the 2-litre Dolomite to fourth overall, grasping class honours.

It had been an exhilarating experience, with lots of sideways motoring, but I found on re-examining my report and *Autosport*'s 47 years later that I had been totally pre-occupied with our Capri class battles and had ignored the overall leaders! Black mark.

THE FINAL CAPRIS ON TRACK

Although I had spectated plenty of times at the Nürburgring, including the 84-hour Marathon de la Route, I had only driven the shorter but very entertaining *Südschleife*, albeit in my favourite RS2600 Capri. On that occasion, added to the unfamiliarity and fearsome reputation of the track was the fact that Ford had paired me up with former Lotus Formula 1 driver John Miles. Subsequently, John became an inspiration to me for his writing (*Autocar*) and at Lotus Engineering, plus some important work on redeveloping the Capri at Tickford, and with the private development of the enlarged 3.2-litre motor (WM Developments). However, on this occasion I was distinctly overawed, and did a scrappy lap or two before watching a master at work.

Nürburgring return in a 2.8i

The 2.8i injection was debuted to UK media with a German trip that included a session on the full-length Nürburgring (14.2 miles). I'd already understood from our road miles that this was the best showroom Capri yet. The injection engine was appreciably smoother than previous incarnations of the V6 – although with notably less mid-range grunt than the now revered earlier 3-litre Essex – and the simple chassis and braking upgrades were much more effective than

anticipated. These qualities were thanks to a then UK-based SVE (Special Vehicle Engineering) team's practical on-road mileage and parts-bin know-how.

Brief Silverstone encounter

Pure track miles in the last Capri body during the '80s included a 3-litre car prepared by Capri loyalist Holman Blackburn for the 1982 Silverstone Tourist Trophy. Sadly, the clutch plate sheared on the starting grid after we had qualified 22nd of 48 entries.

CC Racing Developments 3.0 Capri at Donington

There was a very wet but privileged outing in chassis CC16, a proper CC Racing Developments third-edition 3.0 Capri, at Donington in 1982. Former Ford dealer and committed loyalist Vince Woodman had already won a Mallory Park race when the car carried his traditional Esso/VMW Motors sponsorship livery, but at the time I had it there were just the VMW and CC constructor reminders of its identity. By then, it performed in a back-up role, with 255bhp from a Neil Brown-prepared engine.

I was able to make much more regular and exhilarating use of over 7,000rpm than I had been allowed in Chris Craft's Capri II at Donington in 1977, despite the deep puddles that bedevilled lower sections of the circuit. This was probably the best-mannered Capri racer I encountered.

Racing swansong

Fittingly, my last formal race was in a final-edition Capri 3-litre. Brands Hatch was the backdrop in May 2000, but I had tested the 1980 machine, owned by Colin Potter, in November 1999 at a Snetterton group outing for assorted classics, from Fiesta to Jaguar XJS. Colin, who had some keen sponsors, asked me to share the 3-litre for a two-driver Classic Saloon Car Club (CSCC) race the following year.

We met again at Silverstone for a pre-event test, and I did 22 laps, the best just below 1m 13s, but that was about two seconds short of the target time. My only diary excuse related that 'rear tyres were shot', which made the slowest and quickest corners such an on-off sideways adventure that I thought the deal may be in danger. We then had a third outing, as a pre-race test at Brands Hatch,

▲ This four-way group track test for *Cars & Car Conversions* magazine became preparation for my last race, in that yellow Capri 3.0 of Colin Potter. Also visible are Escort XR3, Jaguar XJS and Mazda RX7, with the Capri's functional race interior below. *Jakob Ebrey*

➤ Appropriately, my last race was in a classic event, driving a 3-litre Capri at Brands Hatch shared with owner Colin Potter. The car was swift enough to qualify on the front row, but sadly we were handicapped by a small fuel tank and overheating that allowed only a top-six finish. *Author*

and were below the 59s lap record, so Colin nobly stuck with me.

A beautiful 13 May Saturday with 26°C ambient temperature at Brands Hatch Indy circuit allowed me a respectable 57.78s lap in the very yellow Ford, making full use of the V6's elevated 7,200rpm limit. That was enough for a close (0.41s) second spot on the CSCC grid's front row, next to the legendary Graham 'Skid' Scarborough in a similar 3-litre Capri.

Sadly, the good news ended there, as the car didn't have sufficient fuel-tank capacity for a full race distance at a winning pace. The owner started and performed well up front, then we had a bit of chaos at the pit stop and I set off with one shoulder harness adrift. I had to slow to fix that and watched the water temperature escalating to 115°C. The full-house 230bhp JW Developments V6 muttered rebelliously below 6,000rpm in third and fourth, really only having enough fuel for a V4, shuddering at the elevated water temperatures.

We did make it home to finish sixth, but the closing laps were horrible, allowing those we'd led to pass by. As I had set second-fastest lap, by an even closer margin to the winning Scarborough/ Jess Yates crew than during qualifying, it was particularly galling.

Only later did I realise the personal significance of that race. In hindsight, I'd been autocrossing, sprinting, rallying or racing since 1968, so it had been a very fair 32-year motorsport life! Others have gone on longer (stand up Mike Wilds, you know you want to!), but I think they would all have struggled to have had so much low-cost fun!

ROAD AND TRACK SMILES IN THE VERY LAST CAPRI

On 26 June 2008, I was unexpectedly reunited with D194 UVW, Ford of Britain's official last Capri 280, as it attended the Western Group of Motoring Writers media day at Castle Combe. The Western Group hired the challenging Combe circuit annually, and invited public relations people along in a neat twist of usual practice, so that motoring writers ended up serving those who normally delivered them client-biased editorial, hospitality and loan cars. It was very popular, as the PRs got to drive rival brands, so attendance was strong.

The Combe track has quite special memories for me, as I held a lap record at a sniff under 100mph on the pre-chicane layout with the Collins Performance Engineering Sierra Cosworth RS500, but, at the other end of the spectrum, I have also driven a red double-decker bus there… Anyway, back in June 2008, the lunch break saw all the serious folk putting away some high-calorie fare, while four of us lined up an assortment of classics for a parade, which featured a Range Rover and an aged Austin. Ford had brought along the final Capri, and all concerned generously allowed me some laps in it. Well, I hung back for a couple of corners, but the temptation was too strong and I let the 280 have a life above 3,000rpm… I enjoyed the freedom to tackle the quicker corners and ease it through one snappy, sharp right/left chicane, one that asks much of the brakes as you arrive at around 85mph, and then need second for the immediate changes of direction.

The 280 was as much simple fun as ever, but even then it was on its way to the high-value classic status it has today. A bit of a shame, because the appeal was generated around affordable and simple transplanted components, a comparatively slippery aero shape, and low weight. Yes, it needed work if you wanted handling and brakes on a par with more expensive prestige badges, but today the values of classic Capris and Escorts are such that they're often sold at higher prices than many of those supposedly superior motors.

I was once again reunited with D194 UVW in June 2009, at the Ford AVO Owners' Club annual show-and-meet day at aristocratic Stanford Hall, Leicestershire.

Before that Sunday mission, I clocked up a 160-mile day, collecting the Capri from Ford's press garage in West London and whistling down to the Kent/Sussex borders to see the remains of a BMW coupé, which even an impractical optimist like me could see wasn't worth saving at the asking price… At least not then, but today it might well be economically viable.

I went on to Cowdray Polo Park and took plenty of pictures there, as I was pretty sure this would be the last time I drove a V6 Capri of any kind. I must say the Capri was a pleasure to drive even 23 years after it left Cologne. I particularly liked the quick and light steering, and the ability to accelerate swiftly enough to blend with the traffic and then to sustain dual-carriageway pace with plenty in hand. The eyeball ventilation on a perfect English summer's day was appreciated, and the seat contours/driving position were as excellent as ever, but I prefer cloth-inlay seats to leather for hard work behind the wheel.

The Ford AVO Club day was one of the best I attended, partly because of the weather, but also because I got to chat with my mechanic heroes from my Boreham days again – Bill Meade and Mick Jones. Peter Osborne and I did another photo session with the Capri 280, this time posing it in company with the club's Ford Advanced Vehicles factory-linked cars, including a superb RS3100 in white, TBL 636M.

▲ I got lucky in 2009 with a visit to a Ford AVO Owners' Club meeting in the last Capri, posed here alongside Vernon Witney's shimmering RS3100. *Author*

➤ In July 2010, I attended a reunion of SVE personnel in – where else? – Essex. Hosted by founder/manager Rod Mansfield, former employees and some sub-contractors assembled, along with these fabled examples of SVE endeavour. The Capri 2.8i in the foreground is my old friend –D194 UVW. *Author*

CHAPTER 9

FIESTA

Ford arrives late at the front-wheel-drive party

Considering that front-wheel-drive automobiles date back to before World War II, and that the 1959 Mini and European transverse, front-engine machines from Fiat and Renault were selling by the million in the later '60s, Ford of Europe's successful 1976 entry with the Fiesta was tardy, to say the least. Yes, Ford of Germany had fielded front-drive Taunus-badged 12M designs in 1962, but they weren't massive sellers and were succeeded by a more appealing rear-drive Taunus generation. The German market was dominated by rear-drive designs until the VW Beetle/K70 made way for a new Volkswagen smash hit, the Golf, which first appeared in March 1974.

Ford of Europe was well prepared for a design to sell across the Continent, and that awkward British island community. Britain adopted the Fiesta as its best-selling car on a rota with the 1980-onwards breed of front-drive Escort and the subsequent 1998-onwards Focus. Front-drive Fiesta, Escort and Focus allowed Ford of Britain to dominate its home market from 1977 to 2020. By October 2022, Ford had sold more than 20 million Fiestas globally, over 4.8 million just in the UK. The Fiesta headed the British market in 1996–98 and again in 2010–20, making it Britain's best-ever-selling badge by 2021. Yet in autumn 2022, Ford announced that its switch to electrification and away from saloons and straightforward hatchbacks would see the Fiesta dropped from production.

THE EVOLUTION OF A BESTSELLER

The Fiesta moniker was applied to seven generations of what began as Ford's smallest road-car offering, and throughout that time the various models retained front-wheel drive, although the gearboxes became increasingly sophisticated beyond the original four-speed manual. The final Fiestas routinely offered five and six manual ratios, plus a seven-speed Powershift-

branded semi-automatic. The later smaller Fords delivered stronger, safety-conscious body and cabin engineering, low-drag aerodynamics, plus extensive digital-screen information. Additionally, there were obvious gains in overall dimensions, inside and out, plus added weight, electronic sophistication and safety features, which gradually moved the Fiesta away from its original range-starter simplicity and comparatively diminutive dimensions.

In the beginning...

Specific Fiesta engine production began in 1975, in Valencia, Spain, and that new factory made complete cars from 1976, although right-hand-drive UK didn't receive its unique market needs until 1977. The Blue Oval had learned much from earlier rivals, so that the Fiesta was a cleanly styled hatchback, boxy by today's raked standards, but straightforward in layout and carrying durable oily bits.

The base 957cc Fiesta in 1976 had but 40bhp and 47lb ft of torque to drag 1,576lb (715kg) from rest to 60mph in less than 20sec, topping out at around 80mph. An alternative higher-compression version of the 957cc four-cylinder offered a claimed 85mph and 0–60mph in less than 17sec. Those stats are just for reference against some of the later performance Fiestas, since the point was competitive, low-octane fuel economy with 40mpg potential and good carrying capacity for modest size.

Right from the start, the company offered a Ghia version, which seems counter-intuitive in an economy hatchback, but it foretold that Ford's business instinct indicated more derivatives to follow. Within months there was an 1,117cc stretch of the simple crossflow (no overhead camshafts here) with 52bhp DIN and almost 60lb ft of torque. So everyday flexibility was markedly improved, and an ability to thrash into the later 80mph zones was conferred, along with 15sec 0–60mph times.

Naturally, those family friends weren't the models that enthused the media and motorsport enthusiasts, although it's worth remembering that all models had a slick four-speed transaxle gearbox, and the handling wasn't just failsafe, but deftly enjoyable, with a ride quality Ford forgot about when the company engineered the following

▲ The author tests some early RS options on a Fiesta at Brands Hatch in the late 1970s, but Ford UK Competitions Department personnel didn't get really serious about the baby Ford until they had to run it at the Monte Carlo Rally in 1979. *Ford*

front-drive Escort. More significant was the 1977 introduction of the punchier 1,298cc motor, adapted from the straightforward Escort 1300 rear-drive predecessors for a front-drive role with 66bhp and 69lb ft of torque. The engine debuted in the Ghia, but would also feature in the Fiesta Supersport, allowing 95mph to be approached, with 0–60mph in the 12sec bracket. These models, under various European badges, affordably squatted below the subsequent XR2 performers.

The path to higher-performance Fiestas was predicted by two distinctly separate 1.6-litre derivatives. Forgotten today, the American version packed an emissions-constricted 66bhp and sold over 300,000 copies, but wasn't replaced when the Fiesta was facelifted in 1983. The Fiesta went back on sale in the US in 2008, but was dropped from four significant markets, including America, from 2017.

Equally overlooked by all but hard-core Blue Oval motorsport followers was a rather more radical 150–160bhp version of the crossflow engine implanted into a chassis featuring replacement front suspension arms and vented disc brakes. That Group 2 Fiesta was dubbed 1600S in the entry list for the 1979 Monte Carlo Rally. The pair of factory entries, driven by Ari Vatanen and Roger Clark, finished 10th and 13th respectively – an acceptable result on such low power – but Ford's international factory rallying path into the '80s wouldn't follow either Fiesta or any other front-drive route.

▼ The Fiesta was available in 1,117cc S trim in the UK from its launch in 1977, although the 'sporty' version was more about matt-black bumpers, door handles and window surrounds and a stripe along its waistline than any performance enhancements. *Ford*

First-generation Fiesta one-make racing

Yet there was still a route to European commercial success for Ford, but it wasn't as obvious as simply developing a 1.6-litre Fiesta for the showroom. It started on the track, almost two years before you could buy a 1.6-litre first-generation XR2. Even the Escort XR3 beat its smaller brother to the marketplace by almost a year.

Boreham staff, from shop floor to adjacent utilitarian offices, knew that the rear-drive Escort's glory days were fading. Ford competition and marketing personnel also realised they didn't want to have the older rear-drive 1.6-litre Escort Sport or Mexico out in any high-profile 'celebrity' or one-make series. The job of such one-make series was to publicise the hardware spectators and viewers could buy in the showroom.

Through their 1978-onwards development work to create an internationally eligible rally Fiesta, the UK competition staff had a head start on reworking the company's baby into an affordable one-make competitor. Boreham had investigated using the 200bhp-plus potential of the 16-valve Cosworth for its Fiesta programmes. That Fiesta-Cosworth package was abandoned for premier-league motorsport, Boreham unconvinced of a front-drive competitor's ability to deliver outright victories, owing to the inherently unsuitable handling and traction qualities for the multiple hostile surfaces that are integral to international rallies.

Boreham then pursued a complex and belatedly ditched turbocharged rear-drive version of the front-drive Escort platform (the RS1700T), which showed great testing pace. Sadly, the car's potential was countered by an insuperably slow transition to the production status needed to gain international homologation. While Ford vacillated, Audi's 4x4 turbocharged quattro became a fearsome international rallying reality.

Back on the Fiesta front, Boreham selected for motorsport development the versatile 1.6-litre pushrod Kent crossflow engine, familiar from the 1967 Cortina through skinny Formula Ford single-seater race cars to showroom rear-drive generations of the Capri and Escort. Some basic engineering development had already been carried out, since this 1,598cc unit had been chosen for the American Fiesta. So, that

MONACO LAUNCH FOR UK FIESTA INTRODUCTION

The belated UK debut of the Fiesta was a personal luxury, as 12–13 January 1977 saw couples invited to Monaco and the fabulously splendid *Hôtel de Paris* in Casino Square for the launch of the right-hand-drive models. Along with many other UK media folk, we punted the debutant around majestically expensive local streets – and over the inevitable twisty trio of Haute, Moyenne and Bas Corniches.

The first RHD Fiestas in sub-1-litre and 1.1-litre forms were a stark contrast to rear-drive rally Escort thrills and the 1979 Group 2 Monte Carlo Rally Fiestas built by Ford Motorsport. However, showroom Fiestas were impressive entrants in an already competitive pan-European small-capacity hatchback market. I liked the failsafe handling and clean design. Even in Ghia guise, the seating and cabin furniture were far from supportive or luxurious, but by this time much of the well-established opposition was stark to the point where you could cut unwary flesh on an unfinished door trim.

As a trip, the UK Fiesta's launch was characterised by getting lost in both Monaco and Nice, but there was also massive compensation in a lunch-time visit to the Mas d'Artigny hotel, and not just due to food of memorable quality, or the privilege of views over some of the most expensive housing in Europe, out to the Mediterranean. For me, the killer touch – based on this hotel's regular media use by companies from Porsche and BMW to Renault and British Leyland – was the provision of individual swimming pools on the terrace outside your state-of-the-art room. At the time, that hotel room, bathroom and sitting area offered more floor space than my seriously mortgaged maisonette in West London!

I did swerve into some deviations from the road test route, to the Monte Carlo Rally territory that I remembered from my 1973 Ford PR work with the works Escort team. Back then, I supported a trio of superbly driven, powerful but sadly short-of-traction RS1600s. I also savoured an illicit run over one special stage where I was allowed to drive a rival Datsun 240Z factory recce machine that had seen a lot of bent-panel action, but remained a fine drive despite its long snout.

That sort of Monégasque entertaining largesse, with intent to influence, has faded these days. Today, the media rough it using economy air travel and hotels, filing frequent stories to meet their daily and nightly online deadlines. Mind you, the generation before my time used to get protracted round-the-world trips on the pretext of looking at the leading British manufacturers' overseas factories!

Oh, and there were gifts given on trips. They could extend to leather coats and such expensive cameras or portable typewriters that I'd leave them at customs, rather than pay the import duties…

ubiquitous and tough 1.6-litre unit became the focus for a Fiesta one-make racer, detuned and less radically modified than those Monte Carlo Fiesta 1600S examples.

XR2: the first performance-orientated production Fiesta

The two strands of development intertwined – one was the Competitions Department and its one-make racer, the other the parent company's desire for a Fiesta performance flagship. So, Special Vehicle Engineering (SVE) within the Ford R&D centre at Dunton, Essex, was assigned the task of developing a showroom product under the XR2 brand.

SVE flourished, and created many more swift Fords, from two versions of the XR2 to Sierra Cosworths. The department began with a nucleus of just 10 personnel back in the 1980–81 era of the Capri 2.8i and the first XR2, but grew to over 50 hand-picked staff deployed by 1985, always under the management of former race driver and established Ford engineering executive Rod Mansfield.

The showroom XR2 by SVE debuted in December 1981, but by then there had been two race seasons of the Fiesta 1.6-litre Championship, co-sponsored by retailers Debenhams and Ford Rallye Sport. The XR2 branding wasn't initially used in the race series, as the showroom item was still under development.

Before these race and sporting road-car

derivatives of the Fiesta appeared, standard versions of the lower-power models were converted, both for track use and as fast road cars. Back in 1977, Ford at Boreham developed a range of parts – from cosmetic to power increases – to be marketed under 'X-pack' branding – an image that had proved charismatic and profitable on the rear-drive Capri and Escort.

The competitions people took the routes we've already mentioned to compete internationally in 1979, but unofficial work on a 1.6-litre Fiesta continued outside of SVE's labours on what became the XR2. When I competed in a 1980 Fiesta Championship round, my car was one such unofficial 1.6 Fiesta, carrying the small rectangular headlamps of the regular production model but flanked by extended X-pack wheel arches, housing the optional Rallyesport/X-pack four-spoke 13in alloys, wearing then-generous 185/60 Pirelli P6 tyres. The engine had been wrestled to 90bhp, and the company's press garage had made up some dual side stripes in light blue to complement a darker shade of blue body.

That 1980 road car was a pretty good prediction of the XR2, although naturally the race cars differed in stripped-out weight, safety gear – including roll cage and fire extinguisher – and the installation of circuit-friendly Dunlop 190 VR-specification tyres hiding vented front disc brakes. They needed that braking power and friendly rubber, as the engines veered towards Formula Ford Blueprint standards, generating over 100bhp.

▲ ➤ Original Fiesta XR2s are rare 40 years since their debut, but this clean example, courtesy of KGF Classics, Peterborough, records the original specification inside (two-spoke wheel, primitive radio/cassette player and striped cloth) and out. The 'pepper pot'-style alloy wheels had proved a hit on the Capri 2.8i. *Allen Patch/KGF Classic Cars*

▲ Appreciation of first- and second-generation XR2 Fiestas has grown in the 2020s. This excellent example of the MkII was nestled amongst a forest of later 'fast Fords' at a winter Haynes Breakfast Club meeting. *Author*

Meanwhile, back at SVE, the major Capri success was followed by the showroom XR2 production programme. Hardware from Ford's giant European parts bin was assembled for assessment, testing and legal validation – a rather longer process than knocking together 20 converted race cars, or the unofficial Fiesta 1.6 performers that I experienced.

Least surprising was the deployment of the faithful 1.6-litre cast-iron crossflow engine again. SVE specified the five-bearing crankshaft and ancillaries from the production US version, but cleverly allied them to European breathing apparatus via the 1600 GT camshaft profile of rear-drive Cortina, Escort and Capri legends, drawing the aged but proven eight-valve cylinder head from the same established heroes. Yet the Weber 32/34 carburettor came from the Escort XR3, working with unique inlet and cast-iron exhaust manifolds, leading to a multiple silencer layout and a modest single tailpipe.

That rag-bag exhaust trimmed power figures, registering 84bhp at 5,500rpm. Yet there was useful mid-range pulling power in this low-weight machine – 91lb ft of torque by an accessible 2,800rpm. You could also flog it just past 6,500rpm, which meant you could steam up to 90mph in third before changing into the modestly overdriven top gear. Utilising 1.6 Escort/original XR3 four-speed ratios and a final-drive taken from the 1.6 Escort (rather than XR3), meant you could maintain Britain's legal 70mph limit at 3,800rpm.

The real trick in the engine bay was that the motor was mounted slightly lower than in mainstream production: only 15mm (0.59in) lower, but it was enough to allow the drive-shafts a straight run, and did the handling no harm either, as subsequent generations of British touring-car racers would prove. For British roads and track adventures, the XR2's handling packed a fun factor comparable to the original Mini: agile and failsafe until the driver ran out of basic talent.

The XR2 components were straightforward, but SVE reworked the Fiesta front MacPherson-strut layout, including increased caster angles and lowered tie-bar mountings that co-operated with lowered spring platforms. Unique damper settings were specified, and no front anti-roll bar was fitted, although the beam back axle featured a Fiesta Supersport bar, along with that model's springs. The reset ride height reduced the overall height at the roofline by 1.7in (43.2mm).

The brakes were uprated by raiding the Escort 1.6/XR3 parts box again, so the XR2 wore 240mm (9.45in) vented discs up front. The rears were much smaller, as per front-drive practice – 178mm (7in) drums – but SVE did take the trouble to employ enlarged hydraulic slave cylinders and 8in Escort servo assistance. As with subsequent Escort RS1600i and XR3i models, the retardation problems centred on a slack brake pedal, legacy of a crossover shaft in the move from left- to right-hand steering.

Volkswagen suffered the same RHD export failing on the first Golf GTI.

External clues to the unique XR2 identity were led by larger round headlamps, supported by auxiliary driving lamps sprouting in front of the grille. Also useful were unique 'Pepper Pot' alloy wheels, echoing the earlier 2.8i Capri in style. These bore generous 6in rims, with routine 13in diameters, and effective 185/60 quality radials. There were also nods to aerodynamics, with a chin spoiler and modest rear blade, both in matt black, as were the bumpers and restrained wheel-arch extensions, which increased overall width by nearly an inch over lower-power Fiestas.

Decoratively, there was a unique side stripe that had seen an amazing amount of team effort from Ford's UK styling studio, carrying the XR2 logo, repeated on the hatchback panel in capital letters in case anyone missed the message. The interior was Spartan, even by the standards of the last century: a 140mph speedometer and 7,000rpm tachometer framed combined fuel-tank contents and water-temperature dials, whilst the steering wheel was an uninspiring two-spoke plastic job, but only three turns were needed lock-to-lock, and the diameter was a manageable 15in.

Today the first-edition XR2 – like its Escort XR3 brother – is becoming a rare sight on British roads, and even at well-attended classic-car meets. Remember, it was conscientiously created and rumbustious fun. Definitely underrated compared with the RS-branded Escorts.

Seven generations of Fiesta

Fiesta is long-lived as a Ford nameplate, in continuous use from 1976 to 2023, and appearing on seven distinct generations of the car. This book is not intended to provide a history of every Ford model, so I have focused on those that stand out in my hands-on experience. Here is a brief outline of the Fiesta's loyal service to the Blue Oval over six decades. As already explained, an

▲ ➤ The first ST variant of the Fiesta (2004–08) brought 148bhp from 2 litres and the capability to dip below 8sec for 0–62mph. The interior was cheered by a couple of supportive sporting seats. I had such an ST with 'racing' stripes for a local-newspaper road test in period. *Ford/Author*

initial and unsophisticated first-generation Fiesta spanned 1976–83, but from 1981 included the XR2 performance model, developed from competition and American experience, with the largest Fiesta petrol engine then offered – 1.6 litres.

A more-rounded successor appeared for 1983, and ran until 1989, able to accommodate diesel and ex-Escort CVH engines, plus a five-speed gearbox. Again, an XR2 variant was offered, this time with the CVH engine mated to a five-speed transaxle.

Because of its performance range-starter pricing, the carburettor XR2s of two generations and two engine types clocked up more impressive sales numbers than their big-brother Escort XR3s. From 1986 to 1989, XR2 sales numbers increased from nearly 16 per cent to just over 20 per cent of all UK Fiesta sales, spiking at 20.25 per cent in 1987, with just over 31,000 units sold that year.

For 1989–95, the third-generation Fiesta was updated to provide a more competitive mainstream product, with the addition of a five-door option, and the faithful old solid-beam rear axle was ditched in favour of a semi-independent torsion-beam layout. Now the XR2 was able to adopt an 'i' for injection, a first for a Fiesta. Initially, the XR2i depended on the eight-valve CVH four-cylinder engine, but it took on the 16-valve Zetec motor in 1992, before passing into history two years later, overshadowed by more able rivals and the stinging insurance rates of the period.

The first Fiesta production RS model arrived in the third generation. The RS Turbo also had the CVH motor in modified ex-Escort RS Turbo trim, although it deployed Ford EEC-IV injection and management, plus a smaller Garrett turbo, to provide 133bhp. That model was later replaced by the Fiesta RS1800 in 16-valve form (130bhp), using the Zeta – renamed Zetec – engine shared with the contemporary Escort XR3i in a lower spec (105bhp).

For the 1995–2002 fourth-generation Fiesta, the rounded front-end look was back, although the

▲ Displayed at Goodwood Festival of Speed in 2010, but never produced, this was Ford's vision for an RS Fiesta, based on the fifth-generation design. The last Fiesta to carry the RS badge in production was the rare 1992–94 Fiesta RS1800. *Author*

◀ Before and after the advent of ST models, the Fiesta could be modified with Mountune power, as with the Zetec S sixth-generation example shown here in 2010. That tuner option still existed in 2023, usually with official Ford blessing on a number of ST lines. *Author*

DRIVEN

Road cars

RHD press Fiestas (three examples) Monaco–Nice, 1977, Ghia and basic, 950 and 1117cc. **1.3 Fiesta Sport/Ghia** Castle Combe and local roads, 1977, Ford press event. **Fiesta 1.6 X-Pack** 1980, LVX 909V; road prediction of XR2 at 90bhp; fun but broke down (distributor). **1.3 Supersport** Hardknott Pass, Cumbria, 1980, UK TV ad filming **Production XR2** Road test, 1982; 84bhp, 105mph, 0–60mph in 9.5 secs, 28mpg test average; cost £5,500. **Fiesta RS Turbo (two examples)** Ford Lommel test track, Belgium, 1989; Tony Dron/author pre-production assessments; early photography with G413 XNO, 1990. **Fiesta RS1800** L157 PVX versus private rear-drive RS1800 registered SWN 817S, published *Ford Heritage* December 1994. **Fiesta ST 2004–5** Road test for Surrey newspaper, 2005, EO54 OTP; economical, efficient via rationed 148bhp fun factor. **Fiesta ST-2 (two examples)** Tested at Llandow track and Bridgend local roads, April 2013; impressive!

Competition cars

1.6 race series Fiesta Oulton Park, RS Rallye Sport Challenge, 1980; 10th practice/9th race. **1.6 Fiesta XR2 (two examples)** Pembrey Beach, TV commercial, 1981 **Mike Smith/Ilford racing XR2** Silverstone track test, published April 1984 *Performance Car*; standard engine and brakes previewed showroom spec.

basic structure was similar to its predecessor and its platform was designed to be shared by Mazda, appearing as the 121. Visibly more Escort-like than ever, the Fiesta proved able to outsell its stable-mate to head the UK sales charts during the late '90s. No obvious performance RS or XR models were offered, but the engine bay benefitted from the arrival of small-capacity Zetec motors and the final departure of the Kent OHV engine.

The fifth-rendition (2002–08) Fiestas had the front-end look of the 1998 Ford Focus mated to a boxier and more capacious interior than its predecessor, and the cabin was distinctly better equipped. Overall improvements included standard anti-lock braking (not the slowcoach system of earlier Escorts) and passenger airbags. No RS performance variants were offered, although a Concept Fiesta RS was displayed at one point. A performance model was provided by the 2004–08 ST (Sports Technology or Super Touring in *Fordese*), packing nearly 150bhp from 2 litres, slashing 0–62mph times to a claimed 7.9sec with decent economy. I did have one to test for a local paper and thought it efficient rather than exhilarating.

In this era, the mid-range pulling power and fuel economy of the diesel reigned under state indifference to diesel emissions. So Ford performers weren't powered by just the 1.6- and 2-litre Zetec petrol engines, as they were accompanied by vigorous turbo diesels. These were often quicker than their petrol counterparts in real-world low- and mid-range acceleration,

▼ Ford lost no opportunity to promote Fiesta ST and its competition pedigree. This photograph showing the ST, taken at Brands Hatch, emphasises the company's rallycross involvement, not just with radical Fiestas, but back to the RS200. *Ford*

utilising TDCi branding and sharing Peugeot-Citroën (PSA) technology.

For 2008–17, the sixth-generation Fiesta, based on the versatile B-platform created for worldwide Ford factory assembly, became a global product, and more recognisable as today's sleeker hatch. During the production run, in 2013, Ford gave this Fiesta its latest corporate front grille style, which it called 'trapezoidal'; the media tended to assert that this new look had been lifted retrospectively from Ford's ownership of Aston Martin! More importantly, the ST variant became a credible halo model, one with award-winning handling and enough horsepower (180–197bhp) to entertain.

Previewed and manufactured in Germany, the final seventh evolution of the Fiesta was not sold in North or South America, or most of Asia, yet the Fiesta's B-platform is an important underpinning to the Puma crossover, Fiesta Active and the now-preferred premium Ford plate, Vignale. ST models were retained, now rated at 197bhp (200PS), with 140mph potential.

In July 2023, the final two Fiesta models rolled off the production line in Cologne, bringing 47 years of continuous Fiesta production to an end.

Footnote *The current (at the time of writing) Puma, based on the Fiesta platform, does offer an ST nameplate and power hikes, plus the prestige of having won the 2022 Monte Carlo Rally, as prepared and entered by Cumbria's M-Sport organisation, founded and run by former Ford Escort top competitor Malcolm Wilson. The Monte-winning M-Sport Puma Rally1 was driven by 48-year-old nine-times World Champion Sébastien Loeb.*

1.6-LITRE FIESTAS, AND A PREDICTION OF THE XR2

Fortunately for your boredom threshold, and my self-respect, my most memorable Fiesta outings had nothing to do with corporate PR – indeed many of my driving experiences upset my former Ford employers considerably. One of my most informative and enterprising outings came in May 1980, beginning at the scene of so many personal Ford adventures – the bleak, brick-walled, single-storey Ford UK press garage.

Heavily defended against the surrounding West London break-in artists, it always contained the cars Ford most wanted to see speeding away profitably from its dealerships… Oh, and often some other extraordinary stock, from show cars to long-term loans for VIPs (mostly male sports stars), and usually the pair of GT40s that the company has always owned.

On this occasion, that barbed-wire-protected and glass-shards-walled compound contained the company's prediction of what would become the XR2, now allowed out to see what the media made of a 1.6-litre Fiesta's performance and sales potential. I had a busy schedule for the converted Fiesta, which still had miniscule, rectangular (rather than round, as would appear on the XR2) production headlamps. The creative and resourceful press garage had even made up its own double striping for the sides, whilst the RS sales operation had provided four-spoke wheels, and the mainstream options list delivered a flip-top glass sunroof.

I sped off on what had become a regular run for me since 1968, West London to Ford's Boreham Competitions Centre. In period, the Fiesta's 1.6 litres seemed particularly willing, as they were housed in a compact, low-weight body, and I subsequently discovered that LVX 909V offered more horses than the production 84bhp. It also stopped pretty well when the vented front discs responded to an elongated pedal prod, jinking with conviction along the more interesting lanes leading to Boreham. I had six interviews lined up for my first Capri book, so the Fiesta was kept busy buzzing between Boreham, Dunton Research Centre and the administrative HQ at Brentwood (now sold off).

Another Fiesta 1.6 featured in my busy weekend on Ford business, as on 25 May 1980 I buzzed up to Oulton Park with my teenage son to drive in an early round of what was dubbed the Debenhams RS Rallye Sport Challenge. Now with over 100bhp to play with in the stripped-out race version (still wearing the tiny rectangular lamps of the production line), there was a lot of action to be had. Most of the heart-fluttering moments were fun, but they were sometimes lurid, as the brake pedal was even sloppier in action than that on the prophetic road 1.6. When the brakes did answer my increasingly desperate calls, they could lock a wheel, or more...

My diary recorded: 'Horrible practice, 10th of 15 after a huge spin at Old Hall.' That wasn't too bad, as subsequent Ford, TWR-Rover and BMW star Steve Soper and the other entries were all packed in a 3.5sec span. The diary helpfully added: 'Great race, really enjoyed myself. Steered round a large accident and finished ninth, lapping a second quicker than practice.'

However, that 1980 day of contrasting fortunes wasn't finished with the Waltons. That subtle dark-blue XR2 prediction stuttered to a halt on the A40 dual-carriageway just outside Witney, Oxfordshire. I think it is the only time I have been stranded on the road in a Ford – back in the day they were simple enough for even a mechanical moron like me to fix. This time, a failed distributor proved beyond my limited abilities, and the special Fiesta was ferried by the breakdown people to a secure facility nearby to await retrieval.

My son was great company during this episode, but my wife wasn't so amused about her 80-mile round trip to retrieve us.

CALL TO LAKE DISTRICT SUPERSPORT LABOUR

An August 1980 Sunday-to-Monday money-spinning call saw me travel to England's most mountainous – and some say most beautiful – area: the Cumbrian Lake District. The advertising agency for Ford UK was Ogilvy & Mather, as was the case throughout my generously rewarded 1980–84 PR/advertising episodes. Yet filming for what would be a 30-second national TV ad was sub-contracted to Rees Harrington, a friendly but strictly timed and disciplined agency whose emblem was an elephant on roller skates. I never did ask why…

Back then, I had my first second-hand BMW, an early 528 offering boxy-outline head-space and plenty of pace. I think that Bimmer must have seduced some key personnel, so I got rehired for UK shoots. That car was a warm refuge on cold filming days and nights, and pretty comfortable too, with a blue-velour plush interior, and it was therefore favoured for accommodating and transporting agency personnel, who found it hidden away from mass-parking areas on Ford-funded shoots.

For the long haul (by UK standards) up from southern Britain to the Lakes, I was asked to pick up an Essex-based personal assistant for the agency producer. We headed for the Pennington Arms Hotel, Ravenglass, part of a lovely sandy resort. No drama on the motorway run up country, but a convivial evening with the assembled

➤ An advertisement derived from the 1980 commercial film we made in the Lake District with the Fiesta Supersport. Only a 1.3-litre engine, but it was a good, agile step towards 1.6-litre pleasures. *Author's archive*

15-strong crew wasn't the best preparation for the routine 4am morning call to filming arms.

That beachside hotel location was a stark contrast to the steep gradients in store for a choice of two Fiesta Supersports to ascend. Titled *Dawn on the Mountain*, the Peter Sugden/Terry Holben script was designed to show off sportier Fiesta features. That meant the 1.3-litre motor, uprated suspension, alloy wheels with low-profile tyres, Ghia-specification seats, spoiler and extra driving lights. The official voiceover strap-line was: 'The new Fiesta Supersport, it shortens straights and straightens bends.' In hindsight, I was intrigued from a writer's viewpoint to see that the briefing headline for the advertisement's main message was 'Man and Machine in Perfect Harmony', which the agency later adapted for multiple Sierra ads.

I hadn't seen a Fiesta Supersport, never mind sat in one, as this hint of the XR using mass-production parts was late to the UK market. Yet the 1.3-litre/66bhp powerplant already existed under other Fiesta badges, so it was hardly a daunting proposition. Actually, these early Fiestas did develop hiccups when 'Action' was shouted (yes, they really did that, with a 1–2–3 countdown), and we were constantly swapping between the pair of them for clean runs up and down the challenging Hardknott Pass's gradients and hairpins.

When it was going, the low-power Fiesta was heroic, with deft handling and good traction during very wet and slippery conditions: I was amazed how much was achieved to crew satisfaction in those early-morning hours. The next, and final, day was a sharp personal lesson in filming life. We sat on the mountainside for three hours waiting for the weather to clear. We tried to get some suitable 'Brrm, Brrm' sounds recorded, but the wind ruined that too. My diary says we didn't start work until midday. Even then: 'Both the bulky movie-camera rigs sticking out from the car, and the usual false number plates, made me nervous on narrow, crowded public roads.'

Oh, get over yourself Walton, most of the driving was fun – and the money was great!

SANDY SEASIDE WELSH XR-RATED EXCITEMENT!

I thought life in late October 1981 was exciting enough, in a week that began with a race in a spinning, but recovered, rapid Autocavan Golf on Sunday, followed by a dash from forested Wales for a pre-RAC Rally date with Hannu Mikkola and Michèle Mouton – Audi's finest quattro drivers that year – then back to London for the Motorfair show, where I had been contracted to provide the show-catalogue editorial.

That opened a week when I drove a pair of Fiesta XR2s on a Welsh beach for another TV ad. As my BMW was now understandably unwelcome in the presence of Ford executives, I collected a Ford demonstrator Granada injection, which proved a totally underrated value-for-money flagship. The biggest European Ford saloon packed the smooth 2.8i V6 powertrain, plenty of storage space (human and luggage) and a comfy long-distance cabin.

Since the Ogilvy & Mather agency people had fought so hard for me to keep this lucrative driving job, and as I also became more involved in sub-contracted Ford business, my boxy old 528 was eventually sold for a new Capri 2.8i in the early '80s.

So the Granada fled down the M4, just past Llanelli to find 21 personnel accommodated at two hotels, and nine vehicles, including a 'hero' primary car and back-up XR2, both in red. Other vehicles included the special-build Caspar camera car that would pick up the dune- and beach-based action at Pembrey Burrows, along with an Alan Mann Aviation-supplied helicopter crewed by pilot Andrew von Preussen and cameraman Peter Allwork.

The first day's sunlit filming was just the sort of unbelievable fun where you cannot believe anyone would pay you to enjoy such automotive play with no public or police worries. The idea was to slalom the XR2s, in downhill-skiing style or like water-skiers performing arcs behind a speedboat.

We did have a couple of mishaps. I twice concentrated too much attention on the camera car, and speared the agile Ford into the sea, but it chugged out under its own power, apparently unaffected. The other setback wasn't my fault: the heavyweight camera car got stuck in a deceptive beach trench and retrieval demanded some skilled manoeuvring, but we were all finished by 5pm, albeit that would have been a full 12-hour day for most of the crew.

The planned final day of shooting became a wet-weather survival course, particularly because the object is always to try and match filmed weather conditions from day to day, otherwise viewers suddenly zoom from sunlit paradise to the reality of British downpours. There were some bad-tempered attempts to get the helicopter sequences synched in with the beach-level action, and we had to stay over another night in Pembrey to expensively add an extra day to the schedule.

That final day wasn't a sunny paradise either, and one of the XRs had hiccups from its previous soaking, so walkie-talkie messaging with the helicopter became ever more strained. We finally completed the shooting in one-and-a-half hours. Despite my experience of other Ford commercials located from the South of France to San Francisco, this Welsh adventure covering those stimulating XR2 beach sessions stood out in my mind as the most enjoyable film driving job I was lucky enough to experience.

I also came away deeply impressed by the XR2's handling and value-for-money character. These Fiestas really deserve the same accolades and strong prices that classic Mini Coopers receive so regularly today.

QUICKER '90s RS FIESTAS

These aren't particularly warm or rewarding memories, but there were lessons for me and Ford Motor Co in a market sector that matured to feature some quality hot-hatchback acts from many mainstream manufacturers, particularly Peugeot, Renault and Volkswagen. In short, Ford illustrated that you can swap nameplates too far and retain uncompetitive engineering too long, even if such hardware profitably pleases the financial departments.

Here were some RS Fords that didn't thrive long in the showroom, and haven't found a warmly appreciative modern-classic audience. The two Fiestas in question are the 1989–92 RS Turbo and the visually similar badge sacrilege (to the hardcore RS Ford community) of the Fiesta RS1800, which had only its nameplate in common with the legendary 1975–80 rear-drive Escort rally warrior.

The better drive was undoubtedly the normally aspirated Fiesta RS1800, but chronologically we start with the RS Turbo and the knowledge that the preceding and sometime contemporary 1989–93 XR2i of around 105bhp, and of changing specification, was the model on which Ford based Fiesta RS derivatives. Both RS types yielded around 130bhp, but offered totally different driving characteristics because two engine types were used – the turbocharged 1.6 CVH for the RS Turbo and the larger 16-valve Zetec 1.8 low-emissions offering (from summer 1992) for the RS1800.

Fiesta RS Turbo

First, the earlier RS Turbo. On 31 August 1989, I travelled to Ford's Belgian test-track complex at Lommel, via Maastricht, Netherlands, a trip not aided by forgetting my passport. However, it was a Ford 'internal flight' and I was able to pass through borders painlessly with a small expeditionary squad, including my contemporary, the late Tony Dron. On arrival, the Special Vehicle Engineering boss, Rod Mansfield, had arranged for a pair of RS Turbo production predictions to be on hand, these carrying Escort RS engine technology in 132bhp guise. Also available for comparison drives were the XR2i starting point, plus the benchmark Peugeot 205 1.9 GTI and the VW Golf GTI 16-valve.

The test sessions and facilities were extensive for that intense Belgian driving day. Despite some earlier and utterly disgraceful off-track Brands Hatch test-day moments in the Janspeed 1.9 Peugeot that Kieth Odor drove (a production-racing season before we shared a Sierra Cosworth), I saucily recommended learning lessons from Peugeot. A cheeky Pug GTI was always an entertaining drive in terms of steering and handling characteristics.

◀ ▼ This clean 1991 example of the Fiesta RS Turbo was sold by KGF Classic Cars over 20 years after it was built. The turbocharged 1.6-litre CVH engine produced 133bhp. *KGF Classic Cars*

However, Ford corporate-engineering DNA drilled in the need to retain safety standards in mass production to cater for a wide range of driving abilities. This was relevant in that I liked the Peugeot lift-off oversteer/tail-out stance on an eased throttle at high cornering forces – a strictly forbidden handling trait within Ford, especially since the original Escort RS Turbo proved such a wriggler in power on/off situations.

The result for the Fiesta RS Turbo suspension set-up was a bias towards safer understeer, but it still had some of the irritations of front-drive restive steering under power. I felt embarrassed about my contribution, which muddied what should have been a more entertaining and still-affordable drive.

Footnote Evo *magazine's editorial was warmer than I expected when looking back on the Fiesta RS Turbo in October 2018. Its verdict: 'For good old-fashioned fun, the RS Turbo is hard to beat. It might not have been able to rival the Peugeot 205 or Renault Clio for poise and engagement, but there's no shortage of exhilarating, hang-on-for-dear-life fun. Lag is limited by the standards of the time, but compared to modern machinery the Fiesta's "pause... whoosh" approach is addictive. The physicality extends to the driving experience, with the unassisted steering proving to be a device you wrestle rather than guiding the Ford through a corner – it's particularly bicep-busting under hard acceleration, where torque-steer adds an extra direction-changing option. Yet like most ageing performance cars, there's a real sense of connection. While the thudding low-speed ride gives way to surprising suppleness and control, although we're talking relatively here – it's no modern hot-hatch, and it was still not in the same league as its Gallic contemporaries either.'*

In 2022 it was reckoned that fewer than 130 Fiesta RS Turbos roamed British roads.

Fiesta RS1800

The Fiesta RS1800 was a much more straightforward journalism job and a much more civilised car than the preceding Fiesta RS Turbo, partly because the body had been strengthened during a facelift of the 1989 third generation, one that contributed a half-step towards the fourth-generation Fiestas of the mid-'90s.

Gone was the reworked Escort CVH turbo motor. Tightening emission regulations ensured that Ford implanted its Zetec 1.8-litre engine with its in-house EEC-IV electronic management for this second deployment of the RS1800 badge, but it retained a decent 130bhp. So, performance remained similar to Fiesta RS Turbo with around 130mph potential coupled to 0–60mph in plus or minus a couple of tenths adjacent to 8sec.

If anything, the earlier turbo had the edge in a straight line, particularly in the real world, as it delivered maximum torque (134lb ft) at just 2,500rpm, whereas the later Zetec RS1800 needed 4,500rpm to harvest a more modest 120lb ft. The later RS1800 Fiesta also incorporated power steering, which damped out much of the pull and fight of the Escort and Fiesta Turbo deviants.

My late 1994 *Ford Heritage*-published verdict stated: 'The Fiesta RS1800 makes a competent modern hatchback, but it is a rotten recipient of the RS1800 legend.' It was a cynical exploitation of a legendary Ford badge, but both the RS1800 and Fiesta RS Turbo are rare alternatives to the usual modern-classic hatches. Today, RS Fiesta choices might serve the hardcore Ford fan well as everyday transport, all at a much more affordable price than the older Fords – or the usual Peugeot/VW or Renault Clio 'modern classics'.

◀ ▼ In 1994, with *Ford Heritage* magazine I created an ancient-and-modern Ford series. Regular contributor Dave Wigmore sets up at Goodwood racecourse with two RS1800-badged Fords: the 1992–94 front-drive Fiesta and the legendary – and extremely rare – road version of the 115bhp MkII Escort RS1800. *Author*

RETURN TO AFFORDABLE FUN WITH FIESTA ST

St George's Day (23 April) in 2013 saw me back on a Welsh Fiesta mission, this time to drive the ST variant on track and local roads – and again it was a thoroughly enjoyable experience. For the first time in many years, I felt a faster Ford again offered financially accessible and desirable driving pleasure. The four-cylinder turbo hatchback was sold at less than £18,000 in both ST-1 and ST-2 trim levels: unlike the contemporary Focus, this ST was only offered as a three-door body.

Here's an edited version of how I felt about that ST for the *www.fromthedrivingseat.com* website in April 2013.

'Faster Fords were always meant to fill the financial gaps that prestige badges could not reach, because the Blue Oval could make more on the "pile 'em high, sell 'em cheap" philosophy. The first whale-tail Sierra Cosworth did the job best, putting £10,000 between it and the far-better-quality, but not much quicker, Porsche 944 Turbo. More recently, the media loved the Focus RS types, but you had to say they were expensively front-drive versus more impressive 4x4 Mitsubishi and Subaru rally spin-offs. [A four-wheel-drive Focus RS subsequently appeared in 2016.]

'Surprisingly, the Ford One plan to get some global sense into the product basics through platform sharing seems to have done us all a favour, with Ford performers slamming back into the affordable world with the much-misunderstood ST branding. Those initials gave Jeremy Clarkson cheap laughs on BBC TV's *Top Gear*, but originally stood for Super Touring [and more recently Sports Technology], a saloon-car racing category Ford had dominated from BTCC to world titles with Mondeo in the Andy Rouse and Prodrive eras.

'Now ST stands for a comeback for the affordable Ford performer. The public seems to be going for the format, with record Focus ST sales in Europe versus the benchmark Golf GTI, and warm reviews in the USA for the big-brother Focus with 250bhp. Then came the Fiesta ST in 2013 with an advertised power of almost 200bhp for the USA and 182PS in Europe. In fact, both motors are the same Ecoboost 1.6-litre unit, and give the same test figures, it's just that the Americans quote the ultimate 15-second overboost stats, which also reveal a maximum of 290Nm versus a steady-state 240Nm torque.

'The results are pretty spectacular for the Fiesta, with nearly 140mph available and a 0–62mph time dipping below 7sec. In other words, the straight-line clout nearly matches the revered BMW M3 back in the non-turbo four-cylinder day. But the 1,163kg turbocharged Fiesta – latest in a line that stretches back to the '80s Fiesta XR2 generations – is about a lot more than performance figures.

▼ The 2013 rendition of the Fiesta ST with 1.6-litre turbo power readily delivered 180–200bhp, with overboost. Seen at Llandow, these cars displayed tidy track manners to match their 140mph potential. *Author*

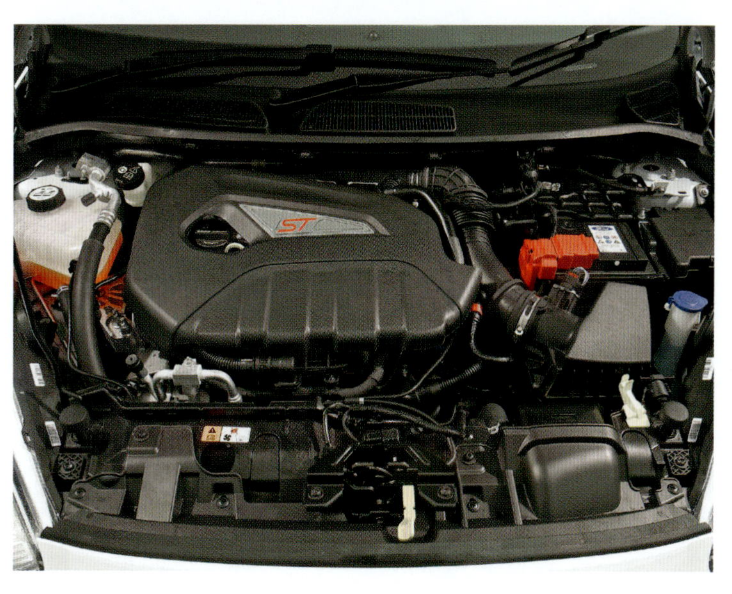

1981 XR2 VERSUS 2013 FIESTA ST

2013 ST figures in brackets

Engine

Configuration: Four-cylinder, eight-valve transverse uprate of 1.6-litre 'Kent' crossflow OHV (Sigma four-cylinder, 1.6-litre, 16-valve DOHC) **Fuel system:** Carburettor (direct fuel-injection, turbocharger) **Maximum power:** 84bhp at 5,500rpm (180–197bhp at 5,700rpm; 15sec over boost) **Maximum torque:** 93lb ft at 2,800rpm (177lb ft minimum at 1,600–5,000rpm)

Transmission

Type: Four-speed manual (six-speed manual)

Suspension

Front: SVE-modified MacPherson strut (Ford Performance MacPherson struts) **Rear:** Beam axle (torsion bar)

Wheels and tyres

Wheels: 6Jx13in (7.5x17in) **Tyres:** 185/60 HR (205/40 ZR17)

Body

Three-door hatchback

Dimensions

Wheelbase: 2,286mm/90in (2,489mm/98in)

Weight

Kerb weight: 800kg (1,163kg)

Performance

0–60mph: 10.0sec (6.7sec) **Max speed:** 102mph (138mph)
Fuel consumption: 28.4mpg (31mpg)

'We drove it on the compact, but challenging, Llandow track in South Wales and surrounding public roads. We were promised a special driving experience and the small Ford delivered. A central Ford Team RS sorts these ST variants out and does a fabulous job within strong production-line constraints.

'The ST packs effective 278mm (11in) front and 253mm (10in) rear discs. Not only did they stand up to tight track use, repeatedly asked to clip 90mph fourth-gear pace to 30mph slot-in second for a very tight 'Bus Stop' chicane, but they didn't protest and were supported by an equally efficient and exploitable chassis.

'Wheels are 17in, shod with 205/40 Bridgestone or Pirelli showroom tyres. Ford states that the ST carries a sports-suspension variation with a 15mm (0.6in) chop in ride height, but the devilish fun factor is in the detail. Unique springs and dampers of course, plus the steering knuckle and a stiffened rear-torsion-beam action, transform already good handling into race-track-friendly adjustability.

'The usual traction control (ESC) software is reworked alongside eTVC (electronic Torque Vectoring Control) – gentle brake intervention to tame the Fiesta in hard cornering, responding to (electrically assisted) steering, braking and acceleration inputs to revive the old MG motto: "Safety Fast". In this application, you can also switch traction control off, which makes the Fiesta even faster though quicker (third-gear-plus) bends, but you'd be better off with the ESC system on in wet and slower going.

'On the road, you can feel that creditable torque tweaking the steering during hard acceleration over bumps: containable, but shows front-drive isn't the purist solution to road running. Safer, yes. More predictable, yes. Mass-production and passenger friendly. Tick those vital 21st-century boxes, but don't expect perfect steering behaviour under full-power duress.

'The interior is uprated well, with giant Recaros and a leather-rim wheel, but the cabin layout and instrumentation is also the usual current Ford fashion for a fussy mess of dials, readouts and controls. The faux-gridded foot pedals are a step back into a fake world that's not worthy of this honest road and circuit ace.

'As a road car, the ST offers a worst official fuel-consumption figure of 35.8mpg (Urban) and we got 27.6 without track use. Prices were very competitive at launch, the three-door-only line (Focus ST caters for five-door hatch or estate) stepped in £1,000 breaks for ST-1 and ST-2 trim, both under £18,000 at the time of writing.

'Recommended as a practical road car – the mid-range pulling power is as accessible as some diesels – and for light track sessions.'

SECOND-GENERATION XR2 SILVERSTONE TRACK PREVIEW SHOWS PROMISE

Even judged by my 55-year motoring media experiences, obtaining a spring 1984 personal encounter with the second-generation XR2 was unusual, and illustrated how desperately competitive the specialist motoring-magazine business had become in the '80s.

I had an offer from former Ford Competitions employee Stuart McCrudden to drive the second-generation XR2, produced at Valencia in Spain from February 1984. This was before right-hand-drive versions hit British showrooms, as UK sales of that second edition suffered while European homologation was sorted. Months of delays occurred as SVE struggled to reset front-suspension tolerances and tame the unique exhaust system, which developed unacceptably high internal temperatures prior to an expensive material change within.

Our 1984 XR2 scoop had the looks of a racing hatchback, for it was sponsored by Ilford Film for then BBC Radio One DJ Mike Smith to race. Yet this stripped interior, single-seat Fiesta relied on an unmodified production CVH ex-Escort engine, brakes and transmission. The venue was also initially unpromising, all right-hand turns on the then triangular layout of Silverstone Club Circuit. Yet, 38 years ago, track operators could be flexible, so I was allowed to probe the likely public-road reflexes of the latest XR, and have the extensive photographic session completed, on the hinterlands of the GP track, including Hangar Straight.

Since nobody was more competitive at getting previews of potentially popular motor cars than Jesse Crosse, editor of *Performance Car*, I offered it to Jesse as part of my monthly freelance contributions. This revamped magazine wore a sophisticated format, born out of a flamboyant *Hot Car* predecessor and destined to fight for a splinter of the profitable prestige and performance magazine market with similar newcomer *Fast Lane*, frequently packing a passing stab at the established polished format of *Car*. In 2023, only *Car* survives of those monthly magazines.

To business… The big statistical benefits of the second evolution of the XR2 were a 12bhp bonus via the 96bhp Escort XR3 overhead-camshaft engine, a widened front track, and a move from four- to five-speed transaxle. I was able to feel those performance and stability benefits with the extended track session, but remained unimpressed with the soggy production brakes.

Customers waited over six months for their ordered CVH-powered XR2s, whereas the press gained an early UK supply. Yet the media had begun to test the smallest Ford performer alongside the Peugeot 205 GTI, and that led to some Ford orders defecting to the available and well-reviewed Peugeot!

I also had to wait for the delayed arrival of this new XR2 in the showroom before judging the interior, as I sat in a mono-seat metallic cell, which lacked even a functional rev counter at Silverstone. When I did get to try the XR2 on the road, I found the chief benefits, besides enhanced performance, were reduced interior noise levels (particularly at a motorway 70mph, just over 3,000rpm) and better-contoured seating. I didn't like the rubbery two-spoke steering wheel, but the price was kept competitive, beneath £6,000, whereas most of the opposition, including the 1.6-litre Peugeot GTI and Golf GTI were over that £6,000 barrier.

I couldn't argue with the performance benefits the reworked XR2 recorded in independent tests. It delivered a 5–8mph bonus on top speed, to hover around 110mph, and a second was sliced from the 0–60mph runs, flickering around the 9sec bracket. That tough CVH motor and five-speed gearbox also brought an overall fuel consumption bonus in my experience, later XR2s recording 31mpg versus 28.4mpg for the first edition.

Despite these strong contemporary benefits of the later XR2 – never mind the XR2i – today I would try and buy the first XR2 for its character and fun factor. It's a philosophy I apply to old and new classics, from Frogeye Sprite to Audi TT: often the original has more long-term financial value and character than subsequent factory updates or restyles.

CHAPTER 10

ESCORT

FRONT-WHEEL DRIVE/FOUR-WHEEL DRIVE

Ford pursues the front-drive hatchback herd

The long-running front-wheel-drive versions of the Escort, produced from 1980 to 2002, including commercials and US models, covered four generations that English-language media have unofficially dubbed MkIII to MkVI. I have used that identification here, although Ford never used it officially, and neither did Ford recognise MkI and MkII for earlier rear-drive incarnations. To confuse us further, there were production 4x4 Escort derivatives in the later run, including a redevelopment of the 1992–94 RS2000 front-drive

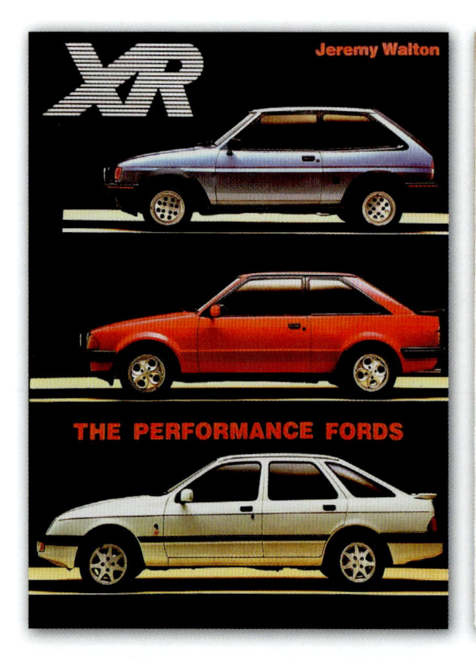

◄ My author contributions to the 1980s boom in XR-branded Fords highlighted two generations of Fiesta and XR3/XR3i Escorts, plus the XR4i/XR4x4 in 1985. The specific XR3/XR3i Escort *The Enthusiast's Companion* followed a year later from the same publisher, but had multiple authors. *Author's archive*

▲ I was fortunate to have several working links with BBC TV/Radio 1 disc jockey and presenter Mike Smith, which included this Snetterton track test of the Escort RS Turbo that Mike used successfully and forcefully in production saloon car categories. *Motorsport Images*

models to carry four-wheel drive in 1995–96, the final RS2000 4x4s morphing into the MkVI body with unchanged engines.

Famously, there was a successful Escort RS Cosworth showroom and competition low-production three-door, which is covered in Chapter 14, as it was built from a Sierra RS Cosworth 4x4 base, wearing Escort identity – partly for technical convenience and because the Sierra would be making way for the Mondeo in 1992, so Ford no longer needed to promote the Sierra.

PRODUCTION AND SALES HISTORY

We had better start by defining the periods of the four widely used Mark (Mk) designations applied to front-wheel-drive Escorts.

The first front-drive Escort: 1980–86 (MkIII)

The first front-wheel-drive Escort – the MkIII – was produced from 1980 to 1986 and was available in many body styles, including three- and five-door hatchbacks, a four-door saloon (dubbed Orion), plus light commercial vans and five-door estate layouts. SVE in the UK and Karmann in Germany co-created an official convertible MkIII – a first for the Escort name – which utilised XR3i power but (initially) not the XR badge.

On the performance front, the XR3 became widely available in 1981–82, and the German-motorsport-department-created RS1600i came to Britain at about the same time as the UK's SVE department had created the later XR3i, which spanned the years 1982–86 in Britain.

The RS1600i had a limited motorsport homologation production life until mid-summer 1983: a hasty RHD conversion had to be hands-on engineered in the UK by Boreham. The RS1600i's motorsport homologation numbers (minimum of 5,000) meant that it lasted only into late 1983 for UK sales. Although the RS1600i became an

⮝ Superbly preserved, this XR3 seen at a Haynes Breakfast Club was smarter than my 1981 staff car. Mine was exploited harshly over 25,000 miles in 1981–83, including international rally coverage. *Author*

instant collectible for RS Ford fans, the factory's fastest performer in its first front-drive three-door hatchback line was the 132bhp Escort RS Turbo, on sale in 1984–85.

Refreshed style, same principles: 1986–90 (MkIV)

The 1986–90 second take on the front-wheel-drive Escort – 'MkIV' for our purposes – was also a commercial hit in the UK and, like its predecessor, offered a wide choice of coachwork from van to pumped-up 'Grand Luxury' Orions. On the performance front, the XR3i and RS Turbo models continued, the XR3i adopting Ford's EEC-IV fuel injection for an official 107bhp, as the '90s and tighter emissions regulations beckoned. The convertible Escort was now offered with 1.6-litre power-unit choices, including the XR3i's 105bhp, but adopted XR3i branding when fitted with that powertrain.

Becoming dated: 1990–96 (MkV)

The 'MkV' designation greeted 1990 with a full range of body styles, but became uncompetitive against the opposition, especially in terms of the car's unremarkable looks and the by-now aged and harsh CVH engine options. Ford did have a 16-valve Zetec motor in development, but that wasn't available for the fifth edition until 1992, when it improved performance and emissions for most mass-production Escorts, though the 16-valve layout and requirement for higher rpm wasn't a recipe for better performance stats in real-world conditions. Because of delays in the Zetec programme, Ford initially had no Escort MkV sports model, save for an underwhelming Escort S carrying the '80s CVH powertrain, spoilers and an alternative 'sports' suspension.

The MkV XR3i line had to wait until 1992 for Zetecs of 105bhp or 130bhp, but its time was up. Whereas XR3i sales had peaked at 25,046 units a year in 1985, the 105bhp model recorded just 1,500 in 1992, and even the attraction of more power failed to entice many punters that year, because the 130bhp version grasped only 1,288 sales.

However, the XR3/XR3i branding was significantly profitable for Ford of Britain during the 1980–92 span, with some 167,971 registered

⮜ Ford in Germany created the first competition-homologated front-drive Escort. Based on the XR3, but with fuel injection and a five-speed gearbox, the RS1600i was the first truly collectible front-drive Escort. *Ford*

in the UK according to my unpublished 1992–93 research, the XR3/XR3i lines accounting for 10.3 per cent of all Escort sales of 1981–89.

By contrast, Ford RS enthusiasts welcomed autumn 1991 for a very competent range filler in the form of the RS2000, which was powered by a specially redeveloped 16-valve version of Ford's tall 2-litre i4 motor. Rated at 150bhp, it should have been the '90s commercial saviour of the Escort's hard-won performance and motorsport reputation. Even with the backing of a one-make race series, it just didn't grab the intended audience. It was sold until 1994, and spawned a 4x4 successor, but by that time the MkV had dated in the European sales race, and the unmodified i4 couldn't generate competitive performance in a heavier safety update of the MkV body, with all-wheel-drive hardware. The Escort to do that 4x4 commercial and motorsport job was the 1992 RS Cosworth, which overshadowed all routine production Escort contemporaries, as we shall see in Chapter 14.

Desperately seeking Focus: 1996–98 (MkVI)

The sixth and final generation of European Escorts was notable for special run-out editions and the use of the heavier (by around 102lb/46kg) body reinforcement of the MkV under a modestly facelifted external style. It was a better-engineered and more competent machine than its predecessors, but it was a million miles below the benchmark its Focus successor would set from September 1998.

On the performance front, the RS2000 continued to attract some custom in front-drive rather than 4x4 format, but the Escort Cosworth remained in its original MkV cladding, so there was no halo performer to seduce those with driving ambition into the final Escort generation.

Perhaps the desperate marketing ploy of badging a few special-edition Escorts in the mid-to-late-'90s with the prestigious Mexico and GTi badges, the latter worthily worn by VW Golf and Peugeot 205 hatchbacks, said it all. There was no pride left in the pre-Focus Escort name… Sad.

We can now put performance front-drive Escorts firmly towards the more expensive end of the post-'80s classic-car market, but which models do I remember now? In my 'Driven' notes we have a long list, but not all remain memorable for the right reasons…

◀ Diana, Princess of Wales became one of the most significant Ford Escort users, covering nearly 7,000 miles before this Escort was returned to Ford. The specially commissioned black RS Turbo reappeared in public with 24,961 miles recorded and its original registration, C462 FHK. It was auctioned for £722,500, including fees, in August 2022. *Silverstone Auctions*

The Mazda Escorts and the final European-built survivors

Entering the '90s, the front-drive Escort's star quality in Britain faded, following a European decline. For 1991, in America and many other non-European mass markets, Escorts became based on Mazda foundations, Ford having acquired a significant ownership stake in the Japanese company. Ford built its Mazda stockholding through the '90s and these Mazda-based Escorts eventually totally slashed links with European engineering, but the Escort also faded from its British bestselling small-car status as the Focus came over the horizon in 1998. The last American-built Mazda Escorts were manufactured in 2001, while the last sixth-edition European Escorts gasped their last in 2000. However, the legendary and toughly effective Escort van staggered into 2002, preceding the advent of the Transit Connect. I experienced a CVH-hauled Escort van regularly in 1992–94, complete with a large sardine adorning its side panels – a noisy but effective transporter.

Footnote *The Escort name lived on into the 21st century overseas, notably from 2015 in China. This Escort was unrecognisable to Europeans, as it had a saloon car's three-box body based on second-generation Focus underpinnings.*

ESCORT MkIII 4x4 RALLYCROSS CARS: THE LINK TO '80s ESCORT SPORTING SUCCESS?

For Ford's competition aspirations, the '80s became a troubled decade. The legends of '70s international rallying victories gained with rear-drive Escorts, and a 1979 world title to end that decade, faded. So did the circuit heroics of the German Capris, which had been brutally curtailed by the 1973–74 fuel crisis, with no alternative mainstream programme accessible, as the Brits wanted all the cash for their Escort antics. Cologne supported the amazing Zakspeed turbo transformations of the Escort and Capri through glorious '80s success against BMW-Schnitzer's blazing turbo opposition, but the style-conscious Capri was expiring in mainland European markets.

For both British and German competition departments, the burning question was: 'What next?'

The back story

As detailed in the preceding Fiesta chapter, Ford UK's Boreham department panted to make front-wheel drive work for international rallying. Ford Motorsport even entered the Monte Carlo Rally, for all to see that this wasn't the magic bullet for world-class success. Yes, SAAB had made it work two decades earlier, but that was in a low-power era, where front-drive steering and traction were undisturbed by less than 100bhp.

Ford management backed two primary programmes for the '80s. The first was a front-engine, rear-wheel-drive Escort RS1700T (dressed in stretched MkIII front-wheel-drive panels) with turbocharged Cosworth BD-series power. The second was a sports-racing car for the new Group C World Endurance Championship, the C100, which packed a 3.9-litre Ford-Cosworth DFL.

DRIVEN

Road cars

MkIII Escort XR3 (two examples) Staff Car *Motoring News* to 26,003 miles, HYM 220W, April 1981–Sept 1983, 29.9mpg; plus test XR3 from Ford, 1980. **MkIII turbo Escorts (two examples)** Both carburettor turbo conversions, 1981; Janspeed four-speed, no intercooler; Ford Boreham's JNO 20Y '1.6L', 125bhp, five-speed, 23mpg, £1,000 RS kit. **MkIII RS1600i/XR3i (two examples)** *Motor Sport* test, 1983; I preferred XR3i; similar performance, 8.6sec v 8.7sec for 0–60mph, both 116mph max, XR3i recorded 30.5mpg v RS1600i 28.3mpg. **MkIII Escort XR3i (two examples)** Staff car at *Motoring News*, 1983–86, UGM 833Y, 7,940 to 36,600 miles, 31.5mpg, bought second-hand for £6,130 and sold for £4,075; also test for 1984 XR book, XR3i demo, A417 UEV. **MkIII Escort RS Turbo** Full road and track test for *Performance Car*, 1984, B932 AAR, recorded 126.8mph, 0–60mph in 8.2 sec, 24.6mpg, £9,250 (+ £470 Custom Pack). **MkIV Escort RS Turbo (two examples)** UK Suffolk press debut, summer 1986 and road test, D771 PVW; I preferred first edition; 1987 CCC feature with Mike Spence Ltd modified version, D331 XGU, 177bhp, 230lb ft torque. **MkIV Escort XR3i (two examples)** Finnish press debut, LHD, A-2277, underwhelmed; test week and 1986 UK rally cover, D396 RVW, 30mpg, not for me. **MkIV Escort 1.6L Estate** Owned from new, 1986–88, 90bhp CVH; covered 30,000 miles without drama. **MkIV Escort XR3i** Turbo Technics, 130/160bhp adjustable cockpit boost, 1988, E77 UDL; excellent. **MkIV Cabriolet XR3i** Road test week, 1988, E907 CVW, 105bhp, power top, £12,000; good rework. **MkV RS2000 (two examples)** Solo drive, Cologne factory and Rhine Valley DIY route, 1991; plus long-term test, 1991, K37 FMC, a lightly facelifted MkVa, appeared on the cover of my book *Sporting Fords: Volume 5* published by MRP in 1994. **MkV XR3i (two examples)** Ford UK press drives, March 1992, 105bhp and 130bhp; unremarkable. **MkVI RS2000** UK Road test week, April 1994, M605 WOO, 0–60mph in 8.2sec, 129mph. **MkV Escort RS2000 4x4** Aug 3–11 1994, L467 RHK, competence and weight up (1,240kg), 0–60 acceleration 0.4sec slower

The C100 showed early pole-position pace, but was never reliable enough to do more than win at German national level, and didn't complete the planned full international schedule in 1982.

Courtesy of Boreham motorsports manager Peter Ashcroft, I saw the C100 race at the Nürburgring. Our chief excitement on that trip was a burst tyre on landing the Ford-owned jet back in Essex – but it was clear that the C100 couldn't deliver the high-level results that Ford required. As with the earlier F3L and subsequent GT70, only handsome American budgets could have pushed through sufficient development to obtain results on the GT40 scale.

The British RS1700T was a sad story too, although that turbo powertrain did help push subsequent RS200 rally-car development through more rapidly, and with good reliability. The beefier appearance of a MkIII Escort on steroids looked set to give all Escort marketing a transferable 'halo' effect. A tough and then fashionable all-white exterior was backed up by some competitive testing results with a range of world-class rally drivers. Yet delivering the mandated production run of 200, to qualify for international event homologation, proved protracted and insurmountable via Ford's European production facilities.

Both C100 and Escort RS1700T were dumped when Stuart Turner resumed control of European Ford Motorsport. Stuart chose the purpose-built Group B rally path for the factory's major effort, via the unique RS200 two-seat coupé, the subject of Chapter 12. In the background, Ford and Cosworth were concocting another master plan for the Sierra in touring-car competition, which would provide exceptional results.

Here, I want to tell the personal tale of two Ford-based rallycross Escorts, which used trick British turbocharged 4x4 engineering that I have always thought could have delivered winning results in the World Rally Championship, but became restricted to European and British titles in less-regulated rallycross. Just three such cars were built and the big difference between the original Martin Schanche version and John

than front-drive (8.6sec v 8.2sec), 129mph, 25.2 urban mpg. **MkVI Escort Cabriolet** Driven to Le Mans anniversary of GT40 winning 1–2–3 in 1966, 13–17 June 1996, GEV 599N, very yellow RS body kit; inevitable Gendarmerie adventures. **MkVI Escort Estate** Used to retrieve stepdaughter from St Tropez, August 1996, N30 GHJ, 115bhp petrol, RS body kit; added 10-day holiday, over 2,000 miles logged and much baggage moved.

Competition cars

MkIII Escort 1.6 Ghia Unregistered pre-production, UK TV advert, Lommel test track, Belgium, 7–11 July 1980. **MkIII Longman RS1600i** Donington six-lap test for November 1983 CCC publication, Datapost, Group A, UK class champion, 155–160bhp, 7,700rpm, 130+mph. **MkIII Gartrac rear-drive (two examples)** Goodwood test, 1983 for January 1984 CCC publication, SRD 395Y, 2.8i road spec, 160bhp V6; Manns Garage/Nick Oatway racer, 3.4-litre Cosworth-Ford GAA V6, 430bhp, 8,000rpm, 154mph. **MkIII 4x4 rallycross** Three tests from 1984–87, for *Motor/Autocar/CCC*, John Welch, UK 1986 rallycross champion, Xtrac 4x4, Zakspeed-derived Cosworth BD, 560bhp turbocharged four-cylinder, 0–60mph in 2.7sec. **MkIII Escort RS turbo (two examples)** 1985–87; *Motoring News* test, Snetterton Willhire 24 Hours winner, Abbott/Smith, cockpit-adjustable boost; Karl Jones/Asquith Motorsport Uniroyal Champion, Mallory Park, 165bhp. **MkIII Escort Zetec** Test at Oulton Park, 17 March 1992, Lloyd Brothers of Stoke 2.0 Zetec-powered club-racing project; quick! **MkV RS2000** One-off drive on 22 March 1992, Brands Hatch, RS2000 Escort RS Series, Ford-owned J723 WOO, 6th overall. **MkV XR3i Lee Allen race** 1993–94 Ford national series champion, raced at Mallory Park, 10 October 1994, published Feb 1995 CCC; practice 4th, race 3rd overall; HT Racing 180bhp CVH. **MkIII 4x4 rallycross** Tested at Bruntingthorpe track and published March 1999 CCC, Euro 1984 Champion, Martin Schanche/Mike Endean, restored to original condition by Xtrac founder; 470bhp, 135mph at 9,000rpm, 0–57mph in 2.2sec.

ESCORT FILMING, AND A WOLFF IN PILOT'S CLOTHING...

The Escort's controversial debut on the UK market came during September 1980, but prior to that I had a profitable July job driving the launch flagship model, a 1.6 Ghia, at Ford's Lommel test-track complex in Belgium. That was for a national television advertisement, and the snag was that I needed all the advertised and harshly delivered 78bhp at 5,800rpm to pull along three actors from a then-popular *Dr Finlay* BBC TV series – all aboard to show what a plucky little luxury item the new Ford had become in its fresh format and an ingenious take on hatchback style with that vestigial boot extension.

Here are some Escort memories from those exotic advertising shoots, for I had been driving European Fords in advertisements since 1979. Some were simple stand-in overnight/day jobs for actors who either didn't drive or were rated unsafe. Bigger-budget outings could take up to 14 days overseas and involve hiring a helicopter for some action sequences (unique images are much more accessible now thanks to cheaper and more-agile camera drones). My helicopter connection was that driving pre-production or freshly debuted automotive products for major manufacturers meant that those with the biggest wallets would regularly hire an ex-Vietnam 'chopper' pilot to cover those inevitable dramatic, lonely-road/mountain-track sequences. Our regular helicopter jockey, Marc Wolff, looked the Hollywood part, in sleek leather jacket and Aviator shades: I swear somebody from Central Casting gave him his name.

In reality, Wolff was a helpful guy, not a Top Gun cliché. Born in Chicago, he spent much of his childhood in Hackensack, New Jersey, before arriving in Britain in 1972 after military service in Vietnam and Germany. As of 1987, he took on dual US/British nationality and accrued a stack of flying and filming qualifications, regularly taking responsibility for aerial or second-unit direction and stunt coordination. Wolff earned *The Guardian* newspaper's 'Alternative Oscars' award for organising and filming the jaw-dropping footage used in the *Cliffhanger* movie, where stuntman Simon Crane zip-lines from the back of a DC9 to a Lockheed business jet, 15,000ft above the Colorado Rockies. Wolff's career bloomed beyond 160 TV commercials; his headline credits cover over 170 feature films and include biggies like James Bond (*Skyfall* and *Casino Royale*), plus some episodes of *X-Men* and *Harry Potter*. He also flew the iconic helicopter sequence used at the 2012 Olympic Games opening ceremony in London, featuring James Bond and 'The Queen'.

So Marc Harold Wolff was a benchmark hero,

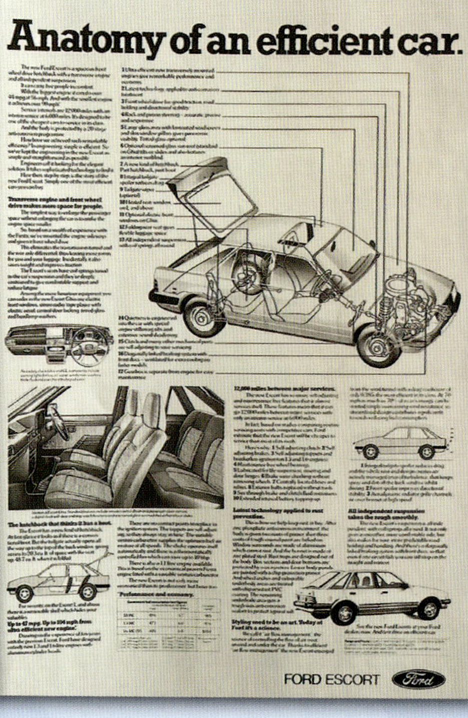

These are the Ford UK-based agency 'Simple is efficient' print advertisements from Ogilvy & Mather, created by Terry Holben (art) and Peter Sugden (text). These accompanied the TV campaign I worked on at Lommel test track. The silver Escort Ghia, with glass sunroof and British plates, looks like the car we used with three actors as cargo. *Peter Sugden Collection*

enviable beyond reason. What were my bit parts in his star-spangled life? Marc, an even braver camera operator, and I, had each other's lives at stake on that Lommel Escort advertising stint. Wolff directed, perching his helicopter a few feet above the banked test track's upper Armco barrier, ready to lift off when collision seemed probable. I aimed the loaded five-door at the side-mounted cameraman, travelling at 70mph with all three grumbling actors aboard.

We did three takes for a director who felt we could shave a bit closer before Marc's threshing departure. I admit it: pointlessly, I ducked down within the wheezing Escort Ghia on the final approach of each pass and become pitifully grateful for Marc's prompt precision. The sight through the hazy glass sunroof of the helicopter's underbelly and cameraman's undamaged legs, safely ascending, was a welcome relief.

For four years, Marc and I were just distant sub-contractors with the common aim of producing a safe spectacle. I would do half-a-dozen domestic and overseas advertising sessions a year. Wolff was only booked for the prestige, big-budget filmed sequences. Away from my small world, Marc racked up missions over deserts, mountains and open water belonging to more than 60 nations.

There was another Wolff thrill before I ceased that profitable, but often boring, precision-driving role for TV ads. We did a prolonged job in Italy, with a dull but spacious saloon version of the Escort that Ford badged Orion in the UK. I drove it around Rome's Villa Borghese Gardens at suitably pedestrian speeds, doubling for actors. We then headed south for some Wolff action, but there was little for us to do.

Marc offered me a lift back along the Italian coast in the rented helicopter, which he would leave in Genoa while I caught a commercial flight home. I was ecstatic to leave the rest of the unit, as we made our dusty departure from the last location, knowing this would be a memorable flight. That trip back from filming in southern Italy, all along the usually sandy coastline to Genoa, was airborne bliss. Marc dropped me off outside the gates of the flight terminal for my return to the UK.

Oh, and I had collected all my unused day money, so my crackling clothing was stuffed full of local lire currency, which amused HM Customs, but that's another tale, without a Blue Oval in sight.

Welch's second example was the latter's use of independent rear suspension.

The first and last of my four encounters with these remarkable machines – three with Welch's, one with Schanche's – came in 1984 and 1999, both times at Bruntingthorpe, which then offered car and motorcycle testing facilities on the lengthy runway and perimeter tracks, plus an improvised infield rallycross layout.

The most detailed and statistically rewarding test came on a cold and misty Thursday with Welch's car. On 28 February 1985, *Motor*'s young and sceptical staff members, David Vivian and John Simister, were sent to the test-track complex at Millbrook, Bedfordshire, to probe the truth about this small Ford's performance – especially the claim that it could run the benchmark 0–60mph time at least as quickly as a Formula 1 car. I attended to write a feature about the proceedings, and to drive this very special Escort again for all but the acceleration runs, which Welch handled while Vivian operated the magazine's then state-of-the-art fifth wheel and its electronic read-outs.

The innocuous white Ford Escort with STP logos delivered the promised blistering pace, despite the reservations that Vivian voiced as they brought 2 litres, two overhead camshafts, four cylinders, 16 valves and a single but beefy Garrett TO4 turbocharger up to temperature.

Their mission? A task that resulted in humbling the contemporary 5-litre V12 Lamborghini Countach, and snipping two-tenths from *Motor*'s record 0–60mph time extracted from a Formula 1 Brabham BT44. Here's how Vivian felt before they ran the first acceleration tests: 'Was this really the car that could humble 700bhp Porsches from a standing start, and shut down a Formula 1 car to 60mph?' Conviction and sensational words soon eclipsed those doubts: 'What was about to follow was history in the making. Despite a persistent misfire at high revs that prevented a full-blooded wheelspin take-off, the Escort catapulted itself into the record books. If the hesitation was brief, the subsequent acceleration thumped my head back with the transient energy of a thunderbolt. Suddenly, sickeningly, I was gravity's prisoner, the relentless force of that incredible powerplant clamping me to the sparsely padded Recaro more surely than the manacles of an electric chair.'

The cold stats *Motor* recorded are in our specifications panel, but the highlight was the two-way 2.7sec average of the 0–60mph runs – the fastest that conscientious weekly magazine had measured, 0.2sec faster than the Brabham, and 2.1sec quicker than the Lamborghini. In fairness, I should point out that the Grand Prix car regained that 0.2sec over 0–70mph, and widened the gap thereafter, rushing 0–110mph in 6.2sec versus 8.7sec for the rallycross Ford.

OK, so drag-racing point proven, the Xtrac-Gartrac turbo Escort formula was built to do a specific competition job, which it did effectively, with Norway's Martin Schanche victorious in the hard-fought 1984 European rallycross championship and John Welch's performances during his 1986 season snatching the British title.

Behind the wheel: a stunningly effective glimpse of what might have been

This text tells how the John Welch rallycross Escort felt to me after my first 1984 outing, followed by some snippets from two other '80s drives. We end with the 1999 privilege of driving the restored Martin Schanche original, when it had fittingly passed back into the hands of its creator, Mike Endean.

'Journalistic endeavour should not be subject to preconceptions, but inevitably you tend to mull over probabilities before you drive a competition car. My reasoning before I drove the latest G4 Xtrac Ford Escort, generously loaned for a day of tarmac and loose-surface experience, was that the rear-drive Escort MkII which we were to use as a benchmark would be easy to drive, and the Xtrac, by comparison would be extremely tricky.'

In fact, I was totally wrong, and the ingeniously effective turbocharged 4x4 Escort not only proved an easier steer on any surface, but was also notably quicker, even for an unfamiliar driver. Back to 1984…

'I had tried to absorb seven fascia dials; twin levers from the transmission tunnel; separate front and rear anti-roll-bar adjusters flanking my now trembling legs, plus boost and brake-bias hand-wheel adjusters, and I was far from confident. John Welch remarked sincerely: "You'll never be short of power and something to play with!"

'Dial in 1.2bar (17psi) and Welch tells you, "That's worth 462bhp and it'll run at that all day, no problem." Stick your nose in further and twiddle up 1.6bar (23psi) and John began to smile: "That's about what you use when you're trying to win. It means 555–560bhp, and that's when the car really comes to life!"

'Then there is the drilled-silver-metal lever, sprouting alongside the five-speed gear-change for the McLaren-TAG style Hewland DGB transmission. Monitored by a hydraulic-pressure gauge in the far corner of the cockpit, this lever provides the other dramatic and instant change that a driver can inflict on this Escort with its four-wheel-drive message. Push the lever forward, and you get a basic 50/50 torque split between front and rear drive, just like an Audi quattro, or with some similarity to the usually strongly understeering characteristics of systems that are grafted onto unlikely machines such as those Porsche 911s, which don't have the benefit of a central differential.

'Rearward lever movement provides a 28 per cent front and 72 per cent rear torque bias. This latter setting provides handling which will be familiar to a previous generation of Escort drivers. Handling, as Tony Fall once grated in dulcet Yorkshire tones, "that a man can understand!" In other words, oversteer that's utterly precise, controllable and available on demand. Handling that does not require a trace of the fabled left-foot braking in order to provoke a slide: you just sit there and build turbo power, then boot it, in the finest tradition of high-powered Ford Escort freaks from past decades.

'John Welch revived my spirits with a run as passenger in the Xtrac. I didn't take a helmet, as I wanted to tape some impressions along the way. Besides, John was also sitting there bareheaded, with the encouraging air of priest ready to hear confession. During the next three or four minutes, I travelled far faster over tarmac and loose surfaces than I have ever done before, yet throughout Mr Welch sat in his panelled grey office, relaxed in calm conversation. Meanwhile the world outside seemed as though it had suddenly been mounted on gigantic elastic bands and was being hurled past our screens in a berserk flurry of fiendish speed.

▲ I was exhilarated to drive the John Welch example of the Xtrac Escort on three occasions. It deployed Xtrac four-wheel drive, a turbocharged BD-series Cosworth engine packing 550bhp, and Gartrac body fabrication. This photo records a final test outing for *Autocar*. By then, Welch (left, chatting to author) had become British rallycross champion. *Author's archive, courtesy of Mike Hodges*

'Welch initially lulled me into some security. His finger stabbed at the gauge cluster: "Water temperature should stay below 90°C. If it doesn't, bring it in quick! Obviously there must be some oil temperature showing before you even think about blasting along, otherwise you simply pop the turbo as you open it up." Next on the agenda was oil pressure: "Should be between 60 and 80psi." Meanwhile, the transmission components zinged with meshing harmony, almost melodious in comparison to the lawnmower cacophony of a Citroën Visa Mille Pistes, for example.

'All the Xtrac's dials are marked with recommended running points, the large boost gauge shaded between 1.0–1.5bar; fuel pressure at 10psi and a tilted oil temperature gauge at 85°C. There is also a large orange light to tell you when you owe John Welch £16,000 for a new engine! An ammeter also flickers over the electrical news in the comprehensive, but vibrating, panel before the driver.

'First gear selects without a clonk in the master's hands, and Welch advises drily: "Don't be coy with the shift. If you want a gear, just make damn sure it knows you mean it. You won't break it." From first gear's position, isolated across the gate, the remaining cogs are a simple push-and-engage operation.

'Warm-up completed, John works his way through the close ratios; 9,400rpm registers as his foot floors the enlarged right-hand pedal. At apparently exactly the same moment as the gear lever hits the other side of the gate, 9,400rpm comes up once again in second gear. From second to third there's not even a break in the soundtrack, just the numbing blare of hundreds of horses regaining 9,400rpm. Fourth to fifth; there is still no slowing in momentum as this incredible Escort gobbles up 100mph, on its shattering schedule to a 120mph maximum, in the same way as mortal cars skate through first gear under full power.

'They say 0–80mph occupies three seconds. It's entirely credible. The fastest road car I have tested did 0–60mph in 5.1sec! Certainly the acceleration is superior to the 510bhp Audi Sport quattro that *CCC* experienced in Wales, but the beauty of this car is that it also handles in any way that conditions demand. It answers driver commands, rather than demanding power-assisted clutch operation and the engineering nonsense of simultaneous braking and power to "set it up". Welch throttled back to 90mph before a tight tarmac right-hander. I told myself to have faith as the Ford charged faithfully into the apex at enormous speed. Inevitably, as full turbo boost arrived, the white Xtrac slewed its flanks sideways at the scenery. Immediately, we were rushed up to maximum revs in fifth gear, but this time with a half-turn of opposite lock applied! No, he hadn't meant to provide quite such an effective demonstration of this fabulous Ford's capability!

'Now it's my turn to sit in the right-hand Recaro seat. The new blue Willans four-point harness snaps shut with relatively unused precision. I survey the controls. Steel grids across the clutch and brake pedals guard against muddy feet. An ivory gear-lever knob atop the massive transmission tunnel and a leather-rim steering wheel will do all the work today as: "The handbrake is too far back to be any use at the moment."

'Depressing the black starter button immediately brings forth a sound like a massed band of vicious vandals kicking hell out of a corrugated iron shed. This is followed by the prompt note of a Cosworth BDA as the engine fires, muffled slightly by turbocharging, but full of vigour nonetheless. From the passenger seat I had been constantly amazed at the speed with which Welch could change gears.

'Now I could enjoy the kind of exquisite power-shifting that Lauda and Prost must regard as a routine chore. In fact, with the benefit of three runs, the shift could be operated almost in the light and graunch-less manner of a Japanese road car, with the bonus of non-synchromesh shift speed, providing one made at least a token effort to synchronise engine and gearbox on down-changes.

'I started off with as much tarmac space as I could find at Bruntingthorpe, and that is more than any facility I have encountered in Britain. Here, one could run at whatever speed the car was capable of, and spare some time to ensure that all the instruments said the right things. However, there are some phenomena that you have to accept as normal in this Escort: two of which are three-foot stabs of yellow flame on the over-run from the howitzer-like exhaust planted beneath the rear panel, and copious clouds of grey smoke to accompany the astonishing pyrotechnics.

'However, the point of the exercise was a couple of quarter-mile loose-surface tracks, which could be linked to form a lap. Some 2,050lb (930kg) of Kevlar-bodied Escort hurled onto a loose surface by so much rampant turbo power would normally just be a wild animal.

'Yet it was only the Xtrac's enormous straight-line speed that gave me any second thoughts about the wisdom of unleashing all the allowed 1.2bar. It's the sort of all-round performance that transforms you from mundane motorist into 100mph mud-flinger, and faster than I have ever seen, or tried, before. There is of course some slippage, as four wheels try to cope with a peak 400lb ft of torque, but this can be minimised with the lever set forward in 50/50 mode.

'Then, if you can afford the time, the lever can be pulled back to provide all the oversteer you want with absolutely no driving tricks whatsoever. There is some typical four-wheel-drive rally-car straight-line 'wandering' at fourth- and fifth-gear speeds, but it is so gentle that a Blomqvist or Mikkola would probably not even consider it worth mentioning. As for me, I was only in fifth gear merely to say I'd been there on the loose, rather than actually needing it to feed an urge for more speed.

'Modern rally cars all have remarkably high standards of braking and ride quality over every surface they are likely to encounter amid myriad World Championship challenges. The Xtrac felt absolutely on a par in respect of riding ability, and only slightly less secure on the brakes, the fronts needing further bleeding on what was still an out-of-the-box competition car when tested. Neither brakes nor dampers are particularly special, although I suspect that the hydraulic expertise shown elsewhere by the Gartrac team provides even better damping than Sachs or Bilstein, because they have tailored the settings exactly to an individual car.

'Switching from loose to tarmac resulted in a noticeable increase in steering effort to pull the car around a second-gear crescent. The steering immediately lightens as the car emerges from its understeering stance – one that previous four-wheel-drive Porsche, Audi and Citroën motoring experience had told me would likely be an eternal ploughing match, on full understeer lock.

'In slower corners, the full 72 per cent rearward bias ensured that the nose would point obediently inward and the tail would slide wide immediately power was applied; but unlike a

A triple take on articles and images covering the original Escort rallycross car created by Mike Endean for rallycross legend Martin Schanche. I tested the amazing G4-tagged Escort in restored state at Bruntingthorpe. *Paul Harmer for Cars & Car Conversions*

THE G MACHINE

It costs £60,000, will out-accelerate a Formula One car to 60 mph and is cleaning up in European Rallycross. What is it? Jeremy Walton explains while David Vivian recounts the day at Millbrook that time stood still...

RALLY TECH

The X factor

Reborn into the 1990s, the 1983 Xtrac Escort won the 1984 FIA European Rallycross title. Jeremy Walton drives the day away with the Escort that allowed British motorsport transmission engineering a genuine breakthrough

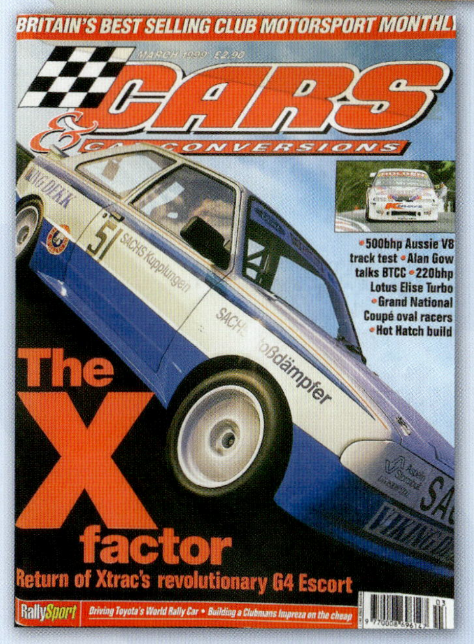
BRITAIN'S BEST SELLING CLUB MOTORSPORT MONTHLY

MARCH 1999 £2.90

CARS & CAR CONVERSIONS

500bhp Aussie V8 track test • Alan Gow talks BTCC • 220bhp Lotus Elise Turbo • Grand National Coupé oval racers • Hot Hatch build

The X factor

Return of Xtrac's revolutionary G4 Escort

RALLY TECH

Mike Endean's X-Files

From Hewland to to rallycross Escorts to Xtrac. How to get ahead in motorsport

Endean's pioneering 4x4 G4 layout, as applied to the Xtrac Escort, broke the major manufacturer ice

conventional rear-wheel-drive Escort, you would also be gaining speed at the rate of a quattro which, I assure you, is an exciting combination of capabilities.

'This understated Escort simply did everything I asked of it, and usually double what I expected, but in half the time. From observing inside and outside, I can only convey how impressed I was by pointing out that in my opinion its marginally lighter weight, plus a bonus of about 200bhp over present Peugeot 205 T16 figures, and the transmission's effortless handling of such enormous tractive forces, would currently place it commandingly in a World Rally Championship-winning position for at least the next 12 months.

'You have only to see what Schanche's original car does to even the best-driven and highest-powered quattros, to know that the Xtrac formula works, and there is actually no reason why the 4WD system shouldn't handle the 650bhp potential of a 2.1-litre BDA in turbo trim to make it quicker still!'

My high opinion of the Xtrac-Gartrac hand-built Escorts never wavered, as our later rewinds now demonstrate.

Although my original '80s reports for *Cars & Car Conversions*, *Motor* and *Autocar* were all with the second of the Xtrac Escorts belonging to British Champion John Welch, I think it is apt to leave this – the most emotive and effective competition Escort I drove – with a couple of extracts from my final encounter with the breed. Published in March 1999 CCC, this front-cover feature had the benefit of hindsight, plus the presence of affable creator Mike Endean, who had just retired from his successful Xtrac transmission company – a topline British specialist that remained a leading international motorsport transmissions supplier into the 21st century.

The 1999 restored icing on the 4x4 turbo cake that freezing January 1999 day was the first Xtrac-Gartrac-constructed G4 rallycross warrior. Yes, as originally campaigned by 1984 European Champion Martin Schanche. There were some key differences to the Welch Escort, including a four-speed gearbox, live rear axle with traditional Ford competition four-link restraints, and the fact that it used Xtrac purpose-built front uprights, rather than the converted Boreham Fiesta components.

We were back at Bruntingthorpe, which featured the kind of super-quick perimeter road swerves that characterised Silverstone and made both venues so suitable for motorsport and high-speed car and motorcycle testing.

Here is what I felt about this faithful restoration 23 years ago: 'This Escort rallycross legend has now been converted from *Motoring News* show-time exhibit 13in-wheel specification to the 15-inchers we'll use for this sortie. We are allowed out between Canberra jet take-off and the Triumph motorcycle testers – now there is a sub-zero hero's task for you, to feel what 135mph at 9,000rpm feels like (a literal knee trembler!). Then it was off to have a skid around an improvised rallycross circuit, more representative of the sub-110mph speeds of the car's heyday.

'The point about the Xtrac Escort was that it allied astonishing turbo performance with agile handling, nippy enough for short-circuit rallycross and club-driver skill levels. I know I drove Xtrac Escorts – and the GM Astra offspring – four times. And the cars never did anything but astonish not just me, but the hardened independent testers on the now defunct weekly *Motor*.

'I had serious track time to reacquaint myself with the Xtrac legend, and it took me all of that time to finally twig one curiosity: that RS-branded steering wheel is actually a Momo unit built for Alpina BMW! This is relevant because you need all the rim leverage you can muster to clear the trailer, or to bully 470bhp into your way of thinking over a track that goes from sheet ice to dirt, to water, to bone-dry concrete and back.

'Compared to the savage action of Welch's Escort on full boost, this original restoration is a pussycat. Neither driver stalled during a day full of restarts, and the camshafts allow you to putter along at 3,500rpm on a very light throttle. Unless you're in first, nothing much will then happen until 5,500rpm, and then it does not really bite your ears off until 6,500–7,000. Then it converts into a 2,000rpm thriller [on the way to 9,000rpm] with a gear-change that is the sweetest and fastest H-pattern I have ever driven in such a powerful competition car.

'Around the rallycross track there were a few chassis surprises. The brakes – period AP

1984 ESCORT MkIII XTRAC 4x4 TURBO

Engine

Configuration: Four-cylinder, Ford Motor Company-supplied Cosworth BDT. Engine and transmission mounted 22in right of centre, front-axle line between cylinders 1 and 2 **Capacity:** 1,860cc **Compression ratio:** 7.8:1 **Fuel system:** Modified Zakspeed turbocharger installation, twin wastegate Garrett T04B turbocharger. Kugelfischer mechanical direct fuel injection with electronically controlled enrichment on boost. Bosch electronic ignition, linked to turbocharger boost and transmission hydraulics. Air-to-air intercooler in front-grille panel **Lubrication system:** External side-mounted dry-sump pump, with rear-located tank. Serck water/oil heat-exchanger mounted above rear differential **Cooling system:** Crossflow water engine-cooling system. Rear-mounted Citroën CX2200 water radiator with twin fans; one thermostatically controlled at 85°C, one controlled from cockpit **Maximum power:** Regular output 456bhp at 6,500rpm with 1.2 bar (17psi) boost; 560bhp at 2 bar (28.5psi) boost **Maximum torque:** 390lb ft in 4,000–7,000rpm range

Transmission

Type: Adjustable 4x4; Hewland DGB gear set in Xtrac alloy casing; Xtrac gear-change mechanism; drive through reduction step-off gears at 45° to a central epicyclical gear-set providing 28/72 per cent front/rear torque split; drive to rear axle through open propeller shaft; drive to front axle through quill shaft **Clutch:** Sachs three-plate **Differentials:** Xtrac front crown-wheel-and-pinion; Xtrac modified Atlas rear-axle casing; both gear-sets 3.44:1 ratio; hydraulic-pressure-controlled slip limiters in front and rear casings, linked to a speed-sensor electronic system. **Front driveshafts:** Xtrac competition-specification equal-length drive-shafts; front MkII Granada Lobro constant-velocity joints inboard of front Group 4 Fiesta-based hubs in Xtrac castings; rear MkI Granada Lobro joints

Suspension

Front: Gartrac adjustable spring-platform MacPherson struts with modified proprietary gas-filled inserts; 170lb/in-rate coil springs; track-control arms and compression struts with rubber inner bushes and adjustable outer spherical bearings; rear-facing, blade-type anti-roll bar, adjustable by lever on right-hand side of driver's seat; anti-dive geometry

Rear: Independent in fabricated subframe mounted on original axle-link locations; modified RS1700T uprights; adjustable-spring-platform Gartrac struts with modified proprietary gas-filled inserts; 160lb/in-rate coil springs – top located via strut turret, lower by unique triangulated arms with rubber inboard joints and adjustable outboard spherical joints; forward-facing blade-type anti-roll bar, adjustable by lever on left-hand side of driver's seat

Steering

Type: Unassisted rack-and-pinion, 2.5 turns lock-to-lock

Brakes

Front and rear: 10.5in-diameter ventilated discs front and rear with AP four-piston callipers **Hydraulic system:** Single-line front-rear dual circuit with Gartrac G3 pedal box and balance bar, no servo **Handbrake:** Inline on/off hydraulic handbrake

Wheels/tyres

Wheels: Revolution 8in x 15in alloy **Tyres:** Surface-dependent; 822 x 15 multi-surface competition

Body

Type: Gartrac G4 – MkIII Escort G3 RWD modified to incorporate Xtrac 4WD system; front chassis rails with under-slung drive-shafts accommodating wheel travel; centre G3 section reworked to accommodate 4WD mechanism; Group 4 Escort MkII/RS 1700T steel roll cage; Kevlar body and passenger-door options **Interior:** Lightweight Recaro seat(s); Willans competition harnesses; Smiths analogue instrumentation; Momo steering wheel

Weight

Race weight: 2,090lb (948kg)

Performance

1985 performance, published 23 March 1985. Measured by Motor staff at Millbrook Driver, John Welch. **Maximum speed:** 115mph **0–30mph:** 1.1sec **0–60mph:** 2.7sec **0–100mph:** 6.6sec **Fuel consumption:** 4.3mpg on intermittent 2 bar boost

11in discs and original four-pot callipers – are magnificent, especially over rutted mud, when it really takes a kick to lock them up, and even then it is a gentle slither. The amount of understeer generated on used Yokohamas and 8Jx15 Compomotive wheels (rather than original Revolutions) was forearm-battering. Once the back end is slithering out of line, the 470bhp is so gently applied and monitored that the fabulous sideways sensation should be part of the UK driving test.'

In summary, it was just a period tragedy from a Ford fan's viewpoint that these Escorts never made the World Rally Championship scene. They looked enough like showroom residents to link with public sales, disguising their astonishing engineering and agile capabilities. Just as for the RS1700T – which Mike Endean saw plenty of during running repairs within his early Xtrac workshop – even this superior Escort all-terrain warrior would have had to circumnavigate the homologation requirement for a production counterpart. As with the RS1700T, this ingenious hybrid, with many ex-Ford components, would have taken too long to grind from stellar conception to Ford limited-production reality.

Shame!

TWO ESCORT TURBO GENERATIONS ON ROAD AND TRACK

Competition-orientated prototype Escort turbo fun: the back story

Ford Germany's 115bhp RS1600i delivered the first front-drive Escort to be recognised for Group A international motorsport, for which a minimum production run of 5,000 examples was required. Yet those inveterate 'cross-country tinkerers' at Ford's Boreham workshops tinkered away in 1980–81 with a much more competition-orientated Escort, boasting a turbocharged 125–140bhp. One such Escort turbo prototype was ready to demonstrate to senior management in 1981, and received a very enthusiastic reception, but that example was based on a carburettor Escort base.

Competition-orientated prototype Escort turbo fun: behind the wheel

The mild 125bhp example (JNO 120Y) of this first Escort turbo I had from Boreham in 1981 was a real stealth warrior, dressed entirely in 1.6 L exterior panels and badges. It was enormous fun, and you could see just why the men in slick suits and ties far up the managerial ladder were so excited about this Escort offspring. The idea was that you would be able to buy an aftermarket turbo installation from the Ford RS dealer network for a predicted £1,000. Such grassroots technology also supported the 1983–84 Escort Turbo Rally Championship, as competitors retrofitted turbo conversion kits.

▼ Back in the mid-'80s I worked regularly as a road tester for *Performance Car* magazine. Here, I'm driving a MkIII Escort RS Turbo demonstrating the meaning of 'power understeer.' This photograph was taken (probably by Peter Robain) at Millbrook test track. *Author's archive*

A first-generation production Escort turbo for homologation: the back story

It took until April 1983 for an official Blue Oval decision to ratify a production Escort turbo for 5,000-off sale to the public, and therefore Group A homologation. Significantly, that same Stuart Turner-chaired Essex HQ meeting also gave the green light for the more radical Sierra turbo (which became the Sierra RS Cosworth/RS500 in 1986–87), but more of that later.

Special Vehicle Engineering (SVE) at Ford Dunton was assigned the task of engineering the showroom Escort turbo using the fuel-injected XR3i as a base. SVE received some Ford Motorsport 'musts' for its 'to do' list: extended wheel arches to cover tyres up to 10in wide, a limited-slip (front) differential and road-car suspension basics (particularly tie-bar front-suspension layout) to handle at least 200bhp in competition trim. Power would never be the problem, as Motorsport directed that a Garrett T03 turbocharger be deployed, rather than the more street-friendly T02, as it had already seen up to 260bhp in its preliminary tests. Adding to Ford turbo-tech knowledge, and providing a near production-ready turbo installation, former Garrett employee Geoff Kershaw sold his Turbo Technics company jigs and tooling designs to Ford, a deal that benefitted both the pioneering carburettor and subsequent Bosch fuel-injected Escort turbos.

The full stories of the intense development of both first- and second-generation Escort turbos are told in my earlier *RS: The Faster Fords* books, so here I skip happily past those 1983–84 years of hard work. However, we have to remember that showroom power/torque was always restricted (an official 132bhp and 133lb ft) to balance against transmission durability and possible warranty costs. The other factor in the public persona of the RS Turbo was the brave decision to select the Ferguson-patented Viscous Coupling (VC) limited-slip differential for use in these cars. This was the first time it had been employed on a vehicle produced by the thousand, and calibrating it to suit then leading-edge front-drive turbo power delivery was a voyage into unknown territory.

XR3/XR3i/RS1600i PERFORMANCE

Sources: These figures are an indicative amalgam of my multiple experiences at Motor Sport, Motoring News and Performance Car – all of them with access to fifth-wheel timing equipment of differing generations. Performance Car was the most conscientious, but figures could often be notably slower than those published by Motor and Autocar, as those weeklies had the best cars and first access.

	XR3	XR3i	RS1600i
0–30mph	3.56sec	2.78sec	2.9sec
0–40mph	5.23sec	4.49sec	4.4sec
0–50mph	7.17sec	6.53sec	6.2sec
0–60mph	9.75sec	9.05sec	8.7sec
0–70mph	14.92sec	12.13sec	11.8sec
0–80mph	16.9sec	15.8sec	15.8sec
0–90mph	23.0sec	21.3sec	20.3sec
0–100mph	32.6sec	30.6sec	28.0sec
Standing ¼-mile	17.0sec	16.9sec	16.7sec
Maximum speed	111.5mph	113.6mph	116mph
Fuel consumption	27mpg	31mpg	28mpg

A first-generation production Escort turbo for homologation: behind the wheel

Looking at my diaries and earlier books, I can see that I drove at least five turbocharged versions of the first front-drive Escorts, two of them simply redeveloped production racers (those of DJ Mike Smith and production-series overall champion

◀ At rest, the Mike Smith Escort RS Turbo production racer. The stooping figure guiding the race-suited author (right) is Stuart McCrudden, a source of Ford track knowledge back to the Anglia. *Motorsport Images*

➤ My favourite action shot of the Escort RS Turbo, which proved particularly fast for production saloon car champion Karl Jones (seen here), and durably quick for DJ/TV presenter Mike Smith and talented Lionel Abbott to win the 1986 Snetterton 24 Hours. *Courtesy of BF Goodrich*

Karl Jones) in the 160–165bhp zone. The Jones-driven Richard Asquith machine, on BF Goodrich TA-R specialist rubber, was by far the best: easy to place accurately, even on the tight confines of Mallory Park, coupled to excellent traction out of the hairpin under full power. As for all Asquith Motorsport's competition cars, it was efficient on a sensible or low budget, and just got the job done with a minimum of fuss. Yet this turbo did have an extraordinarily talented conductor in Jones: mild-mannered and funny out of the car, ferociously spectacular in race mode.

For my public-road miles, I had plenty of smiles. The first production Escort RS Turbo was an RS Ford alright, a bit rough around the edges, but fantastic fun to drive, because it was developed by engineers who loved driving. The first RS Turbo could get ragged under power over crests, and might well get a literal wriggle on as power was applied, or released, over bumpier British lanes. However, these traits were due to both the VC differential and Ford's learning foray into front-drive versus independent rear suspension. Yet, this Escort was exciting and engaging in a way few mass-production cars could match in 1984–85.

MkI RS TURBO PERFORMANCE

Sources: Turbo-1, Performance Car/*author, March 1985; Turbo-2,* Autocar*, 1986*

	RS Turbo-1	RS Turbo-2
0–30mph	2.9sec	3.0sec
0–40mph	4.2sec	4.7sec
0–50mph	5.7sec	6.1sec
0–60mph	8.1sec	9.2sec
0–70mph	10.8sec	11.4sec
0–80mph	13.1sec	14.1sec
0–90mph	17.0sec	18.5sec
0–100mph	21.2sec	23.0sec
0–110mph	30.5sec	–
Standing ¼-mile	16.5sec	18.8sec
Maximum speed	128mph	125mph
Fuel consumption	27mpg	27.4mpg

Second-generation Escort turbo: the back-story

By contrast, the 1986–90 second edition of the Escort RS Turbo was much more of a showroom-biased product, for there was now no need for a competition cousin. It became much more customer-friendly, particularly the handling via a recalibrated VC differential, although it gained weight. As a result, it yielded some straight-line acceleration and lost shedloads of street cred, as the second-generation RS Turbo looked similar to any other MkIV Escort, particularly the now outdated XR3i.

Second-generation Escort turbo: behind the wheel

My remark about the outdated XR3i is not an easy one for me to make, as the earlier XR3/XR3i served me well in a sea of Golf GTI benchmarks. Yet when I drove and reported on an eight-car *Performance Car* hot-hatchback group for a road, track and cross-country test in the later '80s, there was no doubt the XR3i was a slowcoach in a performance field led by marques VW, Peugeot, Toyota and MG.

In some important ways – especially the fitment of ex-Sierra XR4i front disc brakes and a reduction in motorway noise through taller gearing – the second Escort turbo was a far better motor car for the public, but it had lost much of its raw character, and will always retain less 'modern classic' resale value.

▲ The modest media introduction of the second-generation Escort RS Turbo saw me driving on East Anglian roads in the summer of 1986. This RS, D771 PVW, was also the example submitted to *Performance Car* for a full road test. *Tony Dron*

MkV XR3i LEE ALLEN RACE CAR: MALLORY PARK RACING FUN

Sometimes, routine magazine track tests turned into a competition outing for the guest journalist. Uncommon, because it costs significant cash every time you wheel out a competition car – the more sophisticated it is, the more track time tends to be measured in thousands rather than hundreds of pounds, dollars or euros. As the decades rolled by into the '80s and '90s, operating costs shot up, so the importance of media publicity for attracting or satisfying a sponsor and its vital financial contributions became pressing for ambitious drivers and race teams.

Even though the printed word was no match for broadcast TV, we magazine writers with form were, and still are, asked to drive these cars and publish our frontline experiences. The big difference from the '60s and '70s became requests for some cash to accompany your wordy presence.

However, the most common deal in my case – and for many classic magazines today with auction-house advertisers – was for the drive to take place when the car's owner wished to sell for a top price. This wasn't the case with my initial track-test outings, but even that previously described 1970 run in Roger Williamson's Anglia (see Chapter 2) was purely to sell it so that Roger could get on with his single-seater career.

Some 24 years later, on 10 October 1994, I was back at Mallory Park with another smaller four-cylinder Ford, albeit on the longer 1.35-mile track. I did a pretty average job during testing, 2sec slower than the time that 1993–94 Vecta Ford Challenge champion Lee Allen set with his lap record. So it was a surprise when the team (Roland Hayes's HT Racing) asked me to race the car at the end-of-season-final Vecta Championship event, also at Mallory – and they stuck to their word even when the competition Escort was sold just prior to the race for £7,500.

Under the skin: heavily modified by HT Racing

So what did the new owner get for far less than the period price of a new Escort XR3i?

Firstly, an outstanding race record: in the previous two seasons Lee Allen and his Jensen Blue 1993–94 Escort had dominated Class C of the Vecta national series to such an extent that the combo had never finished lower than second in its category, and seized eight of nine possible 1994 pole positions. The 1.6-litre Ford had been crashed three times, usually when trying to oust bigger-capacity racers from the overall order. Gutsy...

Technically, what did that £7,500 buyer acquire? Built from a new MkV body, with an integral TIG-welded multi-tube safety structure from Rollcentre, this UK national-racing saloon was more radically modified than the FIA-regulated Anglias, rear-drive Escorts and Capris featured earlier, which ran to International regulations. HT Racing always turned out exceptional race-converted Ford engines, but this time the entire surrounding vehicle received detailed and fundamental engineering thought and hardware.

The process started with tilting the mass of the engine lower in the engine bay, which also benefitted the driveshaft angles. Unlike the expense-no-object premier-league '90s BTCC 2-litre touring cars, the cylinder head was inclined forwards (rather than tilted rearwards towards the cabin bulkhead). Another moved mass was a centrally mounted fuel tank (its contents carefully monitored for short races). Holding a maximum of just five gallons of 100–102-octane super unleaded, the triangulated aluminium tank slotted beneath the raised rear floorpan was an absolute pain to refill.

Power steering took advantage of an ultra-quick rack. Also utilised were unique wheels, monstrous Sierra Cosworth Brembo front brakes, and intensively tested and revamped suspension. The modifications combined to improve torsional rigidity and relocate weight mass and roll centres to the benefit of handling. For instance, the front track-control arms – always a whippy point on a production Ford wearing classic MacPherson struts – were replaced by solid fabricated items, carrying equally solid rose-joint mounts.

At the back, an RS2000 MkV donated a beam axle, but sorting out the radically modified machine's handling wasn't a weekend's work. Working their way through a 'very big box of springs' and making use of gas-filled-damper expertise from Leda, the HT Racing team installed 350lb/in front springs, coupled to 450lb/in rears. The set-up required utterly different negative-camber settings on either side. The five-spoke 7.5in x 15in Dymag wheels were in expensive magnesium, rather than heavier aluminium, and ran mandatory 225 Yokohamas up front and slimmer 205-section rears.

The car weighed 1,841lb (835kg) on my race day and the Roland Hayes 1,629cc version of the CVH race engine deployed 179bhp at 7,500rpm, to give a power-to-weight ratio of 218bhp per tonne, or almost 110bhp per litre. Obvious aids to power for a CVH race unit without turbocharging were an enormous cold-air collector and twin 45mm Weber carburettors. The engine featured two years of practical development, a reprofiled Kent camshaft, a compression ratio circa 10.5:1, and the ability to leave the start line using 8,500rpm, with a maximum of 8,000rpm thereafter.

Behind the wheel: impressive cornering ability and a podium finish

A tight track that featured both a 127mph right-hand swerve and a 30mph hairpin highlighted a truly different front-wheel-drive Escort experience. The cabin had some lightweight reminders of the road car, including a heater/ventilation knob that was actually a stick-on facsimile! The business bits comprised the firm embrace of a four-point Willans harness, a Corbeau race seat and a Sparco suede-rim steering wheel to twiddle between rapidly achieved full locks.

Gazing through a windscreen emblazoned with HT RACING capital letters, I shifted attention to the 9,000rpm Elliot rev counter and heard the helmet-muffled echoes of four cylinders snorting at idle, and wishing they were living and loving their normal 5,600–8,000rpm regime. I ensured that the adjacent smaller dials told a comforting story regarding oil pressure, and both oil and water temperatures, for there was no cooling fan.

For both qualifying and race, this 48-year-old driver earned some pace from the young champion's Ford the hard way: stretching braking zones and leaning on the cornering ability and agility provided. My opposition had lighter Fiestas, with half-a-dozen regular front-runners equipped with similarly powerful Roland Hayes magic motors. It was also not helpful that my recent experience at Mallory was in a 400bhp TVR V8 Tuscan and a 550bhp Sierra RS500, both emphasising straight-line performance rather than 'last of the late brakers' capabilities.

A little more faith in this Escort's extraordinary

END GAME

The main sponsors and founding club have quit. Despite these traumas, Lee Allen scored his second successive Ford Championship title. Flushed with success, he then agreed to let Jeremy Walton race what must be this country's best handling fwd Escort

This is the brainiest Ford Escort I have ever driven. Not the most expensive: factory Fords like the one I drove in 1993 cost £150,000plus. Ford Motorsport now ask over £200,000 for the Monte Carlo Rally-winner. This Escort was not the fastest sprinter, either: the 1980s turbocharged, 550bhp, Zakspeed-engined John Welch rallycrosser still holds that slot at a 2.7secs for the 0-60mph flash.

Sold immediately prior to my Mallory Park track test and race (that's confidence for you...), Lee Allen's 1994 Vecta Challenge Champion cost the new owner around £7500. For that I believe the punter snatched the smartest - and certainly the best handling - front-drive Ford in Britain. The race records tells you a bit about its prowess. In the bitterly fought Class C of the National 1994 Vecta Challenge series, eight from nine pole positions were gained. It took six fastest laps (three records), seven wins and two second places.

Until I drove it, this racing Escort had never been lower than second place, although it had failed to finish three times in 12 events: two crashes when intruding on Cosworth territory and one ignition failure. Even with this strange stranger behind the wheel, this Mk5 with the winning habit finished third, maintaining its podium pedigree. It looks like most of the Mk5 Escorts that continue to rule the sales roost in the UK, but the Jensen blue hatchback with inoffensive RS2000/XR3i front and rear spoilers hides an uncompromising array of technical tricks. A tilted engine, replacement suspension and central fuel tank, boost its power-steered cornering capabilities to unearthly levels.

Built from a brand new body in 1993/94, the Allen Escort owes a lot of its speed to the people that occupy the same trading estate industrial units as the Allen Family's Truck Align company. A few doors away lies HT Racing - leading supplier of Ford engines to Vecta-men - a set-up particularly strong in supplying 180bhp CVH power units.

Despite the cheerful chatter from Lee and number one racing mechanic David Widdows, you need only look at the car to know these two are thinking racing men.

Even those deceptive Dymags - actually cast at 7.5 x 15in Magnesium, hoping to hit the 750kg class minimum weight - came from supplier/manufacturing contacts on the Charlton SE7 estate, rather than directly from the company's Wiltshire headquarters.

Dymags do a good job, fitted with mandatory Yokohama A008Rs (225/50 front, slightly slimmer at 205 section on the back). However, even such an exotic five spoke presence, plus Kevlar composites for bonnet, bumpers, tailgate and boot lid spoiler replica, will not budge the racing weight. The barest Escort interior in Britain to wear door trims, and a lot of fabricated aluminium, cannot get it below the 835kg we recorded at Mallory that Sunday morning. This meant I enjoyed 218.3bhp a ton, better than the current street Escort Cosworth's rating (175 per ton). Enough to put an Escort on par with a similarly powerful, but smaller, Fiesta.

The Roll Cage Centre provided a steel tube kit that was TIG-welded into the shell by HT Racing. Central to the cornering balance of the car was the fabrication of an alloy five-gallon petrol tank, an elongated triangle in side profile. It fitted where the rear seat passengers would normally feel the footwell. That move provided a low and light fuel load, but many racers would be exasperated by the cumbersome refuelling process that accompanies that location. At eight to nine racing mpg, the process is usually only necessary once a meeting, for the baffled tank picks up every last drop of Super unleaded.

The other basic installation thought was aimed at durability, as well as snatch-free front drive handling. The CVH is tilted forward (nobody knows how much), which allows the differing length standard halfshafts to run straight, eliminating the 1993 CV joint dramas that dogged Lee's previous championship contender.

This race Escort was power steered to take advantage of the far quicker rack provided, but also allowing uncanny precision for low effort. The only penalty is little initial rim feedback.

The under bonnet is also notable for the neat alloy fabrication of cold air ducting to the twin Weber 45 carburated HT Racing engine and the oversize Maniflow exhaust. Boss Roland Hayes reported: "After two years development on the CVH - particularly addressing three specific lubrication problems - we have got them to such a reliable specification that we measure within plus or minus three bhp for every unit, and offer a year's warranty."

When Roland Hayes tells you that the power is; "not the 185; you've been told, more like 179bhp at 7300rpm," you remember it as the first time an engine assembler outside Cosworth has demoted an official power quote. Maximum torque is unmeasured, but Roland estimates some 120lb.ft is available at 6700rpm. He added that the Maniflow big bore layout had added bhp throughout the rpm range (a surprise to everyone). The 20bhp power gain since Richard Longman's Data Post winners of the 1980s has been in extending the rpm range beyond 7000.

Every ounce of Hayes' experience has

● Hardly an engine 'born to race', yet HT Racing have coaxed serious power and reliability from the CVH unit

36 Photography: John Colley & Interschnapz

FEBRUARY 1995

37

cornering capabilities, and a lap time kissing a 90mph average, placed us fourth on the grid, just 0.67sec away from pole position on a tightly packed grid. That 1994 Escort time was similar to my lap speeds in 1991 with the formidable Collins Performance Engineering RS500 Sierra, which I think says it all about the 1.6-litre Escort's cornering and braking abilities.

There was the usual queasy wait for an afternoon race, punctuated by officials trying to persuade us to display some evidence of track manners. There was also much deep thought on how pace could be improved through the fastest Gerard's curve, as I could see 127mph data-logged on entry, and a sweaty 110mph on exit! A brace of S-bends and the heavy braking into the hairpin could yield the fractions I needed…

The start evaporated into a blare of revs and a haze of blue tyre smoke – not bad, but still fourth. Weave, bob and challenge over the opening laps. I watched the trio of Fiestas ahead lifting rear wheels under heroic cornering duress, and admired the blue contrails from equally brave late-braking manoeuvres. I was now closing on the third-position guy: I hit the brakes slightly early and shifted into first to avoid rear-ending his baby Ford, with the bonus of getting the engine running cleanly on the exit of the hairpin. The tactic proved effective, and on the next lap, third place was mine as my rival headed for the hairpin's solid outer reaches.

I made a bid for second place, scratched round a tenth faster than my rival, but reached a personal limit as the Escort shuffled out of line for what seemed elongated seconds on the quickest curve. So, positions remained unchanged to the close finish, with a narrow 0.63sec gap between second and my third place, and the winner a comparatively comfortable 2.26sec ahead.

It had been a British club-racing Escort education, and I wasn't too unhappy on the (loaned) Escort Cosworth road-run home. As a bonus, the editor of a rival magazine had the decency to finish amongst the backmarkers, also allowing me to lap him in practice, earning extra Brownie points!

▲ The last front-drive Escort I used in competition was the Vecta Ford Challenge championship-winning Lee Allen car of 1993–94 with 180bhp. It was a very competitive proposition thanks to outstanding handling. At Mallory Park I qualified fourth and finished third in a 10-lap race. *Cars & Car Conversions*

Ironically, my international and long-term Escort experiences escalated for much of the car's remaining European production life, from 1990 to 1998, until the Focus arrived to rescue the company's mainstream reputation. RS loyalists would have to wait until 2002 before new showroom Focus RS product was offered.

That lack of RS products affected my business, especially as the British no longer dominated the creation of sporty/competition Ford vehicles for series production. Fortunately for my family finances, I had a strong overseas network, born two decades earlier with Germany's Cologne Capris and multiple American visits, plus regular magazine work in France and Italy, so it was no surprise that my later Escort outings encompassed some memorable overseas kilometres.

In 1991 I had an early outing in the revived Escort RS2000, the emotive nameplate reborn in front-drive 150PS format under the misleading advertising strapline: 'The Champ is back.'

Changing times in terms of showroom products were reflected by my 24-hour solo trip to Ford Cologne to assess if this was a worthy use of the RS2000 badge – a name that had affordably entertained so many, including the author, in two rear-drive formats.

I picked up the silver Ford outside Cologne airport arrivals and drove quietly out of town. First impressions were entirely favourable, the styling neat and clean, probably understated in the light of RS-branded Ford traditions. The Recaro seats were a pleasure to occupy, and a

▲ This second-generation (MkV) front-drive Escort RS2000 lived with me on long-term test, serving as a cover car for a book, wedding transport, Ford service test mule and versatile daily performer. *Author*

GENERATION GAMES

The RS2000 first debuted a lazy 2-litre engine in a small saloon back in 1973. Five generations later, Ford still offers the model. Can the same formula still be fun 20 years on? Jeremy Walton finds out. Photos by David Wigmore.

Ford of Britain did not actually invent the RS2000, even though it was the British specialist division of Ford Advanced Vehicles in Aveley which got the job of engineering and manufacturing it initially.

The concept came from two sources: the Ford of Britain dealers and specialists (such as Lumo, Allard and Superspeed) which could only economically offer the public 2-litre sports Escorts packaged with unsuitable gearboxes. These met a definite demand in the days when Ford had a sporting chasm in its new rear-drive Escort range.

At the time, Ford sold either low-cost thrills in the shape of the 1300GT, or the more serious challenge of the Twin Cam.

However, the original RS2000 manufacturing request came from a second source — Ford of Germany. The initial batch of RS2000s in 1973 were all LHD and notched up 987 European sales.

Most of these were made in the UK, which retained a significant press fleet prior to RHD introduction, all of them blue-striped.

Ford of Germany controlled the plant at Saarlouis, which made the original multi-million-pound order and 1162 subsequent LHD models on the first rear-drive Escort body. Saarlouis continued RS2000 manufacture thereafter.

The 2-litre Pinto engine was a tight squeeze in the

➤ In the summer of 1995, I compared a road-test front-drive RS2000 with a private owner's original rear-drive RS2000. The later Escort was losing market-place appeal, and the first edition – as so often with cars – had more emotional pull and lasting value. *Ford/David Wigmore*

rake-adjustable, three-spoke, leather-rim steering wheel was a handy twiddle that didn't look as repulsive as other Escort fitments.

The uprated twin-cam 16-valve engine offered more power (148bhp) and torque (140lb ft) than any previous showroom Escort, including the RS Turbo. Top speed was now independently measured at up to 133mph with 0–60mph averaging 8sec dead, but the weight showed up in the overall fuel consumption averages in the 26.5mpg region. It was all managed so competently by the Ford EEC-IV engine management system, with the powertrain mounted in a much heavier and better-equipped body than previously. Thus, no drama, and competent acceleration was the recipe, rather than a ragged rocket-ship ride.

I headed alongside the Rhine for a relaxed contrast to the city, and then turned onto some of the byroads away from the smooth main road. Actually, the Germans love tidy and unruffled tarmac, even when it is winding up and down through picture-postcard villages. This was the first RS Escort with power-assisted steering, and the feedback was precise rather than informative.

The RS2000 suspension had been tuned to XR3 urban standards, rather than rumpled British back-road twists. Again, well-mannered, but not a stand-out driving machine. However, there was no doubt that four-wheel disc braking, plus Teves electronic anti-lock, were massive improvements over the MkIV Escort's rather cruder disc/drum ABS layout. I liked, rather than loved, that 1991–96 Ford RS2000. Yet the fact was that it had been developed as an XR3 range-leader, with XR+ project coding rather than RS or motorsport-homologation intent. At a debut cost of £13,995, as a more capable and civilised XR Escort, it was fine, but as a raw RS drive of memorable sensations... unconvincing.

However, my next RS2000 outing, on 22 March 1992, saw some high-adrenaline moments at Brands Hatch…

'THE UPRATED TWIN-CAM 16-VALVE ENGINE OFFERED MORE POWER (148BHP) AND TORQUE (140LB FT) THAN ANY PREVIOUS SHOWROOM ESCORT, INCLUDING THE RS TURBO.'

MkV RS2000 TRACK ACTION AT BRANDS HATCH

Following an Oulton Park midweek outing in an experimental 2-litre Zetec-powered club-racing Escort, there came an unexpected call from Ford sports managers to prepare for weekend circuit action. They wanted me out at Brands Hatch in a front-drive RS2000 for the 22 March debut of their seven-discipline Ford Escort Rallye Sport Series, to be run over 18 rounds. This was right in the Ford one-make tradition of good prize money and an example of the chosen model as the end-of-season premier award.

Naturally, those incentives lured some good names out to play. I was on the grid as a one-off to top up numbers, along with a few proper rally drivers who were available and had been attracted to the single event. That dry Sunday, we had former British Champion David 'Dai' Llewellin. Ford also had the honour of future World Champion Richard Burns winning a subsequent RS2000 Aintree sprint outright.

Under the skin: minimal modification

Electronically monitored, the engines ran full catalytic-converter exhaust systems and could only be rebuilt to manufacturer tolerances (or 'blueprinted'). A Quaife Torsen limited-slip differential was allowed, and an inch could be sliced from the ride height. Standard five-spoke RS2000 alloys were employed, but wore trick R1 BF Goodrich 205/50 ZR tyres. In fact, 'mine' was a pretty regular production car, road registered J723 WOO, with an interior retaining production trim save for a tailored roll cage, Willans safety harness and a Corbeau competition seat.

Behind the wheel: enjoyable production-based fun

As you'd expect, the car was straightforward to drive, with notably improved traction and grip over the production car, the power steering gaining feedback from the slightly larger and stiffer construction of the BF Goodrich rubber. The engine felt just like the media-ready regular item, with all the claimed 148bhp present at 6,000rpm, and the 4,500rpm maximum torque delivering an amiable street-wise power curve. All this in a weight that wasn't allowed below 2,491lb (1,130kg); hardly flyweight, as the showroom item was officially catalogued at 2,600lb (1,179kg).

The rally drivers set the practice pace, Ian Gwynne (57.06sec/75.96mph) shading Llewellin by just 0.7sec. Only regular racer/writer Tony Dron joined them in 57sec lap times. In 10 laps I graduated from 59.1sec to 58.41sec/74.2mph to place fifth, a scant 0.08sec ahead of the next of 13 competitors.

The traditional 10-lap fare of a busy BRSCC club-racing event saw Llewellin startled by the rude manners of regular racers, and he fell back into fifth-place warfare with Gwynne. I lost one slot to finish sixth, slightly compensated by dipping into 57sec lap times, while Llewellin's eventual fifth place was accompanied by the fastest lap at 57.08sec/76mph. I had enjoyed it, and Ford's always-razor-sharp press service ensured my local newspaper knew all about it, which did no harm for my local profile.

Footnote *I also had a road test week with the RS2000 4x4 variant, using the usual 150PS powerplant. I had seen various 4x4 transmissions fitted to front-drive Escort variants over the years at SVE, but the final German production version, built on that UK pioneering work, was an extremely civilised installation. So much so, this was probably the safest Escort I drove, but the added weight and unchanged horsepower meant that it came across on dry roads as the 'nice guy' who would not trouble your heartbeat. But… it was certainly the Escort to own if you wanted value-for-money all-weather traction beneath an RS Ford badge. Perhaps perfect for a Scottish vet with strong Blue Oval loyalties?*

SORTIES IN A PAIR OF MkVI ESCORT Si MODELS

By 1996, my freelance life was more fractured, divided between regional television, business TV presenting and printed media, with less book work and more magazine outlets, rather than long-term UK contracts. Although my writing appeared in Italy (*Quattroruote*) and France (*Auto Hebdo*), plus via BMW links in the USA, and Germany becoming significant, my relationship with Ford in Britain improved. Principally, this thaw was due to the fact that since 1983 I hadn't driven in national Ford TV advertisements. Some Ford managers found such appearances so aggravating that I was covered in confidentiality documents with every 1979–83 filming session.

Escort Cabriolet Si

Aside from the glorious privilege of driving the company GT40s on multiple 1995–96 outings, I was also granted the use of 1996 last-European-generation Ford Escorts for two special trips to France. Both summer trips were predictably hot, so I was particularly pleased to have a 130bhp Cabriolet Si with powered top for the June outing to Le Mans, where I would drive the UK company's GT40 in race-morning laps, as described in Chapter 5. We managed to have nearly as much excitement in the Cabby, even though we only took a long weekend away.

Although the British leg of the trip to the ferry was unremarkable, save for the temptation to put the power hood up or down while in motion, more serious trials awaited us in France. Despite using side and main roads away from *autoroute* toll-booth speed cameras, I cleverly managed to speed towards an uphill crest.

'Oh, look at all those men in blue… Oh, are they pointing guns at us?'

Gendarmes were out in force on a Route Nationale. Not such a surprise, but the demand for cash and a consequent visit to the nearest town under close French police surveillance to find a bank machine did delay us. So, the protracted drive down to the château that Ford had rented some way south of the circuit meant we were embarrassingly late for our hosts. Not a good start when you have the enormous privilege of driving a GT40 lined up the next day…

Fortunately, there was a Barclaycard service

at the track on Saturday. Suitably refinanced, I had the exhilarating 150mph experience of the morning parade laps in the GT40. The Cabriolet continued to provide excellent transport in the sunshine, with its late-model Zetec 1.8-litre XR3i powertrain and updated suspension sharing some extensive 1995-announced components with the front-drive RS2000, plus 15in alloys.

The revised and well-engineered powered soft-top had become a standard piece of kit on this popular Escort derivative, which now demanded £16,995 versus £11,930 for the range-starting 1.6i three-door hatchback. Sadly, kerb weight was notably porky on the well-equipped Si version of the Escort convertible, meaning that zipping happily through the five-speed MTX-75 gearbox – also a plus point on the RS2000 – brought the Ford figure for 0–60mph of an uncompetitive 10sec, coupled to a maximum speed of 123mph.

Escort Si estate

Now that other 1996 Escort summer sortie to France. Back to my Escort estate ownership roots, except this time the sixth-edition Si model was definitely Ford property. In an unobtrusive metallic grey, it had covered 8,000 media miles, and wore an RS-branded dress-up body kit to support the Si Zetec 1.8-litre powertrain with the higher (115PS) power rating.

We emerged from the Channel Tunnel for the run south via Reims. According to my diary, we could cruise at an indicated 100mph and 4,000rpm in the tall fifth gear. Or, you could brace the occupants and reach a shade under 120mph with the tachometer registering 4,800rpm. The speed was sufficient to get us to Vitry-le-François and the Hôtel de la Poste in daylight, relaxed enough for a celebratory evening meal.

Well on the way to retrieving my then stepdaughter from a holiday job on St Tropez harbour front, our second day cracked a total of 750 miles via a wet spell. We completed some off-motorway kilometres, then circled Dijon, running back on the *autoroute* to Lyon and the warm welcome of southern sunshine. Our distinctly shabby-chic holiday-apartment accommodation was comparatively cheap, inland of Port Grimaud, about five miles and many more traffic-infested minutes from my stepdaughter's fashionable film-star location. To be fair, she was nobly working all hours in a summer pop-up café and sharing packed worker accommodation with many like-minded Brits and other nationalities.

I'll spare you tales of our nine days of blazing sunshine, frequent swimming, impromptu but tasty meals and the accessible public beaches – attractions that made this Mediterranean coast such a wonderful experience for northern Europeans. That was before the south of France became so expensively fashionable that other destinations, such as Spain, became the favoured venues for mass tourism.

At the end of August we collected an emotional stepdaughter, prised her away from workmates/best-summer-holiday friends and Mr Hot Lips, to pound our way back to Britain. Back in Berkshire, the grubbier Escort had honourably served us a tad over 2,000 miles, the majority in France. Because it had been driven hard – and carried all our stuff and the debris my stepdaughter had accumulated in those summer months away – it never had a chance to approach Ford's flighty official fuel-consumption figures closing on 40mpg. Overall, the Escort Continental workhorse returned just short of 30mpg, and asked for little more than a pint of oil when it had covered 1,200 miles.

The unladen performance certainly felt a match for the Cabriolet we took to Le Mans, with a 0–60mph time of around 10sec, but top speed was definitely below 120mph, probably c115mph.

To me, the Escort estates – from the '70s rear-drive examples to the 1.6L I owned during the '80s, and this hard-worked '90s demonstrator – were undervalued. Their rear-hatch sills allowed easy loading, and they all accommodated surprising payloads, especially the front-wheel-drive versions, which needed no raised floor for a rear axle, and the trio I employed over a 33-year span did a fine job without notable purchase or maintenance costs. Proper workaday Fords!

'THE UNLADEN PERFORMANCE CERTAINLY FELT A MATCH FOR THE CABRIOLET WE TOOK TO LE MANS, WITH A 0–60MPH TIME OF AROUND 10SEC.'

CHAPTER 11

SIERRA XR4i AND XR4x4

Mr Versatility in so many suits

The Sierra, which ran from 1982 to 1993, was bravely styled in the aerodynamic mould, but retained Ford's faith in rear-wheel-drive, a transmission system that had become an increasing rarity amongst '80s mass-market manufacturers. By this point, front-wheel-drive was the norm for high-volume production, and rear-drive was a featured asset for then less-common German prestige marques Porsche, BMW and Mercedes. The Sierra's sleek shape was a brave step forward, especially for the home market, for the British held the now-aged and distinctly three-box Cortina as a national icon, if only for its starring roles in pop culture, and for serving the nation as the sales rep's fleet-market office-on-wheels through preceding decades.

◀ The back of the most radical model update in Ford's European history. I drove these two preserved Fords in 2013, illustrating the major changes, which was educational, but not emotionally engaging. The mass media had a field day of 'jelly-mould' name-calling. *Author*

▲ August 1985 saw me out in Andy Rouse's Sierra XR4Ti at Snetterton for some mixed weather, two exhilarating drives and a lot of banter from fellow motorhome refugee Barry Sheene. *Motorsport Images*

SIERRA EVOLUTION

Even in the technically more sophisticated continental European markets, the slippery three- and five-door hatchback outline of the Sierra was a bold statement, aerodynamically a giant step forward over the Cortina (basic Sierra models had a drag coefficient of 0.32Cd versus the late-model Cortina's 0.45Cd), albeit a bit unsteady in a side wind for *autobahn* use. The independent rear suspension was a significant blessing for those countries like Belgium that retained cobblestones for some road surfaces. Steering earlier performance Ford live-axles in the Cortina and Capri over such sections was a guaranteed knee-trembler, one that left your eyes wobbling.

Over the years, the Sierra was available as the Sapphire in South Africa, with that badge reserved for the 1987 debut of the Sierra Sapphire four-door variant in the UK. In America, the Merkur branding was used for a turbocharged high-performance version (XR4Ti) that had common body roots with the V6-powered European XR4i, both of which we explore in the following driving memories.

The distinctive Sierra design was the product of wind-tunnel analysis and stylists Uwe Bahnsen and Patrick Le Quément, championed by high-profile Ford executive (ex-BMW and GM) Bob Lutz. Bob also backed the sale of XR Merkurs to the USA at BMW prices, and the XR4i's 'stop and stare' styling.

During the 11-year production span of the Sierra, its styling rewarded the courage of its creators and was only lightly facelifted in 1987 and 1990. However, it wore more powertrains and panelwork suits than a sharply dressed man. During its lifetime, the Sierra could have three, four or five doors, in hatchback, three-box saloon or estate bodies, and 2.7 million examples were sold worldwide. With a range of petrol and diesel motivation, a Sierra buyer could choose from 60bhp basic petrol 1.3L drudge to rare 220bhp RS500 Cosworth motorsport special.

DRIVEN

Road cars

Pre-production 1.6 to 2-litres (four examples) 25min filming action, *The Open Road is Calling*, road Sierras and dusty off-road, plus mass-formation sequences, south of France and UK studio, April–May 1982. **XR4i 2.8i advertising launch (two examples)** San Francisco TV and brochure filming and photography, public and US National Park roads, February 1983. **Sierra 2.0 Ghia** TV ad, *Alone on the Road*, freezing Dartmoor action, Feb 1983. **Press XR4i (two examples)** Mulhouse, French-German border, JVX 3Y and test car, 12–13 April 1983; 128mph, 0–60mph in 8.6sec, 18.2mpg. **AVJ XR4i turbo** Road tests, published July 1984 CCC, 1985 *Performance Car;* similar 200bhp performance to Janspeed levels; caught fire at Millbrook, but quickly contained! **Janspeed XR4i turbo** Road test for *Motor Sport*, published September 1983, OWS 33Y; uprated suspension, 205/225 BF Goodrich tyres on Compomotive alloys; 137mph, 0–60mph in 7sec, 18mpg; the best I tried. **XR4x4, 2.8 V6 (four examples)** Ride and handling tests in Essex, plus two UK media drives, 1985–86; C776 FHK long-term test for *Performance Car* magazine. **Modified XR4x4 (two examples)** Road tests, September 1989, published in *Motor Sport;* 2.9 Turbo Technics twin-turbo, 225bhp, 273lb ft; C-plate 2.8i supercharged Sprintex, 250bhp. **Sierra 2.0 GL** Back-to-back test with Cortina MkV, Gloucestershire, 2008; depressing. **Sierra XR4i** Refresher at Salisbury race course, September 2021, A276 RPP, courtesy Allen Patch/XR Owners Club.

Competition cars

Group A Merkur XR4Ti Track test, Snetterton, published *Motoring News*, 4 September 1985; car of Andy Rouse, UK British Saloon Car Championship winner. **XR4x4 production racer** Snetterton 24 Hours, June 1986; slow, finished 19th after one-hour-plus repairs...

It wasn't always a rear-drive machine either, as an efficient 4x4 variant, in radically reworked XR4x4 trim (modest by comparison with the XR4i), was revealed in 1985, and its Ferguson transmission system could also be found in later RS Cosworth four-doors and the Escort RS Cosworth three-door.

So, the initially controversial Euro Ford played many parts, so much so that we've had to separate off the RS Cosworth performers and competitors to appear in a later chapter.

My miles in Sierras had some echoes of the first front-drive Escorts, because my first encounters were for pre-announcement television advertisements. However, there were big budget differences, in that the Sierra filming was a truly massive effort spanning April to May 1982 in England and on the Côte d'Azur. The French location, and the presence of Rémy Julienne's pedigree stunt squad (James Bond movies and all that) ensured a truly memorable experience. And so did our British presence in France during the Falklands fighting, with one national newspaper headline about the torpedoing of the *General Belgrano* showing a rather different attitude from *The Sun*'s 'Gotcha!': the French version translated as: 'English terrorists sink Argentine warship'.

There was also a chance to steer a brace of black XR4i Sierras in and around San Francisco during February 1983, an equally attractive location to me. Sadly, the US weather didn't behave, and it turned out that most of the filming and brochure photography was done in a rush during the closing day-and-a-half sessions of an extended stay.

For our film-crew bubble – and an enormous US TV audience – the major attraction during that period was the final episode of *M*A*S*H*.

THE SIERRA'S BIG-PICTURE DEBUT

This dream 1982 job, starring Sierra the Euro-debutant, became the biggest TV and film production I participated in as a freelance driver. Usually, I drove in productions that took a couple of days and perhaps 6–12 personnel, but not this time. With over 30 permanent crew and management, two five-star hotels on Nice's posh Promenade des Anglais were booked from 25 April to 12 May. Additionally, six people from Rémy Julienne's stunt team were booked for several days. There were two studio sessions, the magnificent roads around Nice on which to deploy seven hand-built pre-production Sierras of vastly differing power and trim specifications, plus a day of rough-road Sierra action for 1981 World Rally Champion Ari Vatanen, filmed with the help of veteran Vietnam helicopter ace Marc Wolff (see pages 208–209 for more on him).

No wonder the rumoured budget was over £1 million (nearly £3 million today).

Just looking back through the kit logged on my call sheet, all needed to record high-quality images and sound, makes me admire the logistics involved, considerably more than I appreciated 40 years ago. For a start, the movie cameras used in 1982 were bulky Arriflexes that needed multiple limpets (suction pads) and minor scaffolding to attach to a moving vehicle. A specialist Citroën-based camera car with multiple mounting points and Citroën's fabulous hydro-pneumatic suspension was used to complete car-to-car action sequences without notable vibration, and an Elemack dolly train that ran on 30ft of track for slow, precisely focused movement tracking was available. Those substantial items give you an idea of the sheer managerial effort required, and why we had to wait half an hour at Nice airport for the camera equipment that hadn't been transported direct to France – try a massive assortment of lenses from 18 to 800mm, 85 rolls of 35mm Eastman colour film, 50 of them carried as hand luggage. Then 20 empty film cans to take used stock, plus eight Motorola walkie-talkies, six rolls of gaffer tape and eight of camera tape… It just went on… and on… But what about the cars?

Actually, we did the first filming day without the Sierras. A 6.48am call (film rule 1: grab every minute of shut-eye), depart from Hôtel Negresco at 7.20am, to find you are also hired as a unit driver in one of five Ford estate vehicles, but not the prop department's Transit. Drive smoothly, but as fast as possible, through what passes for a French deluxe rush hour, leaving the back seats for the responsible and highly paid director and producer to discuss their plan for the day.

From memory, I drove a variety of left-hand drive Granadas for these morning commutes, but that first day we also had an Escort hatchback, one that I shared with ad agency Ogilvy & Mather copy-chief Peter Sugden. We mounted a bumper-camera on it to shoot some empty-road sequences, from rough hillside stuff at Mont Vial, to Gréolières and the smoothest main roads outside Germany. A key sequence demanded La Colmiane's two tunnels – subterranean roads that would feature rather more dramatically in the next 24 hours of my Sierra life.

So that was Sunday and Monday done, but Tuesday was a stand-out day, as a pair of Sierras straggled into view at the tunnel venue, a couple of hours after our 7.15am departure from the hotel. These were the star – or 'hero' – cars. For the UK, there was was a RHD metallic silver job, packing the 2.3 Cologne V6 motor, with a glass sunroof and Ghia trim. To serve continental Europe, the British pre-production tech aces had additionally created a LHD 2.0 GL, also in metallic silver. These silver-dream Sierras did the bulk of the filming.

By 11am, we were all waiting around the tunnel mouth, along with a fog machine. I'd had time to slip into the mandatory film fancy dress: grey overalls over sporty top and trousers, red shirt and matching 'kickers', or trainers as we might say today. Fetching!

You might say it was a short and (bitter) sweet call to action. I was nervous because I'd had serial advice that these Fords weren't my usual disposable fare, but hand-built one-offs of huge value. More significantly, these Sierras weren't immediately replaceable, and a film unit without a vehicle focal point is a very expensive lesson in the true cost of idle skilled labour.

I fussed to a comfortable position in the RHD 2.3, trussed up in my hated uniform. I wasn't a fan of driving trimmed production cars into uncharted situations wearing shiny nylon overalls – those outfits appeared to be the worst kind of flammable material. I had arrived at too many blind corners and crests to find a film-unit van with cameraman vulnerably located, often waiting for me to appear in a style complying with the director's maximum spectacle and available velocity commands.

I sped the V6 Sierra into ¼-mile tunnels, smoke swirled around threatening walls, and we ran straight and true, man and machine emerging sweaty, but happy, into sunlight. I drove back at

'I SPED THE V6 SIERRA INTO ¼-MILE TUNNELS, SMOKE SWIRLED AROUND THREATENING WALLS, AND WE RAN STRAIGHT AND TRUE.'

40mph thinking this was a piece of the proverbial piss! Time to switch Sierras and driving side, into the LHD 2-litre. I asked the rougher-running Pinto four-cylinder for some director-pleasing Pace with a capital P. Mmm… it seemed a bit reluctant, but I pressed on into the smoky tunnel twilight. I think we were only a third of the way in when my perspective changed: surprise, surprise, and major sideways travel. I twirled the steering wheel onto maximum counter-steer lock, and became an agitated passenger for some metres, desperate to avoid hitting anything solid.

Car and driver slewed to an untidy halt, the Sierra apparently in mechanical distress and unable to move without groaning. The rear-drive differential had done its best, but with no oil visible within the casing, it had seized. Both of us were retrieved back into daylight, but there were long faces all round as it was realised this vehicle wasn't going anywhere soon – and there were no spares to hand to effect a repair.

Until the 2.0 GL could be repaired, that meant we'd be at Stage 4, Victorine Studios, in Nice. I got a break from the traditional early-call action days, meeting a few others for some posing, plus twiddling of steering wheels and gear levers to provide the details that car manufacturers love for brochures and advertising. It was an easy day, with time for a sunshine lunch with distant sea views: since you ask, we had pastries, paté, steak and veg, and I indulged my then obsessive love affair with any apple tart.

While I was lounging, or writing a small Fiat X1/9 book in longhand, those with big money on the line were frantically calling up the Ford engineering cavalry. Obviously the local Ford dealership couldn't help, as the Sierra had yet to be released for sale, but our resident ace mechanic from the UK pilot plant worked in the Mediterranean heat for three hours solid, and managed to free and lubricate the differential. That tough 2-litre then ran adequately for the remainder of the filming days.

'I TWIRLED THE STEERING WHEEL ONTO MAXIMUM COUNTER-STEER LOCK, AND BECAME AN AGITATED PASSENGER FOR SOME METRES, DESPERATE TO AVOID HITTING ANYTHING SOLID.'

Our fourth day was one of the best from a driver's viewpoint. Atop Mont Vial's dusty 7,000ft access road lay a meteorological station, and we could plough up and down there for an 8am–4pm series of loose-surface runs. These were fun, as the pair of rear-drive Sierras could be made quite playful even with hastily assembled production powertrains. Besides the helicopter dudes, we had that Citroën-based flatbed camera car with total professionals aboard. They included driver Claude Rossignol and the key camera operator, laid-back but highly skilled Seamus Corcoran. It was one of those sunshine days when you'd pay just to be there…

The following day was given over to the aces: factory Ford rally driver Ari Vatanen and airborne Marc Wolff obtaining the best possible loose-surface footage from the Sierra pair at a location close to Gréolières. I went along with Peter Sugden to watch the heroics, the softly suspended Sierras surviving nobly. We had the privilege of a Vatanen-chauffeured ride back to the hotel, plus a lot of informal information on how the Escort RS1700T test programme was progressing, then a vital topic to anyone interested in 1982 Ford Motorsport activities.

The weekend saw no let-up in the work schedule, but we had to cover some of the ground already visited. The tunnels were reinvestigated, and I was rightly told not to ruin some shots by braking when pitched from strong sunlight to smoky tunnel. Because of a weather change at Mont Vial on that meteorological-station road, some of the previous footage had to be reshot on my seventh day away, Sunday.

On Monday I had the delight of meeting Rémy Julienne's troupe, led by the man himself and comprising a bunch of like-minded French racers and rally drivers, the latter proving particularly adept at switching from tarmac rallying to racing. The Sierras drove, in a broad V-formation and a village sequence, included the now well-used silver 'heroes', complemented by paired continental European LHD or UK RHD cars in five-door Ghia and L ('base') saloon and estate bodies. Owing to pre-production restrictions, there was only one estate available in UK specification.

For me, that formation driving with Julienne's people became another highlight, with plenty of naughty driving between filming locations and some excellent banter to contrast with the anti-British tone of the French media as they reported Falklands forays.

Meanwhile, there were other more immediate setbacks for the film unit. A rented BMW for agency producer Peter Harrison was bent by the official film-unit driver and a more serious head-on occurred outside Valbonne, but fortunately there were no significant injuries. However, potentially worse was a French helicopter pilot taking over from the unavailable Marc Wolff at Gréolières les Neiges. Monsieur managed to clip the ground with a rotor during an action sequence: no problem for me, but director Gerry Poulson had facial cuts, and others had cuts or bruises, including the pilot.

So, there had been several serious incidents in less than a fortnight, and I wasn't overjoyed when my Ford political problems also boiled over with a threat to pull me off the shoot. Why?

Ford UK PR heavyweight John Waddell was far from pleased when he discovered this wretched former PR employee/motoring journalist out there again with pre-production product. Fortunately, there were some skilled agency diplomats, including Peter Harrison and John Banks. I signed yet more paper manacles, and carried on steering my merry way, guided by a series of sparkling French commands from the Julienne squad.

Rémy would get the French aces on full alert, and I got a stunning blonde lady called Brigitte with other commands: *'Geremeee, partir…'* or *'Moteur regulé, allez vite!'* There was more, but you get the picture: it was a great driving job, generously paid and marred only by internal politics. These covered keeping Ford back in Britain and three executives staying in Monaco happy. Plus a potential divorce for a key senior crew member, including messily aired grievances.

The film days rolled by, with mostly outside action runs for me, or boring studio bits, including the traditional routine with car jacked up to record speedometer and rev-counter twitches, these to be intercut later with the action shots. Because of the incident-packed filming and consequent delays, and depending on whom the Rémy Julienne people had available, I graduated to driving some of the other Sierras in the 'formation flying' interludes with the squad.

My diary tells me I did one long day with the green RHD estate and the two 'hero' hatchbacks, plus long-lens main-road sequences, repeated wheel-spin starts, and some fancy footwork with the pedals of both LHD and RHD models.

Looks like I did some ill-advised marriage counselling between takes, but ended that rewarding work day with a champagne goodbye to the departing Julienne magic mercenaries. I returned to the UK and a few filler film calls for that Sierra epic. There were some outstanding driving memories that meant I took the Sierra's merits seriously, while the UK newspaper media assassinated its 'jelly-mould' looks. I also met some great temporary Anglo-French workmates, and had the joy of knowing that at least one couple met and subsequently married, while the potential divorcees patched up their world with a mega-holiday!

▲ The print advertisement that accompanied the release of the Sierra launch TV ads in summer 1982, mostly filmed in France. The massive money and publicity effort Ford of Europe put behind the Sierra debut was unprecedented. *Peter Sugden Collection*

XR4i: EXTROVERT CHARACTER

The Sierra range featured some very extrovert performers. Today, most remember the now exotically priced classic Cosworth three-doors, with their huge rear wings and cleverly (and cheaply) reworked front-end air intakes and bonnet vents, but there was a predecessor with an even stronger inclination to show off. This was the 1983–85 XR4i, which came in a unique three-door body and featured the Cologne 2.8i motor of Granada/Capri familiarity, alongside a five-speed gearbox and uprated independent suspension.

The XR4i shouted its intent with maximum red striping fore and aft, and even around the horn buttons, gear-lever knob and dash-panel surround. A biplane rear wing set off a decorative, rather than functional, outline (not balanced by a front spoiler), and a set of multi-hole RS alloy wheels sported then-generous 195mm-width tyres.

Today, and in period, the XR4i – and a German-American turbocharged XR4Ti cousin – is dismissed as a commercial flop, but my recent research made me think that these flashy Sierras did a better job than their reputation reflected, especially in motorsport.

The following XR4i filming memories are extracts from the copy I submitted in June 2018 to the XR Owners Club magazine, reproduced here with the permission of editor Allen Patch.

'Back before some of you were born, back when a 125mph faster Ford cost a bit over £9,000, I had a part-time job. Tagged "precision driving", it meant that I got to steer performance and mundane vehicles for TV and cinema ads, plus those glossy-brochure photo-shoots that are prime collectibles for Ford loyalists today. Most of my Ford work was on Capri, Fiesta, Escort and Sierra brands, although I did a couple for Granadas too. The XR4i was used for the Californian adventure related below.

'Top-billing should go to the controversial black and red-striped XR4i pair that arrived in San Francisco for a TV-commercial shoot in February 1983. They literally stopped the traffic when we used the Golden Gate bridge, Mustang drivers the most aggressive in asking: "What the hell kinda Ford is that?" Frequently followed by: "When can I buy one?" They hadn't seen the Sierra before, never mind one with biplane wings…

'Our first action day saw a 5am call at Fairmont Hotel on top of Nob Hill (really…), and the first of many commutes over the Golden Gate bridge to Mount Tamalpais State Park. The XR4i was a pleasure to drive through traffic, or even dodging mud slides when let off the leash through uncluttered terrain. At this state park, the brilliant surprise was a closed, slick, black-tarmac road – the smoothest and most photogenic I've seen – contrasted with yellow double lines over the numerous crests and curves for maximum pictorial impact.

'I suppose we ran less than two miles of it before a turn-around for the next shot, but the complete, costly effort – and the contemporary Queen's visit to be met by US President Ronald Reagan – was bugged by fog and heavy rain virtually throughout. So when it came to proper Californian weather on our scheduled last day, the action was frantic, in

▼ The interior of the showroom XR4i was heavily decorated with red-stripe detailing for the gear lever, instrument surrounds and even the horn buttons on the two-spoke steering wheel. *Courtesy of Allen Patch/XR Owners Club*

order to squeeze in enough footage and also the demands of the stills picture guys, who virtually had to wait in line while the massively paid moving-picture crew earned fees that stretched beyond £1,000 a day. Yes, it really was like that back in the "Loadsa Money" '80s.

'As for many of the other commercials I did, I hated turning the cars around in a series of back-and-forth shuffles, so I'd handbrake them around. The XR4i was surprisingly good at this U-turn manoeuvre, but nothing equalled the first-edition XR2s for slalom agility, whether on a sandy Welsh beach or a narrow and rain-soaked Lakeland pass. This trick was accepted when the film unit were in a hurry, but otherwise unpopular with watching Ford hierarchy.'

To finish, Allen Patch asked me a few XR4i questions that you might like to see answered:

AP: 'GJN 718Y was, of course, one car – was the other registered?'

JW: 'Both carried promotional non-dating plates for filming and commuting, as the US allowed private plates anyway. The Essex GJN plate was only for the UK brochure as I remember it. Unusually, these pre-production XR4s were complete, yet a lot of the Fords I drove in advertisements were badged with higher trim levels, and had wider wheels fitted, yet carried basic engines for filming, depending on what could be shipped to meet filming/broadcast deadlines.'

AP: 'How long did the shoot take?'

JW: 'Planned 10 days, stretched to 12 before the 11-hour haul on a 747 home. I did the international press trip to Mulhouse (home to the Bugatti/Schlumpf museum and Peugeot) for the XR4i two months later.'

AP: Were you the only driver?

JW: 'Yes, on the XR4i and most Ford TV ads of 1979–84. I inherited the British-based job from contracted Ford rally-team drivers – who found the work boring – but Ari Vatanen was the lead driver for two days on the main 1982 French-based initial Sierra commercial (extended to 25 minutes for dealers). The Rémy Julienne team commanded their own guys, plus me, for the formation sequences. I later shared the crossover-pattern driving on a Spanish aerodrome Escort and Fiesta TV ad with Bob Constanduros, who subsequently became a Formula 1 presenter.'

8 Suddenly, there it is, looming through the windscreen: a gleaming challenge in tarmac dipping and weaving along the clifftops.

Short straights end in tight bends: you will need brakes that bite hard. Vicious drops border some of the curves: you'll need pinsharp steering and leech-like grip. Sharp crests lurk here and there: you'll need rock-steady stability. And there are dips too: your suspension will have to be good.

To go quickly on this road, surging out of the bends, ignoring the dips and conquering the hills, you will also need an engine that revs as strongly as it pulls, and a gearbox that suits it perfectly.

You stop for a moment and weigh up the scene. What you're seeing is the perfect road, just waiting for the ideal car.

And from what you've already learned about the Sierra XR4i, you suspect that it has the ingredients to meet the challenge.

Its performance starts with its looks – a blend of the beauty and aerodynamic efficiency of the Ford Sierra taken to striking new heights.

Three doors instead of five increase the raciness. Apart from creating a dramatic new look for performance cars, the unique bi-lane tail spoiler from the Probe III barrier-breaker cuts drag and increases high-speed stability.

Cars as powerful as the XR4i need wide, low-profile tyres. To counteract the aerodynamic drag they normally cause, there are clever side mouldings and shrouds around the wheelarches. The cast alloy wheels, visually superb, are also shaped to cut airflow disturbance down the side of the car.

Yet despite the Sierra XR4i's sheer style, lifting the bespoilered tailgate shows that there's still the roomy boot and load space of its five-door sisters. The looks aren't achieved at the cost of practicality any more than they are at the cost of efficiency.

▲ Some of the action sequences on the photogenic roads we used outside San Francisco for the Sierra XR4i's TV and brochure advertising. A memorable outing, but for 85 per cent of the trip it was shrouded by fog with heavy rainfall, so most filming became crammed into two days. *Courtesy of Allen Patch/XR Owners Club*

▼ Reunited with an XR4i in Wiltshire, thanks to Allen Patch of the XR Owners Club. This was some 41 years after I worked on the TV advertisement in California. *Allen Patch/XR Owners Club*

UK CHAMPION SIERRA-MERKUR XR4Ti: TRACK RECOLLECTIONS

By 9 August 1985, I knew Snetterton pretty well, having attacked my first British 24-hour race two months earlier and raced there regularly since 1972. This time it was a relatively short afternoon acquaintance, but a major privilege, as I was to drive Andy Rouse's British Saloon Car Championship-leading Sierra XR4Ti.

The Anglo-American Ford had already won seven of eight championship rounds contested, and would go on to notch up Andy's third of four British Championship titles. I had been allowed to drive Rouse's race cars since his 1972 Escort Mexico series title winner. There would be more Rouse-JW racer access into the next decade, on through his Rover V8 winners, and later the legendary Kaliber Sierra RS500 and the amazing early-1990s Mondeo V6, but this XR4Ti outing was a real racing-journalism rarity, as Rouse's Sierra had yet to complete its UK championship schedule. Usually, you do not risk the media messing up your championship chances with a one-off publicity outing before you've got your title and want to sell off the winning vehicle…

Looking back, I'm not sure this Ford existed beyond the race track, as it combined the European XR4i body that normally housed the Cologne ex-Granada and Capri 2.8-litre V6 with RHD steering. In this homologated Group A racing format, the German V6 got dumped for the American four-cylinder 'Lima'-motor redevelopment of the Pinto 2-litre, displacing 2.3-litres. This was a heavyweight, weighing considerably more than Rover's alloy V8.

Why did they go to all this trouble? Because of just one word that salesmen and advertisers still love, even when applied to household goods, and racers love too: 'turbo'!

From a racing viewpoint, turbocharging was leading-edge technology at that time. In 1983, a Brabham-BMW had become the first turbo Formula 1 World Championship-winning car, and its 1.5-litre four-cylinder motor had been developed from a saloon-car racing application and a production iron block.

MOTORING NEWS, WEDNESDAY, SEPTEMBER 4, 1985

TRACK TEST: FORD SIERRA XR4Ti

Champ's choice

Clever location

173 mph timing trap

Turbo technology

Tyre talk

Body talk

Action

➤ A rare beast, nominally an XR4Ti, technically as supplied to the US market under the Merkur badge. It packed a German-assembled four-cylinder turbo engine and the outward appearance of the six-cylinder XR4i, but as this two-page *Motoring News* track test displayed, this was a RHD car. *Motoring News*

Ford of Germany had featured in the same DTM race series (with Zakspeed-Cosworth Escorts and Capris) as BMW, and was right on the turbo case. The Sierra Cosworth plan was green-lighted in Britain in 1983, but both the German and US Ford factories were just as turbo-aware.

Under the skin: straightforward and reliable turbo power

The performance Sierra that went on sale in the US featured German assembly at Karmann, using an XR4i body mated to the four-cylinder turbo motor that had also been assigned as an option for 1984–86 SVO (Special Vehicle Operations) Mustangs in Fox-coded bodies: I did drive such a Mustang, as recounted in Chapter 4. This German Sierra with 175bhp American motivation was sold in the period 1984–89 through Ford's Lincoln-Mercury division as the Merkur XR4Ti.

The beautifully presented three-door XR4Ti racer had Ford UK backing and was prepared by Andy Rouse Engineering, which Rouse ran with partner Vic Drake, two ex-Broadspeed staff who had kept gathering new expertise long after Ralph Broad had shut up shop. The project also served to prove some competition hardware outside the engine bay for the forthcoming Sierra RS Cosworth, particularly the rolling chassis.

The Rouse interpretation of the XR4Ti performed way beyond the expectations of its creators and backers, and much of that was credited to the engine's accessible grunt. It durably out-manoeuvred numerous rivals by deploying many standard engine components alongside a modest – by race standards – 6,400rpm limit. The large in-line, four-cylinder, Lima-coded unit, with iron block, displaced 2,303cc (with a bore and stroke of 96mm x 79.5mm), but with the international turbo handicap factor of 1.4 multiplying that capacity, the single-overhead-camshaft unit was rated at 3,224cc. That was not far off a standard and contemporary Rover-Buick alloy V8, and the Ford unit matched that engine for mid-range performance, to the surprise of rivals and bar-room experts. Rouse said of the US engine in a 2015 interview published by *Motor Sport*: 'It was quite basic, just two valves per cylinder, but if you turned the boost up, you had more power than the Rovers, and more torque.'

An equal surprise was the reliability it delivered. 'In fact we've only ever had one engine failure, and that was when our original prototype finally let a rod go at Donington after completing all our pre-event testing,' reported Rouse. There was a head-gasket failure at Brands Hatch, but that was put down to an installation problem, rather than an engine defect.

Back in the day, Andy was happy to talk in general terms about this uprated engine for competition. There were obvious differences from the earlier Mustang SVO production installation, including Zytek electronic management of engine parameters, Bosch injectors, plus unrestricted race-spec intercooling cores from Germany tucked away in the production '2.3 EFI turbo' casting within the engine bay.

A cast-iron crankshaft and connecting rods were also production items, albeit carefully balanced to subdue some of the inevitable vibrations within a 2.3-litre four-cylinder that traced its ancestry to Europe's 2-litre Ford Pinto

▲ A unique three-door hatchback body and RHD steering served the Andy Rouse Engineering XR4Ti, allied to major mid-range power from the simple turbo motor and the usual Rouse civilised track manners. The 2.3-litre turbo four-cylinder delivered mighty torque. *Motorsport Images*

▼ Andy Rouse (right) tells the author the background story to an unexpected championship-winning XR4Ti Sierra before the arrival of the Cosworth RS purpose-built machines. The cockpit was a comfortable fit, despite the difference in driver heights, and retained some showroom reminders in the centre-console-mounted handbrake and extended gear lever. *Motorsport Images*

unit. The cylinder head was heavily reworked, and featured more generous inlet and exhaust valves. The critical heavy-duty head gasket sat within grooves in the head and block – if you think these are engineering preparations for a major rise in turbocharger boost pressure from the Garrett T3, carefully coordinated with a compatible compression ratio, you are spot on!

Rouse would not divulge specifics in 1985, but he did admit to 'well over 300bhp, well over! Torque? We have seen 340lb ft at 6,000rpm, but the real point when you are driving the car is that there is almost as much power at 5,000rpm as there is at the 6,400 limit… You can say it has a lot of squirt, or that you can spin the wheels all the way up through third gear!'

The Getrag five-speed gearbox, with first set on a dog-leg, away from the RHD steering, was then conventional fitment for TWR Rovers and Jaguars, and a natural for BMW's 635CSi coupé. However, the Rouse crew had figured out how to reduce the nose-heavy weight bias by using an elongated bellhousing that stretched so far under the cabin that no clunky gear-change linkage was required. For my drive, I had a 3.36 final drive, one of five possible rear-axle ratios, mated to a mildly preloaded ZF limited-slip differential. The gearing choices allowed a 173mph maximum in German (Avus) speed use, or 146mph over more familiar Silverstone terrain.

Slowing down was no problem, with AP-Lockheed Racing's benchmark 13in-diameter front discs of 1.1in thickness. Rears were little smaller (12in diameter and 1in thick), with a knurled-aluminium-disc bias adjuster on the lower-right edge of the dashboard.

The suspension owed some inspiration and hardware to the Sierra's German factory engineering, with a front hub casting encircling heavy-duty wheel bearings, supported by alloy-cased Bilstein gas-pressurised struts. Bottom suspension links were fabricated by Rouse Engineering, serving alongside compression struts and a selection of anti-roll bars, mounted in the Ford production location aft of the front axle.

At the back, there were obvious changes in layout, as Ford had homologated combined spring/damper units to replace the separate units used in production specification. The independent rear trailing arms were of box-section for enhanced strength, rather than using completely redesigned replacement arms.

Wheel and tyre choices were a bit limited when dealing with such strong pulling power,

and 16in x 9.5in wheels were fitted with then slightly uncompetitive Dunlops. The traditional British supplier had just succumbed to a Japanese takeover, which limited UK access to the later Kevlar-case European construction mixes, which were then reserved for 17in rims.

Aerodynamically, the Sierra shape was born for low-drag fuel efficiency, which also allowed notable straight-line speed. However, the race application wasn't so clever, as there was no front spoiler, resulting in a rear aerodynamic bias with those style-orientated double blades at the back. As the race life of this Anglo-German-American Sierra was ruled by the anticipated homologation of the Sierra Cosworth, there was no way fundamental body alterations could be cost-effective for Ford. So, Rouse and German drivers such as Klaus Niedzwiedz (third in the 1985 British GP support race with 'my' test Sierra) just had to live with it, and drive around the high-speed instability.

Behind the wheel: a consummate performer

I had two Friday-afternoon sessions in the Rouse Sierra, the first curtailed by rain, but it served well to familiarise me with the basics. Compared to more modern touring cars, the interiors of '80s Group A saloons were mandated to carry many more reminders of their production origins. So, the stalks for indicators, two-speed wipers and lights remained behind a functional four-spoke RS-branded steering wheel.

The standard dashboard carried redundant heater controls, and showroom mouldings surrounded the implanted analogue Racepart black-and-white dials. These instruments monitored the usual oil and water temperatures and pressures, but a stranger also had to remember the importance of the cooling-down lap(s), as cast-iron exhaust manifold and water-cooled turbine needed cooling from 920–980°C action heat levels to 600°C before returning to the pitlane and killing the ignition.

A digital read-out was used for boost pressure, but blanked off from my journalistic eyes for those sessions! In May 2022 I met Vic Drake, Andy's engineering partner, and he confirmed: 'The boost on the car was set at 1.8 bar and the output was

▲ The cockpit of the Rouse Sierra XR4Ti championship winner was a functional amalgam of production mouldings, analogue and digital instrumentation (the boost gauge masked for my test!), plus the faithful RS four-spoke steering wheel. *Motorsport Images*

330bhp. The engine was capable of more, but the T3 turbo ran out of puff above 6,500rpm.'

Back in the '80s I sat in a softly contoured Bob Ridgard race seat and optimistically prodded surprisingly standard-size foot pedals in bare metal. Gridded replacement aluminium sheet-alloy, fabricated to cover the footwell, ensured no slippage of your heels.

Start-up had showroom familiarity too, as you simply twisted a key, although the deepening snarl from the engine bay and exhaust, as it settled into a 1,500rpm pit-and-paddock crawl, told you this was a Sierra with serious speed attitude. The track session revealed just how accurate that suggestion of high-velocity had been.

Back then I commented of the unfashionable iron motor: 'The outstanding memory of my Sierra afternoon was that engine. It goes! Andy confessed, you could, "light up the rear wheels all the way down the pitlane if you wanted to be a cowboy".' More significantly, I recorded that, 'from the 5,000rpm point at which full throttle is allowed (by limited traction, with just sideways motoring completed in second and third), to the 6,400rpm limit, is swallowed with Chevrolet Z28 Camaro speed. In fifth gear, a solid 6,400rpm (138mph) comes up as you flash by the 200 marker board, this at the close of a shortened, but still significant, main straight.'

Although the unlikely oversize four-cylinder engine earned the headlines, the AP brakes were outstanding on such a treacherous track surface. One trait that Rouse Engineering and a few other

BANTER WITH BARRY SHEENE

That August Norfolk day testing the Rouse XR4Ti wasn't kind to us on the weather front, but we did have the merry motorhome company of then double motorcycle World Champion Barry Sheene, who at the time was in his saloon-racing era with Toyota UK. I knew Barry best from a '70s week at Silverstone driving Porsches around in circles for the Gunnar Nilsson cancer-scanner appeal. We had shared an early 911 Turbo, and the laughter accompanying mutual spins at Becketts hairpin on the Club Circuit gave us both bragging rights, as we tried to negotiate full boost on road tyres.

More seriously, I had believed in Barry's abilities to make the two-to-four-wheels transition *à la* John Surtees. Sadly, back in the late '70s, when I was contracted to supply freelance PR for the Unipart Formula 3 team featuring one Nigel Mansell and the equally talented Kiwi Brett Riley, saloon cars looked the better bet for Sheene. This was judged on Barry's considerable past leg injuries, and the age at which you are expected to shine in junior racing categories.

Waiting out an hour in the Rouse motorhome in 1985 for the rain to abate, there was much Barry banter along the lines of 'lucky journos, don't know the meaning of hard graft, get to drive all these great cars and don't have to suck up to the sponsors'. Fair enough, for I did drive most contemporary Group A winners through the '80s, including the David Brodie Starion (then major opposition to Rouse and Ford) and Tom Walkinshaw's Rover V8 and Jaguar V12 weapons. More to the Ford point, as mentioned previously I had driven Vince Woodman's smaller-size V8 Mustang GT at Snetterton in 1983. So I was painfully aware – as were any Blue Oval followers – that Ford had lacked outright winners in British saloon racing once the 3-litre Capri was out of homologation. So when the XR4Ti project was created, the search was on for Ford to re-establish a winning reputation.

'THERE WAS MUCH BARRY BANTER ALONG THE LINES OF "LUCKY JOURNOS, DON'T KNOW THE MEANING OF HARD GRAFT, GET TO DRIVE ALL THESE GREAT CARS AND DON'T HAVE TO SUCK UP TO THE SPONSORS".'

top touring-car teams managed to overcome was the understandable reluctance for a showroom saloon to exhibit flawless track manners.

In Andy's case, I think I drove half-a-dozen of his race cars over the years, and they all flattered the driver in a near-Lotus manner by offering the best-possible handling. This was especially true on entry to any curve, Rouse's racers always seeming happy to turn into a corner, although the rear-drive layout meant there would be power slides from the apex or later, according to the amount of torque and mid-range pulling power they carried. This Sierra seemed the most flamboyant, owing to the sheer grunt, and a track that wasn't fully dry, until a high-pressure fuel pump problem ended my educational fun.

Even during that abbreviated experience, the Rouse XR4Ti lapped a lot quicker in mixed conditions than I had been able to manage over dry tarmac in Vince Woodman's V8 Mustang two seasons earlier. Since this is often the overlooked Sierra in a sea of Cosworth successors, I should also add that it was one of only two examples that Rouse constructed in 1985. Rouse's engineering and driving craft applied to this XR4Ti brought Ford its first outright British title since 1968 (Frank Gardner, Escort Twin Cam) and the first scored in the UK by a turbocharged saloon.

Missing the opening event, this really was a last-gasp programme, and the team and Rouse took nine victories in the eleven rounds contested. It meant Andy had won four British titles outright, the last three in a row (1983–85) with three different cars (Alfa Romeo GTV, Rover Vitesse V8 and Sierra XR4Ti). The XR4Ti continued to win for Rouse in 1986, but took a class title rather than the outright championship.

More significantly for Rouse Engineering, the BBC televised every championship round, and Ford featured the team and the XR4Ti for a national TV advertisement. So Rouse and company were right on the radar for the 1987–90 heyday of the Sierra RS Cosworth and RS500 programmes, which would see Rouse Engineering expand to employ 30 staff at its premises just outside Coventry.

Meanwhile, I have only admiration for a unique Rouse racing record, and gratitude for the generous way Andy shared his race-car toys.

XR4x4 ROAD, RACE AND RALLY ADVENTURES

During 1985–86, both *Performance Car* magazine (totally revamped from *Hot Car*) and Ford SVE's creation of a practical 4x4 Sierra successor to the extrovert XR4i became 'significant others' in my automotive life. *Performance Car* was now an aggressively good-looking monthly, headed by ambitious technical writer and talented race driver Jesse Crosse. At its best, the magazine was capable of comparison with the redoubtable *Car*, and I had been lucky enough to gain increasing freelance work, with regular-contributor status.

Simultaneously, Ford in Europe became increasingly serious about turning the Sierra into a world-class competition car through the Cosworth programme. During 1985–86 this was proceeding to a two-stage climax (RS Cosworth/RS500) in the background, while the more affordable and everyday-useable XR Sierra went from loud-boy XR4i biplane to Mr XR4x4 Conservative of greater all-terrain/weather ability.

Ford of Britain had form with 4x4s, with the rallycross Capris and the less-glamorous road-car applications within Police-assessed Zodiac/Capri developments of the '60s, but the Sierra project was the first time the company pressed the button to make serious showroom 4x4 quantities. Dunton-based Special Vehicle Engineering (SVE) was assigned a two-year development span to debut the car during May 1985. Ferguson's FFD division created three prototypes that were handed over to SVE. Ferguson patents were involved not just in the 4x4 systems but also for the viscous-coupling limited-slip rear differential employed.

Other technical innovations concerned the brakes, featured the European company's first use of disc brakes all round, picking up on the imminent arrival of vented front and solid rear discs on the Granada/Scorpio range. That retardation would be well managed by the Alfred Teves electronically monitored anti-lock system, rather than the usual Bosch components.

Although the XR4i/Capri/Granada/Scorpio 2.8i V6 was handed over to the Sierra 4x4, along with the five-speed gearbox, the XR4i was the starting point for 4x4 development work, and the modest appearance was a kickback against the XR4i's flash. The XR4i's rear spoilers and red-line body stripes were absent, and the interior was also from the 'calm-down' department. For Britain, a five-door body was the norm, rather than three-door hatch,

◀ By the mid-1980s, Dunton-based SVE had grown following the success of earlier work on the Capri 2.8i and XR2. SVE staff created a dozen reborn Fords from the Escort Cabriolet through the second-generation XR2 and Escort RS Turbo, then the foreground XR4x4 in black. *Author's archive*

but in Europe there was a choice between three- and five-door all-wheel-drive Sierras.

Deploying the same 2.8i V6, and based on a weight increase of at least 245lb (111kg) over the XR4i, you would expect the XR4x4 to show a reduction in performance and fuel consumption compared with its rear-drive predecessor. Yet the transition to 4x4 and another pair of doors, plus the ABS brake components, didn't inflict a significant penalty, as the XR4x4 returned 0–62mph in 8.5sec, rather than 8.4. Both cars were capable of around 130mph maximum speeds – around 5mph up on the Capri 2.8i family, illustrating that aerodynamic progress had been made.

Even official Ford figures recorded a drop from the XR4i's 29mpg, to 26mpg for its 4x4 successor. I returned 21–25mpg over 23 cross-checked refuels, but that 25mpg was a fluke during gentle holiday travel. The overall average I recorded centred on 23mpg during our long-term relationship, with hiccups from a cracked distributor cap the only blemish on the car's mechanical record.

Following the 1985 RAC Rally in an XR4x4

The XR4x4 came to me in the autumn of 1985 as a very fresh unit, one that I promptly ran-in and then booked in for its first service. This occurred while covering over 2,000 miles as a reporter on the 1985 Lombard RAC Rally. The dark-grey Ford came fully loaded with a glass sunroof and the £869 optional ABS braking, and was always worked hard during my 8,400 miles on *Performance Car*'s long-term fleet, not just during that initial rally coverage, but also to probe the potential of four-wheel drive for the 1986 Snetterton 24-hour race, and laden for a family holiday with teenagers who had no complaints about our transport. That was a rare compliment!

Even as the RAC Rally graduated from tortuous driving tests to a full-on '60s spectacle for special-stage tests, where the fastest time is all that counts, it was a special event for those who supported the hardy annual. In 1985, fans – and, whatever the November weather, thousands lined the muddy forest trials that decided the bulk of the results – witnessed the debut of a home-team Group B supercar. Ford's purpose-built turbocharged 4x4 RS200 would not be recognised until the 1986 season, but Austin Rover's British-crewed Metro 6R4s drew the crowds, and pushed TV coverage to another level.

That year, one of the longest routes in the World Championship (2,151 miles/3,465km) was based in Nottingham and twisted out in separate legs, first swerving south to face the Welsh challenges, traditional forest territory that remained the heartland of Britain's World Championship round into 2019. A second leg took the cars north through the toughest English terrain (notably Dalby and Kielder forests), then dipped into Scotland, before a major halt in Carlisle and some unforgiving Lake District challenges.

The five days of competition had few overnight rest halts. For comparison, crews were faced with nearly 80 hours' driving and 33 hours' rest, whereas a more-recent Welsh-based Rally GB round of the World Championship split drive and sleep time more equally at 37 and 32 hours respectively, all over a much-reduced mileage.

The 1985 action started on Sunday with the comprehensively televised public parkland stages surrounding such as Chatsworth House, Trentham Gardens and Sutton Park. Frequently dismissed as 'Mickey Mouse' tests, they often caught out even leading professionals who underestimated their slippery charms and hidden solid obstacles. As Sunday evening plunged into British winter darkness, the tougher Welsh stages took the competitors into unforgiving timed tests over forestry stages.

That year's RAC officially occupied Sunday 24 to Thursday 28 November, but for myself and navigator/concurrent *CCC* magazine editor Peter Newton, never mind the competitors, preparations started much earlier. Our route planning, to file reports and see day and night action, demanded a substantial number of large-scale Ordnance Survey maps, and an eye on accessible public buildings to gain telephone access in those pre-cheap-mobile-phone days. In addition to our journalism work, Peter had a magazine to multi-task through one of the biggest issues of the year.

I had four editors wanting their stories before I cleared off on the rugged mission. A couple of driving days had to be completed away from the keyboard of my early IBM PC, steering a BMW 325i through Millbrook track performance routines, and assessing a Metro 6R4 Clubman. The latter 250bhp Metro mutant became vital to the patriotic editors of sportier motoring magazines, pending the Tony Pond/Rob Arthur and Malcolm Wilson/ Nigel Harris factory-crew full-fat 380bhp Metro 6R4 entries on that 1985 RAC.

We got into Nottingham on Friday 22 November, attended a number of functions – including artist Jim Bamber's pop-up gallery – and Peter got over his disappointment that we would share a Ford instead of the Audi quattro we had used to cover the route in previous years. After 2,000 miles through every kind of weather, and over surfaces from icy to muddy trails, both of us left the understated grey Ford with respect for its capability and comfort.

Our reporting duties for the organisers and our media outlets started with observing Saturday's technical inspections for a healthy 155 entries. Really, that was just a day of making sure we had insider team contacts prepared for our hasty intrusions at stage finishes, service halts, naughty roadside repairs and longer halts. A hard-pressed crew would need to know your faces and how significant your enquiries were, especially in the heat of battle, fighting for every special-stage second, and faced with technical/management problems that their car-manufacturer paymasters and commercial sponsors would prefer were kept out of the media.

The action was immediate, from an 8am start-ramp on Sunday, all in typical British rainfall. For us, there was maximum first-day interest with the presence of triple World Champion Walter Röhrl with Phil Short (yes, a very tall man) navigating a 500bhp development Audi quattro with the pioneering PDK (Porsche double clutch) gearbox. This box of tricks inspired the redeveloped semi-automatic gearboxes found in many of today's showroom performance cars. On this occasion, it was promptly rolled down a Welsh hillside, so that was one of our early stories to report, followed by the retirement of the remaining factory Audis and Peugeots. So, plenty of telephone calls back to HQ before a restart on Tuesday for the northern stages.

That northern foray was harder than ever, and included a healthy ration of snow, and a roll-over accident for winner Henri Toivonen. With their evocative Martini livery, the Lancia Delta S4s dominated the results, these unique 'turbo-supercharged' machines finishing 1–2. There was a reward for hardy British spectators and media too, as the Metro 6R4's debut on a World Championship event saw Pond/Arthur secure third overall, which would be the best-ever World Championship result for the mid-engine Austin Rover – a technical standout amongst the Group B supercars, as it lacked forced induction of the super- or turbocharged persuasion.

▲ I was a privileged passenger to the real Stig (Blomqvist) in Saabs, Audi quattros (Stig was 1984 World Champion for them) and a brace of Fords – this XR4x4 over icy media-launch terrain, and a Texaco-emblazoned Sierra Cosworth (D373 TAR) over mud prior to the 1987 RAC Rally, where Blomqvist was third at the finish, amended to second in the final results. The ace Swede was also a fine tarmac driver. *Motorsport Images*

Snowy XR4x4 miles en route to Wales

There were plenty of adventures for the Sierra, covering over 2,000 miles during RAC Rally week, but they weren't finished after the event. I had an equally challenging outing when asked to test two competition cars at Pembrey, in South Wales, on a day that was snow-laden at home in Oxfordshire. At that time I had a £38,000 test Porsche 928S parked outside while my XR4x4 long-termer rested at Porsche's UK HQ in Reading.

Manipulating the automatic-transmission Porsche, and 300 horsepower that liked to surge forward after a cold start, down an ice-topped, ungritted urban descent was sufficiently tricky for me to run straight over to Reading and get my Ford back! And yes, the Sierra did the job calmly, albeit the delay due to changing cars encouraged me to use 4,000rpm top-gear pace, giving an indicated 100mph and burning a tankful of fuel at 21mpg.

By this stage, the 2.8i motor was ageing and I found myself not relishing life over 4,500rpm. It did free off, but from a 2023 perspective, my estimation is that you might be better off with the late-run 2.9i, which would make a useful all-round load carrier in the equally later-run estate 4x4 body, when the model lacked an XR badge.

I should leave these contemporary XR4x4 road stories by commenting on sales figures. The media verdict on both the XR4i and Merkur XR4Ti was that they were flops, and that was certainly my impression. But, the XR4x4's much longer production run (1985–92) and availability of variants (2.9i and 2.0Si, estate and three-door bodies) saw 23,540 sold across Europe. The abused XR4i (1983–85) sold 25,662 and the equally dismissed Merkur XR4Ti kissing-cousin, with American engine implant, recorded 42,464. The Merkur won racing titles in American SCCA categories, and Ford of Britain benefitted from two seasons of XR4Ti race success before the Cosworth RS deviant was created and homologated.

I think a little more Sierra XR respect should be allowed after living in the shadow of the RS Cosworth for more than 40 years.

XR4x4 racing debutant finishes... just!

How we at *Performance Car* and a plethora of vital sponsors enabled the Sierra XR4x4's worldwide racing debut was a damp tale that started with our long-term all-wheel-drive V6 Ford on yet another demanding outing. Joining a winter 1985–86 BF Goodrich Porsche tyre-test day at Snetterton, I had journalistic driving duties to perform with 911s – from comprehensive 2.2-litre pace and handling tests to trying the ferociously fast and unforgiving original Martini 3.0 turbo.

Smug at surviving, and blissed-out on Bill Taylor's serial 2.7 RS winner, I manipulated a little track time in my long-term Sierra. Now it had become damp, but I had some Porsche reference lap times gathered. This 'out-of-the-box' 7,000-mile XR was definitely on the kind of (wet-weather) pace that might bring a result in the June 1986 Willhire 24 Hours if damp track conditions prevailed.

All motorsport enthusiasts knew that a much more potent Sierra Cosworth would be available for 1987, recognised for racing and rallying within 12 months, but it seemed worthwhile to attempt to establish competition credentials with this 4x4 Sierra. Stuart McCrudden – formerly a Ford Motorsport employee – had his own race prep and entries business, and was already running BBC DJ/TV presenter Mike Smith in the Ilford-backed Escort RS Turbo. Stuart and I had sporadically worked on Ford sporting projects, or races, since 1968, so when he suggested his company could prepare and run an XR4x4 in the June day-and-night event, we came to an agreement, providing £5,000 was raised in sponsorship.

Most of March 1986 was spent dealing with increasing desperation to deliver the cash, while McCrudden extracted a 10,000-mile ex-press-fleet Sierra XR – complete with glass sunroof – from the

➤ This is the stickered XR4x4 I shared with Stig Blomqvist at the 1985 Finnish media launch of Ford of Europe's first mass-production 4WD saloon car. I would use other examples to report on the RAC Rally and to race at Snetterton. *Motorsport Images*

Ford management car park at Brentwood Ford HQ.

The sponsorship and publicity results were ultimately spectacular, with backing from Roger 'Mr Willhire' Williams himself, Ford ad agency Ogilvy & Mather's Peter Sugden (a regular V6 Capri racer), BF Goodrich tyres and *Performance Car*. We were still short of the sponsorship target, when McCrudden did a late but brilliant publicity deal to run TV AM sports reporter Gary Champion and the channel's camera in the car. That enhanced profile encouraged everyone to stump up the sponsorship balance.

That tardy cash-flow meant we had a bunch of drivers in the Sierra and a poor 10 June pre-event test. It didn't help that the clutch plates failed, and that Champion clipped the rear in a low-key accident. The rapid and charming Roger Williams flat-spotted a tyre, and the dry-weather lap times were sobering. Mr Willhire's 1m 28s lap was our best, versus a 1986 24 Hours 1m 21.56s pole position (Ian Taylor, Mercedes 2.3/16). Both Jesse Crosse and I knew that XR time was tardy, recalling that I had practised Jesse's 2.0 Fiat Abarth 130TC at 1m 25.8s the previous year.

As Gerry Marshall accurately predicted: 'You'll need a 24-hour tropical rainstorm and bit of snow to get anywhere with that!'

I cheered myself up with Snetterton commuter miles in a loaned Lotus Excel, but even I could not ignore the fact that heat-wave conditions were forecast. Oh, and the Sierra had only the most basic safety preparation, for instance, a crude alloy plate blanked off the sunroof, and a 25-gallon petrol tank sat above the rear floor, rather than being cut into the floor to lower the centre of gravity and assist cornering capability. This was necessary, as the car would return to public-road life post event.

We ran with the anti-lock braking disconnected and the Sachs damping locked into its hardest settings, in association with an alleged 1in lower ride height. A magnificent Tim Andrew sunlit silhouette picture was published in the September 1986 edition of *Performance Car*, showing the Sierra staggering into the slowest section. You can see the outside wheels and tyres tortured under the load, while the inside front and rear are on ballet-dancer-tippy-toes full droop.

The race wasn't the complete disaster I had recalled over the years. When I came to write this

◀ The Sierra XR4x4 may have been an uncompetitive choice for the driest of dry Snetterton 24 Hours races in June 1986, but we certainly got a mountain of publicity, including this official event poster. *Author's archive*

text 36 years later, I found that we did finish 19th, albeit delayed by more than an hour to replace that heavily loaded nearside-front wheel bearing via a complete strut and hub change. We also had to replace the power-steering pump and belts, robbed from a brand new 4x4 estate that became our cannibalised donor car.

It had been hard work, especially for a team of regular and volunteer mechanics and helpers, changing 16 tyres and a trio of complete brake-pad sets, never mind pouring in Shell petrol by the hand-held container-full, burned at a respectable 13mpg.

There were compensations in my seven hours' racing on a rota with three others. For me, a three-hour 'graveyard' shift during the summer night was outstanding, the Snetterton funfair in full twinkling view as the Sierra tracked around the long right of Coram at 80–95mph. It was also exceptionally stable under duress, as I discovered when a Honda Civic spun across the front of me at some 90mph; the Sierra simply steered around the obstacle in mid-corner.

It had been a bold project, which garnered some useful publicity, but underlined that you should never trust the British weather to help out, or a journalist's judgement!

CHAPTER 12

RS200

Abbreviated action, long shelf life

Like the GT40 and hand-built GT70, the RS200 stands out from the production-based Ford automobile crowd that dominate this book, as it was created for a specific competition purpose.

That purpose was to return Ford of Europe to the front ranks of international rallying during the brief Group B era, after the blind alleys of front-drive Fiestas and the aborted rear-drive Escort RS1700T. With 200 examples built, the RS200 did achieve a podium place in a World Championship rally: Sweden, in February 1986. However, the Group B category was brutally swept away by crowd-safety concerns and headline fatalities, and was replaced for 1987 as the top rallying category by the production-

◀ When Group B cars were banned from international rallying, the RS200 became a popular choice for TV rallycross. Here are two leading forces in the 1986 British Rallycross Grand Prix at Brands Hatch, European Rallycross Champion, Martin Schanche (left) and Swede Rolf Nilsson. *Eddi Laumanns*

based Group A international category, featuring much less powerful cars.

Such a short World Championship life led to diversions into less prestigious events, the RS200 gathering five national rally titles in 1986 and subsequently becoming a force to be reckoned with in European rallycross. Yet, with its *raison d'être* removed, along with any halo effect from top-level victories, selling the 194 production examples to the public required a massive effort, and it took from the initial 1985 public sale and rally debuts to February 1989 to clear the laboriously constructed RS200 stockpile. Incidentally, there were six hand-built preview prototypes of the RS200 from Ford at Boreham and others, prior to the 'production' cars.

▲ The RS200 was not long for the World Championship rallying stages, but did make an early impact with Karl-Erik 'Kalle' Grundel's third overall on his Swedish home event in February 1986. It would be the best World Championship result for the innovative 'Bitsa' Ford. *Ford*

THE PRECURSOR TO THE RS200: THE RS1700T

Personally, and for specialist engine builders, the RS200 story is prefaced by an RS Ford that only made it to the pre-production stage, with fewer than 20 built in competition or prototype showroom form. I attended the Belgian media debut of the Escort RS1700T, alongside a revamped Granada, in August 1981, and was a total fan, my enthusiasm later fanned by a 1982 conversation with Ari Vatanen whilst the Sierra debut commercial was filmed. Vatanen – and fellow Finnish ace Markku Alén – had set some outstanding test times in the RS1700T. As with many other Ford followers, I waited… and waited, with increasing impatience, for this radical RS Escort to reach production and gain the required international homologation for motorsport.

The business of actually building 200 examples of this specific twist on a Ford theme was the major obstacle, as the RS1700T had the right road-car cues to connect with the hardcore enthusiast, and its testing pace in 1981 was sufficient to convince world-class rally drivers of its potential versus contemporary competitors from Lancia and Renault.

Once more, SVE's genius for on-budget delivery was tapped. Yet the complexities, and the absence of serious corporate commitment to creating a front-drive MkIII Escort body implanted with an aluminium torque tube and rear-mounted transaxle, plus a turbocharged Ford-Cosworth redevelopment of the BDA engine, were major obstacles. They retarded any chance of the RS1700T making the showroom on any realistic schedule.

In spring 1983, when it was obvious to the rally fraternity that you needed a purpose-built 4x4 to win at World Championship level, a Ford managerial meeting approved the RS Turbo Escort (front-drive) and the Sierra RS Cosworths, and both RS1700T and the C100 sports-racing prototype projects were terminated.

Now Ford in Europe needed a new purpose-built rally car to win, with even Audi's fabled 1981–84 quattro competition success fading, as it was overtaken by ground-up designs from Peugeot and Lancia that bore a cartoon resemblance to a showroom denizen. The chunky, short-term answer would be the RS200, which inherited a mild redevelopment of the turbocharged BDT engines that had already been assigned (and 21 engines built) to the abandoned RS1700T.

RS200 PRODUCTION TRIALS AND TRIBULATIONS

The following section provides a brief overview of the development and production of the RS200, but if you want more detail than I can offer here, may I recommend the book work of two men who were involved with the RS200 on the frontline. As well as being a prolific author, Graham Robson was an official consultant at Boreham, particularly producing Motorsport parts bulletins as well as running an RS200 as a long-term loan car. Mike Moreton, on the other hand, was a full-time Ford employee who wrote several books about his time at the company. His CV covered the second-generation RS2000, the RS500 evolution of Sierra Cosworth, and the Reliant operation that was handed the task of making 194 of the RS200s during 1985. He was especially successful at getting awkward projects completed within Ford and via sub-contractors.

The RS200 and the '90s Escort Cosworth were the last of the Ford of Europe limited-production homologation performers to go through the old-school creation process, preceded by Boreham-built prototypes of variants intended for motorsport, or the road. Some of the brightest Ford sporting thoughts – such as the previously mentioned RS1700T and the F3L, GT70 and C100 – were killed by mandatory minimum production numbers and the multi-million-dollar budgets required to support their transition from hand-built prototypes to series production. All these projects lacked the serious dollar support of the American parent company.

'SO THE RS200 FACED THE USUAL FORD MOTORSPORT HURDLES IN TERMS OF MANUFACTURING, INCLUDING THE FACT THAT ANOTHER 194 EXAMPLES WERE REQUIRED IN A HURRY.'

There is a lesson here for Ford executives in any of the European offshoots, which is that Ford in Dearborn has to be motivated to the very top level to make a proper job of a global contender. The last I can recall were the Ecoboost V6 Ford GTs that returned to Le Mans in 2006 and beyond, but even then the GTs took class honours rather than outright victories. Building the necessary numbers for the road was a protracted specialist operation, one confined to American facilities.

So the RS200 faced the usual Ford Motorsport hurdles in terms of manufacturing, including the fact that another 194 examples were required in a hurry if they were to face up not just to the established Group B supercars from Peugeot, Lancia and Audi, but also a home-brewed challenger in the form of Austin Rover's V6 mid-engine Metro 6R4. In retrospect, both British designs for World Championship rallying were ingenious, the Metro particularly bold in its lack of turbocharging at a time when all front-runners used it, while the Ford was simply unique.

The key barrier that had to be overcome to

◀ ▼ Interior and the double-spring/damper suspension layout details of the immaculately presented street RS200 owned by Steve Rockingham. *Author*

ensure that the RS200 made it to the international stage was the manufacture of those 194 examples beyond the hand-built initial half-dozen. First they needed corporate cash, and that was granted on 12 December 1984 in a high-level London meeting with Bob Lutz, the chief signatory. Also present in the building were Walter Hayes and Prince Michael of Kent. Lutz signed off the programme with a budget of £12 million (the equivalent of £34.8 million today). The company always insisted on trying to at least break even on the road cars, but I'm told that would have needed an initial retail price of £60,000, not the sliver under £50,000 at which the RS200 was first listed.

The RS200 programme was agreed on the basis of a first hand-built example budgeted at a quarter-of-a-million dollars, and featuring the input of multiple sub-contractors. The only key Boreham-based gent at this stage was John Wheeler, who held the overall technical brief, while many details, and responsibility for the chassis, lay with Tony Southgate.

➤ My first sight of the unique purpose-built RS200 Group B rally car was of examples being individually hand-finished at Boreham. However, they were not the competition cars shown here, but predictions of the public-road specification, harvesting less than 250bhp from the turbocharged four-cylinder. *Ford*

There were many more hurdles to circumnavigate before a mixed squad of Ford and Reliant employees at Reliant's vacant Shenstone production facility began the process of assembling those 194 'production' RS200s. The key features of the design remained: two seats, an integral roll cage, and a unique layout for the transmission components. A short propshaft fed power from the mid-mounted longitudinal four-cylinder BDT engine to the front-mounted transaxle (with driveshafts running direct to the front wheels), which in turn transmitted drive to a viscous differential and on to the driven rear wheels (thus two prop-shafts of differing lengths running parallel), the layout featuring three viscous-coupling limited-slip differentials.

JQF Engineering (working out of the Lordly Hesketh Northamptonshire former F1 premises) had already completed BDT turbo engines for the abandoned RS1700T Escort, so that unit was always at the heart of the RS200 project, albeit uprated in detail. I was disappointed with the performance of the original road cars versus their showroom price, but when they modified boost and engine management to allow a regular 315bhp, that 65bhp bonus brought an RS200 to life, so that the superb handling, traction and braking could be fully deployed.

'THE GEARSHIFT AND THE TURNING CIRCLE ARE THE WORST FEATURES OF AN RS200 AS A ROAD CAR, PLUS THE CABIN IS A HOT AND NOISY PLACE TO DWELL IN HEAVY SUMMER TRAFFIC.'

The gear-change improved to almost tolerable, or I got better at coping with it, while the clutch's engagement was sudden. The gearshift and the turning circle are the worst features of an RS200 as a road car, plus the cabin is a hot and noisy place to dwell in heavy summer traffic.

Other key features for the 194-strong production run included the purpose-designed and created monocoque of bonded and riveted aluminium honeycomb, with box subframe sections in steel, and Aramid and carbon-fibre reinforcements. External panels in glass-fibre and composites retained the roof-mounted air scoops to feed the intercooler. Also incorporated were cut-down and reinforced Sierra doors, trims and windscreen, and an integral steel roll

cage. A very useful feature of the external body panels was that they hinged upwards and open to give exceptional access to the engine bay and suspension.

The eight Bilstein long-stroke shock absorbers (two per corner) and coil-over springs, working in association with wide-base double wishbones, carried over from the competition cars for series assembly. Effective four-wheel disc brakes and heavyweight, unassisted rack-and-pinion steering also survived.

Ford and a growing band of production, administration, practical engineering and purchasing staff initially built up an erratic spares supply of more than 2,000 items, and rudimentary production lines. Two initial production RS200s (007 and 008) headed for Type Approval and the Boreham base in November 1985. Obtaining vital parts that passed Ford quality-inspection standards became the biggest headache, and the team often resorted to driving out to collect components.

It took only four months to assemble the homologation run of RS200s for inspection by the FISA international motorsport governing authority, but the vehicle wasn't recognised for international Group B motorsport participation

▲◀ People and competition car survivors from the 1960s, and contemporaries from the 1980s, at Ford, Boreham. One constant throughout, and of particular importance to the author's access, was Bill Meade (second on right beside the RS200 and also pictured left). Bill worked on pure competition creations and higher-performance road versions of the Escort, Fiesta, Capri and more, spanning Cortina to RS200. *Ford*

until 1 February 1986. Even then, assembly standards for public sale were unacceptable, even if some 27 assigned competition cars weren't such a problem because they would be stripped and rebuilt anyway.

Such left- and right-hand-drive road RS200s that remained when the Shenstone factory was vacated in May 1986 were shipped to Ford for storage in a Boreham hangar. There, a staggering total of 148 RS200s, in various states of assembly –

DRIVEN

Road cars

Hand-built prototype RS200s (two examples) Boreham-based P3/P4 prototypes, media debut, Sardinia, autumn 1985, published *Motoring News*; B888 CHK, P3 low-speed pictures; B690 CEV, P4 John Taylor accompanied dirt/tarmac drives; both hard-worked five-speeds, under 250bhp. **Privately modified RS200** Guild of Motoring Writers/Autoglass Brooklands–Goodwood road run, 19 June 1994, published *Ford Heritage*, September 1994; red rarity, retrimmed interior, 350bhp, 15mpg. **Factory-uprated RS200** Part of Silverstone Ford track-experience day, published *Fast Ford*, spring 1996; RS200 #122, a 315bhp example. **Private road RS200** Published with valid ownership/purchase tips in *Performance Ford*, November 1999; Steve Rockingham 250bhp concours example. **UK press-fleet road test RS200** For *Motoring News/CCC* column comments, 18–19 November 1986; D439 PJN, wet-weather pictures, reliable!

Competition cars

Ford Motorsport stock RS200 (three examples) Ford at Boreham drives in car number #122 in 315bhp uprated road spec, 5 December 1986; later, two more drives including Blomqvist RAC Rally car in Evolution BDT-E format. **Collins Performance Engineering RS200 Evo** Wet, independently timed test day at Bruntingthorpe airfield for publication in *Top Car* magazine, January 1991, F200 XUY owned by Sue Collins; sensational RS200 Evo 2.3-litre Hart-BDT, rated 630bhp, 0–60mph in 4.21sec, 0–100mph in 8.61sec, max 167.6mph.

minus many designated for strip-down and future parts supply – were placed under the care of Bob Howe (sales and administration) and hands-on engineer Bill Meade. Many detailed improvements were made to noise, performance and comfort levels before the last official sales in 1989.

However, much more work was needed to sell a £50,000 Ford that had no slot within world-class rallying, so 120 of the better road examples of the RS200 were selected for a total rebuild and repaint at Aston Martin Tickford from December 1987 to mid-1989. The later official price was £52,950, but the Aston connection offered the chance to select bespoke upgrades, including leather trim for cabin and seats, or individual paint schemes, plus the power hike from 250 to 315bhp. The horsepower boost could be achieved using the components already present, by upping turbocharger boost and tweaking the Ford-branded EEC-IV electronic management. Beyond 350bhp, the factory rally engineers and conscientious specialists would switch the engine management over to more-expensive Bosch Motronic systems.

If we take £50,000 as a typical RS200 sale price in 1987, that translates to £145,000 at the time of writing. The last British public auction result I can find for the breed was £292,500, in summer 2022 from Silverstone Auctions, but that has probably been exceeded in private transactions, and an ex-works rally RS200 would obviously command considerably more. There were other RS200s offered for sale in summer 2022 at prices beyond £250,000, the most expensive, in the USA, equivalent to nearly £283,000, but that 3,000-mile example did not sell.

▲ By May 1986 Ford promoted the RS200 even in its mainstream catalogues. This competition car, chassis 062, registered C200 KHJ, entered nine international events. Most were in the UK for Mark Lovell/Roger Freeman to win the British title. *Author's archive*

PROTOTYPE RS200: FIRST TASTE OF A TURBOCHARGED 4X4 TWO-SEAT FORD

The official media debut of the RS200 and revised Granada took place in Sardinia on 20–22 September 1985. Coincidentally, I spun my driving boots around and flew out again on the 23rd to drive Lancia's supercharged Delta S4 road car. Ironically, the full factory version of that cunning Lancia would win Britain's RAC Rally that year, while the RS200 awaited international competition homologation.

Accompanied drives in the RS200 were the star attractions for my hardcore editorial customers at *Motoring News* and *Performance Car*. Yet, for Ford corporate staff and any media serving an American market, the star was a person, rather than a vehicle. Bob Lutz had become a motor-industry celebrity long before this Mediterranean island media thrash. I had first met him at BMW in Munich, where he was a respected German-speaking presence for three years and a key force behind the creation of a lightweight coupé based on the classic six-cylinder, which became the legendary CSL 'Batmobile'. Those Munich machines would become the bane of Ford in Europe as they despatched the Capri during the 1973 European Touring Car Championship, especially with a mid-season rear-wing that led to that 'Batmobile' nickname.

Robert Anthony Lutz went on to reach top positions within all three American car giants, including a spell as chairman of Ford of Europe. His star quality was enhanced by his status as a former US Marine and fluency in five languages, plus top honours for an MBA earned at Berkeley University of California.

When I met him again in the hotel bar on that 1985 RS200 Sardinia trip, Lutz held the position of Executive Vice-President, Ford International Operations. He had already exerted considerable influence on the performance side at Ford, as a high-ranking heavyweight behind the XR4i, although that wasn't the boost to his career that it could have been. He tended to support all performance and motorsport homologation projects, feeling that the sales halo a successful performer provided for a company's more mundane machinery was often overlooked.

My companion for these RS200 debut miles was John Taylor, 1973 European Rallycross Champion and a successful Ford Motorsport consultant with an eye for talent-spotting the best in future drivers of either sex. I wasn't on a 'JT' list of driver favourites, but I often got seat time when he was in charge of events, my inclusion seeming to rest on my merits as a definitive illustration of how *not* to proceed at speed…

My first impressions of the RS200 were of a neat exterior that cleverly incorporated some production parts (screen and doors from the Sierra) in a totally individual shape that looked compact enough to be agile within the narrow rallying speed tests that would be its competition fare. Both examples at this media debut were hand-built by the Motorsports department in the UK as part of a six-off pre-production medley of assorted specifications from road to rally. The cloth and plastic interior wasn't up to the standard of later production examples, and I suspect the power on a warm Sardinian day was probably in the 230bhp zone on low boost.

One of these RS200s (P4/B690 CEV) was obviously in better photogenic shape than its sister. I manoeuvred it around a local harbour for pictures in company with Gunnar Palm, whose many successes as Hannu Mikkola's co-driver included victory in the London–Mexico World Cup rally. Gunnar subsequently became a very special public-relations operative at Ford, Sweden. That photo session told me that it was pretty hot in the RS200's cockpit – definitely

'I ADMIRED THE RESOURCEFULNESS AND THE ORIGINALITY OF THE HARDWARE DESIGN, BUT THE RS200 DIDN'T FIRE UP ANY EMOTION IN ME, OR DESIRE TO OWN ONE.'

▲ The LHD interior of a pre-production Boreham-built RS200, as driven in Sardinia media previews in the autumn of 1985. The second picture is of an immaculate RHD RS200 production example with red leather applied to the steering wheel rim and 4x4 selector. I can only ever recall choosing rear-drive once, just to see if it worked! *David Corfield and Author*

requiring side windows lowered – and that the turning circle belonged in a truck. Forward vision over the stubby bonnet was perfect but anything aft restricted, as is the case with so many mid-engine masterpieces.

With that pleasant interlude concluded, it was time for an hour or so as passenger, or driving, with Mr Taylor. By this stage, this RS200's interior was quite dusty, and both clutch and gearbox were apparently in a mood. Changing gear was testing, as the lever was floppy, and selecting across the gate was stiffly imprecise. Obviously, former jump-jockey John T had the determination and previous competition form to impose his will on this well-used prototype, and the road miles were shared between us, so that I could get an initial feel.

As we turned onto sandy and curvaceous off-road trails, the RS200 came alive and made some sense. The ride quality was the major asset, absorbent and excellently judged for mild excursions onto loose surfaces. In fact, that quality was to get even better, because these prototypes didn't have the full double-shock-absorber-and-coil-spring rear-end layout, lacking one coil spring per side at this stage. Another plus point was the traction, especially as we had the road-specification Pirelli P700 tyres on those distinctive eight-spoke alloy wheels, rather than loose-surface rubber of chunkier proportions. Exiting low-speed corners was a no-fuss pleasure, until you had to change gear again.

Noise levels in the cabin were intrusive in a harsh, mechanical manner, from both turbo alloy-block engine and the multiple transmission gear-sets whirring and gnashing discordantly. The simple rack-and-pinion steering was well sorted, and a handy RS three-spoke wheel guided the plot from a left-hand-drive position. A quartet of 11in ventilated disc brakes were capable of accurately slowing a near 2,650lb (1,202kg) mass without servo assistance.

I left Sardinia with mixed feelings about the RS200. It really came across as a bit rough and ready versus some road-car rivals. I just hoped it would improve, and it certainly did, but from a competition viewpoint the RS200 was too little too late. I knew it had been created with ingenuity, and with far less time and resources than you would imagine being at the disposal of a major multi-national car manufacturer, but it simply arrived in world rallying too late.

I admired the resourcefulness and the originality of the hardware design, but the RS200 didn't fire up any emotion in me, or desire to own one. I think that was just personal: I was biased in favour of prime performance within identifiable production bodies, such as the RS1700T or the Xtrac-Gartrac 4x4 Escorts.

Plus the price of an RS200 was way above my pay grade. But then that was even more of a factor with the GT40 – and I still love that.

DRIVING BOREHAM'S PRODUCTION RS200

During November and December 1986, Ford Motorsport's now-shared ex-airfield premises at Boreham had become a refuge for the RS200s that had to be sold. Stored in a large hangar, with 120 picked out for total refurbishment before they could be sold at £49,950 a pop, Ford of Britain needed every bit of media exposure it could get to shift that stock of 57 RHD and 63 LHD examples.

So, Ford offered many more driving opportunities than were available at the car's debut. This meant I was allowed experience of steering factory RS200s during 1985–96 in power levels from under the official 250bhp to a remarkable 600bhp Evolution engine mutation for rallycross.

Press-demonstrator 250bhp RS200 road car

Most of my RS200 drives were centred on the company's Boreham test-track facilities but I did have D439 PJN, a left-hand-drive 8,300-mile company press-fleet demonstrator, for an overnight stay. It was the subject of protracted and obstructive photography in pouring rain at Epsom. The brilliant compensation became a fabulous 30-mile session in darkness, composed of slippery local lanes. Here, the compact RS200 excelled, riding bumps beautifully, powering between high grass banks impressively, and showing strong traction even on showroom-issue Pirelli P700s. You could see forward over the stubby bonnet through that Sierra screen without annoying reflections, thanks to the planning foresight of ex-rally international navigator and then Ford Director of Motorsport Stuart Turner.

The only downside was the cumbersome turning circle for tight junctions, although the rapid-response steering itself offered great feedback at the expense of high rim loads. Similarly, the brakes provided excellent retardation, but the pedal felt initially numb on harder applications. I suffered the occasional

▲ A welcome visitor for a sleepover at my home, a press-demonstrator RS200 made a resounding contrast to the second Escort estate I owned and abused. This RS200's appeal and the wet conditions ensured it didn't get much rest, as it proved just the job for back-road overnight outings. *Author*

◀ A then-prominent member of the RS Owners Club, Steve Rockingham always presented his briskly used RS Fords so beautifully that I used a couple for book photography. This is his pristine RS200 in Buckinghamshire. *Author*

➤ This unpublished image of the shabby chic author with the road test RS200 was found in a dusty file, one of a number of photos originally taken by the photographic or editorial staff of *Motoring News/ Motor Sport*. Those in the earlier chapters date back over 50 years, well before the digital age, and are stored in big black books as contact prints and negatives. *Motorsport Images*

clunky gear-change, which still demanded a positive hand across the gate. Faster became the better shift option: swapping second to third or *vice versa* could be obstructive, but a far bigger problem in suburban traffic was the uncommunicative 'in-out' clutch action.

In more regular road use, the cold starts were reliable with the clutch depressed, and the properly managed engine settled into an 1,100–1,500rpm beat. You were advised not to drive away immediately, and not to exceed 3,000rpm until the water temperature hit 80°C. Thereafter, the 'ordinary' 250bhp specification preferred 3,200rpm as a reliable take-off point to avoid stalling, then pulling away cleanly, albeit at a fairly embarrassing noise level.

Although 8,700rpm was the large rev counter's advised absolute maximum of the 10,000 displayed, I found 4,000–7,000 generally sufficed, and third gear was more than enough to break the UK's blanket 70mph speed limit. The natural cruising pace seemed to be around 90mph (140kph) in this low-power, revised road specification. Cabin noise was much lower than in my earlier experiences, the press car wearing an engine cover. Maximum boost on that demonstrator flicked from 0.2 bar at a sluggish 2,000rpm, through 0.5 by 3,000rpm, and a full 0.75 just before 4,000rpm came up on the dial.

I returned the Ford with regret and some good minor-road memories. Yet I still never fancied ownership – even when RS200s dropped into the £30,000 zone during the '90s – at least not in the same way as I lusted after V6 Capris, V8 Mustangs and too many RS Escorts to mention.

'I RETURNED THE FORD WITH REGRET AND SOME GOOD MINOR-ROAD MEMORIES. YET I STILL NEVER FANCIED OWNERSHIP – EVEN WHEN RS200s DROPPED INTO THE £30,000 ZONE DURING THE '90s.'

Fun with a 315bhp uprated competition RS200 (chassis #122)

Back to the Boreham-based blasts, and one of the unregistered RS200s came my way twice in the official 315bhp uprated specification. I drove that chassis (#122) at Boreham and – 11 years later – around Silverstone on a media day, an event designed to show off all Ford's European performers, particularly the Group A Escort Cosworth.

That first encounter with the safely uprated 315bhp RS200 occurred after its last World Championship event: the November 1986 British RAC Rally, which featured too many RS200 retirements and a best finish of fifth. The Boreham encounter was worthy of a black-and-white flashback movie clip. Picture a bleak Essex former airfield in the December twilight with a passable impression of gale-force winds. Accompanied by my former FAV colleague Bob Howe – then a consultant on the RS200 post-production programme, responsible for sales – we unlocked a handy aircraft storage hangar. Yes, there were lots of RS200s hidden away, but I was confronted by a temporary showroom-cum-conference hall.

Five of the two-seaters decorated the entry, backed by another three of the purpose-built beasts in a central selling zone, which had been equipped with audio-visual devices, surrounded by emotive action posters. All this was topped by a podium for Bob Howe, Bill Meade or any other luminary on-hand to spread the enticing rarity/collectability/performance story. This strategy had shifted 20 examples, with five more deposits placed during the month after I called.

Aside from sampling the RS200 in 315bhp trim – the 65bhp bonus then cost £952 and was worth twice that in driving pleasure – I also learned of the many options that could be fitted beyond the 'basic' 250bhp starter model. These included the Tickford (formerly Aston Martin Tickford) total interior revamps, in leather, with colour choices and even radios offered at extra cost.

One of the most important original differences between the road- and rally-specification cars was the absence/presence of the two-wheel-drive selector alongside the gear-change lever, which was reserved for competition-designated machines and provided motive power to the rear wheels only. So were a four-lamp auxiliary-light housing, and black, rather than red, trims to seats and steering wheel. The suspension settings differed too, the rally-orientated RS types sitting 25mm (1in) higher and featuring heavy-duty neoprene bushes and harder damper/spring rates.

I left Boreham with respect for Ford efforts to recoup some cash from a programme that had had the heart slashed out of it. There was more than a fair share of tragic accidents on spectator-fringed high-speed stages, and major fires, which contributed to the bigger Group B safety picture. All of which led to the consequent cancellation of the category for 1987. Ford was lucky, in that its Group B machine had been properly crash-tested, and could be legally sold in multiple markets: sadly, not the USA, but North American collectors knew all the answers, and 47 were subsequently sold either to the States or Canada. Ford also never went to the extra expense of producing a full, flyweight Evolution version: uprated engines, yes – totally revised cars of dubious safety standards, no.

◀ The double spring/damper secrets of rapid travel on wet and rough roads revealed beneath the RS200's upward-opening rear bodywork. It was muddy and wet for my overnight with the hand-built Ford – perfect conditions to bring out the best in the beast. *Motorsport Images*

Reunited with chassis #122 in the 1990s

I waited into the '90s before driving that 315bhp RS200 again. At first sight, the swift Silverstone track wouldn't have been my first choice, but it was raining and that RS200 entertained many that damp day, with only the factory Escort Cosworth, in Repsol colours, capable of firmly outpacing the Motorsport-striped '80s survivor. It was an honour to drive it again, for the proper 'Stig' – Blomqvist – had also run it up the hill at Goodwood Festival of Speed.

I inherited the familiar RS200 after the best thrashing efforts of Ford-contracted rally star Gwyndaf Evans, and 1996 Eurocar V6 champion and all-round talent Neil Cunningham. Gwyndaf commented wistfully: 'I wish I had been up for a factory drive when these things were around. Tremendous!'

I sat behind the left-hand steering wheel and surveyed the ranks of Ford switchgear – four click-and-activate switches were allocated to the lights alone on rally versions! – and a similarly generous spread of round dials. That track day, the large rev counter was equalled in significance by the boost gauge, which had drooped away from the 1.0 bar (14psi) needed for the extra rasher of power. It didn't really matter in those streaming-wet conditions, as the 4x4 element and three clever differentials proved more than capable of handling Britain's remaining Grand Prix track.

The civilised 1,100rpm idle grumbled happily, but the well-used clutch travelled from disinterested disconnection to sharp engagement in the last millimetres of travel. From 2,000 to 4,000rpm, the motor sounded just a bad-tempered note, more akin to the discords of harsh mechanical clatter. First gear despatched, it got a lot more interested in progress from 3,500 to 4,000rpm. Playtime commenced after the lever flopped back into second listlessly, but thereafter the conventional five-speed shift behaved, doubtless terrorised into acquiescence by the previous occupants!

Out in the open spaces of Northamptonshire, I did try operating the engine above 7,000rpm, but there didn't seem to be any valuable extra horses corralled beyond that figure, so I stuck to using the 4,000–7,000rpm band of previous experience. That wet track was a brilliant place to challenge the RS200's chassis and 4x4 deployment. Unruffled, it could handle full power out of 50–100mph corners, where the rear-drive legends were distinctly twitchy, and the front-drive brigade ploughed towards grassy grazing grounds, nose first.

Pushed harder, this abused example sported some brake judder, which I hadn't encountered on any of my previous RS200 drives. This did mean it could lock up all four wheels slowing for sharper corners. I hadn't managed to provoke such unruly behaviour before, even over Sardinian dust, but that may have been more to do with not enraging John Taylor…

RS200 EVOLUTION FROM COLLINS PERFORMANCE ENGINEERING

I found myself at a very wet Bruntingthorpe in January 1991 assessing four tuned 500+bhp Fords for editorial director Chris Wright of *Top Car/Fast Car* magazine, including timed acceleration runs. Three of the cars were various Sierra species: a three-door RS500 Cosworth, a Group N four-door RS Cosworth and a 4x4 four-door. But the star of the soggy, but expertly timed, straight-line acceleration tests was an RS200 Evolution prepared by Collins Performance Engineering, a company that I hadn't encountered before but was about to become a very important part of my life that year, as you will find out in Chapter 13 (see pages 296–297). The Collins RS200, F200 XUY, was the quickest example of the breed I've driven, and looked the part in Ford Motorsport corporate stripes.

The engine for this 2,137cc BDT evolution of the RS200 was originally developed by Brian Hart, Ford's trusted sub-contractor for all things BDA-based, and delivered 630bhp. The rest of the RS200 Evo was pretty special too. That largely unloved Ferguson gearbox had been replaced by a sweet-natured, swift-shifting Xtrac five-speed, the 4x4 layout running a 37/63 front/rear power bias that damp day. Brakes were rewardingly efficient, with 330mm (13in) discs all round, supported by AP Racing four-piston callipers. Sheer adhesion was augmented by oversize BF Goodrich 275/40 and 315/35 tyres on 17in Revolution wheels, with 10in front and 11in rear widths.

The stats, such as 0–60mph in 4.2sec, 0–100mph in 8.1sec and a quarter-mile turned in 12.41sec, were impressive on such a slippery day. Yet acceleration figures and a recorded 167.6mph on the main runway were just a part of the story, for this was actually the best drive I had in any RS200.

Here's how I felt at the time about that Collins RS200 Evo, taken from a recorded conversation with Chris Wright in which we discussed all four of these fast Fords.

'The thrill was the RS200; it was worth coming even in the awful weather, just to drive that. It's very intimidating as a road car, and compared with the others it is like being in a fighter aircraft rather than a commercial-airline Jumbo. I've driven four or five RS200s before, and this was the best. There is enormous grip for the standing start, although you have to rev around 8,000rpm before feeding the clutch in. All credit to the BF Goodrich rubber. It ticks over at around 1,500rpm and you get nothing until 4,000 – then you get more power than you ever dreamed of. It is happy between 5,000–6,000rpm and an absolute dream between 6,000–9,000rpm. When you get the full 1.5 bar boost, it is very noisy in there, like sitting in a jet fighter – and there's just so much grip on these (wet) bends.

'IT TICKS OVER AT AROUND 1,500RPM AND YOU GET NOTHING UNTIL 4,000 – THEN YOU GET MORE POWER THAN YOU EVER DREAMED OF. IT IS HAPPY BETWEEN 5,000–6,000RPM AND AN ABSOLUTE DREAM BETWEEN 6,000–9,000RPM.'

'The Xtrac gearbox is a huge improvement over the Ferguson, but the car had been geared for Oulton Park. You could catch 70mph in first, and while this doesn't do a lot for acceleration, 4.2sec to 60mph isn't bad. The brakes are fabulous, with feel right at the top of the pedal, so you can really brake deep into the corners.'

I had one other experience of an RS200 Evo, a factory car assigned to Stig Blomqvist for rallycross. For some reason I have no personal record of that brief early-evening session at Boreham, so I'm not sure whether or not I actually drove it, but I certainly enjoyed some laps alongside Boreham's John Wheeler.

PRIVATELY MODIFIED RS200 ROAD CAR: BROOKLANDS-TO-GOODWOOD PARADE

Stuart Heath of Body Craft in West Sussex allowed his treasured and uprated red RS200 out for a long weekend tackling the Guild of Motoring Writers (GOMW) Golden Jubilee Run of 17–20 June 1994. The 73.85-mile transit between historic Brooklands and active Goodwood celebrated the 50-year growth of that motoring writers' organisation from seven founders to over 400 members. It was precisely timed and featured a proper 'tulip' road book of the type used in flat-out special-stage rallying, but the atmosphere was pretty relaxed, with the best of British summer weather and an eclectic entry list.

It was always going to be a privilege to drive this right-hand-drive RS200, which, at 350bhp, had a 100bhp bonus over the factory pre-production and media examples that comprised much of my experience. This one was also quite special beyond the 1.3 bar boost and reprogrammed electronics that allowed more powerful playtimes.

Theoretically, each vehicle was allotted the job of representing a year in that 50-year

▲ In 1994 the Guild of Motoring Writers organised a run from Brooklands to Goodwood. Stuart Heath's red RS200 was chosen to represent 1984 and proved a top draw even amongst the many prestige machines attracted to the event. This view was taken at Brooklands, in front of the clubhouse. *Author*

Guild history, but that obviously got harder as the years rolled by. One of my favourites today has become the type of Humber staff car that the RAF used during World War II, and a 1944 example was chosen to lead off our entries, which were honoured on arrival at the Goodwood Festival of Speed with a downhill parade on the hill-climb course and a static display area outside the house. The final, 1994, slot was allocated to a then-current Aston Martin DB7 for Guild Chairman John Blauth. 'Our' Ford was sandwiched between a Maserati Quattroporte and a Lamborghini Countach.

I picked the RS200 up when it had recorded just 5,570 miles, and I added nearly 200 more over that four-day loan. Instrumentation – a boggling eight dials, including a Porsche-style 8,800rpm red-lined rev counter – was certainly plentiful. Owner Stuart had a deadpan sense of humour and advised: 'Run it on four-star or Super Plus – it doesn't go well on diesel!' He also gave me some useful engine-preservation guidance for cold starts: 'You'll get nine warning lights when it has fired on a normal no-throttle start, please don't let oil pressure go above 6 bar, or the revs above 1,500, for the first 30 seconds. You'll find the battery light doesn't go out until 1,800rpm but that's fine.'

This RS200 had more changes than just a power hike, the most obvious being a quality respray in red by the owner's Body Craft company. Rouge themes carried over to pinstriped leather seats and the RS three-spoke steering wheel. Ford supplied road RS200s with red fabric trim for the seats, and complemented that with the usual RS steering wheel in a matching shade.

Another obvious alteration was Mickey Thompson-branded tyres on the showroom Speedline wheels, and there were some thoughtful comfort and convenience moves that Stuart had learned from his ownership. A prop for the rear clamshell section of bodywork was stored above the passenger, there was a parcel-shelf protruding beneath the rear window and overhead cockpit lights had been added.

Besides following Stuart's instructions about starting the engine, I also incorporated some of my usual competition/extreme performance-car precautions, as applied regularly to 277bhp BMW M1s. These included pressing the clutch as the ignition was activated and keeping rpm low until normal water temperature (or oil temperature on more enlightened machinery) had been reached. The official Ford post-cold-start recommendation was 80°C water and no more than 3,000rpm.

I had longer than usual to adapt to this privately owned RS200, and that was all to the good in tempering my initially cool reactions. The car suffered from wheel imbalance and shakes at lower speeds, and I wasn't impressed with the over-size (305/40 ZR16) Thompson tyres. I have to admit, however, that the car looked a little tougher on that American drag-strip-inspired rubber.

The five-speed gearshift was better than on the early examples I tried, but operating the sharp in/out clutch was a real pre-occupation in summer traffic. Fortunately, the water temperature was never a concern, even in such testing conditions. Unleashed on the lumpy Brooklands banking, the RS200 was a revelation. The ride became unequalled for composure and shock absorption, the acceleration addictive from 4,000rpm. That may explain how the occupants got a bit carried away and missed a clearly marked exit slot!

Vowing to behave, we joined the road run, with the skilled crew of that Lamborghini Countach shadowing us along highly enjoyable lanes and byway sections. Yes, the Countach was massively powerful, but it was also a tad wide for British minor-road action, yet it was usually hovering in our mirrors with occupants

RS200 SPECIFICATIONS

Main figures for production RS200, factory 1986 RAC Rally car statistics in brackets

Engine

Configuration: Mid-mounted, longitudinal, inline four-cylinder, Cosworth-Ford BDA series; T-suffix for turbocharging via Garrett T31/4E. Alloy cylinder block with Nikasil liners, alloy 16-valve cylinder head, belt-driven double overhead camshafts. Multi-point fuel-injection managed by Ford EEC-IV (Bosch Motronic) **Capacity:** 1,803cc (1,800cc) **Bore and stroke:** 86mm x 76.62mm (86.0mm x 76.6mm) **Compression ratio:** 8.2:1 (7.2:1) **Maximum power:** 250bhp at 6,500rpm (450bhp at 8,000rpm) **Maximum torque:** 215lb ft at 4,000rpm (361lb ft at 5,500rpm) **Specific power per litre:** 138.7bhp (250bhp) **Power-to-weight ratio:** 196.8 per ton (436.9bhp per ton)

Transmission

Type: Permanent four-wheel drive (selectable two-wheel/ four-wheel drive, front-mounted transaxle, five forward gears. **Gear ratios:** 1st: 2.692 (3.23); 2nd: 1.824 (2.14); 3rd: 1.318 (1.68); 4th: 1.043 (1.36); 5th: 0.786 (1.14); final-drive ratio: 4.375 (4.37)

Suspension

Front and rear: Independent, via double wishbones, twin Bilstein gas shock absorbers and coil springs per side, anti-roll bar. (Replacement components developed for competition).

Steering

Type: Rack-and-pinion, 2.3 turns lock-to-lock (1.8 turns), turning circle over 50ft/16m

Brakes

Type: Dual-circuit hydraulics, no servo assistance **Front and rear:** Ventilated 285mm/11.2in (300mm/11.8in) discs

Wheels and tyres

Wheels: Speedline alloy, eight-spoke, 8in x 16in (magnesium, 6 to 11in widths, 16in diameters) **Tyres:** Pirelli P700 225/50 VR (Pirelli choices according to surface, 245 section used for performance tests)

Body and chassis

Type: Two-door, two-seat incorporating Sierra doors, trims and windscreen, integral steel roll cage, external panels glass-fibre with Aramid and carbon-fibre reinforcement, aluminium Ciba-Geigy 5052 bonded and riveted honeycomb monocoque with box sub-frame sections in steel.

Weight

Kerb weight: 1,286kg (competition weight 1,050kg)

Dimensions

Length: 4,000mm/157.5in **Width:** 1,764mm/69.4in **Height:** 1,322mm/52in (adjustable!) **Front track:** 1,473mm/58in (1,502mm/59.1in) **Rear track:** 1,499mm/59in (1,497mm/58.9in)

Performance

Maximum speed: 140mph (118mph at 8,900rpm limiter) **0–60mph:** 6.1sec (2.8sec) **0–100mph:** 17sec (7.3ec) **Fuel consumption:** 16–18mpg (6–8mpg on rally stages)

obviously having a ball. They proved excellent company at the mid-morning White Horse checkpoint, 25.7 miles into our journey and not far from Midhurst in West Sussex.

Meanwhile, the RS200 proved a remarkable machine: strong traction under full-bore power, with a compact body coupled to accurate steering and consistent brakes, all enabled driver and unique Ford to find a harmonious sweet spot. The unassisted, heavy steering, with truck-like turning circle, could catch you out if you missed a turning, but overall I was so much more impressed with this RS200 than I had been from previous brief encounters.

At Goodwood, the static Ford attracted interest from crowds, but it wasn't entirely the right crowd for Ford lovers. We were greeted after parking in front of the glorious Goodwood House by an onlooker threading his large head through our open window and exhaling beer fumes: 'That has to be the ugliest car there is.'

Personally, I thought it was rather neat, but I prefer it in white paint, planted on sensible road tyres, please…

CHAPTER 13

SIERRA RS COSWORTH

A scintillating Ford sporting icon

The development of a 'Super Sierra' was never a secret; indeed it was a keenly anticipated performance Ford, especially amongst the motorsport fraternity. It seemed to take an age to materialise, because of that 'open-secret' approach. Because there was such an obvious gap in Ford of Europe's saloon-car racing armoury, the company deliberately leaked hints of a Sierra to the media, dealers and hardcore motorsport personnel. Initially, some speculated that the Sierra XR4i or XR4Ti could do the job (see Chapter 11), or even South Africa's XR8, a (very) limited-production Sierra featuring a 5-litre V8 heart.

The real starting point for the Sierra Cosworth was a Stuart Turner-chaired April 1983 meeting that delivered initial project agreement, leading to a 15 July 1986 UK sales debut.

SIERRA RS
COSWORTH

However, the 5,000-off RS Cosworth three-door wasn't homologated for International Group A and Group N events until 1 January 1987. Even then, development of the 'Super Sierra' wasn't finished, for the company had shrewdly authorised and created another 500 road models that qualified under the FIA's Evolution regulations for further development tweaks, such as an enlarged turbocharger and aero upgrades with spoiler enhancements, to produce a killer race Sierra. Well, until Nissan delivered a 4WD Skyline GT-R monster the Australians aptly dubbed 'Godzilla'.

The first 5,500 three-door Cosworths were created as the basis on which to build competition cars. Yet, the amazing 16-valves-plus-turbocharger technology that Cosworth Engineering supplied had a much longer and more civilised life in a subsequent pair of four-door editions – Sierra transformations that took us into the '90s and provided the underpinnings for a final Cosworth transplant, the 4x4 Escort RS Cosworth, which you can read about in the following chapter.

THE SIERRA RS COSWORTH STORY

▲ Track limits, Walton! My first Snetterton 24 Hours in an RS Cosworth Sierra, one usually driven by Dave Morgan, came in 1987 with some top-class team-mates. *Author's archive*

Showroom RS Cosworth three- and four-door development

Once the managerial decisions had been taken and the necessary multi-million-dollar budget agreed, there were practical decisions to be made to speed the process of delivering a turbocharged European Sierra for international motorsport.

That qualifying production number of 5,000 meant that the 'Super Sierra' had to be available for public purchase through Ford dealerships, particularly in the UK and Germany through the designated RS dealer networks, plus selected European neighbours such as France and Italy. The engineering effort for the road-legal cars was split between Ford's in-house SVE team (within the company's main Dunton Research and Development facility) and, for the engine, Northampton-based Cosworth, Ford's long-term partner for Formula 1 and production cars.

Cosworth worked around the basic 98–100bhp SOHC Pinto eight-valve unit, keeping its capacity at 1,993cc, but supplied a new twin-overhead-camshaft 16-valve cylinder head together with an intercooled Garrett T03 turbocharger chosen for race horsepower rather than street manners.

Ford Motorsport knew that it needed at least 300bhp to be competitive in motor racing, and envisaged that at least 200 horses would be the restricted road-legal result. It transpired that Cosworth had a job to constrain the output of the showroom-specification engine to that 200bhp level, and some of the pre-production examples reportedly ran 205–210bhp regularly, with a best of 218bhp. The important point for the public and ultra-critical Ford technicians within the main company – who weren't all delighted to see SVE get all the glamorous higher-performance tasks – became the accessible torque.

The official Sierra Cosworth torque summit was 203lb ft at 4,500rpm, but the engine generated such amiable slabs of twist through the 2,000–5,000rpm range that even mainstream Ford engineers were impressed by the traffic manners of their cut-price supercar. In stark contrast to the rude 'on/off' turbos of the period, the Cosworth engine produced 80 per cent of its maximum torque (equivalent to 163lb ft) between 2,300 and 6,500rpm.

Key features of Cosworth's Pinto engine redevelopment programme contributed by outside suppliers included a preference for Uniroyal belt-driven camshafts, that Garrett AiResearch T03 turbocharger, Mahle pistons, and Weber-Marelli injection and engine management. A compact intercooler was used alongside a reduced 8:1 compression ratio, with boost set at a modest maximum 0.55 bar with warranty considerations in mind.

You can judge the thoroughness of Cosworth's work by the fact that the only totally unmodified Ford components were the pre-RS500 cylinder block, water pump, two pulleys and an auxiliary shaft. Later, even these were modified.

Incidentally, the turbocharged Merkur XR4Ti had initially lacked separate intercooling, relying instead on finned intake manifolding, as did Lotus for its contemporary Esprit turbo. The uprated 'Pinto by Cosworth' engine – with heat treatments applied to forged-steel connecting rods and crankshaft – proved extraordinarily tough in competition, but a Motorsport thick-wall block was required for the race-engine outputs of the RS500 (450–560bhp).

Next on the list of motorsport requirements for the special Sierra, which led to the most controversial appearance of any showroom Ford, were the functional aerodynamics. An SVE regular, Gordon Prout, oversaw this tricky aspect, balancing the pure aerodynamic wish-list against the requirements of pan-European road-safety regulations and manufacturing practicality. The priority was to obtain a degree of downforce – rather than the uplift that afflicted even high-performance road cars of the period. Gordon explained to me some of the ingredients they had to balance back then:

'That downforce was the reason we had such a high rear wing on the RS Cossie, and it had to be in that position: any lower and its effect was nullified, any higher and the drag figure went up to an extraordinary degree. We would have loved to extend the wing rearward for even better figures, but the law prevented protrusions beyond the bodywork. Besides, the expense for a matching extended bumper put that approach out of the question.'

Even during 1984, two prototypes looked like redevelopments of the three-door, 'six-light' (referring to the number of side-glass panes) XR4i, but that was a shape developed without any request or design briefing for motorsport needs, as Andy Rouse's XR4Ti racing experience swiftly underlined (see Chapter 11). It also seems the sales and media reception for that XR hatchback with six side windows, and a lower cost of conversion, ironically led to the decision to adopt the Plain Jane range-starter Sierra L as the foundation for the glamorous Cosworth.

The final body package was unique and has become a modern classic. Features included engine-cooling sheet-metal alterations as well as more obvious aerodynamic appendages. These were finally predicted on a three-door hatchback Sierra L with Phoenix-manufactured body kit. In detail, that comprised wheel-arch extensions, side sills and front bumper/spoiler. The latter was a full-width vertical extension with a flexible rubber blade that cut front-end lift and wasn't as vulnerable as stiffer plastics or aluminium in pavement parking scrapes. The metal modifications comprised twin heat-extraction bonnet louvres that weren't strictly necessary for road use. An air intake between the headlights, which contemporary Sierras lacked, was also allowed. Aside from the basic steel monocoque, polyurethanes were employed, along with rubber body additions.

Back in 1985, Ford was obsessed with

'THAT DOWNFORCE WAS THE REASON WE HAD SUCH A HIGH REAR WING ON THE RS COSSIE, AND IT HAD TO BE IN THAT POSITION: ANY LOWER AND ITS EFFECT WAS NULLIFIED, ANY HIGHER AND THE DRAG FIGURE WENT UP TO AN EXTRAORDINARY DEGREE.'

◀ At first sight, these rear-end views show the transformation that Ford's SVE department carried out from mass-production Sierras to the limited-production three-door RS Cosworth. Actually, the plain-looking Sierra on the left is a five-door 2.3 V6 Ghia – a very smooth motor but not a collector's item like the smart white RS. *Author*

aerodynamics, just like any other mass-manufacturer trying to play catch-up with the pioneering 1983 Audi 100's low drag factor of 0.30Cd. Ford had been caught out in its claims for Escort Cd factors during independent German media tests, so the aero slipperiness of each model would be heavily publicised: thus we learned that the XR4i (and presumably its Merkur cousin) cleaved the air with 0.32Cd, versus 0.336Cd for the first Cosworth.

Those statistics meant that Prout and many others invested in wind-tunnel hours via Cologne Motorsport and Merkenich (Ford's German technical centre) personnel – plus some pioneering sub-contractor time in the UK with Motor Industry Research Association (MIRA) facilities at Hinckley. They effectively balanced the contrasting demands of showroom and circuit, as the Cosworth developed significant rear downforce, compared with strong lift recorded by the normal street Sierra shape. Subsequently, the RS500 added multiple aero aids to improve downforce, and paid a measurable aerodynamic-drag penalty of 0.351Cd.

Many of the basic body measurements, such as the generous wheelbase (2,609mm/102.7in), were more suited to mass-production Sierras than racing. Even in full competition trim, all three-door Cosworth Sierras looked comparatively tall and slim. This impression impacted on me, despite the Group A-required wheel-arch extensions, which could only accommodate the mandated 10in rim widths with difficulty when ultra-low ride heights were required.

In the showroom, the roofline was officially 1,337mm (52.6in) above ground, and the body width was reported at 1,727mm (68in), with the Cosworth three-door official Group A and N FIA international homologation form stating maximum tolerances on the showroom statistics.

Compared with the radical engine and aerodynamic upgrades, much of the rest of the initial Cosworth specification was almost conventional. The transmission featured what was thought to be a well-proven BorgWarner T5 gearbox from the USA, albeit the fifth (top) gear ratio had been replaced for faster sustained speeds at lower rpm in Europe. The BorgWarner (ex-Mustang SVO) gearbox was linked to a viscous-coupling limited-slip differential, with 3.65:1 final drive.

The T5 'box did prove fragile during Ford tests at Lommel at continuous *autobahn* speeds (130mph), when failures were encountered, delaying production readiness by nearly six months. Replacement bearings were specified, and subsequently gearbox lubrication was improved. The T5 remained part of the Cosworth menu through multiple derivatives until the advent of Ford's own MT75 gearbox in 1990.

Even when the pre-production Cosworths were heading to Spain for their media debut, there were background panics at Ford, as the power-steering-pump brackets required replacement with a later specification. The Cam Gears-TRW hydraulic power-steering pump remained in place, along with the 2.63 turns lock-to-lock rack-and-pinion layout, but multiple detail changes were made before production to reduce the twitchiness that was often criticised by the media, and particularly emphasised on the deep contours of Spanish test-route back roads.

The steering geometry was never going to be perfected in the three-door originals, as Ford Motorsport had demanded a raised roll centre (achieved via replacement cast-iron front hubs) to suit a low-riding race car on slick tyres. John Hitchins, a veteran of so many effective SVE chassis settings, kept it simple on spring rates and the preferred Fichtel & Sachs gas shock absorbers or strut inserts.

As with so many suspension and steering details, multiple adjustments were a feature of the run into production, particularly tighter caster settings to prevent the front wheels fouling the constricted wheel-arch space on lock. Additionally, the front anti-roll-bar diameter increased from 26 to 28mm (1.02 to 1.10in), reducing initial understeer. The ventilated front disc brakes were generous in dimensions by period standards, and Alfred Teves ABS was fitted, with a high-pressure hydraulic system that allowed a taut brake-pedal action – a welcome improvement over so many soggy centre pedals on fast Fords.

At the trailing-arm independent rear, much sharing of proven hardware went on, including the XR4x4-derived settings, and the solid disc brakes used by contemporary Granadas/Scorpios. A light, 14mm (0.55in) rear anti-roll bar was installed. Alloy 7x15in wheels were supplied by ATS-Rial with Dunlop D40 tyres (originally co-developed with Porsche) initially specified in 205/50 VR15 sizing. Today, those tyres are unavailable, but I have used Bridgestone successfully for both road and track RS500s. The Jesse Crosse 2021 restored press RS Cosworth runs the same make, with what Jesse describes as 'a slightly harsher secondary ride'.

Footnote *The RS Cosworth three-door, at a debut price of £15,950 (the equivalent of around £42,000 at the time of writing) was notably well-equipped, from built-in tilt-and-slide manual glass sunroof, down to the electronic ABS braking that was listed as a £1,000-plus option on most rivals. Electric assistance was part of the deal for mirrors and front windows, with tinted glass also standard. The cashmere-cloth, Recaro-seated interior, with four-speaker sound system, was another notable bonus versus the asking price of contemporary rivals.*

RS500: rush-job special

The production of a further 500 Cosworths – dubbed RS500s – with multiple race-biased features for road-legal sale, was a vital ingredient of Ford's European and World Touring Car Championship ambitions. Yet, the realities of building those evolutionary Sierra RS types saw yet another Ford homologation-run, limited-production pantomime – one that required an outside sub-contractor and a small team of overworked technicians to execute it under the project management of the long-suffering Mike Moreton.

Mike, the Ford product planner, became the go-to executive who oversaw many awkward-to-manufacture performers from bright ideas to reality. Notably, these included the commercially successful second-edition 'beak-nose' Escort RS2000 and the RS200. Mike also played a supporting pre-production role for the grand finale to the Cosworth Ford Pinto-based cars, the Escort RS Cosworth.

Of course, it wasn't just about Mike M, as the cooperation of many others, from the highest management to production-line workers, was needed to make what had looked a straightforward Cosworth continuation happen. I should mention the engine liaison and homologation expertise of Boreham Motorsport's John Griffiths, and the work of Lothar Pinske (Ford Motorsport, Germany) and Eberhard Braun, who expanded on his original Sierra Cosworth aerodynamic wind-tunnel research to deliver a measurable improvement in the RS500's downforce dynamics.

The mandated extra 500 RS Cosworths were manufactured according to their usual Genk, Belgium routine. When shipped to Britain, they were stored at Ford's regular Thames-side auction-

◀ These two side views show that the RS500 could come in black or white. The black car, the traditional colour for Ford's demonstrators, is the Peter Sugden example I drove in 2022. The white car shows off the extra RS500 aero better and highlights the tail-pipe differences from period original to restoration alternative (due to lack of availability of original) in the 2020s. *Author's archive*

sales and storage facility, the quirkily named Frog Island. Sadly, the bonus Cosworths wilted during the six months awaiting their rebirth. Cosworth had to rebuild and bench test all of their engines before they could be released in revised RS500 format: I do have a note from Cosworth recording that 525 such motors were completed between May and August 1987.

Cosworth co-founder Keith Duckworth had laid down the basic moves required in production to form the basis of specialist upgrades. These modifications would see the revised engines (coded YBD) capable of generating 500 race horsepower, instead of the 350–360bhp of the originals (coded YBB) recorded in 1987. Those production foundations within the Cosworth-coded YBD series began with a strengthened cylinder block, usually referred to as a thick-wall item, that was cast by the Cologne-region manufacturers of the Pinto iron-block mass-production unit, and delivered direct to Cosworth.

Further durability modifications included higher-capacity oil and water pumps, plus enhanced water and intercooler radiators. Generating the extra power was never a problem, but required a truck-size turbocharger (T04 in the Garrett AiResearch armoury), and twin fuel-injectors, often referred to as an eight-injector or twin-rail layout, although the road cars were delivered without the second set of injectors connected.

The cylinder head required enlarged ports for the increased flows of intake mixture and exhaust gas. Elongated bolts that ran though the revised cylinder block to the main-bearing mounts provided additional anchoring for the aluminium 16-valve head. For road use, the exhaust manifolding incorporated constricting sleeves that held power gains down to 20bhp over its predecessor: an official 224bhp at 6,000rpm, coupled to maximum torque of 206lb ft by 4,500rpm.

Now, what about the rest of the vehicle, outside the engine bay?

Many engineering and manufacturing alternatives outside the obvious Ford defaults were considered, but the schedule was restricted by cost and time. SVE was already overloaded, and the Genk Sierra plant couldn't be used for such small, fiddly tasks. It was decided to settle for the Tickford establishment at Bedworth. Above all, there was the imperative to meet a July 1987 homologation date. The road-legal RS500s were successfully inspected, and the car made its competition debut in the World Touring Car Championship at the fast Brno track in Czechoslovakia in August 1987 with a winning 1–2 result.

What went into those public-road RS500s assembled between June and August 1987?

Most obvious were the aerodynamic enhancements and bonus engine-cooling intakes. An additional 30mm (1.18in) lip surrounded the upper rear wing, and a wrap-around extension was fitted to the lipped front spoiler. These modifications were credited with mildly enhancing front downforce, but accounting for a major improvement at the rear. As before, the components for these aero modifications were manufactured by Phoenix in Germany, joining their work on the previous Cosworth's wheel-arch and body-side extensions.

To comply with the competition need for a substantial shift in the rear-trailing-arm-suspension location points, almost 2in (51mm) further forward, the road Cosworths came equipped with U-bolts and nuts to satisfy the authorities that they could be converted to the homologated evolutionary specification.

That the RS500 programme was completed on time, and made a reported $500,000 profit, was a remarkable exception to the record of so many 'homologation specials'. Previously, these tended to be unloved motorsport-orientated love children that offered few perceptible road-use benefits.

'THAT THE RS500 PROGRAMME WAS COMPLETED ON TIME, AND MADE A REPORTED $500,000 PROFIT, WAS A REMARKABLE EXCEPTION TO THE RECORD OF SO MANY HOMOLOGATION SPECIALS.'

More doors, more civilised... and fewer thrills!

On 8 February 1988, the four-door edition of the Sierra RS Cosworth (Sapphire in some publicity materials, no RS branding in LHD markets) launched in Britain at £19,500 (equivalent to a substantial £54,286 in 2023). It met an accommodating audience for its understated rapid-transit potential in a comparatively modest four-door body. According to production records supplied by Cosworth, this was probably the best-selling of all Cosworth Sierras. They accounted for 13,390 of the YBB engines that served three- and four-door rear-drive Cosworth-branded Fords. Big spoilers were out, a big luggage boot and proven 204bhp running gear were in.

Wherever possible, the four-door was implanted with carry-over three-door RS Cosworth hardware, rather than the RS500's special features. The 1,993cc YBB engine, at 204bhp with Garrett T03B turbocharging and Weber-Marelli injection and engine management, remained, as did the T5 BorgWarner gearbox and viscous-coupling limited-slip final drive, offering the same ratios as the three-door predecessors.

A lot of durability lessons had been learned via the production three-door pioneers, and the results were incorporated in the continued production (in Genk, Belgium) of the 1988–89 four-door. Aside from detail changes to the Cosworth engines, the four-door body – trimmed to Ghia levels – brought revised brake-pad specifications from Ferodo and beefier wheel-bearing assemblies (always a Sierra production-racing weakness) from big-brother Scorpios.

The suspension again featured Sachs gas-damping, but spring and damper rates were unique. The diameters of the relocated anti-roll bars were 28mm (1.10in) front (shared with road three-doors) and 16mm (0.63in) rear, rather than the previous 14mm (0.55in).

The alloy wheels were a fresh 7Jx15in Ford design. A colour-matched body kit was pretty well mandatory for street credibility, and the Ghia outline took on a replacement grille/lip spoiler section, body side extensions, and a rear-hoop spoiler, for a quoted 0.33Cd (an improvement over the previous three-door RS Sierras). Ford quoted body weight at 2,750lb (1,247kg), up just 20lb (9kg) on the official 2,730lb (1,238kg) for its limited-production RS500 predecessor. The initial three-door Cosworth weight quoted in Ford press packs was substantially lower than the RS500 or four-door, at 2,651lb (1,202kg). A 1 January 1987 homologated minimum international competition weight for Group N/production categories – ratified by the RAC in the UK and the FIA in Paris – was a skimpy 2,491lb (1,130kg).

Ford's performance claims were right in line with the previous 204bhp hatchback, and when we did a back-to-back at Goodwood, for Janspeed to assess the worth of four-door versus three-door for the 1988–89 racing season, there were only scant tenths of a second between the showroom pair. However, the manufacturers were a bit optimistic when it came to independent media measuring the performance at a test track.

Typically, maximum speed would be in the 142–145mph bracket, rather than the 150mph tag Ford preferred and that the German homologation authorities had agreed (151.4mph/243.6kph). While Ford put the UK benchmark 0–60mph run at 6.1sec, one magazine reported less than 6sec, but most recorded 6.2–6.4sec.

Under full road-test conditions, including performance measurement, the overall fuel consumption would be a little over 20mpg: I recorded 22–27mpg during 5,000 miles of company use in 1990–91, but that was after considerable sorting at an official Ford RS dealership, because that 40,000-mile example hadn't been kindly treated.

From a Ford company viewpoint, the popular four-door production successor to the extrovert three-doors also absorbed some of the 15,000 YB engines originally contracted at the rate of 5,000 a year. Although Sierra Cosworth and Escort XR3i/RS Turbo sales were savaged by their reputation as victims of rampant theft and consequent escalating insurance rates, Cosworth eventually manufactured 38,334 YB engines (from the 1985 YBB through to the final 1995 YBP used in late RS Escorts) over a 10-year span, some for competition applications only.

▲ In 1990 the 4x4 version of the four-door Sierra RS Cosworth was premiered for the media to drive in left-hand-drive form. Assembled in Genk, Belgium, but still with UK Cosworth turbo power, it deployed trusted Ferguson-patented all-wheel drive and viscous-coupling limited-slip differentials. *Paul Wilson/Author's archive*

The final Sierra RS Cosworth: 4x4 – but prematurely obsolete

In February 1990, the final Sierra RS Cosworth, with four doors and four-wheel drive, arrived on the UK market at £24,995 (the equivalent of £60,534 now). Although it was an excellent vehicle, the 4x4 sold at a slower rate in the UK than its rear-drive predecessor for all but its 1990 debut year. The total UK sales of 3,661 were actually less than any other Cosworth Sierra or Escort derivative.

Particularly suited to British weather and rumpled roads, the model only lasted until December 1992, killed by the continually escalating insurance premiums due to theft (I had one stolen in 1991) and by the imminent debut of the Escort RS Cosworth.

An effective, but weighty, Ferguson-patented 4x4 system featured a viscous coupling again, with a fixed 34/66 front/rear apportionment of power, via epicyclic gearing. Another major update was an all-new gearbox.

Ford had developed its own MT75 manual five-speed, far from perfect in my estimation, having a long and floppy gearshift throw, but my 2022 experience revealed the previous BorgWarner T5 was pretty stiff on the selection front, unless thoroughly warmed. That said, I preferred the T5's precision and shift speed, and it felt more durable to me as well, but I have no facts to prove that theory.

The YB engine family had moved along to the YBG or YBJ for 4x4 use, both up 16bhp on the rear-drive Cosworths. Both contained conscientious detailed upgrades that covered inlet manifolding/fuel rail, cast-iron exhaust manifold and alloy sump, plus revised oil and water pumps, and yet another head-gasket specification.

These later engines were distinguished by red or green cam covers, and were produced in two different emission specifications. The 'green-top' was compliant with US 1983 emissions regulations and permitted 220bhp at 6,250rpm, along with torque of 214lb ft by 3,500rpm, with a catalytic converter. The 'red top' had its exhaust breath passed to less-restrictive Euro 1504 requirements, yet official power figures remained the same, except that 6,000rpm was the quoted peak without a catalytic converter.

Outside the engine bay, detailed modifications were the key to the 4x4 specification. SVE project leader Ray Diggins told me: 'There is a stronger mounting point for the rear dampers, and brackets to tie between the rear cross-member and the turrets. The inner wings are thicker than a standard Sierra, and the complete body takes slightly longer to build than usual, so the modifications are restricted only to the Cosworth Sierra 4x4. The benefits from a stronger C-pillar are associated with use of composite glues to stitch things together, while the front bulkhead and inner wings have detail strengthening that you cannot see at a casual glance. It all helps complete the integrity of the body in road and special-stage use, when it will be equipped with an integrated roll cage of course.'

The road-car suspension was overhauled, particularly at the back where the older three-door RS types displayed extensive flexing. Positive wheel cambers were visible evidence that 4x4 characteristics had arrived, but the Sapphire RS front and rear spring rates were retained. These were reported as a lot stiffer at the back than all but the hardest progression offered with the variable-rate system at the rear of the earlier Sierra XR4x4.

Gas-damping from Fichtel & Sachs was employed again, but for this RS Cosworth the anti-roll-bar diameters were increased by 2mm (0.08in) front and rear. Much of the credit for this Cosworth's uncanny adhesion was attributed to effective 4x4, revised suspension, and the replacement of Dunlop D40 rubber by bespoke Bridgestone ER90, created for SVE.

The 4x4 system, manufactured by the GKN-ZF partnership at Viscodrive, under an agreement with Ferguson and Ford, remained fundamentally the same as for previous SVE applications from the XR4x4 onwards. That meant a power split of 34/66 front/rear, delivered via epicyclic gears, and further modified on slippery-surface demand by viscous couplings. Unlike the Group A works rally RS 4x4s, no front limited-slip differential was used for the road cars, or Group N. The viscous couplings were set up quite 'loose' in the road car, to avoid interfering with the standard-fitment electronic Teves ABS braking.

The centre differential was housed in an aluminium transfer box, which replaced the standard MT75 gearbox extension housing. Other SVE-developed components abounded, most from the XR4x4 programme of 1983–85. It's worth noting that the front-drive was delivered from the central differential via a multiple-row chain, which drove a short propeller shaft running to the front differential. The front right-hand drive-shaft was conventional, with CV joints at either end, but the left-hand driveshaft passed through the oil sump to a support bearing. The result was an unequal-length driveshaft system that resisted torque steer. Supporting the front-drive components was a very important cast-aluminium subframe, which replaced the usual Sierra pressed-steel structure.

The rear-drive system used the propshaft and differential system from the previous RS Sierras, again with limited-slip differential.

Braking was upgraded to a quartet of vented discs of fractionally under 11in diameter. However, retardation remained a weak point in the Group N rally car. Seriously heavy road use could require uprated friction materials and fluids, but I didn't find that a necessity in 26,000 miles of driving in three 4x4 RS Cosworths that included RAC Rally coverage.

The braking system kept the pedal towards the top of its travel with none of the disconcerting increases in pedal movement that the original hatches exhibited over bumps at track velocities.

◂ Nice thought: this Sierra Cosworth 4x4 photo is from an anniversary card I received from Ford Special Vehicle Engineering boss Rod Mansfield. Dated 1 February 1990, it marked 10 years of SVE projects – 17 models and 600,000 units produced. *Author's archive*

By 1990 standards, this Ford performer was well-equipped in showroom format. Electronic anti-lock braking, central locking, four electric power windows and electric mirrors, plus extensive stereo equipment, was taken for granted. There was both lumbar and height adjustment for the driver's Recaro seat in the UK.

The only early options for all markets were metallic paint, air-conditioning (not previously offered with a Cosworth engine), CD player and leather upholstery. The air-conditioning option didn't become a reality until Cosworth incorporated the appropriate pulley on the water pump and an extra belt drive, remounting the power-steering pump to rejig the front-end power-take-off points. That work was complete in time for the 1991 Ford model year (starting September 1990).

Looking at the 4x4 exterior, the most obvious sports identity lay with those 7Jx15in alloys of previous Sapphire rear-drive RS parentage. Test cars were all shod with the ER90 Bridgestones of 205/50 ZR15.

The body was little changed externally over the previous four-door Sapphire RS, but the devil was in the detail, with the 'blacked-out' rear-lamp lenses of the 1990 Sierra, plus a 'white-out' effect to the front flasher lenses. The extra body-kit panels and mirrors continued to be colour-keyed.

For the final year of production, smoother alloy wheels and a stiffened dashboard design were incorporated from mainstream models, along with a dull replacement steering wheel. Other RS body details that spanned the Sierra RS's production life included the colour-coded body kit (auxiliary lamps in the polyurethane front spoiler/bumper assembly and raised rear spoiler) and a return of the bonnet vents that had been deleted from its rear-drive predecessor.

Official performance figures for the 220bhp 4x4 were good and when I tested the car at Millbrook I was within a tenth of the claimed 6.6sec for 0–60mph. I was more struck by my 0–30mph time of 1.7sec – impressive by any standards. Again, we were nowhere near the Ford-reported 150mph maximum speed, and on Millbrook's banking our best equated to 141.6mph.

I got 21–27mpg from 4x4s that I either tested, used to follow the RAC Rally, or had as a company car. We did once have the opportunity after a *Motor Sport* road test to refill at the test track, and found the press-fleet 4x4 had managed 8.1mpg!

DRIVEN

Road cars

Pre-production three-door RS Cosworth (two examples) UK media debut drives, 5–6 December 1985, Spain, C240 HVW and C241 HVW; fabulous! **Pre-production three-door RS Cosworth** London–Mugello, Italy, 17–22 July 1986, for publication in September 1986 *Performance Car*, C232 HVW; test with BMW M3s; track and 2,300 road miles; 23.35mpg overall, 17.7mpg on-track; repaired twice – alloy welds to sump, my fault; exhaust manifold/downpipe separated at track. **Production three-door RS Cosworth (two examples)** Jesse Crosse long-term at *Performance Car*, 1986–2022, now restored by Crosse and featured in *Autocar*, 2021, D990 PVW; Steve Rockingham car, used for photography and guidance for RS Cosworth book, 1993, D538 NOK. **Press fleet three-door RS500** Used for five days of high-rpm thrills, September 1987, E201 APU; 18.9mpg. **Pre-production four-door RS Cosworth, 2WD** SVE late development vehicle to gauge media reaction, Essex roads, September 1987, D806 VEV. **Modified RS500s (two examples)** 1988, Mountune 304 and 360bhp road upgrades; Mountune also engineered Robb Gravett's 1987–90 champion Group N and A Cosworths. **Modified four-door RS Cosworth, 2WD** 1989, Brooklyn Garage/Dennis Osborne; £2,662 switchable 280/295bhp upgrade; excellence from British Touring Car Championship and Sierra Rally Championship-winning preparation team. **Production four-door RS Cosworth, 2WD (two examples)** Road test and 1991 company car, F322 ODX; also driven by François Delecour, Boreham snow fun. **Modified four-door RS Cosworth, 4WD** Tested for *Fast Lane*, 1990; superb £32,178 Rouse Engineering 304R; T25 small turbo; 0–30mph in 1.7sec(!), 0–60mph in 5.6 sec; 0–100mph in 15.7 sec, 18.7mpg. **Production four-door RS Cosworth, 4WD (two examples)** May 1991–July 1993 *Motor Sport* staff car for 22,895 miles, H335 GRT, 'Gertie', cost £23,000; press-fleet demonstrator, November 1993, used to cover RAC Rally, stolen from hotel guarded compound mid-event! **Production four-door RS Cosworth, 4WD** *Motor Sport* replacement for stolen road-test car, December 1992, H622 KHK. **Mountune RS500s (two examples)** Tuned road-car-day drive for CCC magazine, 1994, wet, wet thrills! **Restored RS500** Driven on lanes and dual-carriageway, 6 April 2022, Linfoot restored, 97,000 miles; leased then owned by Peter Sugden, former Ford of Europe copy director; charismatic!

Competition cars

Group N race three-door RS Cosworth Snetterton 24 Hours, June 1987; wet start, fast (3rd overall best) but DNF; drivers author, Rod Birley and Sean Brown (quickest); regular Uniroyal driver Dave Morgan spun me as a pre-event passenger, twice! **Group N Janspeed three-door RS Cosworth** May–October 1988 racing season, 1- to 24-hour races, support Kieth Odor; best result 3rd Oulton Park; 265–280bhp; same Cosworth and Odor won 1989 Group N Championship. **Group A WRC three-door RS Cosworth (two examples)** Driven at damp and cool Boreham airfield, 26 May 1988, published in *Motor*, properly timed and allowed plenty of playtime, D372 TAR, stunning!; also rode as passenger to Stig Blomqvist in D373 TAR sister Cosworth (3rd 1000 Lakes 1987), pre-1987 RAC Rally Weston Park media day. **Group N UK, three-door RS Cosworth** *Motoring News* RAC Rally preview, tested Pembrey, 20 October 1988, with Gwyndaf Evans and 4x4 Mazda 323; 265–272bhp, 'forest/tarmac' boost curves. **Group A Rouse RS500** Driven at Silverstone for *Performance Car* feature with Sytner BMW M3, January 1989; 1,100kg (2,425lb), estimated 0–60mph in 3sec, 0–100mph in 6sec, 158mph; 170mph at Bathurst 1988. **Group A RS500, Karl Jones** Snetterton track test of Duckhams-sponsored three-door against Honda CRX for *Motoring News*, 1989; under-developed versus Rouse RS500; Jones bravery spectacular! **Group N UK, four-door RS Cosworth** Coaching session, Donington, spring 1989; Chris Lord/Evans Halshaw, F111 XCP. **Group A/N works four-door RS Cosworth (three examples)** Boreham open day, February 1990; passenger to Colin McRae (spun) in a second life for rear-drive D372 TAR; drive and passenger with Jimmy McRae, 4x4 Group A, G687 BTW; drive with Gwyndaf Evans, Group N 4x4, G97 CHK, no brakes and plenty of Mountune horses! **Group A WRC four-door, 4WD (two examples)** Snowy Boreham airfield drive for *Autocar & Motor* and four other magazines, February 1991; passenger with François Delecour, G94 CHK, 3rd place Monte Carlo 1991, 295bhp, Q8 colours, seven-speed gearbox; drive in Malcolm Wilson's 7th-placed Monte Carlo car, H319 MTW. **Race-spec RS500** Vecta Ford 1991 UK club-racing season, G819 JWN, by Collins Performance Engineering, Congleton; 500–550bhp, Bridgestone 'road' tyres; one outright win, four lap records, five second places and one third place.

PRESS PREVIEW OF THREE-DOOR SIERRA RS COSWORTH

Even before mass manufacture of the Sierra RS Cosworth began, pre-production examples were allowed out for exciting overseas forays.

Although mainstream production of the first Sierra Cosworth would not begin until 17 June 1986, Ford of Britain decided to let the media loose in 15 examples (10 of them UK RHD) on 5–6 December 1985. These cars provided pretty accurate predictions of a package that would be launched at slightly under £16,000. Since Ford was claiming 150mph, coupled to remarkable acceleration and outrageous appearance, plenty of sensational press ink and widely broadcast words were anticipated, and received.

On a cash-for-dash front, Ford and its SVE department were sure of a welcome, as the next contemporary car offering similar performance statistics cost at least another £10,000, and wore a Porsche 944 turbo badge. More rivals would appear, but Ford retained the initial UK publicity advantage. Its chief race-track rival, BMW's 200bhp showroom M3, didn't make much sense for the UK, as it was only offered tardily with left-hand drive, and at prices upwards of £30,000. Additionally, for the performance/competition hardcore, the BMW didn't feature the dial-up horsepower boosts of a turbocharger.

So, what were those early UK-specification RS Sierras like to drive at their premature media debut? From our 7.30am arrival at Heathrow, through to Jerez at midday, they were a riotous succession of whirling white Spanish snapshots, for we were back in London by mid-afternoon the following day.

I don't recall any of the predicted major accidents, just a lot of very hard driving in two of these C-plated early Cosworths. I had the chance to swap Sierras and complete photography in company with technical expert Martin 'Sharp-Eyed Man' Sharp. I had worked with Martin at Ford AVO, but he was then the go-to source for what factory competition engineers were up to in world-class rallying.

I wrote about that Spanish trip for *Motoring News*. My story included description of a high-speed convoy that fellow participants often talked about in later years: 'The deserted Seville–Jerez motorway basked in winter sunshine, its concrete-

▼ This pre-launch view emphasises how well-equipped the 145mph Sierra RS Cosworth was. Glass sunroof is prominent, but it also delivered ABS four-wheel disc braking, power steering and electric windows and mirrors, plus Recaro seating. All plus 204bhp for £15,950. *Ford*

block surface now briefly asked to bear the weight of 20 plump German Dunlops. Briefly, because the five Ford Sierra Cosworths flying in formation, southbound, were cruising at a relaxed 137mph… The assorted drivers within – from 50-year-old company executive to engineers and journalists in their 30s – obviously felt equally relaxed about the phenomenal cruising pace. For you could see conversations continuing as wind noise remained restrained.'

So the 'Super Sierra' made a lasting impact beyond the whale-tail and associated 'racer for the road' design messages. My most vivid initial impressions were of how cheaply and cleverly they had managed to deliver an efficient interior, with features such as a compact XR3i steering wheel, plus twin front Recaro seats and a boost gauge (one that was invisible with UK right-hand steering!) borrowed from the Merkur XR4Ti. Never mind, it all worked well, as the SVE development engineers also loved driving.

Not so clever at the debut were harsh engine vibes – at that time worst passing through the 4,500rpm zone – twitchy handling and light steering. The latter two traits could combine to give you an unsettling country-road ride. Most of these features were ameliorated in the 1986 (UK D-plate mostly) production examples, and they remained fabulous performance value for money that is equally prized by the classic-car community 36 years on, even at the £60,000+ dealer asking prices of 2023.

SIERRA RS COSWORTH vs BMW M3: THE 1986 ITALIAN JOB

Although my 1969 first foray into Europe for work was to drive a longer distance, with a ferry trip to Sicily and back via Turin, this epic multi-car trip of 17–22 July 1986 was the most action-packed 2,300 miles I've ever covered. We were to drive from London to the fabulous 3.3-mile Mugello track, set in one of Italy's most beautiful regions not far from Florence.

Tackled in an early (1985) three-door Cosworth (C232 HVW), the objective for *Performance Car* magazine was to bring Ford's 1987 European Touring Car Championship contender into head-to-head contact with the prestige opposition: BMW's first-edition M3. That rivalry would escalate with a first World Touring Car Championship title also up for grabs in 1987.

Back in 1986, editor Jesse Crosse hadn't just wheedled a 14,000-mile pre-production Sierra Cosworth out of Ford, but had persuaded BMW's press people in Munich that it would be a good idea to have the white whale-tail invade its international press debut! Yes, the confrontation would be of intense interest to both sides, but it was still a 'big ask'.

Both Ford in Britain and BMW Motorsport in Germany were fascinated that we were taking our Correvit timing gear along for side-by-side track acceleration runs of the similarly powerful showroom M3 and RS Cosworth (the M3's 200bhp played an official 204 Ford horses), plus indulging in some limited track time for the Ford alongside the M3.

Making it happen involved a lot of tolerance from BMW, but both sides agreed that we should hide this scoop test from the regular European media by agreeing an arrival time over a long weekend in between sessions for the mass media, who would be exercising over 30 immaculately presented M3s. In any case, by the time the already thrashed and travel-stained Ford arrived, there was no danger of confusing it with a pampered press machine.

Initially, the trip started off well, in sunshine. I met art editor Paul Harwood and staff photographer Peter Robain at the magazine's Middlesex offices, where they appeared alongside a long-term BMW 325i the magazine operated. The Bimmer was packed to the boot and back-seat maximum with a large quantity of rather valuable camera and timing gear, worth thousands. Thereby hangs a tale…

I installed myself in the familiarity of a Ford cockpit and remembered that Jesse had endorsed its status with the casual comment: 'It's obviously had a hard life, but it goes really well.' Lackadaisical performance was never an issue with the original Cosworths, and for long road spells this one could idle off-boost, keeping the laden six-cylinder BMW company, particularly in heavily regulated Britain and France. Overall consumption, including the track sessions, averaged over 23mpg, specifically measured over 2,200 miles. I then did at least another 100 joyous and selfish UK miles that were unmeasured.

Travelling disaster!

We were in a Calais hotel by midnight and were off to a deceptively bright and early French start. I doubt we did 20 miles before the Ford-BMW convoy was halted, not just by police but gendarmes acting on behalf of French Customs. One look at the BMW boot, and a casual question followed on the value of all that gear loaded within, with a swift demand for documentation. As we had no paperwork, we were escorted to the nearest cash machine to pay a fine equivalent to over £500.

The long day turned into a series of setbacks, some severe. We had lost half a day with the customs and police, Paris and the Périphérique were log-jammed, and I was the only one with access to a credit card, then very limited to the amount you could withdraw at cash machines. The Visa card did keep us topped up across numerous French halts, and by evening we were running in 11pm darkness in Hautes-Alpes ski-resort country between Grenoble and Briançon. I enjoyed winding the Cosworth through twisty sections, until on an uphill left I hit one of many fallen rocks. Swiftly after that unwelcome 'clunk', the oil light blinked. Race-trained to know that absent oil means imminent consequences, I stopped. What now?

◀ Photographs from our eventful 2,300-mile Italian Job to pit the pre-production Sierra RS Cosworth against BMW's new M3. From bug-splatting to an unscheduled pit stop at a Renault garage, the pictures tell the story. *Peter Robain/ Performance Car*

Fortunately, Robain was a Land Rover man of cross-country resourcefulness. The Cosworth was jacked to allow limited under-engine access in the torch-lit partial darkness. The leaking cast-alloy sump was mollified with repeated chewing gum implants, and we limped the Cosworth close to the nearby Serre Chevalier resort at about 2am. I chose the village high street of La Grave because I could see a small roadside Renault-branded garage. I knew that Renault had specified major alloy engine parts for decades.

That we found only a single hotel room for a trio of Englishmen at that dark early-morning hour was a minor miracle. Much laughter followed the next morning from the self-styled 'Ali Baba' garage proprietor and his youthful assistant, on hearing the 'three-blokes-in-a-bed' tale. We set off gingerly in mid-afternoon sunshine for the nearby Italian border and, remarkably, the Sierra sump stayed sealed for the rest of that tough trip. However, there were other RS Sierra components that would throw tantrums in the face of so many BMWs gathered at the picturesque track backdrop.

The Italian trek south was harsh, especially for

the occupants of the heavily laden 325i BMW. We got to Turin around 8pm, heading on down to Brescia, Parma, Reggio Emilia and Bologna. Now the Cosworth cleaved through heavy rainstorms, turbo motor happy with a water-cooled life, but visibility was limited and I felt for those following in the smooth, but now distinctly less responsive BMW. The thunderstorms did ease as we reached BMW's Tuscany press-launch base, checking into Montecatini Terme's Bellavista Palace, a five-star hotel, at 1am on Sunday morning.

Clandestine comparisons and a hard bargain

Now time became precious, for the deal was that the very temporary collusion between the lone Ford and BMW's media fleet had to be completed before the next media group arrived. That meant premium-quality photography, performance timing and the full M3 experience all had to be bagged – and out of sight. That was the plan, and I'm proud that Harwood, Robain and I completed that mission, but there are relevant details I omitted from earlier accounts.

Even in the rush to get *Performance Car*'s Cosworth-versus-M3 morning trials completed, particularly the full performance figures and side-by-side action photography, I was still able to drive three M3s (I had full reports to write about the new BMW for other media outlets) on local roads as well as around the Mugello track. The M3s were all properly developed, with and without catalytic-converter exhausts. Some track sessions were hard on them – and the accompanying Uniroyal tyre team. We had an informal competition around the fabulous turns and twists of the heavily contoured circuit that saw the best man on 2m 37.3s versus my 2m 38.0s. Many beers were consumed that evening to salute those fun-packed laps, but first there was a small Sierra RS snag to overcome…

Messrs Robain and Harwood had been out clicking frantically through the breathtaking Tuscan backdrops, returning with some excellent village scenes featuring the two rivals. Sadly, the Cosworth's exhaust system simply dropped off when asked to indulge in track life. Much head-scratching, as a turbocharged car without a sound exhaust system is a badly wounded performer.

▼ *Performance Car*'s two early covers – February 1986 and September 1986 – to headline the arrival of the Sierra RS Cosworth. I apologise for the childish doodling on the white C235 HVW, but remain proud of our efforts to welcome Ford of Europe's significant circuit performer.

There was also another obligation to fulfil: BMW had some hardcore BMW Motorsport engineers present – and they desperately wanted to experience the RS Cosworth out on track.

A deal was struck: BMW mechanics would repair the stricken Ford but the price was to allow an M3 development engineer actually to have a drive in our pre-production Cosworth. I had verbally promised Ford UK press personnel that I wouldn't allow any BMW people to drive this early production car... so, promise broken – I'm not proud of that.

The hardcore BMW mechanics taught me a lot of colourful German swear words as they welded up the exhaust system but they did a great job. So solid was the result that it lasted not only the track laps with the BMW Motorsport engineer aboard but also the entire trip back home. It took the Ford press garage back in UK some equally expletive-laden labour to separate the track-repaired iron manifold-to-downpipe unions.

Some background about BMW's motorsport-orientated rival

That enforced track time at Mugello with the BMW Motorsport engineer taught me just how thoroughly Munich had redeveloped its M-format 3 Series for the road, starting rather earlier than Ford had done with the Sierra Cosworth. BMW Motorsport GmbH began work on the M3 for public sale in July 1981 and planned 15,000 examples. At Ford, the first planning meeting wasn't until March 1983, and SVE was handed the project by September of that year.

Ford initially intended to manufacture a total of 5,500 Sierra RS Cosworths to meet the homologation requirement, but that was only true of the three-door first edition, as the company then carried on and built more Sierra Cosworths as four-door derivatives. As it turned out, BMW also made more M3s than originally planned. Including all evolution versions, but not the M3 cabriolet, some 17,184 M3s were built and more than 500 competition-only M3 kits were sold profitably. Ford totalled over 30,000 showroom Sierra Cosworths, but the difference was that these came in three- and four-door bodies, and also with an 4x4 version.

At Mugello, Dieter Quester demonstrated a low-power prototype of the race-spec M3 that underlined equally detailed preparation for the track. Nevertheless, it could be seen, even before any competition confrontations, that the turbocharged Sierra Cosworth formula was always going to have a power surplus compared with a normally aspirated M3.

Apparently, BMW senior management had dismissed the turbo route for its performance 3 Series. Ironically, their vision at the top of the famous four-cylinder building looking over Munich's 1972 Olympic Park was clouded by the fact that Mercedes and its deadly 3 Series rival, the 190E series, already had a 16-valve 2.3-litre performer (2.3/16) to sell to an eager public. That engine's cylinder head and some ancillaries were the work of… Cosworth!

SIERRA RS COSWORTH THREE-DOOR vs FIRST-GENERATION BMW M3

Source: Performance Car comparison test, back-to back average of two runs, same driver/passenger, dry Mugello track, published September 1986

	Sierra RS Cosworth	BMW M3
0–30mph	2.4sec	2.3sec
0–40mph	3.4sec	3.5sec
0–50mph	5.2sec	5.4sec
0–60mph	6.7sec	6.9sec
0–70mph	9.6sec	9.6sec
0–80mph	12.0sec	11.8sec
0–90mph	15.3sec	16.0sec
0–100mph	19.3sec	20.5sec
Standing ¼-mile	15.4sec/91mph	15.3sec/88.7mph
50–70mph/fifth gear	7.1sec	9.9sec
Maximum speed (official)	149.8mph	145.9mph
Fuel consumption (road/track)	23.35/17.6mpg	26.9/18.1mpg
Maximum power	204bhp at 6,500rpm	200bhp at 7,200rpm
Power per tonne	167.2bhp	169.5bhp
Drag coefficient (Cd)	0.34	0.33
Debut price	£15,950	DM68,000 (£18,550)

The journey home

Monday morning, and time for the pictorial people's final photography in 5.30am soft light in Montecatini old town. Both cars filled, we made a noon departure for a superb coastal run up to La Spezia and Genoa, then via Turin to exit Italy through the Susa valley and Colle delle Finestre. From the near-12,000ft summit, we phoned ahead to book into Michelin-listed Hôtel Million in Albertville, a Savoie regional destination in France that we reached without incident by 8.45pm that Monday evening.

Really, the rest of the trip, after an 8am departure from the fabulously tasty Million fare, was an unremarkable high-speed run along the French *autoroute* network via Chambéry and Mâcon, although the Cosworth and its totally seduced driver did get into an unworthy dice. Some Moto Guzzi bikers were attracted by that whale-tail and an unfamiliar Ford, with a result that proved turbocharged four-wheelers then had a top-end velocity advantage over contemporary two-wheeled Italian legends.

We ate south of Paris, and were in Calais by 6.30pm for the 7.45pm ferry back to Dover. By the time the bug-spattered C232 HVW, in grime-white, was tracking round the M25 in relaxed twilight 85–90mph illegality, I had fallen totally in love with it. I'd always held another plain-white Ford, the original Ermine White Escort Twin Cam, as the Blue Oval love of my life, but now the Sierra RS Cosworth's lure of accessible speed, and the promise of many more adventures to come, presented serious challenges to my loyalty.

As it turned out, the Sierra RS Cosworth would deliver on every bit of its long-term promise, adding a bonus for me as a frequent track companion. Yet I never owned a three-door for more than temporary transitions between road and race, and I failed in bidding for two early road examples. Sad, but the memories of trips like these are some compensation.

RIDING WITH THE MOUNTUNE RS500s

Ironically, Dave Mountain, a youngster with excellent pedigree at leading Mini preparation specialists Swiftune, became the 1979 founder of ultimate-engine-tuning concern Mountune – an established force amongst winning Fords by the time I trekked out to its Maldon, Essex, premises to drive a pair of RS500s. Over 100 international championships had been won, including hard-fought British titles with Robb Gravett's Trakstar RS500 facing off against similarly equipped Andy Rouse. However, the cream of the prestige crop was the Escort Cosworth 1994 Monte Carlo victory. When this section was written, I realised and respected just how consistently successful Mountune had remained over the decades, still a leading engine supplier in the massively competitive BTCC series, and provider of Ford-approved performance packages.

I spent a very damp winter day in 1987 driving some muscular road RS500s for *CCC*. Despite hostile weather for the most powerful RS500s I had then driven on public roads, the company of the cheerful Mountune proprietor, mostly surrounded in challenging spray as we played the Essex version of 'Convoy', meant a fair dose of laughter, along with some serious driving experiences.

Dave Mountain was equally candid about some of the basics they conquered on the way to understanding how to release reliable public-road power from an RS Ford that was only ever expected to provide the foundations for a world-class race saloon. Here are some of my period observations…

'Some initial snags were discovered in the RS500s released to the public. For instance, the under-piston oil-cooling sprays had to be realigned to hit their targets!

'Mountune also discovered the hard way that some 204bhp Sierra Cosworth iron blocks had to be scrapped, as they were outside manufacturing tolerances for bore measurements. The sleeves placed within the exhaust top manifolding obviously strangled power potential in pursuit of reduced warranty claims.

'Brake-horsepower test results from Mountune revealed that the earlier 204bhp [pre-RS500] units tended to give more power than officially claimed, closer to 210bhp than 204. In contrast, the customer RS500s produced less than advertised, just below 220bhp. Back in 1988, Mountune listed three power levels for RS Sierras: Stage 1 offered plus 40bhp for five hours' labour and a VAT-inclusive cost of £454.25. Stage 2 came in at a bonus 80bhp, covering eight hours' work and some minor parts, such as cockpit-adjustable boost (very fashionable!) at a retail £1,144.25. The third stage was really still being developed during my visit, thus the chance to drive two RS500s on one exciting day.'

Stage 3 was specifically aimed at the RS500 clique and was much more elaborate, occupying eight days on site. This connected up the eight-injector system of the Group A racing layout, allied to a total engine rebuild (including lightened and balanced reciprocating components) with low-compression (7.2:1) forged Mahle pistons. The engine management consisted of the usual 204/224bhp Weber-Marelli system, with the rev limiter deleted and a 'piggy-back' injection/ignition management system co-developed with Microdynamics.

The Full Mounty...

The turbocharging side of the Stage 3 package from Mountune was unique, as it logically involved swapping the giant competition-biased Garrett T04 unit for the usual production T03B found on 204bhp Sierra Cosworths, with the aim of restoring public-road, lower-rpm pulling power. 'The T03 is a gnat's too small to do the higher revs you occasionally use on a road car, so we do enlarge its internal airflow capacity. We think this gives the best of both worlds, with low-end power and top-end boost,' felt Mr Mountune.

Aah yes, boost – always a hot topic in Sierra Cosworth performance and competition circles. With 9.25psi above atmospheric pressure used as a production boost level, Mountune's adjustable systems offered 6–18psi. Dave Mountain's chuckling contrasted with grim greyness swirling as heavier rain fell outside our cosy-office chats. 'You'll find that development car can be twizzled to 20psi,' he asserted, with laughter, 'but if you really don't care anymore, pull the adjuster out completely and try 30psi for max fun.'

Instead of a rare road RS500's standard 224bhp at 6,000rpm and 206lb ft of torque by 4,500rpm, after 15 months of development that Mountune Stage 3 package yielded 350–360bhp at a usefully lower 5,600rpm. 'It's amazing, but that cylinder head needs no further flow work to achieve 400 horsepower,' reported Dave Mountain, adding gleefully, 'just more boost!'

There was an obvious improvement in mid-range pulling power, and some welcome lower-rpm advantages that we discovered that damp day, but no printable torque figures were offered. A by-product of that eight-injector conversion, and Mountune work involving remapping the electronics against a 256-point grid, became improved part-throttle fuel economy. This in comparison to standard Weber-Marelli-managed Cosworths that were mapped in the mid-range to run around 8 per cent rich. That secondary injector system activated beyond 10psi boost, the boost controlled by an adjustable control adjacent to the steering column.

Back to the byways of Essex in the Stage 3 Cosworth, and it was quickly apparent that these engine conversions delivered their power in the most civilised way, realising the objectives of making the RS500 more flexible, as well as perceptibly more powerful. Whether dialled for 6psi or 20, tickover remained stable at 750rpm. Progress was evenly delivered from a chuffing 1,500rpm to the arrival of significant boost by 3,500rpm and on to full boost in third, fourth or fifth. Dialled up beyond 18psi, I sensed, rather than felt, some pulsing under full throttle.

The chassis was under-developed back then, so I had a busy time sawing at the steering wheel and hoping for prompt brake-pedal reactions over standing water and back-lane hummocks. Back then, the RS500 for the road was just a mule for homologation race-tuning, so it would be much

'THERE WAS AN OBVIOUS IMPROVEMENT IN MID-RANGE PULLING POWER, AND SOME WELCOME LOWER-RPM ADVANTAGES THAT WE DISCOVERED THAT DAMP DAY, BUT NO PRINTABLE TORQUE FIGURES WERE OFFERED.'

improved by today's ready supply of aftermarket coil springs and dampers, a proper laser-beam check of all suspension geometry, and superior brake friction materials.

It was then time to try and get some sensible feedback from the 'Full Mounty' eight-injector conversion. I began within the downpour, thinking some social responsibility was in order. Forced to adapt quickly, I soon twiddled from 10psi boost to 15psi. From that point, speed-crazed withdrawal symptoms intruded at anything less than 18–20psi and the attendant lashings of lovely horsepower. An armful of steering-correction lock became the new norm.

A brief flit onto a local motorway's assortment of long curves and undulating straights flicked up illicit speeds at the same frenetic rate as the boost and rev-counter needles twitched. Such speeds on soaking roads showed that the RS500's extra aerodynamic appendages also allowed a measure of stability at motorway speeds, not something evident in lesser Sierras.

It was educational fun, learning to live with that much accessible power, even if it could spin the wheels in fourth and fifth, given enough demand for acceleration and a deeper kind of puddle. 'Educational' because such Essex experience provided me with fast-track knowledge that proved useful when asked to drive the road-legal 500bhp+ Collins Performance Engineering RS500 at a wet test track… and the first race of that Collins season was wet too (see pages 296–297).

The only problem with those Mountune road RS500s, especially in eight-injector format, was that they just became totally addictive!

RESTORED RS500: COPYWRITING CHIEF'S REWIND ROAD-RUNNER IN 2022

Inevitably, the name Peter Sugden crops up in my earlier chapters covering TV advertisements. Peter was an enduring creative force behind the Ogilvy & Mather multi-media words for so many years between his 1972 return from Australia to his official 1997 retirement as a Ford of Europe Creative Director. Even then he was called back into action whenever continuity was required, such as for Ogilvy & Mather's celebration of 30 years holding the Ford account, or the prestigious Shell-Ferrari press and film ads that marked their 50-year road and track association.

So Peter was a durable and effective elite copywriter, particularly when partnered over 30 years with art director Terry Holben. However, our focus here is on the outstanding pedigree of the showroom RS500 Sugden has retained ever since it was his brand-new Godfrey Davis leased company car. Subsequently a 1987 advertisement hero in its own right, appearing alongside the headline 'Eleven Rounds with the World Champion' (a reference to the 11 rounds of the 1987 World Touring Car Championship, with the manufacturers' title going to the Eggenberger Motorsport RS500), Peter's black RS500 was dolled up with appropriate Texaco decals and the wheels photographically transformed into race rollers.

Peter was generous enough to share this restored 97,000-mile vehicle with me on a memorable March 2022 day. He has had properly exotic cars over five decades, and a wealth of competition experience, including with Jaguar and Ferrari marques, but he also experienced the rarity of his employers backing a series of production-based Ford race saloons. In hindsight, he recalls he started with: 'Fiesta, then a Capri and finally a Cosworth RS in the Uniroyal Production Saloon championship. The highlight of the season was always the Willhire 24-hour race, and I did seven of those with a best result of fifth.' With all that competition experience, plus serial Marathon runs from 1970 London–Sydney onwards, you can take it this man is a hands-on veteran of the delights and pitfalls of competition and classic cars.

So why did automotive connoisseur Sugden elect to buy this RS Ford after four years of leased life, then keep it on through more than 30 years and the inevitable hassles (yes, it was stolen!) of regular long-term and high-speed use?

Prior to this RS500's restoration, Peter recalled for the editor of *Modern Classics*: 'On the road, the RS500 remains the favourite. I drove it all over Europe in the days before speed was deemed the eighth deadly sin, and you could cruise down the *autoroute* at 130mph with impunity. Back then, the French towed caravans at 100mph! It was my everyday car. I commuted to work in it. I went skiing with it (the front spoiler made a great snow plough). And while other people carried all their race equipment in Transits, I folded down the back seats and used the RS500, hard. At the time it was affectionately known as "The Van".'

It was on such a race outing that it got stolen from the Silverstone paddock, as Peter recalled: 'At the insistence of my insurers, I'd fitted a tracker and, to be fair to the cops, they found it three hours later hidden in a cul-de-sac in Birmingham. The thieves must have been doing some doughnuts in it, because the tyre on the back-left wheel had shredded and torn a hole in the rear bumper.'

When it had clocked up a varied 96,000 miles, Peter charmingly recalled: 'To be honest it became a little tired and emotional. When driven hard, numerous lights appeared on the dash, warning of impending failure of every vital organ. The engine developed a worrying misfire at high revs. The discs were warped and shook the steering wheel. The suspension wasn't as supple as it used to be. It was still fun to drive, but no longer inspired confidence that you would arrive at your destination.'

◄ ▼ Peter Sugden's RS500 road car spent a spell dressing up as the World Championship-winning Eggenberger Texaco Star Ford-backed machine. Here we see the Ogilvy & Mather advertisement for 1988 and the surviving windscreen sticker, an honoured slice of patina today. *Author*

Peter's RS500 was transported to Yorkshire, where established race and road RS500 specialist Paul Linfoot Racing tended to its ills. Shrewdly and nobly, the Linfoot unit resisted Peter's attempts to try and uprate power, along with other non-period updates. The result is a true, original time-warp; down to the faded and

SIERRA RS500 COSWORTH ROAD CAR

Engine

Configuration: Front-mounted, longitudinal, inline four-cylinder. Cast-iron thick-wall block. Aluminium-alloy cylinder head, four valves per cylinder, DOHC, belt-driven camshafts, inlet tracts 65mm (2.6in) instead of 56mm (2.2in) diameter. Garrett AiResearch T31/T04 turbocharger, enlarged intercooler. Weber-Marelli engine management, eight-injector layout (only four injectors connected on road cars). Ancillary engine changes for durability/accessibility/homologation reasons. **Type:** Ford Cosworth YBD **Capacity:** 1,993cc **Bore and stroke:** 90.82mm x 76.95mm **Compression ratio:** 8:1 **Maximum power:** 224bhp at 6,000rpm **Maximum torque:** 206lb ft at 4,500rpm **Specific power per litre:** 112.39bhp **Power-to-weight ratio:** 180.65bhp per tonne

Transmission

Type: Rear-wheel-drive, viscous-coupling limited-slip differential. **Gearbox:** BorgWarner T5 **Gear ratios:** 1st: 2.95; 2nd: 1.94; 3rd: 1.34; 4th: 1:1; 5th: 0.80; final-drive ratio: 3.64:1

Suspension

Front: Independent by MacPherson struts, lower track-control arms, coaxial coil springs, Fichtel & Sachs twin-tube gas damper inserts, 28mm (1.1in) anti-roll bar. **Rear:** Semi-trailing arms, Uniball joints, coil springs in front of axle, separate telescopic Fichtel & Sachs gas dampers, 14mm (0.6in) anti-roll bar. Unique alternative trailing-arm mount installed for racing homologation

Steering

Type: Power-assisted, Cam Gears/TRW rack-and-pinion, 2.6 turns lock-to-lock

Brakes

Type: Servo-assisted, Alfred Teves electronic anti-lock **Front:** Ventilated 283mm (11.1in) discs **Rear:** Solid 273mm (10.7in) discs

Wheels and tyres

Wheels: 'Cosworth' Rial alloys, 7x 15in **Tyres:** Dunlop 205/50 VR15 original fitment

Body

Type: Three-door hatchback Sierra L with Phoenix-manufactured body kit, plus RS500 modifications, rear wing to Ford design with 30mm 'Gurney' lip and lower RS parts secondary spoiler with cut-out for upper-wing pylon. Usual 'Cosworth' wheel-arch extensions, side sills, front bumper/spoiler, with extra RS500 air intakes (five in total, including official deletion of auxiliary lamps inboard of flashers) and front hard-plastic lip/extension spoiler blade. Twin bonnet louvres, air intakes between headlights and bumper

Weight

Kerb weight: 2,734lb/1,240kg (up 77lb/35kg on original)

Dimensions

Length: 4,458mm (175.5in) **Width:** 1,727mm (68.0in) **Height:** 1,377mm (54.2in) **Wheelbase:** 2,609mm (102.7in) **Front track:** 1,450mm (57.0in) **Rear track:** 1,470mm (57.9in)

Performance

Maximum speed: 153mph **0–60mph:** 6.2 secs **50–70mph (5th gear):** 10sec (turbo lag...) **Fuel consumption:** 20.6mpg for urban cycle; 33mpg at constant 56mph; 24.5mpg at a constant 75mph

cracked remains of the Texaco screen sticker it wore for the 1987 press advert, and its original jaunty, red-spotted, cloth-covered seats.

Yes, there is one obvious exception to the showroom-spec rule: that big-bore exhaust. Peter says dismissively: 'We just could not get the production item. At first it had a rude exhaust note, but you'll find it quite civilised now.' True, but it still had that hard-edged 'let's do the business' note as it flicked to rampant turbo boost in the 3,500–4,000rpm zone.

Sitting in a showroom RS500 cockpit with all the black-plastic trimmings and red-needle dials after a 34-year absence reminded me what a functionally efficient environment it became after a few production upgrades. The obvious updates over lesser Sierras – bar the foundation RS Cosworth – are the comfortable-yet-retentive front seats, and the compact leather-rim steering wheel.

The sturdy five-speed gear lever seems to sprout overwhelmingly towards the passenger, but falls within a hand-span at the wheel. Foot pedals seemed closer than remembered, but were particularly handily spaced when dealing with the notably stiff BorgWarner T5 shift. The suspension felt awkwardly bump-sensitive at 20–35mph country-lane pace. Narrow, rippled and potholed back-road 2022 terrain brings out the worst in RS500 engine, transmission and ride manners. Yet the brakes remain absolutely responsive, with no-slop promptness, which was reassuring in somebody else's expensive vehicle. I can remember back in the day getting quite rude with the Cosworth engine rpm and cornering speeds, to be rewarded with a truly exciting and entertaining Silverstone homeward-bound road trip in a press-fleet RS500.

A few miles are needed for the shift to ease up, and after a gap of three decades for me to think, 'Aah, I remember this.' I wasn't going to revert 30-plus years to back-road bully tactics, but Peter ignored heavy passing showers and allowed me to discover RS500 life on a rural dual carriageway. That meant I had the visceral pleasure of feeling the Pinto-energising 16 valves, twin overhead camshafts and that oversize T04 turbocharger waking up beyond 5,000rpm. The speedometer twitched from lane-lounger to maximum speed limit with satisfying alacrity.

As we settled onto the soaked carriageway, I had mixed feelings about the heavy-lipped rear spoilers. I was grateful for the stability – not a comment you'd make of every lesser Sierra – but conscious that rearward vision was limited. However, the Ford's pace was such that not much lane-changing was needed, just a watchful speedometer eye!

SIERRA COSWORTH THREE-DOOR WORKS GROUP A RALLY CAR: 1988 CORSICA RALLY WINNER

By 1988, my acquaintance with race and rally incarnations of the original three-door Sierra Cosworths was expanding rapidly. On 26 May 1988, I added to that experience by greedily driving the only Boreham-based factory Group A Cosworth to win a World Championship rally. The car, D372 TAR, was safely back on its home test track with John Wheeler, the engineer who coordinated the car's creation. At 2pm that day, I would be even farther east, at the Snetterton race-track, to join Kieth Odor for a Janspeed test session for the forthcoming Snetterton 24 Hours. We would compete in a production-racing (Group N) Cosworth that we had taken to a third place at Oulton Park the previous Sunday.

The Corsica Rally #8 works Ford was still wearing green-and-white Panach-Ford Motorsport livery, with the names of its winning crew, Didier Auriol/Bernard Occelli, signwritten on a front wing. It was a famous win, convincingly defeating two previously unbeatable Lancia Delta Integrales and a Prodrive BMW M3, with fifth place taken by the other factory Ford of Carlos Sainz.

We had a cool, damp morning to take electronically measured acceleration figures according to the usual two-way-average routines of *Motor*'s test staff. Helping those statistics along were undisclosed 'experimental' high boost (over 1.5 bar/21.3psi) and the presence of John Wheeler at the wheel, working a superb Getrag five-speed gearbox. However, we also suffered increased rainfall and the usual pitfalls of feeding high turbo power through rear drive, albeit those plump rear Michelins were the best-possible factory fare. Additionally, there was a proper limited-slip differential, and the reworked independent rear suspension permitted levels of standing-start grip unknown back in the Ford Escort/Capri live-axle days.

'BY 1988, MY ACQUAINTANCE WITH RACE AND RALLY INCARNATIONS OF THE ORIGINAL THREE-DOOR SIERRA COSWORTHS WAS EXPANDING RAPIDLY.'

Let's peek into the past, and drive in alternate shifts with former Porsche engineer John Wheeler, who joined Ford in 1980, and was the hands-on draughtsman/engineer for the Escort RS1700T, RS200, a succession of Sierra Cosworths, and the ultimate Escort RS Cosworths, in a near 20-year reign at Ford Motorsport, Boreham.

Under the skin: no-compromise rally thoroughbred

Looking back, the first impressions of this Cosworth were of how right it looked in the 'tarmac racer' role. Most rally cars sit up with SUV ride heights to tackle rough terrain, but this had nearly the full racing stance, on costly featherweight magnesium wheels and low-profile tyres. Hell, it even sported a racy side exhaust and more stripes than an ice-cream van.

▼ A moody, memorable and totally magnificent day at Ford Motorsport, Boreham in Didier Auriol's 1988 Corsica-winning factory Sierra Cosworth RS. *Mike Valente*

Inside the Sierra's normally spacious cabin, there was much competition clobber, including then state-of-the-art roll cage scaffolding, rough emery-paper-style surface finishes to floor and foot pedals, plus extensive Lifeline fire-extinguisher systems – particularly important in the Group A era, following fiery deaths that occurred in the previous '80s Group B supercars.

The navigator's comprehensively equipped office was to the right, with many dials (seven!) and read-outs to monitor both car health and territorial progress. The navigator also had charge of the horn, as well as the complexities of the programmable Terratrip elapsed-time-and-distance calculator, plus continuous digital read-outs of turbocharger intercooler temperature.

Even though it was less than a month since this Sierra scored World Championship gold, some seating changes had been made. I could not claim to have sat in Auriol's close-combat position, as two matching Recaros now stood at a respectful distance from the dashboard, the navigator dropped floorwards and radically to the rear, as has become the rallying norm.

Both of us were harnessed by Sabelts, and ahead of the driver's seat was a chunky Alcantara-rimmed three-spoke steering wheel, ironically carrying Stig Blomqvist's signature! The engine literally started on the button – that's where the Ford marketing men got the idea for a starter button as a pose on some showroom production cars in the 21st century.

Behind the wheel: an outstanding experience

The cosseted 16-valve turbo motor settled into an even 1,100rpm tickover. Despite the number of bare-metal panels inside the cockpit – and the extensive weight-conscious diet that removed any trace of soundproofing or rustproofing – we could still chat with the engine at idle, or before it switched to 'Terminator' mode at 3,000rpm.

Compared to the previous Sierra Cosworth engines I'd experienced, the big impression was of accessible power. When I drove, the original regular use of 7,500rpm had become a memory. In my words: '7,000 was rarely transgressed. That was because there was a great spike of pulling power between 3,500 and 5,500rpm that made it a lot easier to get the plot moving with exhilarating pace. Slowing and turning into the tighter corners, on a much bumpier Boreham than had been my experience back in 1968, was also promptly executed, thanks to AP Racing's beefier discs, and power assistance for the quick (a couple of turns lock-to-lock) steering.'

You then had the excitement of emerging from the initial understeer without spinning the green-striped machine in the rapid transition to power oversteer on such potholed, damp and occasionally gravelly surfaces. From 75 to 105mph in the third-to-fifth gear slots, sheer fear kept us on a modest corrective lock. Encouraged to try a lot harder in the transition from 100mph straight, to first- or second-gear pylon-indicated twists, both of us managed to rotate the car

▲◀ I am the lucky driver in red jumper, with engineer John Wheeler. The cockpit was a place of wonder and thunder 35 years ago. *Mike Valente*

1986 SIERRA RS COSWORTH ROAD CAR vs 1988 WORKS GROUP A CAR

Data collated 2022. Showroom three-door versus Boreham-prepared World Championship Group A tarmac Corsica Rally winner, D372 TAR, in brackets

Engine

Configuration: Front-mounted, longitudinal, inline four-cylinder. Cast-iron block. Aluminium-alloy cylinder head, four valves per cylinder, DOHC, belt-driven camshafts. Weber-Marelli injection/engine management (Hand-built at Terry Hoyle with Graham Dale-Jones; 1.2–1.5 bar boost; unique ECU) **Type:** Cosworth YBB series **Capacity:** 1,993cc (Group A, FIA 1.4x multiplication turbo formula = 2,790.2cc) **Bore and stroke:** 90.8mm x 76.95mm **Compression ratio:** 8:1 **Maximum power:** 204bhp at 6,000rpm (315–340bhp) **Maximum torque:** 203lb ft at 4,500rpm (Test on raised boost and prototype revised intercooler, c400lb ft)

Transmission

Gearbox: Five-speed BorgWarner T5, synchromesh (Five-speed Getrag, no synchromesh, dog-leg shift-pattern first) **Final-drive:** 3.64:1 final drive, viscous-coupling, limited-slip differential (Group A viscous-coupling limited-slip differential and choice of final-drives: 4.89:1 tested)

Suspension

Front: Independent by MacPherson struts, Fichtel & Sachs gas damper inserts, coil springs, anti-roll bar (Totally reworked with replacement castings, hubs, springs, anti-roll bar and Bilstein competition gas damping) **Rear:** Trailing arms, coil springs, separate telescopic Fichtel & Sachs gas dampers, anti-roll bar (Totally reworked with replacement castings, magnesium trailing arms, hubs, springs, anti-roll bar and Bilstein competition gas damping)

Steering

Type: Power-assisted, 2.4 turns lock-to-lock (Competition rack, power-assisted, two turns lock-to-lock)

Brakes

Type: Servo-assisted, anti-lock braking system (AP Racing, cockpit-adjustable bias) **Front:** 11.1in ventilated discs (330mm/13in diameter x 32mm/1.3in thickness ventilated discs) **Rear:** 10.8in solid rear discs (285mm/11.2in x 28mm/1.1in thickness ventilated discs)

Wheels and tyres

Wheels: 7x15in (Ford Motorsport/Speedline magnesium 8x16in front, 7.75x16 rears) **Tyres:** 205/50 VR Dunlop D41 (Michelin S1 slicks)

Body

Type: Three-door hatchback, fully road equipped, sunroof, five seats (Built from bare shell, multiple lightweight panels and components, two seats, extensive TIG-welded T45 steel roll-cage fittings)

Weight

Road test weight: 1,219kg/2,688lb (Kerb weight, 1,104kg/2,433lb)

Performance

(Independent road test, based on Motor results for both showroom and competition. Ford Motorsport chief engineer John Wheeler at Boreham airfield drove D372 TAR. Motor staff man Ian Sadler recorded fifth-wheel monitored results on a cool day/high boost at 7,000–7,500rpm gear-changes)

	Road car	Rally car
0–30mph	2.4sec	2.4sec
0–40mph	3.6sec	3.1sec
0–50mph	4.8sec	3.9sec
0–60mph	6.0sec	4.8sec
0–70mph	8.5sec	5.9sec
0–80mph	10.5sec	7.5sec
0–90mph	13.2sec	8.9sec
0–100 mph	16.3sec	10.8sec
Standing ¼-mile	15.4sec	13.3sec
Maximum speed	146mph	115mph
Fuel consumption	22mpg	10mpg

'THIS FACTORY CAR WAS JUST WHAT YOU'D HOPE FOR AS A DRIVER AT ANY LEVEL. A CHARACTER AND A TOP-CLASS WINNER. OUTSTANDING!'

to an untidy halt, in my case with a lot of self-conscious, false laughter.

John Wheeler always pushed personnel and crew to give their ultimate, and he didn't spare a visiting journalist either. Showing rare trust, I was allowed out over the jumping blocks that the Boreham testers utilised to assess stability under extreme duress. Here, I hurled the Sierra flat out in third, fourth or fifth at uneven-height concrete ramps: 'Logic says the car must tip over. Yet it lands with such gradual shock absorption that only the pale face of the looming lens hero [Mike Valente] tells of the short flight of a 2,443lb (1,108kg) automobile.'

I loved all my Cosworth Sierra and Escort drives – except losing racing ground to Paul Bailey's 'RS2000-500' (see pages 151–153) – but this factory car was just what you'd hope for as a driver at any level. A character and a top-class winner. Outstanding!

SIERRA RS COSWORTH 4x4 FOUR-DOOR WORKS GROUP A RALLY CAR: 1991 MONTE CARLO CAR

On 6 February 1991, I commuted nice and early in my 40,000-mile Sierra Cosworth four-door to a sub-zero and snow-dusted Boreham. Exciting, as I would meet François Delecour and accompany him in the four-door works Sierra RS Cosworth he had very nearly given a debut win on the Monte Carlo Rally a week earlier.

Backed by Q8 Oils, this tarmac-biased low-rider Sierra, G94 CHK, was a £100,000 factory-rendition of the 4x4 Cosworth, packed with the trickest seven-speed gearbox and an official 295bhp from Mountune. I then drove the similar factory Ford, H319 MTW, that Malcolm Wilson piloted to a fine Monte seventh.

We also had a naughty slippery session, when we decided to test chirpy François's top-flight reactions in my Teesdale Publishing staff right-hand-drive Cosworth road car. He didn't usually drive on the British 'wrong' side of the road and I forgot to add that my car was 'only' rear-drive, until he broadsided us into a mighty spin at the first sharp corner.

I should provide a précis of the epic tale that saw Delecour so nearly win on his Ford and the

◄ We media were allowed a couple of 1990–91 outings at Ford Motorsports, Boreham, in the Q8-backed Cosworths. This LHD factory 4x4 (G273 BPU) appeared in a variety of liveries in 1991–93 in Portugal, Spain and the UK. *Ford/Q8*

4x4 Cosworth's Monte debut. François was lying a handy third behind the established Group A winners from Toyota and Lancia going into the final day, when he pressed on magnificently to overhaul Carlos Sainz's Toyota Celica GT4, and Lancia's Miki Biasion wilted under snow-laden special-stage pressure. The Frenchman commanded a big advantage ahead of the final 14-mile night stage and realistically only needed to complete it at a steady pace to be assured of victory. Sadly, however, he ran off the road and severely damaged the rear suspension, picking up punctures in the process. Sainz snatched his victory gratefully, finishing 4m 59s ahead of Biasion's Lancia, while Delecour salvaged third place.

Behind the wheel: seven speeds and impressive traction

My turn. Most technically interesting was my first encounter with the Sierra's then freshly homologated Group A seven-speed MS90 gearbox. In the homologation form, it appeared as an H-pattern six-speed, first gear marked 'L' as it was officially a crawler initial gear opposite reverse. The FF-manufactured non-synchromesh gears clacked through the second to sixth ratios with satisfying speed, but I did find the isolated first a bit of a fiddle.

The seven-speed was designed to meet the requirements of a very narrow power band, as the Cosworth engines suffered tightening international intake-restrictor regulations. Yet all the road and competition examples I drove, from 204bhp to 500, were a lot easier to keep within pulling-power revs than any '60s/'70s tuned carburettor car, mainly thanks to excellent electronic management systems, as I discovered for myself with Malcolm Wilson's factory sister car.

The hybrid injection/engine management electronics were credited on the homologation form to Ford EFi/Bosch and Weber-Marelli (covered all bases for scrutineering!), but much of it was the work of local UK specialist Pectel. The Monte motor ran up to 2.0 bar (28.5psi) boost and a 7,600rpm limit to pull along 2,654lb (1,204kg). It was credited with 295–300bhp at 6,500rpm, coupled to 322lb ft (400Nm) torque by 4,250rpm.

For me, the major Boreham test-track memory was of resounding mid-range pulling power, which I exploited between 3,000–7,000rpm during that slippery session. There was significant slow-speed understeer turning into corners at third-gear or lower speeds, but the bonus was proper 4x4 traction on the way out, characteristics that no competition rear-drive rallying Ford had ever exhibited.

Running 8.15x16in OZ wheels and 225-width Pirelli tyres, the factory Ford was geared to give a high (by rallying standards) 134mph maximum at 7,500rpm.

It had been an enormous personal privilege to drive these outstanding Cosworths and meet Delecour, who would have his Monte Carlo Ford victory revenge in 1994, but that is a story for the later Escort Cosworth chapter.

DRIVING THE WILLHIRE 24 HOURS IN A PAIR OF SIERRA RS COSWORTHS

My first track experience as a passenger in a previously successful production-racing Sierra Cosworth at Snetterton was a tad over-dramatic, as the regular driver spun me out of Coram's fast and apparently endless right-hander during a demonstration of his prowess on a wet media-preview day for the Willhire 24-hour race in 1987.

My *Performance Car* link now stretched beyond retained writer, and the magazine placed me as a team driver in four touring cars over the 1985–88 runnings of the Willhire event. The vehicles were a Fiat Abarth 130 TC (1985), Sierra XR4x4 (1986) and front-running Cosworths (1987 and '88). With the Cosworths, it was always good to be in a car capable of leading, rather than cowering off-line in a slow machine, as had been much of our experience in the XR4x4. Although neither Cosworth lasted the distance, their fun factor was fabulous, as all us 'boosty boys' tried to balance exhilarating speed with finishing pace. The 1987 Cosworth turned a 1m 21.8s lap during a media day/test day on modest boost (1.0–1.2 bar), with 240–265bhp available from its Mountune £1,500 race engine.

1987 Willhire 24 Hours: TCCS Morgan three-door Sierra RS Cosworth

For the 1987 event, I was teamed with established tin-top racers Sean Brown, Dave Morgan and Rod Birley. We got no track time on the first day of practice because our rented TCCS Morgan team Cosworth didn't arrive. Preliminary practice, and what is now called qualifying, came on what turned out to be a very wet Friday. As a result, the afternoon practice sessions were cancelled, so my first outing was in the wet evening, with just three laps to make sure I got a race. The heroic, sensational performers in these soaking sessions were, naturally, the less-powerful front-drive cars, Golfs featuring well within the outright front-of-grid positions.

The race was all drama – mechanical, electrical and personal. Although the track was still damp in the opening laps, by the time I got strapped into the white-and-red RS it was dry, and our quickest drivers had got the competitive Ford in amongst its fellows. After some three hours I could barely believe a pit board that said we were running third with yours truly aboard.

It seemed prudent to switch on the headlights, as the car was so quickly into each string of traffic. I don't think that had much to do with what happened shortly after 3½ hours, as the engine just shut down as I entered the corners beyond the start/finish straight, no electrical life present. I drifted into a convenient trackside bay and opened the bonnet to an oven-ready engine compartment, more in hope than expectation of finding something obvious detached. I prodded and probed, and inspected the mass of relays and fuses, knowing that I did have some spares stowed in the cockpit. I felt ever more useless and hot in my racing pyjamas as the June sunshine faded. Curious spectators gathered, along with an observant official, intrigued by what had happened to a front-runner.

After 10–15 minutes featuring an embarrassing lack of progress, I surveyed those onlookers and saw the most welcome face I could have recognised in those circumstances, Paul Wilson. A former senior at the Ford pilot-plant (prototype and pre-production creators) and now Ford UK press garage, Paul stepped up, beaming and ready to offer advice. He couldn't directly help while the car was on the circuit – we'd be immediately disqualified – but he moved as close as he could, and verbally ran me through a logical sequence of electrical checks.

After half an hour or more had passed, we got to an obscured 20-amp fuse that completed the live circuit for one of many electronic modules. On changing the fuse and trying the starter for the umpteenth time… there was a growl, spit, and clatter as the motor revved and settled to an imperturbable idle as if nothing had happened.

I stuffed my balaclava and helmet on in a rush, and drove back to the pits. The team swung back into full-action mode to recover all that lost time – and nobody swore at me! To my surprise, they let me out in it on the regular night rota, but come the morning the Sierra had suffered many of the common failures of the Cosworth in its first race season. It circulated at the tail end of the field,

▲ At my first Snetterton 24 Hours (1987), in the Dave Morgan RS Cosworth Sierra, we spent too much time in the pits, as in this handover, where I am about to step in. *Author's archive*

after some typically British mechanical ingenuity and fortitude to keep it running. But at least the Cossie still lapped very competitively in between the setbacks.

Great driving experience – but no cigar!

1988 Willhire 24 Hours: Janspeed three-door Sierra RS Cosworth

My Cosworth drive in the 1988 Willhire 24 Hours looked even more promising. That was because I was already ensconced in a regular drive and test programme with the Janspeed Sierra RS Cosworth, shared with 1989 Firestone production-racing champion Kieth Odor.

I had helped source the ex-management-fleet Sierra RS three-door at Ford. Additionally, we had healthy sponsorship from BF Goodrich and many others, headed by local Salisbury support from P. Arnold. The result was the smartest Sierra I raced, in metallic dark red and blue, with white detailing.

Here are some edited highlights of what I wrote for a three-page feature entitled 'Our Own 24' in *Motor Sport* in September 1988:

'The weekend after motorised Britain was suffused with pride for Jaguar's Le Mans victory, the 1.9 miles of Norfolk's Snetterton circuit played host to the United Kingdom's only 24-hour race. The contestants were from the booming National Production Saloon Car Championship. Miffed by an absence of success, and further turbo-boost controls, Colt Car Company had withdrawn its support for the Starion, leaving eight Sierra RS Cosworths the favourites for victory in this safely organised BRSCC contest.

'The obsolete three-doors were boosted to rather less than their usual production racing 270–285bhp in the interests of durability. The "Cosworths", as they always seemed to be referred to, were still comfortably the fastest straight-line propositions in the field along Snetterton's answer to Mulsanne – the Revett straight. The Fords, which finished 1–2–3 against all predictions in 1987, faced a field of potentially more reliable, but slower, BMW M3s, Escort RS Turbos (a surprise winner in 1986), Golf 16V GTIs and a lone Mercedes-Benz 2.3/16. Plus sundry representatives of Peugeot, Honda, Lancia, Opel, Fiat – and the flock of GTi Suzukis, which dominate the sub-1,300cc division.

'This was the ninth edition of the annual fixture, which started in 1980 as a mixed sports and saloon event, but the 1985 swop to a saloon car championship race marked the beginning of an escalation in spectators and competitors that was quite a surprise to anyone who remembers Snetterton only for its windy isolation. The June weather is usually kind, and this year was no exception. Unfortunately, that warm weather isn't a blessing for the hard-worked 205/50 road-legal rubber, which twists beneath 2,650lb (1,202kg) of Cosworth through the many hard rights such as Coram's endless 100mph corner.

'This year's car was a pristine example, prepared by the people who first trusted me enough to support my saloon-racing side-line: Janspeed. I didn't really need reminding that it had been 16 years since we tackled 29 novice-season races and came home with a class title. This time, instead of driving for Jan Odor senior, I had agreed to accompany Kieth (yes, that is the Hungarian spelling) Odor, who had been less than 10 years old when I drove for his dad!

'Now Janspeed has seen its turbocharging technology fully employed on the Sierra Cosworth, as the Ford races under strange technical regulations in Britain. A 26mm (1.02in) air-restrictor on the path from intercooler to injection attempts to strangle the Cosworth's overwhelming horsepower advantage. Yet there is no control placed on the boost, so competitors run more than double the Ford-recommended figure of 0.55 bar (8psi). A fair degree of technical

'THIS YEAR'S CAR WAS A PRISTINE EXAMPLE, PREPARED BY THE PEOPLE WHO FIRST TRUSTED ME ENOUGH TO SUPPORT MY SALOON-RACING SIDE-LINE: JANSPEED. I DIDN'T REALLY NEED REMINDING THAT IT HAD BEEN 16 YEARS SINCE WE TACKLED 29 NOVICE-SEASON RACES AND CAME HOME WITH A CLASS TITLE.'

▲ This atmospheric picture of a Snetterton 24 Hours pit stop captures the refuelling, wheel changing and 'Can-you-fix-it?' interludes that accompanied all my Sierra 24-hour missions. *John Overton*

expertise with electronics and turbochargers is called for if the poor old Garrett AiResearch T03 (which spends its life working virtually flat out against the restrictor) is to remain serviceable and the head gasket undamaged. Generally, you can reckon that a winning RS Cosworth in this category will have 280–285bhp, enough to slam it to 135–140mph on many British circuits.

'That road-tyre-smoking power level was my first racing experience of the Janspeed Ford. In late May, Kieth and I shared the driving in a one-hour event around Oulton Park's captivating crests and cunning double-apex corners, finishing third overall. So we went to Snetterton with some optimism, even though all of us knew that this complex car would be far more severely tested over 24 hours.

'Just as you would expect, we took the precaution of running lower boost to place less stress on not just the engine, but also upon replacement running gear. Included were a new BorgWarner gearbox, viscous-coupling differential, brake discs (we bedded in six sets of Mintex M200 brake pads among the testing chores) and Koni gas-insert dampers. In a formula where you cannot change any material part of the suspension, especially the coil springs, the replacement dampers and abused tyres (standard 205/50-sized BF Goodrich TA-R1 for us) end up as the only obvious handling upgrades.

'As ever in "showroom" racing, it was the subtleties that mattered just as much… The Sierra proved incredibly sensitive to ride height, so that it felt terribly nervous when riding on new springs, which were fitted to avoid falling below the specified legal-minimum ride height. As the springs settled beneath the load of a full 120-litre fuel tank, its contents consumed at close to 10mpg, the Sierra's cornering capabilities also stabilised. Unless upset by heavy braking, the RS Sierra turned into an apex on understeer lock, followed by power oversteer.

'Some three hours of practice are provided, half of it the mandatory night session in which every nominated driver (between two and four per car) must complete at least three laps. We intended to drive two-handed for the majority of the race. From the delayed 4.15pm Saturday afternoon start to 10.31am on Sunday, that is exactly what we did, but former Porsche Champion (and Snetterton 24-hour race winner) Bill Taylor stepped in for what was destined to be the last stint.

'Practice left us on the third row of the grid, seventh quickest courtesy of Kieth. In fact, as for the night session, our strategy was simply to do the minimum possible, the car now fully race-prepared so that any excess practice mileage was pointless. To put our lap times in perspective, it is worth reporting that pole position went to the Dave Pinkney/Robb Gravett RS Cosworth in 1m 18.58s (87.82mph), nearly a second faster than eventual winners Lionel Abbott/Graham Scarborough, who would cover a record 1,964.9 race miles at an average 81.85mph, over 20 laps ahead of an M3 and a Golf GTI 16-valve. In race trim, our car recorded 1m 19.43s (86.88mph), the fifth-fastest lap. The winners also established the fastest racing lap, 1m 18.54s, fractionally faster than the best in practice.

'I had the honour of starting the car in a 2½-hour racing session, but we normally ran closer to 2¾ hours, with brake-pad changes scheduled every second stop and tyres thrown at the car during most stops. The daytime temperatures did give us tyre problems. As Kieth put in laps below 1m 20s, we suffered the tread separating from the carcass and consequent deflation. But there was plenty of warning, and it is worth saying that temperatures in excess of 250°F were recorded, even when the covers had reached the pits. Other leading brands suffered the same problems when fitted to the faster cars.

'The rolling start was a daunting third-gear experience, the herd of Cosworths congregating on Riches, the opening corner, in a dusty haze of blue tyre smoke, over-run exhaust flames and searing paint schemes. The initial laps were spent with the Cosworths gradually subduing the interloping Escort RS of David da Costa. I settled in behind the Firestone Firehawk RS Cosworth of Jerry Mahony.

'Dry and dusty night, or sun-soaked day, lap times and cornering patterns remain the same, altered only by the need to accommodate sudden strings of traffic. Some 25 of 37 starters completed 24 hours; only 24 were classified though, the second-placed RS Cosworth disqualified for using a tyre lacking road legality.

'In the RS Cosworth, you need drop no lower than third gear at this circuit, the superb torque a marvellous assistance to avoiding tyre-destructive wheel-spin. Fourth is required to dive into the aptly named Bomb Hole dip that introduces you to the delights of Coram. All through this eternal right you listen to the tortured near-side front tyre wailing, your body braced upon a giant alloy footrest and by the embrace of your Willans harness. You straighten the steering and give the tyre an easier time at the expense of your heartbeat. Prompting the big Ford back into 100mph line for the left/right flick of Russell is a precise process, otherwise the car bites back and clips the forbidden zones of the low kerbs.

'Past the pits, fourth can be exchanged for fifth again at the designated 6,000rpm limit, the side exhaust booming off Armco and backmarkers with exhilarating vigour. Top is only required again on the Revett back straight. The car I had last year reached the rpm equivalent of more than 130mph at the fastest point, but this year's more-conservative specification brought only 121mph. That wasn't enough to cope with the other Sierras in the opening charge, but the Firestone car was running exactly the same lap times at that stage, and I knew Jerry Mahony/Mark Hales had been a leading force in this category before. Thus I wasn't too disheartened by my first-hour placing of seventh, 3sec ahead of the Firestone machine, a position that strengthened into fifth at the close of the second hour.

'Unfortunately, our pitstop at 6.44pm took 13 minutes, after a misunderstanding during a brake-pad change popped a piston from one calliper. That delay summed up our event – we would just get the car back into the top 10 when another mechanical gremlin wormed into view. Amongst the evil ones were: a burst oil cooler, total brake failure after a calliper distorted (I took to autocross on the infield to slow the thing down on the fourth-gear approach to Riches), that tyre deflation, intermittent clutch operation due to a fractured steel component in the operating mechanism, plus a series of electronic-engine-management misfires after the alternator was damaged by a surprise shear in the massive bolt which locates it.

◀ Despite the repeated technical problems I had in two 24-hour outings with Sierra RS Cosworths, the car personified fast Fords, and featured the benchmark fun factor I enjoyed throughout my racing life from 1968 to 2000. *John Overton*

'At 12.52pm on Sunday, the electrical shortcomings eventually mismanaged the engine to the point where the head gasket wept. The travel-stained Ford rolled silently towards my Renault caravan base with less than four racing hours left. Yes, I was disappointed. Yet, 10 hours and 25 minutes of racing in four sessions gave a personal satisfaction that was denied the hard-working Janspeed pit crew of team manager Frank Swanston, engine ace Norman Clancy (who had looked after my Escort Sport a decade previously) and Kieth's regular mechanical minder, Shaun Arnold. All of them demonstrated that they could whip the car back into contention, after each frustrating delay, in professional good humour.'

I enjoyed my hours of sweated labour at the Cosworth coal-face, but there were still two longer-distance races for Odor and I to tackle in 1988: a one-hour event at Donington on 18 September that we failed to finish, then a good-quality following weekend at the Brands Hatch finale to the Uniroyal series. This race featured over 300km (186 miles) of the modified Grand Prix circuit. That final resulted in an encouraging fourth overall on the fabulous circuit. Oh, and my premature declaration that I would retire? Actually, my last race was in May 2000 at Brands Hatch… in a Ford, naturally!

Footnote *There was a season's end non-competitive November outing at Snetterton. In aid of the British Heart Foundation, it involved taking passengers in our regular Uniroyal 24-hour racer. I had the privilege of many laps, and for the first time I managed to get a seat for Norman Clancy, who had put in the hours at our day and night of Snetterton, besides so many races we attended in the 1972 Escort. Despite my inverted racing past, and the immense impact when you initially experience Cosworth acceleration, Norman appeared as calm as ever when he climbed out of the Sierra. I also got to take three potential sponsors for rides, which was important, as Kieth would fight (successfully) for production championship honours in the following 1989 season. On such a cold, but dry, winter day, freed from 24-hour-race durability versus boost worries, the Sierra absolutely flew along the straights and had no tyre-preservation handicaps during full-load cornering. I timed Kieth as being right on our pre-race practice pace… with startled passengers on board!*

1988: ANDY ROUSE KALIBER-SPONSORED BTCC SIERRA COSWORTH RS500

At a Silverstone circuit paddock simply packed with Kaliber low-alcohol beer guests on 4 October 1988, my missions were to drive an Andy Rouse Kaliber-backed RS500 (as used for the 1988 British Touring Car Championship final) on a media-friendly 460bhp, plus an ex-Jim Clark Cortina Lotus (see Chapter 3) with at least 310bhp less. The results of this assignment would be published in the January 1989 issue of *Performance Car*. However, the commission from *Performance Car*, and subsequent feature, omitted the Clark car and went for another BMW-versus-Ford feature, as I also drove Frank Sytner's 1988 championship-winning M3, as prepared and entered by Prodrive, which contrasted the Sierra's sensational straight-line squirt with stunningly effective brakes and head-stressing grip around the jinks and swerves of Donington.

Actually getting into the Rouse RS500 took most of the day, queuing up behind such Z-listers as Joan Collins's umpteenth husband, 'Bungalow' Bill Wiggins, nicknamed during their 11-month marriage. Needless to say, it was worth those hours of standing in line, for this – as with so many Rouse Fords – was simply the best of the racing RS500 breed that I steered.

Just how commercially rewarding the RS500 became for Andy Rouse Engineering (ARE) can be gathered from the fact that up to 30 employees built over 100 race engines and 30 competition RS500 cars, each example demanding over 1,000 hours of labour. Subsequently, ARE also built 'nearly 90' Sapphire Cosworth four-doors, coded 302R and 304R, for road use, so it truly was a versatile and thriving business.

Under the skin: the ultimate race RS500

Andy recalled in 2015 the competition changeover to the RS500 during 1987: 'The RS500 version, with a much larger turbo and bigger injectors, different rear-suspension geometry and a bigger rear spoiler, was quite challenging to drive, because we were making 520bhp and over 400lb ft of torque. Ford paid us to run the Sierra development programme originally, but then the RS500 took off, and in 1988 there were up to 16 of them on the championship grids. So they didn't need a works team any more, and instead they ran a bonus scheme, with £5,000 if you won a race. I won nine of them, so that was quite good!'

In period, prices to race at the front varied from £65,000 from ARE, to Eggenberger advertising their winners for more than £100,000 in 1988–89. Today, you would pay a lot more than that to own a pedigree race RS500, but as a guide, the kind of Rouse RS500 I tested was offered as a Continuation series of three, built from stored Ford Motorsport bodies, at £185,000 by CNC Motorsport AWS for race-ready delivery in 2022.

Such Rouse RS500s were in the top three contenders for 1987 World Championship honour. Yes, I know Eggenberger actually won that title for Ford, but Andy memorably beat Steve Soper at Brands Hatch in an epic confrontation with the Texaco star.

Oh, and Rouse, sharing with Alain Ferté, won an epic Silverstone TT in 1988, where Dick Johnson came over from Australia with the charismatic and demonstrably ultra-rapid Shell-backed red RS500 racer. Eggenberger, Johnson and Rouse machines duked it out for ultimate RS500 respect at that Silverstone TT. The Johnson/John Bowe car proved a worthy runaway leader until it hit water-pump trouble. Then it was Andy on top, ahead of an Eggenberger 2–3–4 trio, but two of those three Eggenberger factory-backed Fords set fractionally faster qualifying laps than Rouse, and Johnson's flyer took pole position to predict what we would see in terms of race pace.

No wonder, then, that I waited my turn patiently at that breezy but dry Kaliber media day, in the face of public-relations people obviously trying to placate us all. For this major sponsor – Kaliber reportedly spent £1 million annually on motorsport sponsorship – the publicity accruing from BBC *Grandstand*'s audiences peaking at 6.5 million, and VIPs, was naturally more important than hardcore motoring media.

So what was so special within this and so many other ARE Sierra RS500s? I think it was the driven, hands-on presence of Rouse and business partner Vic Drake. Drake had a vital and unique input to the programme, as this was the only winning team in the UK to build its own engines at this premier-league level, and Rouse provided not just race-winning pace but also the kind of direct track-to-engineer feedback that decades of front-running experience and a quartet of British titles underlined.

Since the majority of competitive RS500s shared many Ford Motorsport components – including the start-point bare '909'-code body – the 100-hour in-house engine builds gave these ARE RS500s an innate advantage over buying in sub-contracted units. Another important factor in the pursuit of reliable turbo power on the ARE RS500s was Zytek electronic engine management. Andy told us in July 1988 that the team spent 600 hours on race-engine analysis, along with engineers from Derbyshire-based Zytek, reprogramming the electronic control unit (ECU): 'When you consider that the ECU computer receives data on a huge number of parameters, including engine revs, turbocharger-boost pressures, a number of engine-temperature readings and the throttle position, you can see that getting the best from the engine is a complex task. But programming the computer is one of the secrets of getting the engine to work correctly.'

Another turbo-engine soothing operation was to cool it during operational temperature peaks with exhaust-gas temperatures reaching close to 1,000°C. This demanded not just the obvious high-efficiency water and air radiator and intercooling, but also two engine-oil coolers, uprated water and oil pumps, plus coolers for gearbox and differential oil. Oh, and the driver got some additional cockpit cladding to reduce the heat transfer from the turbocharged exhaust gases passing underneath.

A lot of rubbish was talked about peak power beyond 550bhp, although that was available, but not internationally legal, as Andy explained in 2019: 'The most power we had was 530bhp, which was only used for qualifying. In race trim we used about 500bhp to get good reliability. This was with a legal-spec turbo, which had to pass scrutineering. It is possible to make more power by using a bigger compressor, but

▼ Andy Rouse (arm out of the Kaliber Cosworth's window) acknowledges the fantastic battle he has just had with Steve Soper's factory-backed World Championship Eggenberger RS500 around Brands Hatch GP circuit in 1988. *Author's archive*

that would be outside the Group A rules.'

By contrast, the suspension and associated components sounded simple, with Bilstein gas-damping (Spax dampers were also used on the sister RS500 of Guy Edwards), massively uprated coil springs, 13in AP Racing ventilated disc brakes and BBS 8.5x17in centre-lock alloy wheels, wearing race-only Pirelli tyres at the regulated 10in widths. However, getting them all to sing in unison took track miles and a hefty reference to Rouse's previously accrued experience, especially from the XR4Ti Sierra-Merkur programme. It was repeatedly highlighted in period that race RS500s had the power of '70s Formula 1 cars carried in more than double the kerb weight, at 2,420lb (1,098kg). Since the production-based Ford couldn't have a Formula 1 car's ground-hugging ride height, with a proportionately wide track, or anything more than production-based aerodynamic appendages, racing to win against equally determined opponents was bound to demand the coolest head a driver could summon, and equally demanding track development miles.

Theoretically, the RS500 rested on Ford's MacPherson-strut front and trailing-arm independent rear layouts, but radically revamped using Ford Motorsport and third-party specialist parts. The rear suspension became highly modified, relocated to the limit of regulation tolerances via the revised mounting points that Ford-Tickford incorporated in the RS500 road cars.

Behind the wheel: the quickest and best-handling RS500

Mid-afternoon, and the early-morning fog has cleared. It is my turn to drive a machine that Rouse reported as swallowing 0–60mph in 3sec and 0–100mph another 3sec later. It should be a monster to a stranger allotted just five laps around Britain's regular Grand Prix track layout, right? Absolutely not.

That laboriously built bare body, with the integral roll cage around you, plus the visible screen-surround tubes, weaving forward and aft in support of heavy suspension loads, provided reassurance that you could not have more obvious protection. The cabin had reminders of this Ford's production past, including the fascia/dashboard surround and grey door trim.

The ergonomics were some of the best I have experienced in a race car, with the routine RS-branded four-spoke steering wheel raked to suit the high-sided race seat, itself enveloping the driver's shoulders and torso, with a bolster to align thighs to foot pedals, and an enormous alloy foot-brace – all within comfortable reach of the pilot when belted up in a four-point safety harness.

The engine fired up on a flick of the switch and a push of the starter button. I was impressed by the race-tuned four-cylinder's amiability: the heavyweight competition clutch could be released at 1,500rpm in a first gear that could carry you to 65mph at the thunderous 7,500rpm red line. I also registered the presence of seven-dial instrumentation, as there were low-speed photographic passes to complete, to the accompaniment of the usual competition-car transmission-gear gnashing. The eye-catching dial was a 250mph speedometer for the benefit of the onboard cameras. More functional were a VDO-gauge monitoring boost to a 2.0 bar (28psi) maximum, and a Stack 0–9,000rpm tachometer that remembered exactly the maximum engine rpm reached, rather than what the driver was prepared to divulge.

Apart from the isolated first of five forward gears being a stretch away for a right-hand-drive pilot, the £3,000 Getrag synchromesh gearbox was sweet-natured simplicity to operate. Aside from that 65mph first gear, Andy predicted that the Silverstone gearing would deliver 90mph in second, 112mph in third, 132mph in fourth, and a weather-and-wind-dependent 152–158mph on the fifth-gear Hangar Straight.

Over 30 years later, my searing recollection

'THE ERGONOMICS WERE SOME OF THE BEST I HAVE EXPERIENCED IN A RACE CAR, WITH THE ROUTINE RS-BRANDED FOUR-SPOKE STEERING WHEEL RAKED TO SUIT THE HIGH-SIDED RACE SEAT, ITSELF ENVELOPING THE DRIVER'S SHOULDERS AND TORSO.'

is of the muscular acceleration. There was no perceptible boost until 4,500rpm, but asked to deliver between that point and 7,000rpm, each Getrag shift was rewarded with a muffled retort from the side exhaust and more instant velocity than any other Cosworth device I drove outside those rallycross 4x4 Escorts described in Chapter 10. The obvious difference between rallycross and circuit racing becomes the fact that you keep on accelerating, hard, beyond 100mph, on a circuit.

Boost gauge twitching for an easy 2 bar on such a cool day, third gear soared us over 110mph to reach for a closely stacked fourth, which took over ready for a fifth-gear rush to 145mph on these exploratory laps.

As the Hangar Straight faded in favour of the incoming quick rights of Stowe and Club, I was pathetically grateful for those generously dimensioned brakes. I did find the quickest cornering flick left – fifth-gear Abbey on the 1988 layout – a 135mph skipping trip over the ripples and bumps. Rouse just grinned at the end of my stint and commented calmly: 'It gets better the harder you drive it in there.' Even Andy acknowledged that such power on comparatively modest rubber meant that his race RS500 must be 'driven with smoothness and some restraint', to avoid long sideways moments in second or third gears.

This was definitely the quickest and best-handling track RS500 I drove, albeit only briefly. It was a startling contrast to the BMW M3 of Frank Sytner, who took the British title that year after Andy suffered a 150mph tyre blow-out in the final. The Ford emphasis fell on amazing straight-line speed, with an advantage of up to 180bhp reported over the M3. The BMW demanded more of a single-seat driving style, blaring between 7,000–8,800rpm through six gears, dependent on fabulous chassis and braking capabilities for impressive lap times.

I was destined never to race an M3, but there were more RS500 experiences in store…

1988–89: KARL JONES DUCKHAMS-SPONSORED BTCC SIERRA COSWORTH RS500

I didn't expect to drive more RS500s on track after the Rouse Kaliber car summit, but there were more explosive laps in 1989 and 1991. First, Karl Jones's Duckhams-sponsored RS500, prepared by former Ford Motorsport engineer Richard Asquith, was a 1988–89 regular British Touring Car Championship contender. Perennially underfunded, Jones and Asquith had shown what they could achieve with a second place using what amounted to a Cosworth with traces of factory-rally-car DNA in an unsophisticated specification. Body and suspension/brake components were supplied from Asquith's former employers at Ford Motorsport, Boreham.

It was 'mine' for some Snetterton test laps alongside the Jones-Asquith Honda CRX, then appearing in a spectacularly rewarded but a mite brash-and-crash one-make series. Since I then had a CRX coupé as a road car, I was quite interested in the competition version, but back to our Ford priority.

The automotive odd-couple track test appeared in *Motoring News*, but didn't draw in what driver Jones and preparation ace Asquith really needed – more cash. Their RS500 was neatly presented, a credit to the sponsors, but to be more competitive, the team badly needed the kind of race budget that the front-runners attracted.

There was the trademark explosive Cosworth surge in power, but not in the same league as the Rouse Cosworths. Plus, the handling featured near tarmac-rally settings, no match for radically revised and patiently fine-tuned circuit capabilities.

It demanded all of Jones's Escort RS championship skills (he has been seen more recently shining in classic and current Jaguars) to tame it. That RS500 proved – if proof were needed – that you can only go so far in modern motorsport without a competitive cash pile to complement an ace driver and preparation talent.

1991: RACING AN RS500 COURTESY OF COLLINS PERFORMANCE ENGINEERING

When I went to Bruntingthorpe in January 1991 to assess four tuned 500+bhp Fords for Chris Wright of *Top Car/Fast Car* magazine, the star of that soggy day, as I explained in Chapter 12 (pages 256–257), was an RS200 Evolution prepared by Collins Performance Engineering. Standing in the gloom after our timed acceleration runs had been completed, Sue Collins – company boss and serial creator of 500–550bhp RS Cosworth conversions – stood thoughtfully. She gestured to a road-registered white RS500 Cosworth, G819 JWN, that was cooling down.

'I want to do some of that Vecta Fast Ford Championship in my RS200, but I'd like that Cosworth out more regularly. It's a bit of an animal in the wet, just can't find anyone to drive it…' Sue's voice trailed off wistfully and all eyes swivelled at me.

Marilyn, my new partner (later my second wife), was with me. She had worked in motorsport media and was keen to get involved again. As I digested the implications of Sue's comments, Marilyn beat me to a reply: 'Jeremy could give it a go, how about that?'

Sue looked slightly surprised. She paused, and her reply was a counter-offer: 'OK, if he can drive it around the quicker perimeter track here in a reasonable time – and do us some infield cornering stuff, all without spinning in this weather, I think you could be on.'

Now those previous Mountune RS500 sessions in an Essex downpour surfaced in my dusty memory bank. I knew enough to get the plot moving by changing gear whenever the boost threatened to unhinge road tyres, and when we had it running through the fast swerves

▼ My favourite image of the 1991 Collins Cosworth RS500 came from Brands Hatch. Unfortunately, I don't know which trackside photographer created this cracking close-competition image. *Author's archive*

▲ A couple of personal Cosworth Sierra racing memories. Top is the Janspeed production racing RS that I shared with Kieth Odor. The white road-registered machine below is the Collins Performance Engineering RS500 raced in 1991's UK Vecta Challenge with multiple second places, fastest laps and pole positions. *Performance Car/ Author's archive*

of the ageing airfield perimeter track, there was nothing but pleasure to be had as it cleaved through the air, kept increasingly secure by the RS500 aero extensions.

There were a few modest smiles when I returned, happy not to have spun or skated off on the quicker sections. Sue said: 'You've got it if you want it. I'll run my RS200 occasionally, but there's a season of the Vecta Fast Ford series to do, and it's important for this Sierra to appear and publicise my business.' It became a tough season of slow starts to short races, and then recovering for too many second places, countered by strong qualifying and race pace.

This roadworthy RS500 featured a YB-series turbo motor to full Group A specification, rated at 550bhp. The engine was mated to a Getrag five-speed synchromesh race gearbox, sending drive to a ZF multi-plate limited-slip differential. Stopping power was provided by 330mm (13in) front and 315mm (12.4in) rear cross-drilled discs with four-pot AP Racing callipers. Bilstein took care of the gas-assisted damping, working in alliance with coil-over rear springs and metal joints throughout the high-stress mounting points. Revolution 16in-diameter wheels, 7in front and 9in rear, served those road-legal Bridgestone RE71 tyres, in 225/45 ZR and 245/45 sizes.

I loved the thought of driving the Cosworth in a race series that visited some great tracks, including Brands Hatch, Donington and Oulton Park. As it turned out, I missed just one round, and retired from another. The car was always presented pristinely, with that 550+ horsepower (adjustable and unregulated boost) present and exceptionally correct.

Another plus factor rested within the Bridgestone RE71 road tyres we employed. They never wilted, even under the pressure of setting record laps on hot days on Brands Hatch Grand Prix circuit (90.18mph/1m 43.79s) and pre-chicane Castle Combe (95.44mph/1m 9.40s).

Low points of the year were spinning on the opening lap of the wet Oulton Park first round (recovered to seventh overall/third in class), and a start-line collision at Lydden Hill via a 4x4 Sierra Cosworth that was always ultra-quick off the mark. On the latter occasion, not amused, I had a good red-mist run for several laps to catch the miscreant, but nobly resisted the temptation to ease him further into the barriers atop the loop on that compact track. I didn't need to chance a non-finish. I have no results records for that one, but I think I managed to retrieve podium points.

I was also unsatisfied with a third overall at Snetterton behind perpetual winner Paul Bailey's mighty RS2000-500 and Sue's RS200 Evo, but subsequently learned I had been comprehensively blown away with some freak transient-horsepower manoeuvres. I became more content when I found that my best lap was only 0.3sec adrift of Sue in the 630bhp RS200 Evo, but Bailey's best was 1.2sec faster than both of us.

Highlights of the season were a win at Mallory Park, four lap records, and five second places. The Collins car was often seconds faster than anyone over a qualifying lap, but it was cumbersome to get it off a standing start via the dogleg first/second gear-change. It usually took me until half distance in a typical 10-lapper to get within striking distance of the front-runners.

Overall, it was a fabulous opportunity to drive the RS500 legend frequently. It was the most powerful car I raced on a regular basis.

CHAPTER 14

ESCORT RS COSWORTH

The last Ford homologation special

Last of the truly leading-edge European showroom performance/competition Fords, and the final creative gasp of Ford of Britain's insular Boreham-based clique, the Escort RS Cosworth was packed with World Rally Championship technology. The turbocharged YB-series engine and 4x4 transmission, coupled to Group A-only seven-speed gearbox, owed most to the painstakingly developed and highly effective Sierra RS Cosworth hardware, albeit tightly packaged in handier and more promotable Escort panels. The Sierra had become production history when the 1993 Mondeo line appeared, bidding for world domination, leaving the Escort as Ford's RS Cosworth successor.

◀ I spectated on the 1998 Network Q Rally of Great Britain to see the last factory-backed Escort WRCs in action, operated by M-Sport. #7 (R6 FMC) for Juha Kankkunen/Juha Repo finished second, #8 (P9 FMC) for Bruno Thiry/Stéphane Prévot third, and the Gazprom-sponsored #24 (S13 FMC) for Sebastian Lindholm/Jukka Aho, shown here, fifth. *Author*

▲ It was very wet when Brands Hatch management let us out between British Touring Car sessions in 1992 for a track session in an Escort RS Cosworth production road car. *Author's archive*

ELONGATED CONCEPTION, SMOOTH PRODUCTION

The '90s Escort RS Cosworths were the last of the Ford of Europe homologation limited-production performers to go through the old-school creation process, preceded by Boreham-built prototypes of vehicles intended for motorsport or the road. The good news, compared with all low-run predecessors, was that SVE developed and redeveloped the showroom cousins. Also welcome was that the mass-production Ford plant at Saarlouis, which is just in Belgium, but controlled by Ford of Germany, assembled those road cars with

➤ As with the Sierra Cosworths, Ford did not stint on showroom equipment, especially if you took the pricier Lux option. If the body was destined for motorsport rigours, then that pictured glass flip roof became surplus to requirements. *Ford*

ESCORT RS COSWORTH TIMELINE

February to August 1988: Ford Motorsport at Boreham, guided by overall project leader and design engineer John Wheeler, create a concept idea. They identify implanting Sierra Cosworth 4x4 competition and showroom technology within the contemporary Escort outline.

June to August 1988: Authorisation and realisation of the first running prototype by sub-contractor TC Prototypes, Northampton. Escort panels have the appearance of the contemporary (1986–90) model. Forward planning on the '90s mass-production Escort begins, under code CE14. Embryonic Escort RS Cosworth initially coded ACE-14, then simply ACE. Management approval given to proceed after successful tests.

November to December 1988: Fundamental body design work executed with over 400 new components and a high priority on body stiffness/safety. Project ACE submitted for usual European standard safety tests: torsional stiffness roughly twice that of 1990–94 production Escort. Cologne wind tunnel delivers front and rear downforce aerodynamics with 0.38Cd; without wings (final 1993 production), 0.34Cd quoted. Body dimensions stretch as a result of crash safety and accommodating Sierra longitudinal engine and gearbox hardware in what was designed as a front-drive Escort body. Just contemporary front-drive Escort's roof, bare tailgate and doors survived unmodified for production. December 1990, Karmann selected to assemble showroom Escort Cosworth. At least 2,500 examples required to qualify for international motorsport.

August 1989: Ford's proven internal SVE department assigned the job of engineering the production car into multiple prototype and 4P pre-production stages. SVE hand the showroom product specification to Karmann for assembly.

January to September 1990: First road example assembled after American parent company approved road and rally planning and budgets. Within a month Sierra RS Cosworth 4x4 begins production, a major source of hardware for the showroom Escort RS Cosworth (RSC). May 1990 sees the significant MS002 forerunner of the Group A Escort Cosworth rally car created by former Boreham employee Gordon Spooner, then a Motorsport sub-contractor. That publicity ensures significant media exposure. By 12 July 1990, some 17 Escort RSCs exist, six fully functional, more as clay models or interior studies. MS002/H930 JHJ, registered in September 1990, becomes a winner in a Spanish national event (Josep Maria Bardolet): it ran 300bhp with a seven-speed gearbox, and was also employed for endurance and heat trials.

August 1991 to January 1992: Passed fit for production, 10 functional road Escort RSCs manufactured for early publicity and evaluation. By January 1992, Saarlouis prepares for mass production. Tasks for 76 cars reported as: two for Geneva Motor Show; 25 media/press cars; 31 for marketing, testing and homologation; 18 bare bodies for Ford Motorsport UK.

27 April 1992: Karmann production starts, target 25 per day.

5 May 1992: Pan-European launches.

22 May 1992: UK sales begin at a basic £21,380.

December 1992: FIA international motorsport inspection checks manufacture of 2,500 examples, qualifying for Groups A and N. Sierra RS Cosworth 4x4 ends production.

1 January 1993: Escort RSC officially homologated for international motorsport, Groups A and N.

21–27 January 1993: Factory Group A Escort RSCs finish second and third on Monte Carlo Rally debut.

2–5 March 1993: Factory Escort RSCs finish 1–2 for first World Championship win, in Portugal.

March to May 1993: Revised Escort RSC with smaller road-suitable T25 turbocharger and deletion of high rear wing/extended front splitter debuts at Geneva Show, available in UK May 1993 at prices from £22,050 to £22,590. As for many motorsport homologation cars, lots sold at discount in period.

1993–98: As a Group A car, Escort RSC wins eight World Rally Championship (WRC) events 1993–96, including the 1994 Monte Carlo Rally. It then competes in the world series in a revised WRC format and wins two more qualifying events during 1997–98 before the Ford Focus WRC replaces it.

10 January 1996: Ford UK reported final total Escort RSC Karmann production at 7,145 as at 1 January 1996.

January 1997 to November 1998: M-Sport takes over running Escort WRC from Ford at Boreham for final two seasons: two victories in 1997 (Greek Acropolis and Rally Indonesia) are the last for the Escort badge. Redeveloped WRC Escorts, now equipped with Xtrac six-speed sequential gearboxes, secure competitive second and third places on their final RAC Rally of Great Britain in November 1998.

enthusiasm and pride in their work.

Sadly, public appreciation across Europe was slow to pick up, the key UK market slewed by the constant Cosworth dark side of theft and conspicuous accident rates. Naturally, such incidents viciously spiked British insurance premiums, and deterred many who could just about afford a showroom Escort RS Cosworth but not the loaded insurance premiums.

In the more prosperous and larger German market, the battle for patriotic and performance/competition-conscious customers was far more hard-fought. The BMW M3 and Mercedes 190 2.3/16 were battling on road and German championship track with their four-cylinder road-racer machines. Alternatively, there were always varieties of Audi quattro as showroom seductions, or cousin Porsche and its lower-cost four-cylinder 944, then 968, entry-level models. The Blue Oval carried notoriously low prestige in Germany's heartland motor-manufacturing society. Ford branding struggled down the pecking order, lower even than the similarly American-owned GM-Opel multinational. However, such obstacles were usually overcome, and after a protracted development period, the Escort with a Sierra RS Cosworth heart became a production reality in spring 1992, and a homologated international Group A and N competition car from January 1993.

Some key dates in that process are provided on the facing page.

DRIVEN

Road cars

Prototype RS Cosworth (two examples) Pre-production drive Cologne–Luxembourg–Saarlouis–Cologne, 7–11 August 1991, facilitated by John Wheeler; two Escort RS Cosworths, one Sierra RS Cosworth, plus Lancia Delta 16v and BMW 2.5 M3 Evo Sport; most miles in red (SV#716/H724 LOO) and white (SV#717/G277 FOO) cars. **4P production RS Cosworth (three examples, one RHD)** Luxembourg media launch, 29–30 April 1992; drove UK-registered RS Cosworth, plus Cologne-supplied K-AR1822, one of 58 production preview models. **Production RS Cosworth (two examples)** Media drive in Scotland, 13 July 1992; RHD, most miles in J996 BPU; my diary entry: 'Much better than I expected.' **Production RS Cosworth** Full road test, published October 1992 *Motor Sport*, J988 BPU; pictured at wet Brands Hatch BTCC test session; Millbrook 0–60mph in 5.7sec, 138mph (own performance figures). **'Small turbo' RS Cosworth (three examples)** Media drives, May 1994, final road specification, LHD; 20 engine changes listed, including T25 turbo replacing T3 and Ford EEC-IV electronics, high rear wing and front splitter deleted, most kilometres in K-DC 8710 and K-DC 8706; UK road test week, 6–14 October 1994, M412 UHK, covered 513 unremarkable miles. **Brooklyn Ford RS Cosworth (three examples)** Road test, autumn 1994, published in *Fast Ford*, December 1994; Monte Carlo 200-off edition #092, M27 JWP, uprated to 315bhp at 5,600rpm, 312lb ft at 3,500rpm; showroom, M775 KWP, T25 turbo and full wing set; plus 320bhp example. **Brooklyn/Len Halson RS Cosworth** Road test, published *Fast Ford* and CCC, 1995–96 **Private 2.4-litre RS Cosworth conversion** Road test, February 1995, LEN 55; 2,381cc, significant pulling power from 1,700rpm, 370–380bhp. **Graham Goode Racing RS Cosworth (two examples)** Road test, spring 1996, published June 1996 *Fast Ford: Cosworth Special*, K64 KAR, Graham Goode Racing 1992–94 demonstrator with 335bhp, 340lb ft, 163.2mph, 0–60mph in 4.3sec, 22 mpg; road test on public roads/ Bruntingthorpe track through cones, 28 September 1997, published *Cosworth2* one-off magazine, WDB 662, rebuilt for a Malaysian customer, Motorsport (no sunroof) body, 420hp, 7,250rpm limit, 420lb ft at 4,500rpm. **Power Engineering RS Cosworth** Road test, spring 1996, published June 1996 *Fast Ford: Cosworth Special*, M316 WHK; small turbo, low wing set, custom carbon finishes, 8x17in TSW wheels, Wilwood 4-piston front callipers, 278/273mm (10.9/10.7in) Grippa discs; 310bhp at 5,900rpm, 325lb ft at 2,900rpm, claimed 'over 150mph/0–60mph in 4.5secs', 22.2mpg.

Competition cars

WRC works RS Cosworth Driven at Boreham airfield, 21 July 1993, published in *Autocar & Motor*; 1993 Corsica Rally winner (François Delecour), K832 HHJ; 0–60mph in 3.8sec, 136mph at 7,600rpm.

THE END OF AN ERA

The Escort RS Cosworth became the last Ford in-house factory competition car operated directly by Boreham at the highest levels of international rallying, pursuing outright victories. A similar Ford corporate pattern had been seen much earlier with top-level touring-car racing in Europe, Capris the last models directly built and entered by Ford Cologne or Boreham, in 1974–75.

There were other customer-biased projects for the remaining Ford at Boreham management and technicians, including the rare and underestimated Ford Puma Racing limited-production model. For rallying, Malcolm Wilson's outstanding M-Sport operation in Cumbria was then subcontracted (initially in 1996) to represent the company with a final Escort season, then Focus and subsequently Fiesta and Puma ST in World Championship rallying.

M-Sport continues to perform with honour to date, particularly with the second-edition Puma crossover, winning the 2022 Monte Carlo Rally, albeit much more tightly constricted by the realities of budget than in-house Ford systems had allowed.

The showroom Escort was destined to live on until 1998 in Europe (into the 21st century for some markets, including China), when the Ford Focus – and international motorsport regulations – would fundamentally change Ford corporate attitudes to performance and competition vehicles.

Senior British-based Ford of Europe engineering executive Ron Mellor – an enthusiast who had rallied an RS2000 at club level – had the vision to spread attractive performance-car qualities to a wider customer audience. Thus, the mainstream Mondeo and Focus offered premium handling and a much-improved mainstream-product driving experience for many more customers. Availability of specialist hardcore Ford Motorsport qualities in limited-production vehicles – and the RS brand – were put on the back burner from the Escort RS Cosworth's era until the first Focus RS appeared in 2002.

For me, the last Cosworth-branded Ford also marked the end of an era. No longer would I have the privilege of driving Boreham-based World Rally Championship cars, and from 1998 onwards there would be no UK-based SVE advanced road-car engineering department to visit. SVE's 1981–1998 lifespan, involving 33 wildly assorted projects – including two showroom editions of the Escort RS Cosworth – faded from media view from January 1996, as the mainstream engineering team enhanced its mass-production cars.

Confusingly, SVE in the UK did work on rather less glamorous projects after 1996, and there were two more managers after high-profile Rod Mansfield, but the glory days were over. The American Ford parent company had its own Special Vehicle Team (SVT) and subsequent SVE (Special Vehicle Engineering) and SVO (Special Vehicle Operations) outfits that tackled Mustang upgrades and more.

Similarly, there were no Ford European RS cars to write about until the 2002 front-drive Focus. By then, my working life had moved on to depend on writing about other brands, particularly BMW M products in Germany and the USA. I did continue to experience Ford performance products for the showroom, including the first two front-drive generations of the Focus RS, and assorted ST incarnations of the Mondeo and Focus.

My sporting loyalty to the Ford Competitions and subsequently Motorsport department at Boreham was tested in 1997. I was asked to meet a senior British executive at the company's offices at 4 Grafton Street London W1 (now closed), a man with whom I'd worked on an equal basis during my '70s employment, but by the late '90s he had risen far beyond our old pay grades.

After some pleasantries, he asked how I would react to a plan to close the Boreham operation. The corporate financial reasoning was logical, I was told: 'It's costing us more than Formula 1 now it's fully unionised.' That meant consequent cost escalations on the unsociably long hours put in by the dedicated staff, particularly mechanics and engineers. I could see the management conclusions were inevitable, but said I could never support such a closure for sentimental reasons. For it had been the personnel of that Boreham department that had given me my initial, and many subsequent, opportunities to access their stories and hardware.

Quite correctly from a commercial viewpoint, Boreham was closed down, in stages. I was upset by the loss of a base that had given me a chance, but a living had to be earned… During the '90s the majority of my sport/writing time was occupied by BMW, Audi or Lotus, with many other brands such as Alfa Romeo and Renault-Alpine to consider, as most of my media/book contracts were by then in the USA, Germany, Italy or France.

PRE-PRODUCTION RS COSWORTH ROAD CAR: EUROPEAN COMPARISON TESTING

My first serious pre-production encounter with the Escort RS Cosworth took place during 7–11 August 1991, on a trip from Cologne to the Saarlouis factory via the Nürburgring. I was invited to give driving feedback via the goodwill of John Wheeler, chief Escort RS Cosworth Ford engineer on the race and rally project. We had two left-hand-drive Escort Cosworths (prototypes SV 716 in red and SV 717 in white), each serving to assess particular parameters, including NVH (Noise, Vibration and Harshness) and seat comfort in the white machine, which had been specifically created to assist Karmann in its production planning. The cars would be run back-to-back versus an original Sierra RS Cosworth, a rare rear-drive BMW 2.5 Evo Sport (600 made) and Lancia's showroom version of the 4x4 Integrale 16v.

At various points in the trip, I had company from Cosworth engineer Paul Fricker and SVE chief Dieter Hahne (who succeeded Rod Mansfield in January 1991). Dieter really believed in the product, and he retained a unique last example of the smaller-turbo Escort Cosworth.

As an aside, Dieter sold his unique run-out Escort RS Cosworth in spring 2022. His comments revealed it deserved that 'unique' tag: 'The car was made in 1996 as an L-specification with T25 turbocharger. It was the very last one ever built in Auralis Blue. As Karmann had run out of 1991 Escort bonnet parts from Ford-Saarlouis, they used a 1993 bonnet. Spats and splitter were missing from the front of the vehicle. I had the car foiled black later and the upper spoiler fitted, which was not part of the original L-specification. Naturally, it was LHD for my German use – 97,000km on the clock when my wife, Inge, and I decided to sell.'

A couple of representatives from Karmann were also present on the Cologne–Saarlouis trip. Karmann didn't just press unique panels (Osnabruck) and assemble the complete Ford (Rheine), but had also accepted the challenge of providing an alternative showroom seat to replace the expensive Recaros, thus saving on production costs.

A 6.35am call, and I flew into Cologne at 7.30am. I was installed in the Lancia by 8.22am after a briefing from Wheeler and Hahne, plus basic checks for tyre pressures and fluid levels. I had always liked Lancias and had plenty of experience in the Integrale, from multiple road editions up to a Sanremo factory car I tested for *Motor* after a WRC event. Sadly, this 16-valve showroom edition was on its fourth Ford comparative test, terribly tired in the braking, steering and trim zones, finally developing such brake and steering vibrations that it was withdrawn at 1.26pm on our first day. However, I knew enough to realise that the Integrale was a worthy rival, particularly in road trim, with an absorbent ride, good handling balance, and within such boxy and compact dimensions to make it a first-class country-road or rally weapon.

The 238bhp BMW stayed the course, and was the best road opposition to judge against, as the Escorts were running 220–230bhp. My notes said the stretched 2.5-litre (the biggest M-branded four-cylinder) was also the roughest that BMW had ever let the public buy. The other aspects of those M3s – aerodynamics, grip, brakes, ride and handling – were a class act, assembled to standards that prototype or production Fords weren't going to emulate.

▼ I was fortunate to complete some of the pre-production Escort RS Cosworth testing, via some comparison testing across Europe to the Saarlouis assembly plant. This South of France test-track image shows John Hitchens in one of the road Cosworths (H724 LOO) that I drove in back-to-back sessions. *Peter Hitchens*

➤ Former Ford SVE manager Dieter Hahne sold this unique run-out Escort RS Cosworth in spring 2022, with 97,000km on the clock. *Dieter Hahne*

The original three-door Sierra RS Cosworth provided for comparison was certainly under 204bhp, and had to be worked much harder to keep up with these early Escorts. Nürburgring was shrouded in late-afternoon fog at 4.20pm, but we were able to check out the driveability of the Ford EEC-IV management planned for use on later T25 turbo road Escorts (competition cars stayed with Weber-Marelli for the larger T35 turbo models) and I was quite impressed with the 2,500rpm arrival of full boost, even with the T25 layout.

As that summer evening passed, we drove a couple of hours of hardcore kilometres between 5.30–7.30pm that saw some *autobahn* miles fly by at a steady 200kph (124mph), peaking at 220kph (137mph). The sheer chassis grip and vehicle stability was in a totally different league from the original 'whale-tail' Sierra Cosworth, as we not only had front and rear downforce at significant levels, but also the 4x4 element. By contemporary standards – and those of today – it was an outstandingly stable saloon car, particularly resistant to side winds at speed. Both workhorse Escorts retained benchmark handling and adhesion on country roads, although I must reiterate that German minor roads have a smoothness, and constant-radius corners, that do not feature in other European twisty terrain.

By the evening debriefing, within handy morning range of a rewarding Saarlouis factory visit, we had packed in some intense miles, and it wasn't all good news. My noted dislikes included the noise levels at *autobahn* speeds, and I wasn't convinced that the Karmann seats were yet of sufficient sporting character to be compared with a Recaro alternative. Karmann personnel did persist and did a good job, but I – and many others – still have Recaros in my modern road car.

I completed a three-sheet report on our 1991 Cologne–Saarlouis outing. For me, the headline plus points were: 'The chassis abilities – which I am classifying as cornering grip, aerodynamic stability and braking. I believe them to be world-class. I can think of cars that approach some of these parameters in some areas (big-braked, lightweight vehicles have obvious retardation advantages over a steel-bodied saloon that has passed all contemporary safety tests for example). Yet, the overall balance achieved between stop, go and overall adhesion over multiple surfaces and speed ranges is unparalleled in my 24-year experience, setting new series-production-saloon car standards. My congratulations to all concerned.'

Looking back, I'd like to name some of those personnel, starting with Mick Kelly (overall chassis team leadership), John Hitchins (major contributor and SVE loyalist) and Colin Stancombe (leading Fiesta racer/SVE employee). Ford officially credited higher-profile inputs from Jackie Stewart and contracted rally drivers Malcolm Wilson, Gwyndaf Evans and Jimmy McRae. The showroom chassis was undoubtedly tautened up with a particular eye on production-based motorsport and those moves were credited to Terry Bradley, a long-term Ford Motorsport employee whose understated engineering inputs over the years are too easily overlooked. These handling improvements were achieved via reinforced trailing arms, plus braced mounting points for the rear cross-member and tougher front suspension bushes.

My main 1991 comparison beef centred on the Escort Cosworth's power-assisted steering, with my report suggesting: 'In case any smugness develops over my chassis comments above, you can always work on steering response and feedback to provide at least the turn-in response of a production Sierra 4x4.' I strongly felt the feedback was insufficient and that the original Sierra RS Cosworth was chattier. I now put this down to the particular Escort Cosworth geometry demands of 4x4. That steering rack did a very accurate job, and a regular owner-driver should be perfectly happy with steering precision, unless they owned a vehicle with Lotus-quality handling. Even the notoriously abrasive, but respected, Georg Kacher of *Car* magazine concluded in a September 1992 comparison test (VW Corrado V6, Vauxhall-Opel Calibra turbo coupé and original Escort RS Cosworth): 'Escort wins this sports-car contest. It is such a driving machine, responding more to your thoughts than your actions.'

The entire road and factories exercise was an unrepeatable total privilege for an outsider, and a significant part of my development-engineering ride-and-handling education. I was able to draw on this knowledge again in the period 1994–96, as a regular guest driver when the Lotus Elise was in underfunded, rapid development, all thanks to former Ford SVE boss Rod Mansfield and former high-performance-tyre development engineer Tony Shute.

PRODUCTION-PREVIEW (4P) RS COSWORTHS: MEDIA ROAD-RUNNERS

Although the pre-production and manufacturing process for the Escort RS Cosworth proceeded in a much more orderly and pre-planned manner than all predecessors, save the original three-door Sierra Cosworths, it was still a protracted programme. This can be seen from the fact that we were still driving production (4P) previews of the finished product in April 1992. The official start date for Karmann mass-production was 27 April, coincidentally two days before I attended the main Luxembourg media launch which centred on those 4P previews. British sales would start a month later, at a launch price of £21,380 (equivalent to about £42,000 at the time of writing).

For me, the Luxembourg sortie marked the return of my driving licence, after getting a speeding fine and a month's suspension period from a Bicester court for going a little too quickly on the M40. I was in good company, as Damon Hill was also up before the magistrates: he paid a much larger fine than me but served a quarter of my suspension time… My plea in mitigation was that my vehicle that day was left-hand drive with a kph speedometer, but I was actually fortunate, as I had given a Mercedes 500E V8 its motorway head after a brusque conversation with a respected editor, who justifiably felt I shouldn't be late for appointments.

▼ Luxembourg in April 1992 was the venue to test these Cologne-registered previews of the showroom cars. I was impressed with the solid and stable performance, but it would still be 1993 before the car was homologated for international competition. *Author*

Back in the land of Cosworth Fords, I drove three examples, sharing with Graham Jones, former *CCC* colleague and later a Ford PR executive. The first was a transfer from my staff four-door Sierra RS Cosworth to a British-registered Escort Cosworth for the run out to Stansted airport. Next, two 'K' for Köln-registered LHD examples in full-wing/bigger turbo trim, which indicated that the extensive seat development Karmann carried out had emerged in public just for the lowest-price 'entry' models, Recaros sprouting up everywhere else.

Our German drives went well. Graham – previously a deeply knowledgeable race reporter in North America from his homeland Canadian base – was an excellent companion and we got right on the Cosworth pace. We also got lost, regularly, so much so that we even penetrated the nearby Belgian border and put in some touring laps of the old long-form Spa-Francorchamps circuit.

In July 1992, Ford organised a day trip to mark the mainstream arrival and promotion of the RHD Escort Cosworth models, which involved flying from Heathrow to Glasgow, followed by some interesting mileage via Loch Lomond, on to more fabulous tarmac, and returning on a late-afternoon plane.

I drove a couple of the UK-specification Escort Cosworths that sunny day and found them much improved in terms of reduced noise and harshness levels compared with those earlier prototype drives. SVE engineers had done an excellent job of civilising the car, and I had acclimatised to the numb, but accurate, steering. I also continued to appreciate the all-round grip and poise benefits of the Escort over the Sierra.

However, I still feel the four-door 4x4 Sierra RS Cosworth remains underestimated for its more accommodating cabin and understated appearance, with much of the Escort Cosworth's performance. I had at one point reserved the black Sierra RS Cosworth 4x4 demonstrator that got stolen on the RAC Rally, so I could have put my money where my editorial bias had been…

▲ ➤ A tale of two interiors! The LHD machine depicts the period white-faced dials, beloved of that generation, whilst the RHD version shows what you gained with the leather option, but has sombre black-faced instrumentation. In both cases the COSWORTH script was carried prominently in the upper fascia. *Ford*

RS COSWORTH PRODUCTION ROAD CAR: TESTING TIMES AND CUNNING CRIMES...

Just a month after the Scottish trip, in August 1992, an Escort road adventure arrived that coincided with a crime that made national and motor racing press headlines. It started with the routine arrival, on 21 August, of J988 BPU for a road test week, the results of which would be published in the October 1992 edition of *Motor Sport*. By this point, the pricing for the Escort Cosworth had been confirmed: the basic cost was £21,380, with three models, all equipped with catalytic converters, stepping up in prices through £23,495 for the plusher Lux in cloth trim (as tested, Diamond patterns on Recaros) and £23,976 with leather-clad cockpit. At this stage, the only options were heated seats and CD player, then restricted only to Lux models.

The criminal element

The Ford and RS branding – especially the Blue Oval plus RS Cosworth stamped on the rump – always attracted thieves and those who live on the dark side of society. I have had three Fords stolen over the years, including a Sierra RS Cosworth, so I know how that feels. Yet this sortie with the *Motor Sport* 1992 test car marked another level, as it involved international illegal drugs transportation.

We chose Monday 24 August to photograph the test Escort Cosworth at Brands Hatch, admitted courtesy of the British touring car organisation, TOCA, which had a major test session in full swing. We were able to do our tracking and action shots in between the frenzied 2-litre touring cars, a leading team being VLM (Vic Lee Motorsport). I knew Lee through his earlier '80s drives in Golf GTIs, plus his VLM outfit winning the 1991 British Touring Car Championship (BTCC). The team then represented BMW UK with its 318iS two-car assault on the 1992 series, Tim Harvey on the driving strength as a front-runner, and subsequent champion.

In addition to the fastest showroom Escort, we also took along a road test BMW 318iS, to try and picture it in with the VLM racers. The VLM boys were in no mood to accommodate us, and there were rival journalists lined up to test drive one of the team's race BMWs, so we abandoned that idea, took the Cosworth out in the soaking wet and departed. I was slightly puzzled, as I didn't recognise some of VLM pit staff, and Vic's usual 'I may not be quick, but my aftershave do smell nice' banter had been replaced by stony faces all round.

On 8 September 1992, I attended a Range Rover launch in Wiltshire. Just before I checked in, I heard the national radio headlines, reporting that Lee had been arrested with a female companion for importing 40kg of cocaine, worth £6 million, hidden in a transporter and bulky gas bottles. No wonder there were stony faces at that Brands session. Apparently, the customs authorities had tracked the Lee entourage for several months, including surveillance in The Netherlands and at Brands Hatch that day...

After a spell inside, Lee was caught for another drugs offence in 2005, this time in association with well-known Sierra Cosworth competitor Jerry Mahony. I was reminded of the advice I received as an apprentice race reporter from the *Motoring News* Sports Editor in 1969: 'Just remember, Jeremy, half the grid have been in prison and the other half should be!'

On with the road test week

This Escort Cosworth test contained many home-grown statistics. On the Friday 28 August, the Escort went through the full track performance-measurement routines at a dry Millbrook. When I ran these performance tests in the '90s, much of the thoroughness and timing to the hundredth of a second – rather than a tenth – came from methods established at *Performance Car* a decade earlier.

Our 0–30mph run in 1992 was a rarity, for it demonstrated the 4x4 turbo traction and accessibility perfectly, taking under 2sec (1.91sec). The contemporary *Autocar* 1992 result averaged 2.3sec. Naturally we (Mark 'GP reporter' Hughes was then monitoring our times) ran a two-way average of the benchmark 0–60mph in 5.78sec, versus *Autocar*'s 6.2sec. Our 0–100mph occupied

➤ The 220bhp turbocharged heart of the 1992 Escort RS Cosworth was much more reliable than the original 204bhp Sierra RS Cosworths. *Ford*

16.48sec, versus 17.4sec from the industry-standard magazine. I don't have a comparison figure for *Autocar's* 0–110mph, but we took 21.32sec before running out of banked run-off area in one direction, although the run back was uneventful.

Because no skill is involved, our 50–70mph time of 10.8sec in fifth is a worthwhile comparison with *Autocar* and the test health of both Cosworths used: *Autocar* managed 9.7sec for that increment, so there was nothing wrong with that Cosworth's boost settings! They also had us beaten on an overall average 20.2mpg fuel consumption. Including Brands Hatch (some leisurely car-to-car tracking stuff below 50mph), we recorded 19.31mpg. Maximum speeds are very much at the mercy of the weather and track conditions, so I don't read so much into our 138.2mph average for the two miles around the Millbrook bowl, where *Autocar* published an average of 137mph.

'OVERALL, I REMAINED IN AWE OF MULTI-NATIONAL FORD'S CREATION OF SUCH A RADICAL TWIST TO ITS FADING BESTSELLER, BUT THAT DOESN'T MEAN THERE WEREN'T DOWNSIDES TO THAT ROAD TEST WEEK.'

Overall, I remained in awe of multi-national Ford's creation of such a radical twist to its fading bestseller, but that doesn't mean there weren't downsides to that road-test week. The first was a bugbear that I suffered with the Audi quattro and the Porsche 968, and that I continue to dislike. We had a simple, but terminal, puncture at the test track. Just like my Audi performance-test experience in the '80s, there was no spare. We were OK, because Ford despatched a man on a near 200-mile round trip with an instant replacement. Today, I carry a full-size spare on longer trips in any vehicle that isn't designed to carry such an item: I feel an idiot all those incident-free times, but I know how boring it is waiting for assistance...

Another test experience reminded me of earlier RS Fords: the 1992 test Escort Cosworth cut out completely in a hard-fought second-gear corner. It was annoying, but occurred rarely enough to ignore 20 years later.

Footnote *I was assigned a sister Escort Cosworth, J995 BPU, in white, that year for another test week. That was also a cover car for* Fast Lane *with Mark Hales driving. From memory, I did some subsidiary driving in this car for a* Fast Ford *multi-test that was bylined by editor Andrew English.*

ESCORT RS COSWORTHS TUNED BY BROOKLYN FORD

From 1994 to 1996, various tuned Escort RS Cosworths came my way. To be honest, I don't remember all of them, and I am a bit frustrated that I cannot find details of a Castle Combe day I did with a bunch of excellent higher-performance Brooklyn customer Escort Cosworths, so they aren't in my 'Driven' log.

However, I can remember a generality about all the Sierra and Escort Cosworth upgrades I tried that may be relevant today. A sensible increase in boost – in the range 1.2–1.5 bar, in association with some equally sensible supportive electronic and hardware moves – not only uplifted performance sharply from the low, warranty-conscious boost levels of production ersions (1986 Sierra RS Cosworth, 0.55 bar; 1992 Escort RS Cosworth 0.8 bar), but also soothed that iron-block four-cylinder into some distinct sweet spots.

Consequently, the elevated turbocharger pressures could make not just acceleration more entertaining, but also allowed long-distance cruising speeds to become perceptibly smoother. I'm not talking about miraculously transforming the four-pot iron-block Ford into a suave, high-rpm wonder of the contemporary Honda or BMW ilk, but the modifications did allow the Fords to become considerably more civilised for road use, and downright exhilarating on track.

Brooklyn-tuned Escort Cosworth Monte Carlo special edition

My first taste of these upgrades was with the work of family-owned Brooklyn – then a massive Ford dealership in Redditch – on a 315bhp version of the Escort Cosworth's rare (200-off) Monte Carlo special edition, plus a production model with the larger turbo (T35) and full wing set, featuring a slightly better exhaust system, that was rated at 320bhp. Brooklyn had been RS dealers since 1983, with a unique separate workshop to deal only with the Ford sub-brand, under the technical wizardry of Dennis Osborne and Andy Fisher. I was familiar with their Sierra RS Cosworth modifications, and knew they had been skilled enough to prepare C240 HVW, the express-fleet Sierra Cosworth, for Phil Collins to win the one-make rally championship in 1986.

Brooklyn's reworked RS Cosworths underlined that strong reputation: I've driven at least four Escort examples, as well the original three-door and rear-drive four-door Sierra variants. All were properly detailed for the road, and cheaper than the many alternatives that were simply high-boosters heading for head-gasket and other embarrassing mechanical failures.

A few words on the Monte Carlo edition. Its cockpit was quite distinctive, with Motorsport-branded cloth and suedette-covered front seats, and alloy bright finishes applied to both gear-lever knob and handbrake button. Just as for limited-run BMW M3s, there was a plaque to record special-edition numbers, and the Brooklyn car I drove was 092/200: it is believed that fewer than 80 Monte Carlo right-hand-drive cars were sold in the UK.

For me, one of the most consistent providers of modified Escort RS Cosworths was Brooklyn Ford at Redditch. They were careful with road-car power outputs, preserving or enhancing torque over sheer horsepower. A 200-off rarity was the limited-edition Monte Carlo package, with 315bhp and maximum torque by 3,500rpm. *Author's archive*

Externally, the big distinctions were quadruple headlamps (an RS option fitted to this car), aping the works rally-car illuminators, and alloy OZ Racing-badged wheels in standard 8x16 sizes. The Brooklyn test car was finished in the Violet Jewell metallic paint that classic-car dealerships now assert is the rarest (estimated 10 to 12 RHD): I can see a Ford-owned demonstrator (L702 RTW) and another Ford Heritage-owned example (L130 NWC) also came in this burgundy shade, but without the quad headlamps fitted to 'my' tuned example.

Brooklyn 2.4-litre RS Cosworth

By February 1995, courtesy of Manchester-based owner Len Halson, I was able to try his personally registered (LEN 55) car featuring a 2.4-litre long-stroke stretch of the YB motor, which would cover 24,000 sound miles before I lost track of it. This engine enlargement also came from Brooklyn's workshops. Details supplied by Martin Lawton showed that most reciprocating components had been totally re-engineered at a 1995 cost of nearly £6,200. The capacity measured 2,381cc, with 92.8mm x 88mm bore and stroke (standard production 1,993cc; 90.8mm x 77mm), and produced 380–390bhp, although the most obvious gain was in mid-range torque, which was beyond 400lb ft. Again, the engine was properly managed, with a sub-1,000rpm tickover, and extra torque apparent from 1,700rpm. The big bonus over a production unit, or tuned 2-litre, came between 2,000 and 4,500rpm.

GRAHAM GOODE RACING TUNED RS COSWORTH: GOODE BY NAME, GOOD BY NATURE

Another company I came to respect and work with through the Sierra and Escort RS Cosworth eras was and is Graham Goode Racing (GGR) in Leicester. I knew of Graham's race and immaculate engineering credentials from Broadspeed apprenticeship through to watching his race Anglias, but that was just the start. He backed and drove many fast Fords, through to the stand-out Listerine-sponsored RS500s in the British Touring Car Championship. Actually, Graham's greatest race successes came overseas, when he redeveloped his UK Sierra RS500s to a lightweight peak, below 1,100kg. Add in 2.0 bar boost for 540bhp in tropical temperatures, and that was sufficient to win in every Asia-Pacific series appearance against the feared twin-turbo Nissan Skyline GT-Rs.

GGR's chief engineering ace, Alastair Mayne, is my knowledgeable and most regular GGR contact, and I'm well aware that their Ford prowess extended to modified Focus RS models, and to other makes, particularly the Subaru 4x4 turbos that many enthusiasts graduated to when Ford abandoned the RS brand between the last Escort RS Cosworths and the 2002 front-drive Focus RS.

My first chance to drive GGR's modified Escort Cosworth arrived in May 1996. Wet Leicestershire lanes greeted Alastair's delivery of the thoroughly reworked K64 CAR demonstrator. Carrying nearly £7,400 of VAT-paid upgrades, the hard-worked GGR package returned supercar performance within handy Ford Escort dimensions. The engine bay became the star turn, offering 335bhp at 6,300rpm and 340lb ft of torque at 4,100rpm. That 120lb ft peak torque bonus was extracted with the help of sophisticated electronic management add-ons that allowed, reported Alastair from the passenger seat, '1.6 bar for a couple of seconds, then falls back to 1.3 bar.' Mild over-boosts continued to be delivered if the ambient temperature in the engine bay was cooperative. Laden with 210kg of humans and electronic timing gear, this red road-runner rapped off 0–60mph in 4.3sec and recorded 163.2mph. Headline stuff, but of more everyday use was the ability to slash in-gear acceleration times, so that the benchmark 50–70mph in fifth took a significant 2.4sec less than the standard model.

Not surprisingly, given GGR's pedigree, the rest of the Escort Cosworth recipe had also been overhauled. Most obvious was the step up in wheel sizes to 8.5x18in from Rondell, with 225/40 ZR Yokohama A008Ps. This allowed some gains in the gearing, so that a 163mph maximum was within the rebuilt engine's safety limits, at 6,800rpm. Behind those effective wheels lurked AP Racing front 330mm (13in) discs with four-piston callipers, the rear discs left as per production.

Naturally, the suspension had been revamped too, operating around Koni damping, with uprated coil springs, a 20mm (0.79in) ride-height reduction and replacement mounting bushes to toughen up responses. Standard 28mm (1.10in) front and 22mm (0.87in) rear anti-roll-bar diameters were retained. There was a telling period comment I made in a *Fast Ford: Cosworth Special* June 1996 magazine: 'The GGR Cossie gave a stunning display of safe speed across wet and challenging roads. So much so that passengers asked to stay in for the action-photography sequences. A unique request in my working life – although it was very wet and cold out there! The action sequences also proved that the ride heights hadn't been chosen with looking-good priorities in mind, for there was no contact between tyres and arches, even at full power in second gear on tight junctions.'

The only thing that prevented me giving it a perfect road-car score was the huffing and hissing deliberately engineered for overrun operatic interludes, via the open air cleaner sitting beside GGR's dump valve. For me, this GGR conversion ranked with Brooklyn's similarly thorough preparation, and underlines that competition participation can and does improve the product a customer receives.

POWER ENGINEERING TUNED RS COSWORTH: NO LONGER JUST ABOUT ENGINES

I had dealt with Fords converted by Power Engineering in West London before borrowing this Essex-registered (M316 WHK) demonstrator in the spring of 1996. The owner was the aptly named David Power, but his technical partner, Dick Prior, was even better known to me as a leading rear-drive Escort preparation ace over three decades. Dick had built race Escorts for David Brodie, especially the ill-fated Group 2 silver-and-blue 1973 warrior. Dick also prepared winning rally cars for Tony Pond, initially at the Norman Reeves Ford RS dealership in West London. Naturally, Dick also assembled TFR 8, Pond's winning 1975 Tour of Britain RS2000. That was an obsolete first-edition RS, but talented Prior also did much of the basic safety and regulation preparation on my second-generation RS2000 in 1976.

So the credentials, were right, but how was Power's final-edition Escort Cosworth?

The car carried over £7,000 of modifications, many of them designed to showcase the company's wares. It was very smart and show-ready. Carbon-fibre weaves appeared extensively, including the front grille. The compact T25 turbo engine bay was full of a broad, aluminium bracing bar, more carbon composites, crafted aluminium tank (for water-injection research) and many aircraft-industry-standard braided connecting lines.

The more road-friendly T25 turbo had been coddled, with a liquid intercooler to back up the standard unit. The remapped Ford injection system was supported by a separate electronic controller, plus a high-pressure fuel regulator. A 75mm (3in) stainless-steel exhaust fed an enhanced-flow-rate catalytic converter. Reported power was 310bhp at 5,900rpm, with a torque boost to 325lb ft at 2,900rpm.

The dynamics of the understated silver

▼ One of the most varied Escort rally days I have had was in 1995, when Ford ensured a Kentish rally stage buzzed with first- and second-generation rear-drive Escorts, as well as a contemporary Escort Cosworth. This photograph shows the engine bay within a Malcolm Wilson M-Sport Escort RS Cosworth. *Author*

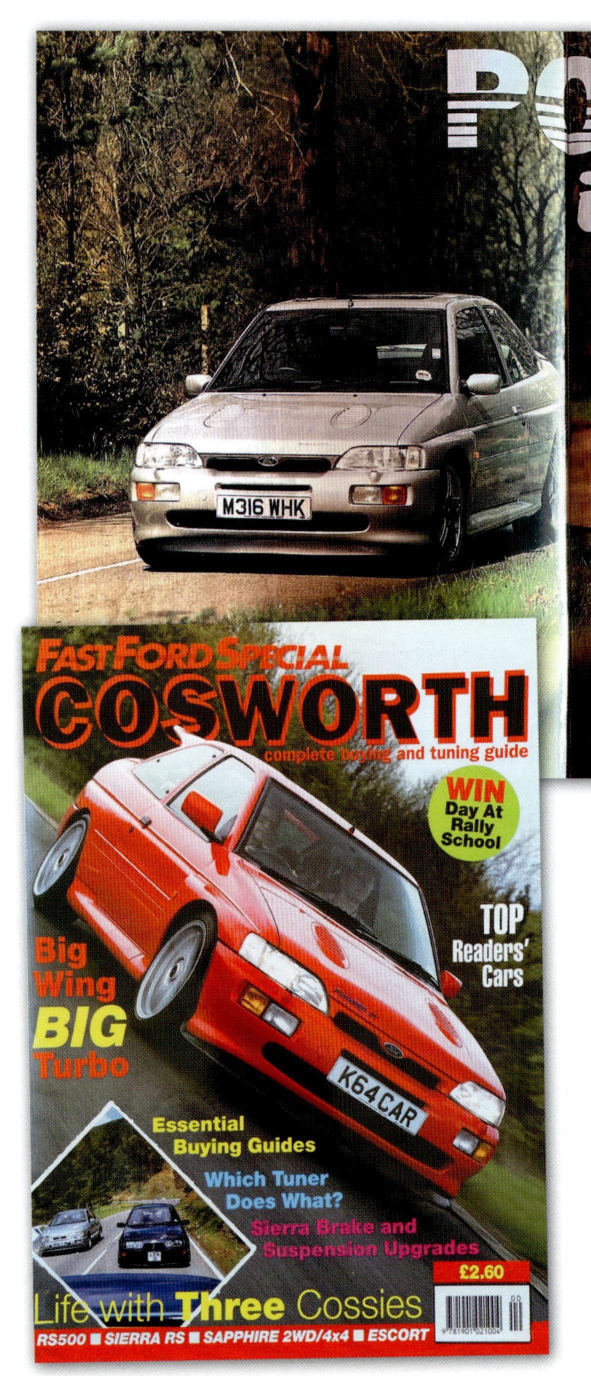
POWER in hand

Not one to do anything in half measures, when Power Engineering decided to build something a bit special in the Escort Cosworth line, it quickly became one of the major players in the converted car industry. But the best is yet to come...

Although Power Engineering is renowned for motors with more street and circuit muscle, the company's 15 staff and demonstrator are testimony to the fact that this company knows what makes a Ford tick outside the engine bay.

We drove a Power-converted RS Cosworth Escort cross-country and did more miles than are usually offered by the converted car industry. We were dogged by a misfire, but the car was such an accessible power and handling revelation that we forgave it its tantrums.

If you need to ask what Power Engineering is, then you haven't understood the current Ford specialist business, for it is one of the major players. Its west London premises teems with up to 25 cars a day — 90 per cent of them the Fords that David Power and Dick Prior concentrated on when they set up as an independent 10 years ago.

Our guide to the test car was Steve Haggar, who reported: "The RS Escort has been our mainstay. We reckon to have seen 4000 of the 19,000 in the UK. A big reason for those satisfied customers is the way that we can put Mervyn Power's experience of our rolling road into rectifying problems or improving performance.

We don't often get the chance to feel pride-of-ownership with a

Words: Jeremy Walton **Pictures:** David Wigmore

➤ By the summer of 1996, Escort Cosworth hardcore interest had hit fever pitch, as this *Fast Ford Special Cosworth* magazine cover illustrates. Inside, the Power Engineering tuned car drive-and-tell story was typical of the many I contributed to similar publications. *Dave Wigmore*

Escort Cosworth had been revitalised too. Most obvious were the 8x17in TSW five-spoke wheels and Yokohama tyre set. Cross-drilled Grippa discs (305mm/12in front and 273mm/10.7in rear), packing Wilwood four-piston front callipers, were effective, in company with the production Teves anti-lock. The suspension had been lowered and uprated along similar lines to the Graham Goode Racing example, Koni dampers at the heart, plus some Sierra RS Cosworth technology, with 25 per cent stiffer spring rates, plus hardened mounting bushes throughout. Initial take-off impressions were clouded by an obstructive Sachs development clutch, but I covered more road miles in a day than were usual for a demonstrator, running through Middlesex, Hampshire, Berkshire and Somerset. Although the use of the smaller turbo inherently led to a more accessible power curve for road use (the 2,000–5,000rpm zone was impressive in any of the usual five gears), it was the chassis that grabbed my attention.

The ride was rougher and harsher than the production car up to 30–40mph, but beyond that it became less jarring. Those replacement brakes rewarded the driver with bonus retardation, and a firm pedal. The usual high grip levels remained, augmented by reduced roll, and taut feedback that made those UK road miles unravel seductively, especially where changes of direction were rapidly required, whether at roundabouts, or crests with subsequent sharp corners.

The Power Engineering demonstrator wasn't as dynamically flawless as the Brooklyn or Graham Goode Racing products I tried, as we were haunted by a slight transitional misfire, and the development clutch was obstructive. Yet it was certainly the smartest and best-braked, and had a well-modified chassis.

WORKS ESCORT RS COSWORTH: 1993 CORSICA RALLY WINNER

I was at Boreham on 10 June 1993 to test the WRC Corsica Rally-winning Escort RS Cosworth (K832 HHJ) for *Autocar & Motor*'s edition of 21 July 1993. Steered by François Delecour and guided by co-driver Daniel Grataloup, the car took a convincing victory over the best rivals from Lancia and Toyota, and marked three Escort Cosworth WRC wins (Acropolis, Portugal and Corsica) from the four opening events that season. Yes, it should have been four from four, as Delecour and Miki Biasion ran in 1–2 formation on the most important round of all, Monte Carlo, but both Escorts succumbed to last-minute fuel-delivery hiccups (and a remarkably powerful charge from Didier Auriol's Toyota Celica GT4), leaving them second and third.

Any experience of a factory Ford, pared to low-weight/low-ride-height tarmac specification, and fresh from victory on its previous outing, had to be special. And so that sunny but thunderstormy Thursday proved. I not only dealt with key management men, including RS200/Cosworth Godfather John Wheeler, but also picked up a drive in the Fiesta race supporting the main BTCC event on the following Sunday!

◀ ▼ A couple of on-event images from François Delecour's victory run on the May 1993 Corsica Rally, including the man himself (note the significant door crash bars) and the tight-turning, flame-spitting crowd pleaser. *Motorsport Images*

I didn't know it then, but this would be the last time I would have the privilege of driving a factory rally car on an exclusive basis – just as the 24 Boreham-based employees in 1993 would not have known that their World Championship Escort RS Cosworths would also mark the end of an era, as the subsequent WRC Focus assaults on the world series would be prepared by Malcolm Wilson's M-Sport operation.

It was a fitting finale. For media purposes, the headlines came in the phenomenal statistics: 360 restricted horsepower sounds average in a 2022 world of 500–1,000bhp road cars, but the phenomenal grip allowed 0–30mph in a scant 1.2sec, and 0–60mph in a two-way average of 3.8sec, coupled to a blink under 10sec for 0–100mph. Credit for those carefully and independently measured times must go to 1973 European Rallycross Champion John Taylor, who performed the acceleration sprints to *Autocar & Motor*'s exacting standards perfectly, while I operated the magazine's timing equipment.

Having been lucky enough to attend the Corsican WRC round, and subsequently to have a passenger ride with Renault's fiery home-grown talent Jean Ragnotti in a 5 GT Turbo on one of those rock-lined, car's-width, twisty stages, for me it was the thought of running up to 136mph at night in this 'Cossie' that deeply impressed… Boy, did they need those vast water-cooled brakes!

Behind the wheel: a pure thoroughbred

Here is the gist of my submitted test text in 1993, before it was edited for publication:

'The four-eyed Ford, worth in excess of £160,000, squats and awaits our command. Inside the stark, metallic cabin there are two dashboard dials and seven forward gears, clattering sympathetically in response to a constant 1,150rpm grumble. The engine noises were emitted from the most potent 2-litres to sit behind an Essex Ford factory registration plate.

'Outside, white and swirling metallic-blue paint are powdered black by the imposing side exhaust, which has spent a day puffing out high-boost gases. From the multitude of fuses, switches and buttons that cling to the "centre console", you need only switch on the ignition and stab the black starter button to set the Cosworth mumbling. There is no need to use the throttle during or after the start. Ford Motorsport-modified Weber-Marelli management takes care of reliable ignition and a stable idle speed, complete with a soundtrack like distant small-arms fire.

'Strangled by the 38mm (1.5in) air restrictor imposed by the sport's authorities, Escort competition #3's menacing version of the Traffic Light Snuffle is delivered contemptuously. For this car has stood three days of unreasonable demands from a Frenchman. One who regards the 7,600rpm limiter as the start of gear-change negotiations, and rock-faced goat tracks as suitable terrain to exercise a maximum of 136mph. "Our" 700-mile Corsican rally victor wears its front-end scrapes (a gravel spin aptly described by the hard-working mechanics as a "scuffette") with insouciant pride.

'Now, K832 HHJ is hardly going to beg for mercy as a mere motoring journalist climbs aboard. The cabin comprises a competition collection of bare metal and T45 HG-steel scaffolding supporting the roll-cage foundations. The integrated cage is now so comprehensive that it is nearly as difficult to drop into the left-hand seat as to slide into an American stock car. The driving position is that of a contorted garden gnome, perched way in front of the laid-back navigator, but the "tools of the trade" are excellent. An anti-sweat suede rim for the three-spoke OMP wheel, enveloping Sparco seat, and six-point Sabelts, are backed up by gridded foot pedals. These positively seduce slim feet, encouraging the most intricate of left-foot braking, or heel-and-toe manoeuvres, without a second thought. In fact, they seem to do the job for you. The superbly crafted footrest looks as though it could serve as an airframe spar, while the co-driver is braced by a footwell filled with Kevlar composites, materials that are also (very expensively) used to encase the fire extinguisher.

'I pause to admire the neat twin roof hatches that provide the majority of ventilation on a thunderous 20° day. Each hatch is spring-loaded to notch through a variety of air-intake positions, individually tailored to suit driver and co-driver, with a choice of draught patterns achieved via an internal metal slide for the co-driver. Yes, Ford was the company that brought a '60s market (initially in the Cortina) 'Aeroflow'... The electronics age made the traditional rallying competition dash panel, one suffused with black-and-white dials, obsolete. Just two analogue instruments survive in front of the driver. Yet Ford has not gone for the digital displays that adorn leading BTCC racing saloons. Pectel Control Systems – an innovative UK company – provided an eight-mode digital read-out. During my drive, this was switched to monitor turbocharger boost, rpm, air and water temperatures.

'Cinch the belts a little tighter, and dip the clutch through most of its travel... cccluuunk. With the gear-lever collar felt, and raised, you can access the only dogleg path in the 'H+2' gate. The non-synchromesh gearbox offers the first of seven speeds. On the move, you barely need the twin-plate AP clutch; the gear-change is the fastest and sweetest I have used in any kind of saloon. The only danger is that of getting lost within three main alleyways to the shift pattern. Subsequent practice proves that a light throttle and just 1,200 Stack-indicated rpm is enough to set the plot rolling.

'Even with a constricted engine air supply, the 1993 Escort runs 0–60mph a full second faster than Didier Auriol's rear-drive Sierra managed in 1988. Today's factory Ford equals the eye-rotating 1.2sec stride for 0–30mph exhibited by the full-blown RS200 of the Group B era. Fabulous acceleration statistics aren't the primary point, which remains the manner in which this Escort shrink-wraps around you, hauling in the farthest tarmac horizon – preferably one a dozen dirt-surfaced bends or more away – with instant zoom-lens magnification. Manipulate pedals, levers and wheel in the right micro-second sequences, and there is no faster way to traverse tricky terra firma.

'Even at 3,000–4,000rpm low speeds for photography, the steering feedback promises instant gratification: more power-steered pleasures than a man can decently accommodate. Although the regulations lopped an inch from overall tyre widths this season (now 9x17in), the quick rack, one fractionally under two turns lock-to-lock, prods this Michelin-slick-shod Escort onto lock with the alacrity of a Caterham Seven snorting steroids. Considering there is the aforesaid power assistance, 58 per cent of vehicle

◀ Courtesy of Rod Dyble at Ford, I attended the February 1997 Swedish Rally, where M-Sport fielded a pair of Escort RS Cosworths for drivers Carlos Sainz and Armin Schwarz. The close-up spectating sight of 100mph over snow laden forest trails was worth the trip alone. *Author's archive*

weight over the front wheels, plus a trio of viscous couplings to serve the 4x4 system and its epicyclic centre differential, such alert reactions are rather more commendable than for the simple (Caterham) Seven.

'The complete chassis and its suspension system were rejuvenated. Sure, there are MacPherson strut and semi-trailing-arm principles, yet the rules freedom allowed in selecting the optimum mounting points, plus the replacement of key components with magnesium (ie, said trailing arms) and alloy castings that support a big-brake layout, ensure the package travels sensationally to a new destination. The modifications transform the already impressive Cosworth Escort road car into a two-seater competitor with racing-car grip (the G-forces leave you with a sore neck as a passenger) and flawless 4x4 traction. Its ability to claw tarmac is soothed by a Bilstein-damped ride that only deteriorates below 35mph on concrete surfaces.

'As John Taylor whips through 25, 62 and 81mph in the first three gears, threading the Escort through the complex infield of Boreham at fourth-gear speeds up to 90mph, we gain an insight into the sheer breadth of forces generated and absorbed in such spectacular progress. Water-cooled brake callipers, damper cases and turbo intercooler (all

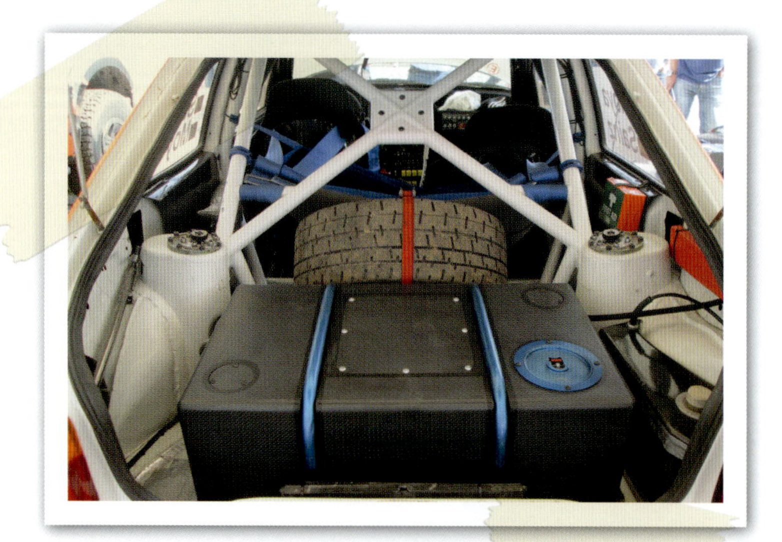

▲ ➤ Just a bit of bracing needed! Open hatch and engine bay show how the factory specification reinforced a body that was considerably stronger than the company's earlier competition Cosworths. This factory and privateer Escort RS carried the P6 FMC registration (seen on at least three chassis numbers). Drivers included Carlos Sainz senior, who scored an outright win, and Stig Blomqvist for more recent Goodwood special-stage appearances. *Author*

with separate boot-mounted containers) shrug off our attentions. They can maintain consistent temperatures in 40°C ambient, underwriting six water, oil and air radiators packed behind those quadruple lamps.

'The 7,500rpm shifts, and the constant need to flick the alert Escort from 90mph fourth gear to 25mph first, highlight two more outstanding attributes, and one defect. There is a superbly managed engine, and a motorcycle-quality snappy gear-change, but also a propensity to understeer, dropping off boost when shuffling back-and-forth to first (officially a crawler gear on the homologation form) from second in tight corners.

'Boost, more than thrice the showroom allowance, was managed flawlessly. Between 3,000 and 4,000rpm, the works Escort packs a 3 bar (42.7psi) pressurised punch – a blow that flings the Ford forward with addictive dependability. Such turbo-boost software, developed by Ford Motorsport staff in association with Pectel and hardware assemblers at Mountune of Maldon, has been remapped since that first (1993) win in Portugal. Now, the accent is on 3,000–7,500rpm drivability and premium access to bundles of torque. The official figures culminate in a minimum of 360bhp at 6,800rpm and a whopping 550Nm (404.4lb ft) torque reading by 4,500rpm. These for an engine that would cost £18,500 (ex-VAT) to duplicate, with its trick camshafts, pistons and lightweight-steel crankshaft. A "seriously modified engine" in Chief Ford Motorsport Engineer John Wheeler's words.'

For me, it was the total performance package of this winner that was most memorable. Race-car steering feedback and responses, fabulous 4x4 grip, neck-tweaking acceleration, countered by head-butting brakes and the feeling of supreme confidence in the car's capabilities. In previous eras I had steered Audi's fabled factory/David Sutton quattros, and Lancia's factory Delta Integrale, never mind the original rear-drive Escorts and the Corsica-winning 1988 Sierra RS Cosworth, but this Escort remained supreme in my lucky experience.

I found this heavily regulated Escort RS Cosworth was a triumph over the restrictive regulations, and showed genuine improvements over the past legends mentioned. Yes, the Audis were more brutally powerful, but the Escort RS Cosworth, as created and entered by Boreham personnel, was the best all-round rally car I was fortunate enough to experience.

Put it this way, offered a choice from my 50+ years' driving other people's competition cars, that Corsican star would be the rally car I would want to take home. Naturally, an older-generation classic Escort, tended by legendary Boreham personnel, would be a wonderful sentimental ride… and more practical and accessible to somebody of my 76 years, but I really saved the ultimate and best factory rally Ford Escort for last.

1992 ESCORT RS COSWORTH ROAD CAR vs 1993 WORKS WRC CAR

Showroom RS Cosworth versus WRC tarmac Corsica Rally winner, K832 HHJ, in brackets

Engine

Configuration: Front-mounted, longitudinal, inline four-cylinder. Garrett AiResearch T3/T04B turbocharger, two-stage water/air intercooling. Multi-point fuel-injection managed by Ford EEC-IV (Weber-Marelli/Pectel engine management) **Type:** Cosworth YBP (YBT) **Capacity:** 1,993cc (1,994.5cc) **Bore and stroke:** 90.8mm x 77.0mm **Compression ratio:** 8:1 (7.5:1) **Maximum power:** 227bhp at 5,750rpm (360bhp at 6,800rpm) **Maximum torque:** 220lb ft at 2,500rpm (404lb ft at 4,500rpm) **Specific power per litre:** 113.9bhp (180.6bhp) **Power-to-weight ratio:** 175bhp per tonne (305bhp per tonne) **Turbo boost pressure:** 0.8 bar/11.4psi sustained, 1.2 bar/17.2psi mid-range over-boost (3.0 bar/42.7psi to overcome mandated 38mm/1.5in air-intake restrictor)

Transmission

Type: Permanent four-wheel drive, epicyclic centre differential, 34/66 per cent front/rear power split, viscous-coupling centre and rear limited-slip differentials (Alternative power splits from 58/42 to 40/60 front/rear) **Gearbox:** Ford MT75 gearbox, five forward gears (MS93, seven-speed) **Gear ratios:** 1st: 3.61 (2.07); 2nd: 2.08 (1.529); 3rd: 1.36 (1.2); 4th: 1.1 (0.995); 5th: 0.831 (0.792); 6th: (0.654); 7th: (0.536); Final-drive ratio: 3.62 (4.63 to 3.25:1, plus axle and reduction-gear homologated alternatives)

Suspension

Front: MacPherson struts, Sachs gas damping, lower track-control arms, 28mm (1.1in) anti-roll bar (Magnesium uprights, alloy fabricated track-control arms, compression struts, replacement coil springs with water-cooled spring-seats, Bilstein gas damping, adjustable caster and camber, cockpit-adjustable anti-roll bar) **Rear:** Semi-trailing arms, coil springs, Sachs dampers, 22mm/0.87in anti-roll bar (Homologated ±20mm/0.79in mounting-point tolerances. Cast-magnesium semi-trailing arms, concentric coil springs and water-cooled Bilstein dampers, toe-in/out control links and cockpit-adjustable anti-roll bar)

Steering

Type: Power-assisted, variable-ratio rack-and-pinion, 2.45 turns lock-to-lock (Competition power-assisted 'quick rack' 1.9 turns lock-to-lock)

Brakes

Type: Servo-assisted, anti-lock braking system (Cockpit-adjustable bias front to rear) **Front:** 278mm/10.9in ventilated discs (AP Racing six-piston callipers and 355mm/14in ventilated discs, water-cooled) **Rear:** 273mm/10.7in ventilated discs (AP Racing four-piston callipers and 355mm/14in ventilated discs, water-cooled)

Wheels and tyres

Wheels: Ford Design, Ronal alloy 8x16in, eight-spoke (Tarmac-spec: OZ 9x17in) **Tyres:** Pirelli P Zero 225/45 ZR16 (Tarmac spec: Michelin S1 slicks)

Body and chassis

Type: Unique three-door, pressed-steel monocoque. Few production Escort panels, augmented by parts and assembly via Karmann. Multi-section body kit in polyurethane: spoilers and rear-wing set, side sills. Twin airbags, high-tensile steels and UK anti-theft measures (Ford Motorsport, reinforced body welding, plus integrated TG45 steel roll cage and extensions)

Weight

Kerb weight: 1,325kg (Competition weight, 1,200kg)

Dimensions

Length: 4,211mm/165.8in **Width:** 1,762mm/69.4in **Height:** 1,425mm/56.1in (Adjustable, tarmac/forest). **Wheelbase:** 2,551mm/100.4in (FIA tolerance, ±1 per cent)

Performance

Maximum speed: 138mph/142.5mph minus wing set (155mph at 7,600rpm) **0–60mph:** 5.7sec (3.8sec) **0–100mph:** 17.4sec (9.9sec) **Fuel consumption:** 22.8mpg average (Corsican stages: 6mpg)

CHAPTER 15

MONDEO

A car for the world

My experience of 'Mondeo Men' embraced many motorsport competitors, from multiple World Champions Jackie Stewart and Michael Schumacher to British Touring Car quadruple champion Andy Rouse and Kiwi charger Paul Radisich. Yet none of us would have experienced the pleasures of showroom or circuit Mondeos if it hadn't been for the creative engineering team led by Richard Parry-Jones, who realised that fine handling and a quality ride could be brought to mass-market Ford customers, not just a niche performance market.

WHO WAS 'MONDEO MAN'?

The phrase 'Mondeo Man' sprung from slick Labour Party strategists as they targeted their potential voters in the 1997 General Election. Lo and behold! Tony Blair acceded to the Prime Ministerial throne with support from equally demeaning demographical targets, 'Basildon Man/Woman' and the often Ford of Britain-linked 'Essex Man' (or 'Woman' or 'Girl') with bonus clichés applied. The 'Mondeo Man' tag was particularly damaging to showroom sales of performance Mondeos compared to perceived prestige rivals. The 'Mondeo Man' label is sometimes quoted as the reason why Mondeo ST values remain underrated today compared with earlier Ford XR, RS, Cosworth, and more recent ST performance models.

Ford's Mondeo planning targeted a much bigger picture than county clichés and political machinations. For the Mondeo, as you'd guess from the fact that it is derived from the word for 'world' in various languages (such as *monde* in French and *mondo* in Italian), was yet another shot at building a common car to meet worldwide tastes, needs and aspirations. In fairness, over five generations (1993–2022) Mondeo basics did make it to many parts of the globe. Under other names

▲ How the Mondeo should be driven at Brands Hatch's Paddock Bend: double World Cup winner Paul Radisich in full flight. In 2022, Paul was still driving race Mondeos that classic collectors had gathered in New Zealand. *Author's archive*

– Taurus, Fusion and Contour – the Mondeo covered vital markets way beyond Europe, such as the Americas, China and the Middle East.

Global domination the aim… scattered production the game!

Ford as a multinational did its best to market the Mondeo globally, yet regional sales zones developed. Bearing the previously mentioned four additional nameplates beyond Mondeo, its underpinnings provided the basis for multiple vehicle platforms. From a European perspective, the Mondeo badge applied through a 29-year span, the first generation going on sale in March 1993, and the last example manufactured in Valencia, Spain in April 2022.

In the USA, primary regions (including Canada) were served by the 1994–2000 Ford Contour and the later (2012–19) Fusion. Ford's Mercury division, selling the car with the obscure Mystique tag over the same six-year period, backed up the mainstream Ford Contour version of the Mondeo. The nameplate Mercury, a Ford sub-brand, was dropped in January 2011, having survived since 1938. Relevant to this book, Mercury – twinned with Ford's still-surviving Lincoln luxury brand in 1945 – facilitated some brave imports for the US market. These included two successful generations of the Capri between 1970 and 1978, and the less popular (Sierra XR4-based) Merkur covered in Chapter 11.

What amounted to a fourth generation and total rebirth of the Mondeo on a replacement platform became Fusion in most of the Americas between 2012 and 2019.

Worldwide, that wasn't the end of the regional variations, as a fifth-generation variant was revealed in January 2022, manufactured and sold only in China. It was bigger all round, with the rear seat space particularly emphasised, as is traditional for this market. China had taken the Mondeo in its second-generation form since 2002, and it's worth highlighting that China has grown into the world's biggest car market. Incidentally, the Ford Taurus nameplate resurfaced in April 2022 as a reworked Mondeo for Middle East regions.

The V6 and ST200/220 factors

The first-generation Mondeo in European CLX, GLX and Ghia guises omitted specific performance variants, offering just a nod to the more enthusiastic driver via 136bhp Si clothing. Ford in Europe, especially Britain, always favoured a high-performance 'halo' effect for its often mundane mass-production/fleet sellers, a habit that can be traced back to forebears such as the Cortina (GT/Lotus/Twin Cam) and Sierra (XR and RS Cosworth).

However, the Mondeo was a much more capable and safer car than its mass-market predecessors, right across the model range, achieving a fine balance of handling, ride and steering alongside the fundamental switch from rear-drive (as on the Sierra) to front-drive. Higher-performance Ford saloons for the showroom now eschewed the RS Cosworth route.

Ford tried its utmost – as demonstrated by the purchase of Jaguar Land Rover and Aston Martin – to acquire the gloss of a prestige badge through better products, even if the Blue Oval's long-established identity means that the company has generally found it tough to elevate itself to more prestigious levels. Aiming at more profitable niches occupied by the likes of Audi, BMW and Alfa Romeo, Ford deployed for the Mondeo a blend of V6 horsepower and comprehensive equipment levels, from air-conditioning to tailored interiors, and exterior styling upgrades centred on wheels and aerodynamic additions.

As for the V6, the first thing to establish is that it was a brand-new design, with no hangovers from Ford's previous European-market V6s – with cast-iron block and pushrods – found in range-leading Capris, Granadas and Sierras. The 1994 arrival of Ford of Europe's V6 Mondeo engine was credited to appropriately multi-national US, German and UK teams, but there was an important link with the past: the high-speed casting process used for the major aluminium components was the first use of that established Cosworth technology for mass production.

'THE MONDEO WAS A MUCH MORE CAPABLE AND SAFER CAR THAN ITS MASS-MARKET PREDECESSORS, RIGHT ACROSS THE MODEL RANGE, ACHIEVING A FINE BALANCE OF HANDLING, RIDE AND STEERING.'

The new V6 engine's aluminium block and quadruple-camshaft 24-valve heads were complemented by a forged-steel crankshaft. Observers have pointed to Ford's Mazda links as an inspiration, but Ford in the USA – and the architecture of the engine, with chain drive to those overhead cams – refer to an earlier modular V8 4.6-litre engine, one deployed from 1992 within the acreage of a Lincoln VIII coupé, and originally envisaged with a smaller V6 cousin.

Official Mondeo V6 power figures reported 168bhp at 6,000rpm (170PS at 6,250rpm in catalytic-converter trim) coupled to 162lb ft (220Nm) of torque at a comparatively elevated 4,250rpm. Ford claimed 0–60mph in just over 8sec and a 139mph maximum for saloon and hatchback bodies, but the estate was confined to 133mph; even worse performance statistics were admitted within Ford sales brochures for optional automatic transmission. They reported losing 9mph from maximum speed and being nearly 2sec adrift on a manual gearbox over 0–60mph, never mind the real-world fuel consumption. Naturally Ford glossed over that, but the fact was you were looking at 21mpg or worse in urban conditions.

Ford's logic was that a special performance version wasn't required initially. Yet the advent of the 2.5-litre V6 engine for the 1994 model year – and the natural desire of the marketing and sales force to have something special to talk up – virtually coincided with the debut of the UK's mutant V6 Team Mondeo racers. There was very little in common between the track star's frenzied 2-litre Mazda-Cosworth 300bhp V6 and the amiable 2.5-litre/168bhp street Mondeo. Yet, simply badged as V6 24v models in four- or five-door bodies, carrying the option of Ghia trim, the Mondeo V6s did profitable business in Europe from their 1994 introduction, especially in Britain.

In 1996, the Mondeo was facelifted, using the corporate oval grille and fisheye headlamp clusters, although the foundations remained those of the previous platform. This outline would serve until

2000, when the link with 2-litre touring car racing was severed, as subsequent Mondeos weren't used by Ford in motorsport. Showroom 21st-century models were dominated by diesel engines, until the electric/hybrid revolution began to sweep mass-produced performance petrol engines away into ultra-niche markets.

In 1996, a European line of ST24-badged performance Mondeos appeared, using the same 2.5-litre aluminium V6. Ford's 'ST' identity meant either 'Sports Technology' (in glossy colour brochures) or was intended to convey an association with 'Super Touring', the 2-litre touring car racing category that spread across Europe to become a global phenomenon into the later '90s.

During 1997, Ford of Britain emphasised the links with Super Touring racing with a five-door ST24 Mondeo, promoted alongside the debut of the unsuccessful five-door BTCC challenger. For the road, that meant some external body cladding, embracing front and rear bumper revisions, allied to extended side skirts and a rear spoiler, plus sportier seats and more individual interior.

That ST branding extended corporate use of the ST label, along with increased horsepower (thus ST200 and ST220) and an enlarged and palpably much better 3.0 version of the V6, which occupied ST220 engine bays from 2000 to 2007. This later engine had many detailed upgrades that made it a better second-hand buy for the durability and maintenance aspects of private ownership.

DRIVEN

Road cars

Production 1.6- and 2.0-litre Mondeos (two examples) St Tropez media debut, 14–16 January 1993, 1.6- and 2.0-litre engines; impressive showroom cars, performance potential unknown. **Production Mondeos (two examples)** Scotland-based Mondeo drive, Gleneagles Hotel, 4–5 May 1993; one drive with triple F1 World Champion Jackie Stewart, published by *The Guardian* newspaper. **Production Mondeo 2.0Si (two examples)** Road test and commute to Silverstone, British GP, 9–16 July 1993; ride with Michael Schumacher around GP track in 2.0Si; a second example also road tested in 1993. **Mondeo 2.5 V6 (two examples)** Road test week, North Wales, 24–25 June 1994; preview of 168bhp V6 engine; estate 24v good enough to want to buy; second test from 16–19 August 1994. **Mondeo ST200** Drive of facelifted Mondeo, 2 August 1999; fitted with 200PS (claimed power) 2.5-litre V6; still not overwhelmed. **Mondeo ST220 (two examples)** Scottish-based media previews, 2005; drove both 155bhp TDCI and 3.0-litre petrol V6 variants in facelift four-door and estate bodies; excellent.

Competition cars

Super Touring Mondeo 2.0Si-V6 Track test at Brands Hatch, 18 February 1994, published in *Autosport*, 28 July 1994; Andy Rouse Engineering-prepared winner for Rouse/Radisich. **Mondeo four-cylinder race project** Silverstone track test, 27 February 1997, published in CCC, May 1997; 295bhp/8,500rpm Roger Dowson/Mountune; Malaysia-based Asia Pacific & Australian races.

MONDEO LAUNCH: ON THE SCOTTISH HIGH ROAD WITH JACKIE STEWART

On Monday 4 May 1993, we travelled beyond first class, as Ford arranged that one of its executive company aircraft fly us away from Stansted's exclusive Business Terminal and up to that pearl amongst Northern European cities, Edinburgh. There to meet us, a fleet of Zetec-powered (later Zeta) petrol Mondeos, from a 90bhp 1.6-litre model to a 136bhp 2.0 Si at this launch stage. The American-developed petrol 2.5-litre V6 and 1.8-litre turbo diesels followed within a year. An unremarkable 45-minute drive up to Jackie Stewart's favourite media-event hotel, Gleneagles, occupied the motor noters next. We met the triple World Champion and former shooting ace that evening in the conservative luxury of the Braids Room, for cocktails and dinner.

However, J.Y. Stewart wasn't the sole attraction, as Ford knew that chief engineer Richard Parry-Jones had a good-news Mondeo story to tell. So, plenty of chatting and drive-time could be accessed with the former RS2000 club-

DRIVING WITH A TRIPLE F1 WORLD CHAMPION

The Mondeo launch event in Scotland wasn't my only Ford-mounted outing with Jackie Stewart. When he retired from F1, in 1973, we shared an RS3100 for part of his retirement tour, including BBC South studios and some 1973–74 outings in the Formula Finesse national series. Formula Finesse involved balancing a ball in a bowl on the bonnet of a Cortina and driving through a tightly coned course, timed to the tenth of second, without dropping the ball. I never did see Jackie's times beaten. Acting as a chauffeur to racing royalty was nerve-wracking, but Stewart was very polite, as was newly crowned Emerson Fittipaldi. I had that honour conferred when I had to ferry 'Emmo' from Heathrow to a central London party, one celebrating the Cosworth-Ford DFV engine that he had used to win the 1974 world title in a McLaren.

'FOR ME, STEWART WAS THE SMOOTHEST AND CLEVEREST OF DRIVERS AT ASSESSING INCOMING MESSAGES – FROM OUR 5–25MPH FORMULA FINESSE DRIVES, TO THAT FAMOUS 1968 GERMAN GRAND PRIX.'

Jackie (today Sir Jackie) was famously photographed with the Capri RS3100 in a widely used Ford PR black-and-white photograph. I was also driven – with many other hacks – by JYS around Silverstone in an Elf-backed Capri II. We later used the same track for three-at-a-time passenger rides in a production Mondeo with Michael Schumacher conducting. I haven't highlighted that experience, as the outing could not allow Schumacher to demonstrate the talents that would win seven World Championships, although his precision in a heavily laden showroom saloon was outstanding.

For me, Stewart was the smoothest and cleverest of drivers at assessing incoming messages – from our 5–25mph Formula Finesse drives, to that famous 1968 German Grand Prix held in an atrocious foggy blanket at the Nürburgring. On that day in the Eifel Mountains, Jackie won by a truly massive margin of just over four minutes, averaging 86.82mph, in 100m visibility at best, in his Matra-Ford V8. And he also drove with a healing broken wrist in a cast...

◀ At the 1993 British Grand Prix, emerging Benetton ace Michael Schumacher carted media cargo, including me, around the full Silverstone circuit in a Mondeo 2.0Si. *Ford*

rally driver, who led a team of 3,000 technicians across Europe to unlock the fine-handling, mass-production Ford that was the Mondeo. Mainstream Fords such as the Fiesta, Focus and Puma 2, are still gathering plaudits for their ride and handling qualities 30 years later, but the Mondeo arguably provided the foundations.

The background, gleaned through shared Mondeo 2.0Si chats over 1h 20m 'laps' of the challenging roads around Gleneagles out to the Trossachs, was illuminating. The insight reflected quite a lot of the Mondeo pre-debut development miles that occupied 15 months for Jackie's involvement. Allowing so much time and in-depth conversation with Stewart and Parry-Jones was unique in my experience of major motor-manufacturer access, so *The Guardian* also got a unique 600 words for £150 in 1993.

Jackie Stewart was the headliner of course, in his low-profile 29th Ford-contracted year as a consultant and ambassador – 'our silent asset' as Parry-Jones described him. Sometimes he ruffled the feathers of notoriously fickle parent-company politicians, but in Europe his input to late-programme handling sessions for the RS200, Sierra Cosworths and Mondeo were well publicised.

Jackie's feel and feedback for steering, brake-pedal progression and throttle reaction/calibration – or any of the other dynamics that matter to public or professional drivers – was simply unmatched. Naturally, his time was very expensive, so the company engineers and the contracted consultant drivers did by far the majority of development days and the heavy lifting of endurance mileage. Ford utilised its own purpose-built Lommel facility in Belgium for the bulk of testing work for most European markets, including Britain and Germany. One engineer commented: 'I knew Lommel track a lot better than my home by the time we finished!' Ford US used its own facilities, and the suspension and steering settings were obviously markedly different to Europe, biased towards a softer ride for longer average journeys in a wider range of hostile conditions.

After all this multi-million-dollar fine-tuning talk, and thousands of man hours, Parry-Jones had a cheerful put-down for the international engineering programme: 'After we spent all those millions, my wife says she doesn't give a damn about all that handling bullshit. How could I have developed a car without a button on the boot?'

◀ In 1993 Ford brought out their big sporting guns to support Mondeo publicity. Triple World Champion Jackie Stewart was a proper tutor at the rather fabulous Gleneagles Hotel, which meant that I could write a 'one-on-one' road-drive story for *The Guardian* newspaper. *Ford*

REFLECTIONS ON DRIVING MONDEO PERFORMANCE ROAD CARS

Today, I can look back and review the 1994–2022 life of 'performance' Mondeos, from the 2.5 V6 to the final ST-Line. Available with a variety of petrol, diesel and hybrid powertrains, none packed more than 190PS, and at the end of their production life all were priced near or beyond £30,000.

There is a strong case for writing that – in a modern-classic market gasping at the value of more obvious RS and Cosworth-branded Fords – ST variants of both the Mondeo and Focus are undervalued. This was the case for earlier XR2/XR3/XR4 lines, until values of the Blue Oval's earlier charismatic performers went ballistic. To cut a long story short, expect the 21st century's ST220, already touted by media and auction houses, to be the next ticket on the Blue Oval money-train express. The ST220 certainly became worthy of wider recognition for its blend of performance, practicality and winning links to touring-car racing.

Time for reflection on my Mondeo V6 and ST road-car driving experiences over the years.

Mondeo V6 estates

It was the summer of 1994 before I drove V6 variants in four-door saloon and five-door estate format. My first experience was a disappointing overnighter, in the era when Ford's sporting performers were judged against its outstanding Escort RS Cosworth. Although the V6 Mondeo's suspension system provided fine handling, Ford UK could be mean, and in this case you were confined to steel wheels with 195/60 VR tyres.

'THERE IS A STRONG CASE FOR WRITING THAT – IN A MODERN-CLASSIC MARKET GASPING AT THE VALUE OF MORE OBVIOUS RS AND COSWORTH-BRANDED FORDS – ST VARIANTS OF THE MONDEO AND FOCUS ARE UNDERVALUED.'

The alloys were the same 15inx6J dimensions, but were allowed 205/55 VR lower-profile and wider rubber.

A second, longer family outing to North Wales in the 24-valve estate underlined the point of a V6 in public-road, four-up and baggage mode. I really liked the smooth running, an extra rasher of acceleration (although I missed the old 3-litre V6's mid-range punch) and the Mondeo's tidy chassis manners. Since I had bought my own Ford estates before (rear- and front-drive Escorts), I would have been up for that V6 Estate, if post-divorce financial cramp hadn't intervened.

Mondeo ST24

As noted earlier, there was a 1997 ST24 variant of the Mondeo theme, but it carried the same power as before and really existed to reinforce the fact that the first-edition Mondeo had been revised with an oval grille and wraparound headlamps. It also served to highlight that a five-door variant had been made available, with hints of the (sadly lacking in success) British Touring Car programme of that year. As with the previous model, the ST24 closed on a 139mph maximum, covered 0–60mph in 8sec and ran suspension set 15mm (0.59in) lower than less-powerful Mondeos. The alloy wheels grew slightly, with 16x6in measurements accommodating 205/50 ZR tyres.

Even over a brace of carefully monitored UK drives I didn't warm to the ST24 variant, especially as other brands offered more attractions for similar money.

Mondeo ST200

In August 1999, I had the chance to spend time with the mildly facelifted Mondeo in ST200 guise. That primary body shape would figure in Prodrive's clean sweep of 2000 BTCC drivers' and manufacturers' titles. For the performance public, it was probably more important that they had cheered up the 2.5-litre V6 to a claimed 200PS at 6,500rpm and 225Nm torque at a peaky 5,500rpm. The Ford-declared moves to enhance peak power

◄ ▼ Ford ensured we had every opportunity to road test the 168 and 200bhp-rated Mondeo V6s. This is the ST200 I had in August 1999. I still like the original design, but the subsequent ST220 was a much better all-rounder and remains on many modern-classic-car media wish-lists. *Author*

covered: polished porting ('extrude honed' were the buzz words), enlarged and high-flow air-filtration/throttle body, replacement lower-weight (hollow) intake and exhaust camshafts, and a low-pressure system for the traditional twin exhausts.

The suspension had also been overhauled, with some components shared with Ford's Cougar coupé. The wheels and tyres were upgraded to 17x7in alloys carrying 215/45 ZR rubber. Internally, the trademark Recaros remained, and externally Ford referred to the favoured optional paint as 'Ford Racing Blue Metallic'.

Sadly, the brakes hadn't kept up with what had become pretty peaky performance from the V6: I noted a 'soggy pedal', which must be the automotive equivalent of Mary Berry's 'soggy bottoms' on TV baking shows. I felt 5,000–6,800rpm was needed for the reworked V6 to 'feel sharp'. The suspension changes had also left the low-speed ride in tatters for British country lanes.

The delivery driver managed 24.7 careful mpg, but I returned 23.5mpg overall and remained unconvinced about the public-road performance 'improvements'.

Mondeo ST220

By March 2002 (UK sales of the ST220 begun in May 2002) I had a personal handle on the ST versions of both the Focus (ST170) and Mondeo (ST220). That access came via a media preview of both the Focus and Mondeo, driving in Southern France. As before, the ST-specification applied across four-door saloon, five-door hatchback and estate versions, but the bigger news was the upsizing of the aluminium V6 to 3 litres, rated at 217bhp (226PS) at 6,150rpm, plus a useable 275Nm of torque, albeit with an elevated peak of 4,750rpm.

Ford claimed 151mph, and a subsequent 155mph maximum, coupled to a sub-7sec 0–60mph dash. Slightly over the 7sec barrier was more realistic, in line with Honda's fabled Accord R (now a benchmark for modern-classic values)

➤ The clever 2002 marketing and showroom twist to the ST220s – and other ST Mondeos – was that you could opt for five-door saloon or estate bodies. Both carried modest external distinguishing clues and shared sports alloy wheels. *Ford*

and the contemporary V6 Vauxhall/Opel Vectra GSi. Contemporary rivals nominated by Ford in a media briefing were Accord R, at £21,495, Alfa's tempting 156 V6 Veloce at a more affordable £20,285 (poor service and resale values), and £26,870 of BMW 330i, BMW's then calling-card, a suave straight six.

It is worth noting the June 2003 fitment of a fine Getrag six-speed gearbox for the Mondeo ST220, to replace the Ford MTX five-speed. The Mondeo's V6's engine expansion was accompanied by growth in the standard wheel diameters to more credible 18in alloys wearing 225/40 R tyres, and an honestly redeveloped sports suspension. A restrained body kit was also introduced, with subtle lip rear spoiler watching over twin exhausts. Favourite Ford metallic paint, referred to as Performance Blue, remains sought-after, but there was also a Machine Silver ST220 edition. Just 538 examples were assembled in late 2004, registered in Britain in 2005, and carrying a build date and limited-edition number atop the 3-litre V6.

I shared that double media preview with then *CCC* editor Steve Bennett; the ST170 Focus was particularly significant to that magazine's hot-Ford-attuned readership. Yet we also gave a full workout to the ST220 in the nearby Monte Carlo Rally terrain, both on and off the suggested routes.

My Mondeo notes carried a score sheet, marked out of a 10 maximum; lowest was a 4/10 for the cabin (Recaro seats the redeeming feature) with white-faced instrumentation. I still find later-model STs messy on the instrument clarity, with too many colours and small dials, some

⮟ Often cited as a modern classic in the saloon format, the ST220 is particularly valued in this Performance Blue paint. Twin exhausts and a modest body kit accompanied the 217bhp V6, plus a set of 18-inch alloys, and a six-speed Getrag gearbox from June 2003. *Ford*

subsidiaries looking like they've been plucked from a back-street discount warehouse.

ST220 ride and steering attracted 8/10 and 9/10 scores, and there was another 9/10 for the sheer grip exhibited on Uniroyal Contact T2 tyres: 'Squeaky but effective' was the comment that probably applied equally to one of the drivers within! The V6 motor's soundtrack scored 7/10, with 8/10 for performance, benefitting from a stronger 1,500–3,000rpm range than the earlier 2.5 V6. Safety features attracted 9/10 from me, but there was the proviso on both STs that the brakes had distinct limits in hard use over tight twists.

My verdict was that the ST220's chief programme engineer, Jürgen Gagstatter, had supervised a team that yielded a 'very well developed' end product. That team included former SVE engineer Geoff Fox on his last mission after 40 years of Ford UK graft. Handling development benefitted from Nürburgring sessions that aimed to balance enthusiastic driving qualities, whilst retaining many of the fine ride characteristics that were present in the contemporary Mondeo and Focus with lower power.

I think that our verdict was justified over the years. In 2019, respected motoring writer Tony Middlehurst introduced a *Piston Heads* piece on buying an ST220 with these words: 'Folk who have been through the whole Mondeo ST thing, from ST24s through ST200s to the ST220, will usually agree that the 220 is the one to have. It's not just the fact that it had more power than any of its predecessors, or that it had all the equipment you'd ever want as standard fit, including big, supportive electrically adjusted, heated leather Recaro seats. It's the all-round package of performance, handling and, especially in the most popular hatchback version, or the estate with its self-levelling suspension, surprising practicality.'

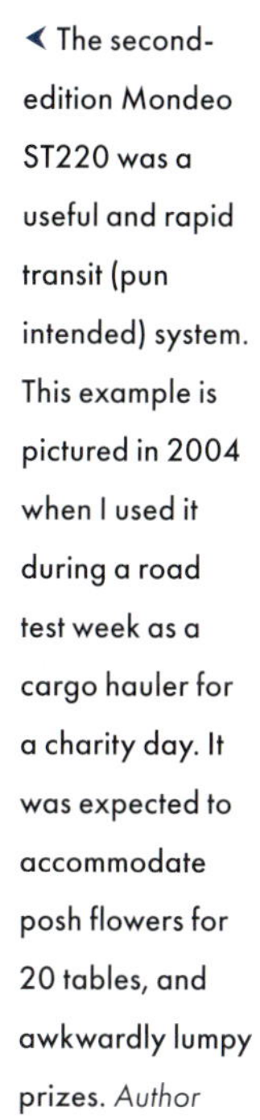
The second-edition Mondeo ST220 was a useful and rapid transit (pun intended) system. This example is pictured in 2004 when I used it during a road test week as a cargo hauler for a charity day. It was expected to accommodate posh flowers for 20 tables, and awkwardly lumpy prizes. *Author*

Later Mondeo ST

From a personal viewpoint, the 2006-previewed (Paris Show) Mondeo generation of 2007–12 is beyond my experience, along with the subsequent generations. As the ST5 turbo engine became the performance choice, I experienced that charismatic five-cylinder turbo in Focus ST and RS Focus performers, not the now much-heftier Mondeo, which by that time shared the Galaxy and S-Max platform.

In 2007, Ford previewed the revised Mondeo at the Paris motor show. I found it remained a good-natured middleweight, but always cried out for more power to match its increased weight and size. *Ford*

THE CHALLENGING BIRTH OF THE BTCC/SUPER TOURING V6 MONDEO

The arrival of the front-drive Mondeo in Si format to join the 2-litre British touring car circus wasn't part of some macro-managed corporate plan, far from it. The Mondeo's race- and title-winning form went from initial Rouse-era euphoria to expensively confused four-season loser, before it was brought back to British prominence by a three-car Prodrive assault. During a less-competitive era, the Mondeo's last corporate fling, in 2000, resulted in a steamroller 1–2–3 final BTCC championship result.

Initially, even the layout of a race Mondeo, from front to back, was confused, for there were multiple engine and transmission choices. Abundant finance wasn't offered promptly to research those technical possibilities, as the UK branch of Ford's multinational operations was already eating corporate cash at a spectacular rate, supporting the creation of the Escort RS Cosworth and its competition programme. Eventually, with a change of top European management and the enthusiasm of Ford of Britain's marketing men, a budget was released. A decision came so late (December 1992) that the Mondeo project missed almost half its debut 1993 BTCC season, joining the battle at round seven of fourteen. However, that delay wasn't all due to sluggish Ford corporate inertia.

A rear-wheel-drive false start, and that V6

Andy Rouse told me in 1994 how they had invested serious labour in creating a pair of *rear-drive* Mondeos from the bodies of two pre-launch front-drive road cars. This exercise was based on the premise that rear-wheel drive had continued to prove successful with BMW's 1991–92 championship titles. Besides, rear-drive was the natural format for a race car. Cosworth was selected as a supplier, and the company's earlier consultancy role had investigated possible base Ford and Mazda four- and six-cylinder power plants. They revealed a 2-litre version of the 2.5-litre Mazda V6, seen in the contemporary Ford Probe coupé, as the most likely to take power beyond 300bhp, running at a higher rpm than the four-cylinder opposition in the BTCC's 8,500rpm regulated format.

Looking back, in July 2019, Rouse explained some of the background to that unique (for BTCC racing) selection of a V6 motor: 'Ford's four-cylinder engine at the time wasn't suitable for racing, because the bore was too small and the valves were too close together. The V6 Probe engine had plenty of valve area and a big bore because it was designed as a 2.5-litre engine that was destroked to 2 litres for racing. Once the engine was fully developed, it was very good and was probably the first Super Touring engine to make over 300bhp. However, as the regulations evolved, and engines were being placed behind the front axle, the bulk of the V6 became a handling issue that was difficult to overcome. We presented Ford with an improved engine layout using their four-cylinder engine to try and convince them to develop it for racing, but that didn't work out at the time.'

For further background on four cylinders versus V6, see 'An alternative four-cylinder Mondeo racer' (page 334).

Andy took up the early rear-drive testing tale: 'First, there was an obvious shortage of sheer horsepower to make the rear-drive gamble work. We tested at Snetterton, Brands, Pembrey and Silverstone, but Snetterton was the significant one. Using a speed gun on rival cars and our own, we discovered we were 8–9mph down on our rivals. It was pulling the cornering speeds OK, but the 285bhp we had initially was just not enough to roll the lap speeds we needed.'

There were also a couple of inconvenient BTCC regulations that stood in the way of converting a front-drive Mondeo into a rear-drive winner. That Snetterton lack of top speed was directly down to the power losses suffered in taking a gear-drive through a 90° angle from a mandatory (original front-drive) transverse-mounted engine to feed the rear. As it transpired, the Rouse equipe was fortunate that it didn't pursue that backward route, because Ford schedules for making 4x4 versions of the Mondeo – needed for Rouse to meet BTCC regulations for the rear-drive variant – were optimistic, so the rear-drive Mondeo would have been rendered illegal anyway.

Rouse continued: 'When we got to the last week of April of that year [1993] with our rear-drive project, it was obvious that if we wanted to race by our target [British] Grand Prix July date, we would have to go the front-drive route. We took one of the two rear-drive cars we had made and turned that into the front-drive prototype. All done in-house... in 14 days! Basically, we just used whatever we had around to make it work, to get it running so that we could see just how quick a front-drive Mondeo could be. We used a gearbox that we had from another project, adapted strong drive-shafts to suit and so on.'

A new start with front-wheel drive

Talking about the first run of the front-drive Mondeo, Rouse recollected in early 1994: 'It felt good at Snetterton straight away. In fact it was quicker than the rear-drive car straight away, so we knew it had potential.' He added with the clarity of hindsight: 'We discovered from the start that the Mondeo was much too heavy. So Ford opened up the [Genk, Belgium] production line on a Sunday and sent 30 special shells down the line. They reduced the metal thickness on all the panels, and took out anything that we didn't need – all the crash-strengthening in the doors, lots of brackets and stuff.'

A number of Ford executives in Britain and Belgium thus took a big gamble with their careers, including key Mondeo marketing player and newly arrived UK executive Nick Palmer. Authorising these special-builds was a brave move, because the hassle within the Genk factory was considerable. The mass-production Ford had to be removed from the production line at an unorthodox point and every build requirement thought through on a weight/strength basis, rather than with the usual emphasis on high-speed mass-production and showroom goodies.

Rouse added in my '90s interviews that these Mondeo race bodies: 'Don't have things you don't need for racing – brackets to hold trim, underseal, even undercoat paint. All were deleted from the build process at the plant. The paint may sound minor, but a coat of primer adds 4–5kg. So, we eliminated much of the production [painting] process, and the cars were raced in three coats, including etch-primer, undercoat and the colour you see. The only strengthening is that the Motorsport bodies received double spot-welds. The additional strength a car needs to go touring car racing really comes from the changes we make here at ARE [Andy Rouse Engineering], especially in the purpose-built roll cage.'

Another major hassle was getting early delivery of Cosworth power, as the Northamptonshire concern had Grand Prix obligations to fulfil. Andy remembered in a recent interview: 'We had our own engine-building department with three technicians under the control of my business partner, Vic Drake. Initially, the plan was simply to get the car up and running on our version of the V6 motor. Our first motor ran in the third week of January [1993]. A much bigger achievement than redeveloping the RS500, since we had to design, freeze specification and order basic parts like the steel crankshaft and rods. In fact every internal part was redeveloped, leaving just the basic head and cylinder block from the road car.'

It emerged under contemporary questioning at Cosworth and Andy Rouse Engineering that the Northamptonshire motor magicians had originally contracted with Ford to deliver the first power units in August 1993. That schedule would have effectively written-off any serious 1993 outings, never mind a July British Grand Prix support-race debut. Constant corporate and preparation-team pressure was applied, and Andy Rouse Engineering received its first compact Cosworth-Mazda V6s in late May.

'IT FELT GOOD AT SNETTERTON STRAIGHT AWAY. IN FACT IT WAS QUICKER THAN THE REAR-DRIVE CAR STRAIGHT AWAY, SO WE KNEW IT HAD POTENTIAL.'

➤ Although conceived, assembled and tested under extreme time pressures, the front-drive Mondeo, with Cosworth reimagined Mazda V6, proved immediately competitive – good enough to win against the best. *Motorsport Images*

⮟ Packaging that unique V6 engine and its ancillaries in the Mondeo was a challenge, as rivals tilted simpler four-cylinder units rearward and ever lower. *Motorsport Images*

Those early units delivered around 285bhp, but during the summer run-up to their race debut at Pembrey – a month before the July Silverstone GP target – Cosworth was 'able to concentrate on our engine supply and performance,' added Andy in late 1993. He continued: 'During the year I would say it gained about 10bhp, but it's always had good top-end power. In fact it would be fabulous if we could run it to 10,000rpm, but the BTCC 2-litre rules [which Andy helped frame!] stop us at 8,500. Initially, the V6 was very 'top-endy' and really liked life above 8,000. Further development in association with our preferred engine-management system from Zytek filled in some of the early gaps. Standing starts required replacement electronic programming and they filled in the 7,000–8,000 band a little more strongly.'

Peak V6 pulling power was always competitive in high-rpm theory, but wasn't accessible over the wider torque band that the four-cylinder opposition relied on in these comparatively heavy race cars. Incidentally, the BMW straight-six demonstrated the same narrow-band torque as the Cosworth-Mazda V6, when raced under 2-litre FIA international regulations.

Outside the engine bay, the front-drive Mondeo had racing advantages that contributed to faster laps than the legendary RS500s recorded, even at quick tracks like Thruxton. Visually, a spectator could spot the big plus points: it was much lower and more slippery through the air than most of the white-hot opposition's machinery, with a wider front and rear track that allowed the absorbent suspension to ride the kerbs. Part of that kerb-riding prowess could be attributed to Dutch Pro-Flex shock absorbers with remote oil reservoirs. Spring rates were high: 900lb/in fronts and 1,000lb/in rears.

The principles of MacPherson-strut front suspension and other key showroom configuration details were retained, but almost all were modified. That fresh-componentry approach applied particularly to the AP Racing disc brakes, units also proved on the rallying Escort Cosworth: 14in-diameter ventilated front discs, clamped by six-piston callipers. At the rear, 12in units were deployed with four-piston callipers. At the Silverstone GP supporting race in 1993, where he had just finished third, Paul Radisich educated me to appreciate the race Mondeo braking power: 'From about 125mph, the brakes are something else on the Mondeo. A good hard pedal, but the retardation is stunning.'

ARE made the twin-bearing hubs and stub-axles (accommodating a single-wheel-nut and four-peg Dymag alloy-wheel attachment points), along with the cockpit-adjustable blade-and-link anti-roll-bars. Utilising a bell-crank action and Bowden connection cables, a 40 per cent change in anti-roll-bar stiffness could be dictated by the driver in action.

Radisich commented: 'When we switched to front-drive for the Mondeo... boy, was it ever hard work! We had no power-steering and getting all the power and torque down to the ground and steering it, well, that's a tough job!' Power steering arrived, but not immediately. Since a front-drive layout must cope with steering and power-transmission tasks, the action of the differential was also critical. The original multi-plate mechanical differential was replaced by the Ferguson viscous-coupling limited-slip action.

The tyres had to be selected to deal with the critical track demands. Initial races were on Yokohamas, but they didn't like the heat of extreme summer battles, and Michelin tyres were found to suit the Ford's set-up best, a decision speeded by the deliberate omission of any contractual tyre deal.

Footnote *Looking back on the heyday of 2-litre BTCC, Rouse reflected in 2019 on where manufacturer and importer millions were spent: 'This was an international championship contested on British race tracks, so the manufacturers were all keen to do well. Teams expanded too – in Williams's case [Williams ran the 'works' Renault Lagunas] to 60 people. The cars became more sophisticated and expensive, and engine builders were going to extreme lengths to win the horsepower race. Drivers' salaries went ballistic, and team presentation became an expensive issue. The end result was that the cost exceeded the return, and not every manufacturer could win, so Super Touring burnt itself out after a period as an outstanding motor racing spectacle.' That process had occurred by 2000, when Prodrive took the Mondeo to that 1–2–3 championship result, just Honda and Vauxhall running as official factory opposition.*

SUPER TOURING MONDEO 2.0Si V6: TRIUMPHANT WORLD CUP DOUBLE WINNER

The wheedling and manipulation done, via an extraordinarily helpful Brands Hatch track office on 18 February 1994, I'm braced for 13 minutes and 40 seconds of data-logged lunchtime lappery in the front-running Team Mondeo BTCC racer, as campaigned by Paul Radisich and prepared by Andy Rouse Engineering for a track test which would be published in *Autosport* on 28 July 1994. Recently, I caught up with Radisich in his New Zealand home territory, giving me the benefit of hindsight on those rushed moments, as he had raced generations of Mondeos from 1993 to 1997.

Back then we could see that a winged Alfa Romeo 155, driven by amiable football fanatic Gabriele Tarquini, would dominate the 1994 British Touring Car Championship. However, Radisich and the duotone corporate colours of the Mondeo would win a second World Cup against allcomers at Donington in October (having won the previous year's event, held at Monza). Andy Rouse retired from driving at the close of 1994 as a quadruple British saloon car title-winner.

At the time I drove the 1994 Rouse-assigned Mondeo, instructed by Paul Radisich, Paul had won his first Mondeo World Cup title, against the strongest-possible Alfa Romeo opposition at Monza – and finished third in the 1993 UK title hunt, despite missing the first half of the season.

In 1994, it wasn't just the Mondeo that suffered at the hands of Alfa Romeo, but all the massed opposition: Tarquini's 155 dominated the season, while the Williams-Renault team, with

➤ ⮟ Pre-season 1994, I had the chance to attend a Silverstone test day for the Ford-backed Andy Rouse Engineering Mondeo 2.0 V6 Cosworth. Drivers were Rouse (right) and New Zealand's Paul Radisich (left). *Author*

its radically re-engineered (first of the extreme slant and lowered engine layouts) Laguna for Alain Menu was second in the 1994 series points. Radisich took third after scoring two of the team's three victories. The other win was notched up to Rouse at Donington – his final, and 60th British Saloon Car/BTCC victory (scored over 22 seasons – for a long time the outright record).

Under the skin: a distant relation to the showroom cousin

Appropriately, the race Mondeo in '94 Ghia format had experimentally adopted power steering when I drove, fitted for much of 1994 for the benefit of eliminating extreme steering-wheel-rim loads through the slowest corners and Thruxton's high-speed bumps. For 1994, the 'showroom specification' of the race cars was officially changed from Si to Ghia, primarily because Team Mondeo was allowed to delete the Ghia's auxiliary fog lamps and implant cool-air feeds to the brakes, supported by those vaned Dymag wheels. The cars lost the 1993 rear showroom Si-specification spoiler in the process, but aero-testing had told them it had no value anyway!

For the *Autosport* test I made some scene-setting comparisons between the showroom Mondeo Ghia V6 (pending, when I wrote the test) and hand-built racer: 'The Cosworth rendition of the V6 harvests 120bhp more than the most-powerful Mondeos scheduled for sale [actually, that materialised as a 132bhp bonus on official 1994 Ghia Mondeo V6 statistics when the range-leading Mondeo went on sale]. A racing Mondeo, complete with Xtrac six-speed sequential transaxle, weighs 25 per cent less than its showroom siblings. It is faster too, limited to 8,450rpm, the Mondeo hit 157mph at Monza.'

Footnote *For more under-the-skin details of the BTCC Mondeo, see 'The challenging birth of the BTCC/Super Touring V6 Mondeo' (page 328).*

Behind the wheel: my final Rouse-engineered competition-car drive

Andy Rouse subsequently confirmed 144mph as the rev-limited maximum at the Silverstone GP circuit, but observed drily: 'If the wind changes, we have a 20-minute job to change sixth gear, it is *that* critical.' The test data-logging also told me that the ratios I had at Brands Hatch allowed 69, 90, 108 and 120mph at the rev limiter in second through to fifth, with 127mph coming up in sixth on my sub-8,450rpm entry to Paddock's cambered, downhill charms. The equivalent figures for Radisich and Rouse were 133 and 132mph.

I commented on my Brands experience that: 'The exotic nature of the Cosworth BTCC V6 is underlined in the pits. Rouse's Mark Dorrans completes the full start and warm-up procedure, and instructs us to maintain a 3,000rpm "tickover".' The sunny test-session divided into one 30–40mph lap behind photographer Jeff Bloxham's Golf, then: 'Off the leash, I tug back for higher gears, snick forward to downchange. The slap changes [no clutch needed after engaging first] are the best I have driven in a saloon, with or without synchromesh. Neutral is a sod to find, barely a breath between first and second.'

Aside from that V6 engine – an aristocrat that revved so smoothly between 6,000 and almost 8,500rpm, versus a sea of four-cylinder opponents – outstanding Mondeo memories were of braking and cornering capabilities: 'So much has the Rouse team chiselled away at the handling, that you only feel it is blatantly front-driven when there's a need for full power in second at Druids or at Clearways/Clark Curve. The subsequent restlessness doesn't stop until fourth is home and pulling beyond 100mph.'

Even at daunting Paddock Hill Bend, the Mondeo gave a stranger enough confidence to 'skim downhill, off-camber with almost no further steering lock needed after the apex.' I also researched the fabled kerb-crawling abilities: 'I put a couple of wheels over the inside kerbing at fourth-gear/90mph Surtees. The Mondeo just strode through with barely a millimetric shrug from the chosen line.' The brakes were equally impressive, so much so that over-braking was my significant failing.

I hadn't looked back on this Brands Hatch episode in more than 25 years when I came to write this. I now realise what astonishing access I was granted, as these Mondeos were precious in nature and value (£100,000, rolling chassis only). I can only publicly acknowledge the assistance I received from Brands Hatch circuit office, Radisich, Rouse and Vic Drake that sensational day.

A fitting finale to my 1971–94 test drives in Broadspeed/Rouse vehicles, I just hope I have written worthy words to match their extraordinary abilities.

TOURING CARS

PHOTOGRAPHY BY JEFF BLOXHAM

Mastering the Mondeo

IT MAY NOT HAVE MATCHED THE ALFAS SO FAR THIS YEAR, BUT FORD'S BTCC MONDEO LEAVES **JEREMY WALTON** BREATHLESS. THESE CARS HAVE COME A LONG WAY...

TRACK TEST: BTCC FORD MONDEO

In stark metal, the 1994 Team Mondeo British Touring Car Championship contender exudes racing purpose. It crouches with low-riding menace, a shimmering contrast to the travel-stained Mondeos that munch motorway miles on soft springs and Yorkie Bars. You have to be very special to order the Ford Mondeo's £100,000-plus rolling chassis from Andy Rouse Engineering, never mind the shrunken 2-litre Cosworth version of the Ford's aluminium V6.

The Cosworth rendition of the V6 harvests 120bhp more than the most powerful Mondeo scheduled for sale. A racing Mondeo complete with Xtrac sequential six-speed transaxle weighs 25% less than its showroom siblings. It's faster too. Limited to 8450rpm, the Mondeo hit 157mph when it capped a fantastic debut season by beating the cream of the tourers in the FIA World Touring Car Cup at Monza last year.

First driving impressions are of the growing cockpit sophistication found in the BTCC. The integral T45 steel rollcage barricades the driver into a cabin where everything seems to be adjustable; 'in-flight' electronics spinning an invisible umbilical cord between car and computer crew.

Within that stark cabin, a Stack analogue rev-counter squats adjacent to multiple-mode digital readouts. Each lap time is displayed for eight seconds alongside oil and water temperatures, plus oil pressure.

Such is the sophistication of the data-logging system that Andy Rouse Engineering partner Vic Drake presents me an electronic recall of my magic 1m50s at the wheel – one photographic lap and eight 'off the leash' at the end of the day.

The test car runs the experimental power steering that earned approval of its feedback from both drivers and was subsequently adopted on the race cars before the Oulton Park round in May. The lower effort required at the rim is particularly appreciated at the end of a tough race, or when balancing about 290bhp against the traction improvement efforts of a limited slip differential over the short Brands Hatch Indy circuit.

The six-speed sequential lever appears to be surrounded by two handbrakes. In fact, these slim alloy rods control front and rear rollbar stiffness. Team Mondeo star driver Paul Radisich volunteers that it might be worth 'running the rear bar full soft until the tyres warm up'.

Too true. This is the most sensitive saloon to small adjustments I have ever come across. Within five laps I have to take the risk of altering that back bar setting. I pulled it up a stop, and the Mondeo clicked into an ever more neutral stance that, steering rim twitches aside, defied front-drive clichés.

Assorted foot pedal sizes reflect the racer's edge philosophy at ARE. A massive foot rest and brake pedal are accompanied by a diminutive clutch pedal: 'Just use it to set off,' advises Radisich. The accelerator almost brushes the brake pedal to aid heel-and-toe changes.

The exotic nature of the Cosworth BTCC V6 is underlined in the pits. Rouse man Mark Dorrans, who completes the full start and warm-up procedure, instructs us to maintain a 3000rpm 'tickover'.

Now I was free to clack the gear lever forward for departure. Minimal throttle movements bring an instant 7000rpm. The lap behind VW Golf-mounted photographer Bloxham was hellish, the powertrain unhappy at the dawdling 30-40mph pace.

Off the leash, I tug back for higher gears, snick forward to downchange. The slap changes are the best I have driven in a saloon with or without synchromesh. Neutral is a sod to find, barely a breath between first and second its minimalist home.

The telemetry for Brands reveals that 8450 rpm-limited gearchanges provided 69mph in second, 90mph in third, 108mph in fourth, 120mph in fifth and a solid 127mph at 7500rpm into Paddock in sixth. ▶

Track tester Walton gets the low-down on the Mondeo's cockpit from Radisich (right). The Ford Probe V6 (left) provides 300bhp, which means plenty of front-drive 'restlessness' all the way down the hill from Druids (top)

◀ I seized the February 1994 chance to drive the Rouse-engineered Mondeo during a Brands Hatch test day lunchtime – a wonderful experience but not published in *Autosport* until 28 July that year. *Author's archive*

AN ALTERNATIVE FOUR-CYLINDER MONDEO RACER...

Silverstone South Circuit is one of many infield layouts that has supported the track's full Formula 1 configuration over the decades since the inaugural World Championship GP of 1950. For me, this track provided the opportunity for a distinctly individual racing Mondeo test on 27 February 1997. This variation on the Mondeo competition theme was a joint project between Mountune and the late and revered Roger Dowson. Based at Silverstone, Roger had created a Mountune-motivated four-cylinder Mondeo, as opposed to the Mazda-based Cosworth-reworked V6 motive power used by Ford-backed Rouse, and subsequent preparation companies, in the BTCC.

To the best of my knowledge, the Ford four-door, carrying custard-yellow and Petronas sponsor's green livery, never raced in Britain. It was trailered away overseas the night after my test! It was destined for Gianfranco Brancatelli to steer with intermittent success in the Asia Pacific series, and appeared there in 1997, and twice in Australia, usually driven by 'Branca'.

This privately funded Mondeo project still had links with Ford racing strategies, but ironically a Ford marketing policy would ultimately prevent the four-cylinder from gaining factory recognition. The rolling chassis around the Cosworth YB* four-cylinder engine was that of an earlier Rouse V6 machine from the 1995 season. A comparison between four- and six-cylinder engines was prompted by the swift drop in competitive form of the official Rouse Mondeos after the world-beating 1993 and 1994 seasons.

Then Ford Motorsport boss, New Zealander Peter Gillitzer, had discussed the four-cylinder racing alternatives with Mountune as far back as 1992 (pre-debut of the Mondeo V6 racer). By May 1995, the Essex engine men had done enough initial development comparisons versus the Vauxhall and Renault racing units (respectively prepared by Swindon Racing Engines and Sodemo in France) to be convinced that this was

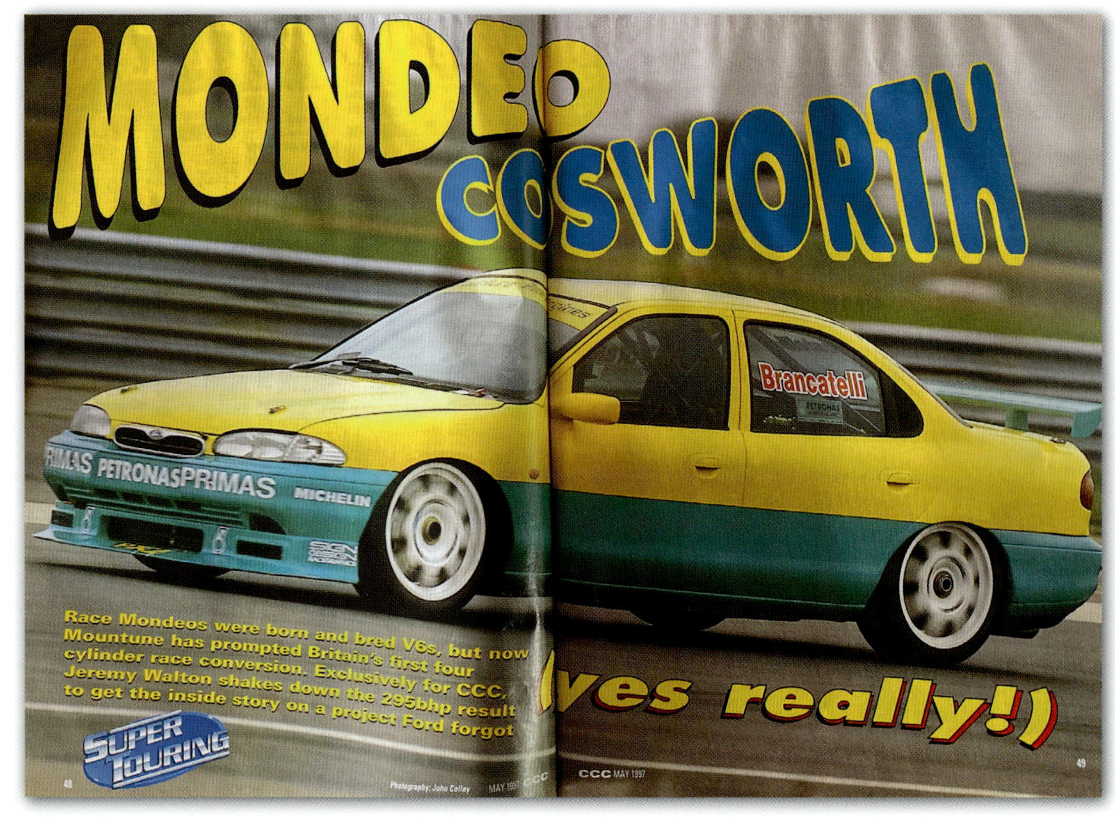
MONDEO COSWORTH

(yes really!)

Race Mondeos were born and bred V6s, but now Mountune has prompted Britain's first four cylinder race conversion. Exclusively for CCC, Jeremy Walton shakes down the 295bhp result to get the inside story on a project Ford forgot

SUPER TOURING

48 Photography: John Colley MAY 1997 CCC

CCC MAY 1997 49

➤ I snatched a half day at a wet Silverstone to drive this four-cylinder Mountune variant on the Mondeo race-car theme, before the car travelled overnight to race in Malaysia and Australia. It was a Rouse V6 Mondeo originally and still existed with Conrad Timms in New Zealand in 2022 beneath a corporate paint scheme. *Author's archive*

a path worth following. Commercially, Mountune knew it could build competitive Ford-based units to meet BTCC regulations (8,500rpm limit) for a fraction of the costly Cosworth V6 race units. Even so, the YB-based alternative path had cost Mountune a reported £40,000 by the time I tested it in 1997.

* *Although the 'YB' designation was nominated in period by this engine's creators, its specification corresponds more closely to an official non-turbo Cosworth YAA. Bore and stroke remained as for production YB-series Sierra Cosworths and mass-production Pinto engines at 90.8mm x 71mm.*

Under the skin: a powerful rival to the 'works' V6

Mountune had created many four-cylinder motorsport engines for Fords running without turbocharging. By November 1995, Mountune had a unit rated at 274 reliable horsepower and 280lb ft of torque. More labour-intensive work yielded bonus grunt, based on the cylinder head's enormous flow-rate capacity and the inherent strength of the YB-series base. Fresh engineering thinking – including a water pump slow-running trick, offered a 6bhp bonus. By June 1996 they could see 294bhp coupled to 291lb ft of torque by 7,000rpm. Time for a back-to-back bench test.

Later that summer, an independent assessment by Nicholson McLaren Engines saw the Mountune unit deliver a peak of 289bhp at the mandated 8,500rpm, and 184lb ft of torque at 7,000rpm. It wasn't enough to persuade Ford to change its official British and German team entries over, as the Cosworth V6 would just surpass 300 horses. The V6 produced less torque and horsepower than the four-cylinder unit through all but the last-gasp 7,750–8,500rpm range, but professional race drivers and teams must exploit that last-gasp advantage.

There was also the powerful Ford marketing argument that its top-of-the-range showroom V6, utilising an American 2.5-litre design, from a different family to the Mazda architecture that the Mondeo Super Tourer used, was harder to sell. However, the showroom Mondeo V6 became more profitable per unit than four-cylinder Mondeos. So, the V6 element in a Mondeo racer was a persuasively strong Ford of Britain sales aid.

MANSELL MONDEO MANIA

More than a decade after the event, Andy Rouse recalled the Mondeo race outing that attracted most mass-media comment, and it was nothing to do with the regular British Championship. For Nigel Mansell's October/November 1993 touring-car debut in the seriously competitive TOCA end-of-season 'Shoot Out' event at Donington was worldwide news. During a 2015 interview with *Motor Sport* magazine, Andy explained: 'He didn't really know what he was getting himself into. I think he thought it was just going to be a celebrity race. We tested for a couple of days [in October 1993] and to begin with he couldn't get the hang of the Mondeo – he thought the braking distances should be the same as a Formula 1 car's – and he went off two or three times. We'd brought two cars, and we put him in the second one while we repaired the first one, and went on like that. But gradually he got the hang of it; he was improving all the time. And he was brilliant to deal with.

'It must have been the best publicity Ford ever got out of going motor-racing. The hoo-ha started on the Thursday before the race, stories in every paper, and on the day 70,000 people paid to come and watch. He turned up with all his minders, and a TV crew constantly sticking a camera up his nose, and all these people hanging around him. It was bizarre. He spent hours signing autographs – and of course in the race he crashed the car. Really wrote it off. That was headlines across the world: "Nigel Mansell in Mondeo crash!" After that, car clubs and charities came on asking us to give them damaged panels from the car that they could auction. When we had given away all the panels, we painted a big red 5 on any other bent panels we could find and sent them those!'

All a stark contrast to Jim Clark's modestly reported championship-winning season in the Cortina Lotus...

◀ Nigel Mansell attracted massive media attention – and 70,000 spectators – when he drove an Andy Rouse Mondeo at Donington's TOCA Shoot-Out in November 1993. *Motorsport Images*

THE MONDEO RACE STORY CONTINUES...

I maintained contact with the Mondeo race story, but only as a spectator. Some of it was downright hard to stomach, particularly 1995's World Cup at Paul Ricard, which I attended as a Ford media guest. As for the BTCC season, the Mondeos – now with brave charger Kelvin Burt paired with Radisich – were way off their front-running 1993–94 pace. I watched Germany's Audi and BMW teams dominate proceedings.

Technically, the weight, radiated heat and placement of the V6 motor hit competitive lap times. Some fundamental race-team rethinking was obviously required, especially as the Ford Motorsport managerial moves became so clumsy and excessively expensive in recruiting preparation and race-car constructor outfits separately. Here's how Andy Rouse recalled his last link with Ford being severed: 'For 1996, Ford took their deal away from us and gave it to Dick Bennetts and West Surrey Racing. They'd had a change at the top at Ford. The new motorsport manager was a New Zealander, and Dick Bennetts is a New Zealander. I don't know if that had anything to do with it. Anyway, it made no sense to me, and I was a bit knocked by that after all the work we'd done for Ford. We'd been on the podium 36 times with the Mondeo. West Surrey didn't build the cars themselves, they went to Reynard to get that done, and I think in all the time they ran those cars they only won one race, so it was a bit of a disaster for Ford.'

In fairness, I must write that I have dealt with Dick Bennetts/WSR ever since his Formula 3 days spawned subsequent 1998–99 double F1 World Champion Mika Häkkinen, never mind the WSR record for winning with MG and then BMW during an ever more intensely fought BTCC after the turn of the millennium. WSR, Bennetts and the BMW 3 Series remain a benchmark team to beat in BTCC as this is written. Rouse was right – Ford and its sub-contracted Mondeo (variously built by Reynard, or Schübel in Germany) were disastrous decisions... not just for Ford, as it marked an uncharacteristic low for WSR too.

Come 1999, and some sense prevailed, hiring the Prodrive pedigree. The Bicester professionals ran '80s race-winning 2.0 BMW M3s in BTCC, and the all-conquering 1994 Alfas, but had even more winning form in rallying with Austin Rover, Subaru, BMW and Porsche. The skilled and smooth-talking Martin Whitaker (a former colleague at *Motoring News*) was at the Ford Motorsport helm now. Ford found some serious money to hire two of the best drivers in the BTCC business: Alain Menu and Anthony Reid. Results improved, Menu scoring a fine win and the team looking solid with some podium placings – but the championship-table points tallies were still dismal (tenth and eleventh overall).

The 2000 BTCC season would be the last for the Mondeo – which had been campaigned longer than any rival – and marked a switch from Cosworth construction of the ex-Mazda V6s to that of Prodrive, in-house. The five-door hatchback outlines were structurally stronger than ever, thanks to ever-advancing roll-cage and extension-structure designs. These cars contrasted sharply with the Rouse-era winners, in that a full set of aerodynamic aids was allowed, from front splitter to raised (and therefore more effective) rear wing.

Ford and distinctive yellow Rapid Fit sponsorship cash contested the BTCC Super Touring series – now much weaker, with only Vauxhall and Honda as official factory teams. Ford drew on the talents of three winning drivers: Menu and Reid continued, backed by Sweden's Rickard Rydell, a previous star with TWR-Volvo. Predictably, they all won races, Menu elected champion driver, and Prodrive Ford annexing the manufacturers' title.

▼ After some truly lean racing years following the 1993–94 Rouse era, the Mondeo returned to form in 1999–2000, under the preparation of multi-talented engineers at Prodrive. By 2000 Prodrive replaced Cosworth as engine builders. *Ford*

So what was outside the rebel four cylinders for my magazine test on a damp winter day? A lot of remedial work. Roger Dowson told me with a wry grin: 'We really did the job twice.' He explained: 'First we got the car ready to do the Malaysian Motor Show, then it was shipped back by sea. We had to start again for the real thing... and clean off all the rust!' Conversion costs for a non-factory project dictated the retention of the (original Rouse-specification) Xtrac six-speed sequential gearbox, but the engine mountings had to be fabricated from scratch, as the YB motor was 3in longer than the V6. The unit was also taller, and sat upright in the engine bay, rather than the laid-back and lowered motors that Williams pioneered for the effective Renaults. That slanted layout was copied by the four-cylinder majority, but required an expensive and complete revamp of the front-end layout to optimise the lap-speed advantage. For a factory Ford team wedded to the V6 strategy, just not worth it...

Initially, the plan had been to simply substitute the four-cylinder unit for the original six-cylinder motor, but as delivered, the ex-Rouse rolling chassis needed a revised wiring loom and rebuilt gearbox. A replacement Pectel engine-management system also became necessary, plus totally uprated oil-, water- and brake-cooling systems in order to meet the severe stresses of Malaysia's tropical climate.

Behind the wheel: a friend with aero benefits

My test was traditionally British, from weather to RHD steering. I was trustingly assigned the task of warming up the engine at 2,250rpm beyond 40°C oil and water temperatures, so difficult in tropical-cooling trim on a British February morning. Tricky too, was moving the 18in Dymag wheels and Michelin wet tyres under low-riding wheel arches, as I had to manoeuvre them via unassisted steering lock. A short, snorting 2,500–3,000rpm drive in first and second gear from Dowson's Silverstone industrial unit ensued, to find the South Circuit layout on the 700-acre site. Journalist colleague Mark Hales, out in the TWR Arrows-Yamaha F1 car, dominated the track, and some fine left-right twists shared with the contemporary GP track, with a simple run back from Abbey corner.

Time to wriggle deep into the supportive Sparco seat and fasten the six-point Simpson harness. Although the gearbox had been roundly cursed in the workshop, I experienced the amiable side of its non-synchromesh speed from first-gear static engagement to sixth. Snapping sequentially through the first three ratios at 6,000rpm was accompanied by characteristic barking from the iron-block Cosworth-Ford, and a very solid, exploitable power curve. A surprise was the lack of front-drive steering weave and kickback; it just got on with the acceleration task, no bother. Also rock solid was the steering feedback in quicker corners, along with messages from the suspension, all the settings simply estimated for the heavier iron-block engine.

I liked this Mondeo variant, because it looked after me on a winter day, feeling stapled to the quicker bits of track. In hindsight I realised this was due to the addition of front and rear wings and splitter, versus the 1993 Rouse car with no obvious aero add-ons. Although aero devices raise cornering speeds notably, it is also the feeling of stability that enhanced aero gives a saloon/hatchback shape that provides a visiting conductor with a confidence boost.

This Mondeo's flawless power delivery, via a unique engine-management system and its calibration, was a welcome surprise. It was obvious from my previous experience that Mountune's ambition to fill in the power troughs of the official Ford-Cosworth race V6 had been successful. The Mountune four-cylinder was particularly strong at 5,000–7,000rpm, and I reckoned this would have been the unit of choice for anyone paying their own bills, and worthy in its own right on tighter tracks.

'I LIKED THIS MONDEO VARIANT, BECAUSE IT LOOKED AFTER ME ON A WINTER DAY, FEELING STAPLED TO THE QUICKER BITS OF TRACK. IN HINDSIGHT I REALISED THIS WAS DUE TO THE ADDITION OF FRONT AND REAR WINGS AND SPLITTER.'

CHAPTER 16

FOCUS ST AND RS

The last mass-market fossil-fuel performance Fords?

The 1998 Focus replacement for the European Ford Escort marked a new era in performance and mass-production Fords. The Focus benefitted from characteristics such as enhanced fuel economy, safety and electronic equipment, admirable front-wheel-drive handling, and benchmark dynamic qualities. In particular, the impressive ride and handling, previously more often found in Ford's pioneering RS homologation/motorsport models, filtered down to the full Focus model range.

The future of the Focus name looked terminal as this book was created. Ford had officially announced the 2025 closure of European Focus production at Saarlouis, and that decision followed the cessation of American Focus manufacture in 2018. China continued manufacturing enormous numbers of Focus-branded regional variants, but the push for electrification could change all that beyond 2025.

THE FOCUS ST AND RS STORY

An instant success

From a Ford perspective, the first-edition Focus was a global hit, indeed industry sources credited the model as the best-selling car worldwide in 1999–2004, and again in 2013–14. The numbers were boosted by Focus sales in America and China, an achievement unmatched by previous smaller Euro Fords.

For Ford in Europe, the Focus halted drooping sales caused by the Escort's faded charms. It did a particularly good job in some of Europe's biggest markets – Germany and Britain – and even won German acclaim for durability: a refreshing change! Of less showroom value, but massively good for corporate egos, the Focus also claimed the UK's Car of the Year 1999 – especially gratifying as it saw off a new Astra generation from General Motors. The Focus also grabbed the award for North American Car of the Year 2000.

As a driver, the Focus was significant for

▲ Teaser action from September 2015 Goodwood Festival of Speed, promoting the third and final 4x4 turbo edition of the Focus RS to potential customers. *Ford*

the moves made under the freshly styled skin: certainly the blend of fine handling and compliant ride became a new benchmark for Ford, and the Focus was then often employed as the yardstick in media comparison tests. The key to the Focus's excellent suspension performance could be traced back to the Mondeo estate's requirement for a space-saving multi-link layout that performed as well as the expensive and space-hungry double-wishbone types found in high-performance applications. The successful Mondeo estate solution was adapted, simplified and lightened for the higher production rates and lower cost of the Focus, and was branded 'Control Blade'. It packed multiple links and thin trailing arms (hence 'blade') to deliver strong body control and excellent steering feedback under duress, coupled to quality bump absorption, way above expectations. The front-end layout remained on MacPherson-strut principles, but was more refined than its predecessors.

The launch range of engines certainly couldn't match the handling breakthroughs, and because of governmental encouragement, turbo-diesel motors sold particularly well in the UK and Continental European markets. Owing to the stronger torque of the European 1.8-litre turbo-diesels, official performance statistics for the range-topping 130bhp petrol engines (0–62mph in 9.2sec and 125mph) weren't much superior to the TDCi diesels (0–62mph in 10.7sec and 122mph). Factor in that the economy of the diesels was totally superior to the petrol engines, and that diesel fuel was significantly cheaper in some key markets such as France, Italy and Germany, and it becomes obvious why diesel engines had come out of the shadows. In '90s Britain, diesels would score strong sales until awareness of their troublesome emissions – particularly Nitrogen Oxide (NO_x) – strengthened in the wake of the 'dieselgate' scandal uncovered in 2014.

Second-generation Focus

The Paris motor show in September 2004 was selected for the first public display of the second-generation Focus. The updated Focus wasn't initially offered in North America, where a reworked first edition continued, but large production runs for the MkII Focus were achieved because of China's massive factory outputs in the period 2005–14.

As ever, the second-edition Focus grew dimensionally and packed in more electronic features, plus alternative gearboxes: six-speed Durashift, automatic four-speed or basic five-speed manual, joined by a semi-automatic Powershift system in 2008, when the European models were mildly but extensively refreshed. The facelifted cars wore a majority of new panels to provide a cohesive range look alongside the Mondeo, with some cues from S-Max and Galaxy MPVs.

The following year, customer indifference to four-door saloons in a world of hatchbacks and increasing crossovers led to the three-box Focus being dropped across Western Europe. However, it continued to be made in big numbers in China, Eastern Europe (including Russia) and South America.

The acclaimed and proven suspension layout was carried over to the second-generation car, tuned to suit its heavier and expanded form. There was a good reason for those gains in size (over 6.5in longer) and weight, as the Focus now rested on a shared platform with models from period Ford stockholding interests at Volvo and Mazda, plus a worthwhile in-house spin-off in the form of the Focus C-Max MPV.

Aside from the ST and RS variants that we explore separately, the mass-market Focus line offered a bewildering choice of motive power, from the 1.4-litre Zetec petrol engine in base models at 79bhp, to the 2-litre petrol Duratec at 145bhp. Again, the TDCi-badged diesels could offer convincing acceleration and top speeds. Even the Focus ST's five-cylinder petrol turbo engine (ex-Volvo) failed to match the 2-litre turbo-diesel's 251lb ft torque, although the 2.5-litre petrol ST reversed the contest, fielding over 220bhp versus the TDCi's 136bhp.

The interior received a significant upgrade in quality, as the original Focus had taken brickbats in the media and showroom when compared with the benchmark fourth-generation Volkswagen Golf.

Third-generation Focus

Yet another twist on the original front-drive, hatchback and saloon Focus themes was forced into the 2010–11 'One Ford' global plan, successfully imposed to save the company under the chairmanship of ex-Boeing high-flier Alan Mulally. These were desperate times for Detroit's mass-marketers – GM and Chrysler lining up for US Federal Government cash, while Ford preserved its independence from obvious state aid.

North America and Europe debuted their third Focus edition simultaneously in spring 2011, but there was a public preview of saloon and five-door hatchback bodies at Chicago's 2010 North American International Auto Show, with a clean new look.

There was a genuine move towards global assembly, with seven countries and their factories involved. Primary manufacturing sources were based in the USA, China, Germany, Russia, Thailand and Argentina.

The third iteration of the Focus offered the usual progression in equipment and trim uprates. Ford's first efforts at the double-clutch Porsche-VW-Audi style of semi-automatic gearbox were afflicted with the shudders, but there were alternatives from five- and six-speed manual to full automatics.

The petrol engine line-up, outside the performance ST/RS variants discussed later, expanded to embrace a 1-litre Ecoboost three-cylinder of 99bhp, with 1.5-litre Ecoboost branding on a 148bhp- or 180bhp-rated unit – one that reported 130–138mph maximum speeds, coupled to 0–62mph claims of 8.8 or 7.9sec with notably clean breath. Diesels remained as important as ever in non-American markets and displaced 1.6 or 2-litres.

The foundations of the third-generation Focus were revised, and shown at the 2014 Geneva motor show. Externally, it had a sharpened outline with less depth to the headlamps and a sportier overall look. A component of this was a corporate grille (shared with the Fiesta), featuring transverse bars that the media dubbed as an Aston Martin echo: personally I liked the clean look, and since Ford was one of many former Aston owners/investors, I thought they had some right on their side. All mass-production Focus versions gained chassis tweaks to move back towards the acclaimed handling of the original.

Fourth-generation Focus

From 2018, some markets received a revamped Focus that was still in production as this was written. The revised car offered lower kerb weights, with uprated body strength and more capacious accommodation, thanks to a wheelbase extended by fractionally over 2in. North America no longer manufactured the mainstream Focus types after 4 May 2018. How secure the Focus future was in the wake of an electrification revolution, buyer preference for SUVs/Crossovers, and Ford USA's desire to eliminate conventional cars – bar the Mustang – from its line-up, had yet to unfold.

However, the Focus story continued to be told in important markets, such as China, where, from 2012 onward, Chongqing became the largest Focus assembly facility, offering three different plants in the same city. At the time of writing, Europe was served by the Saarlouis plant, and Australasia was patchily offered ever-slimmer Focus model choices. Typically, Australia went its own way, by 2022 taking just the Focus ST-Line, which emphasised sporty appearance, not more powerful engines.

FOCUS ST AND RS DOWN THE GENERATIONS

Performance versions of the obviously agile Focus took four years to arrive. When they did, in 2002, we got a dynamic duo: the ST170 and the first RS model since the Escort RS Cosworth. Both retained front-wheel-drive. Even so, the original showroom Focus RS has become a valued modern classic. Some of the earlier ST-branded Focus models – especially the five-cylinder models – also won respect for a mighty soundtrack and impressive performance, plus affordability coupled to depreciation-beating resale prices.

Sadly, production of the last of three distinct RS Focus generations ended in 2018. Ford confirmed in a corporate statement of 20 April 2020 that plans for a 'hybrid-e' fourth generation had been abandoned: 'As a result of pan-European emissions standards, increased CO_2 taxation and the high cost of developing an RS with some form of electrification for a relatively low volume of vehicles, we are not planning another RS version of the Focus.'

The first two showroom Focus RS models arrived in front-wheel drive, the final third-edition examples finally offering 4x4 traction. The expanding Sports Technology (ST) line was now better established, and could be traced back to the ST24 Mondeos of February 1997 in the UK. However, 2002 again marked a significant rebirth, this time for ST-badged variants of the Focus (ST170) and the respected Mondeo ST220 covered in the preceding chapter.

An RS badge returned to Ford showrooms in October 2002 after a six-year slumber. A limited production period running into 2003 at Saarlouis delivered 4,501 examples of the Focus RS in Europe, nearly half to starved UK fans. There were limitations – this was a front-drive hooligan, not a 4x4 echo of contemporary World Rally Championship technology, as offered by Subaru or Mitsubishi. The Focus RS also had a five-speed gearbox (a redeveloped Ford MTX unit) rather than the six-speed found in the ST170 or many rival hot hatchbacks.

Yet the Focus RS had its own strong character, inside and out, becoming a valued collectible amongst the Ford fraternity. This customer loyalty, despite the usual challenges with the ready availability of factory replacement/maintenance parts, never mind the significant cost of most unique RS Ford ingredients.

A second Focus ST – and a feisty performer

▼ It took six years to reintroduce the RS badge in the transition from Escort to Focus. Just over 4,500 first-generation Focus RS models were built. The 212bhp front-drive Focus, with strong torque from its turbo 2 litres, cost £19,995 in the UK in 2002. *Ford*

Focus Cossie

We knew that Ford wouldn't be able to resist giving the new Focus the Cosworth treatment for long and the good news is that they haven't. The bad news is that currently the only Cosworth Focus is a concept car which the marketing types won't even confirm is going into production.

Still it does give us **CCC** types the chance to salivate over a turbocharged, intercooled Zetec E engine potentially producing over 200bhp. The car unveiled at the Detroit Motorshow runs the standard gearbox but Ford have added a viscous limited slip diff between the front wheels. For the street wise element Ford provide 235/40 rubber on 18in multi-spoke rims. Mind you, they needed big wheels to accommodate the 355mm front discs and four pot calipers. This impressive list of fantasy specifications is topped off with adjustable front and rear anti-roll bars and shock absorbers.

Chuck in all wheel drive and another 100bhp and you really will be able to emulate Colin McRae down your local high street.

◀ When I first glanced at this prediction, I thought it was another case of media over-enthusiasm to get a scoop. Then my files revealed Ford had sent out details as an official press release, including the Cosworth link. It didn't happen, at least not in the showroom. *Author's archive*

– appeared at the Geneva motor show of 2005. It packed a charismatic five-cylinder turbo engine, reworked from a Volvo base. It materialised as the first Focus rated as capable of 150mph, thanks to 222bhp at 6,100rpm and a strong 236lb ft pull from 1,600 to 4,000rpm. It sorely needed the traction of a limited-slip differential, actually deleted from the standard spec. There were three ST-coded levels of equipment for three- or five-door bodies, all running 18in wheel diameters. A run-out edition of 500 for the UK featured Panther Black paint, silver detailing and red-leather cabin trim, and carried top ST-3 levels of equipment. Production ended 2008 on a creditable 22,361 manufactured for what had been a niche market.

A second-edition Focus RS materialised for spring 2009 and was gone during September 2010 after some 11,500 had been created at Saarlouis in front-drive only. Based on the ST's turbo five-cylinder it offered a sniff over 300bhp at 6,500rpm and 325lb ft of torque between 2,300–4,500rpm. Ford claimed the second Focus RS breached the 160mph barrier, with 0–60mph rated a tenth below 6sec. Chassis equipment included a unique

▲ ▶ The run-out RS500 model of the second-generation Focus RS appeared in 2010, in stealthy Panther Black, and certainly looked moody. It was a pretty wild front-drive ride, uprated to 345bhp by Mountune. *Author/SMMT*

'RevoKnuckle' front-suspension layout, Quaife ATB limited-slip differential and 19in alloys, accompanied by the traditional Recaro seating. There were four paint choices and a tacky layout for the cabin instrumentation. A 500-off RS500 edition was created, with Mountune 345bhp engine, and just 101 of these Panther Black run-out models were sold in the UK.

A third-edition Focus ST, often referred to as 'ST250', packed 247bhp from a more-conventional four cylinders, and appeared for 2012 in three specification levels – ST1, 2 and 3. Restricted to five-doors, or an effective estate, it carried a 2-litre, four-cylinder engine, plus a BorgWarner division's KO3 turbocharger. It was just as swift as its predecessor, with 0–60mph in just 6.2sec, and a 154mph maximum. Just as noteworthy was that it improved significantly on the thirsty fuel-consumption of the five-cylinder.

The third and final petrol Focus RS was an ultimate 2016–18 rendition, debuted at the Frankfurt motor show in 2015. It switched to 4x4 traction and was propelled by 345bhp, extracted from a 2.3-litre stretch of the previous four-cylinder engine, and twin-scroll turbocharging under the Ecoboost brand.

Torque was accessible and muscular – 324lb ft between 2,000–4,500rpm, or 347lb ft for 15sec of over-boost. This shared ST/RS basic engine, modified from its earlier Mustang application, appears to have been the most fragile ST/RS unit offered: there were reports of explosive failures in both modified ST and standard RS. Any power upgrades need to be carefully balanced by durability measures and proven technical advice.

Ford claimed 0–62mph would flit by in 4.5sec, with top speed in the middle 160mph band. It was available in five-door hatchback configuration only, and had all the goodies, including big Brembo brakes (350mm/13.8in fronts), totally reworked and stiffer suspension, with 19in alloys, plus Recaros.

A run-out Focus RS Edition was manufactured in a 300-off limited run, finishing in March 2018 and available in North America as well as Europe. These cars were usefully upgraded with a Quaife mechanical (as opposed to electronic) limited-slip differential. Just two body colours were offered: Nitrous Blue and Racing Red – plus roof-spoiler and mirror-shell embellishments in gloss black, and bonus RS identity evident on the forged 19in alloy-wheel centre caps.

A last UK-only hurrah in April 2018 saw 50 'Heritage'-tagged Focus RS models, adorned with every extra and implanted with Mountune's 375bhp evolution of the 2.3-litre Ecoboost. There were numerous special editions, including a breathtaking £83,000 three-off finale, RS 50.

DRIVEN

Production Focus 1.6- and 2.0-litre first-generation (two examples) Media launch, France, 14 September 1998; 1.6 Zetec and 2.0 Duratec engines in LX and Ghia trim; noted 'massive' improvement over final Escorts. **Focus ST170** Media launch, France, 22 March 2002. Driving debut of Focus ST170 and Mondeo ST220; 0–60mph in 8sec, 130mph; fine handling, six-speed Getrag gearbox, good accommodation; RS version needed, as Honda Civic R/Mini Cooper S brought sparkling competition. **Focus TDCi Sport first-generation** Test week for *Haslemere Messenger* local paper, November 2003, ET03 ZWE; 115 Duratorq 1.8-litre turbo-diesel offered most power; 0–62mph in 10.8sec, 120mph, claimed 51.4mpg; economical fun. **Focus three-door Zetec and TDCi second-generation (two examples)** Scottish media drives, 9-10 November 2004. **GGR Focus RS first-generation** Road test at Bruntingthorpe airfield and on public roads, 19 May 2004, R5 GGR; uprated RS with 285bhp; impressed on track and loved it. **Focus Coupé/Cabriolet** UK media drive, spring 2007; CC-3 trim, Duratec 2.0-litre, 145PS; fashionable 2-in-1 cars with metal power-tops, heavy and no luggage space, liked civil versatility. **Focus RS second-generation** Regional media drives in Wiltshire, October 2009, AF59 JXC; 2.5-litre, 5-cylinder, 305PS, 0–62mph in 5.9sec, 163mph, 21mpg urban. **Focus 1.6-litre second-generation** MPG Marathon tuition, Berkshire, 28 July 2009; BP Ultimate-sponsored; five-door Econonetic 1.6; urban 47.76mpg versus 55.69mpg of instructor Anthony Sale. **Focus RS500 second-generation** SMMT test day, track only, May 2010, EJ10 OBC; Mountune uprated, interior plaque PR 01; spectacular front-drive mover. **Focus RS second-generation** MPG Marathon, and Economy Run, 19 September–4 October 2009; Ford-owned, registered CAR 1; returned 38.5mpg, third in class, improvement over urban 21mpg in official test; car also used for SMMT Millbrook track day in May 2009. **Focus ST250 estate (two examples)** SMMT test day, Millbrook test track and surrounding public roads, May 2012; also driven only on public roads near Reading, Berks.

➤ The later ST Focus derivatives were a fine blend of practicality and performance, particularly this 2015 estate body. *Author*

The fourth generation of Focus ST five-door hatchbacks and estates arrived in 2019. It was still on European sale in 2023, although the end was in sight, as the Michigan Assembly Plant had rolled out the last Focus for the North American market on 4 May 2018.

However, in Europe, particularly in Britain, a keen following made economic sense for a five-door-only hatchback. The informally labelled ST280 (for 280PS, actually 276bhp), introduced in 2019, inherited the previous RS Ecoboost four-cylinder stretch to 2,261cc, but it was constrained to yield less than the RS engine's 300+ horses. Standard features included a six-speed manual gearbox, incorporating an electronic limited-slip traction aid. Ford reported 0–60mph in 5.7sec and a 155mph limited maximum. Trim levels drifted up, including 19in double-spoke alloys with later Style packs, plus part-leather Recaros.

➤ Although some searing colour combinations were produced, the seats were excellent. *Author*

The Focus ST Edition of August 2021 looked like run-out fare, with unaltered power, and a body in metallic-blue hues. Yet it also offered a seven-speed semi-automatic gearbox. Hidden away inboard of Michelin Pilot Sport 4S tyres and either 18x8in or 19x8in alloys, were revised suspension components, reflecting another Nürburgring handling session. KW Automotive delivered a twin-tube damper-and-coil-spring set-up that was set 10mm (0.39in) lower than the previous ST, and was adjustable over a 20mm (0.79in) range, plus 12 compression and 16 rebound damper adjustments.

More hotter-than-hot hatches from Ford to come? A remote possibility when this was written, as Ford had confirmed in June 2022 that the fourth-generation Focus line would end Saarlouis production in 2025. As already noted, the US had stopped Focus manufacture in 2018. Judging by previous form, China could continue its Focus variant manufacture, but that would be an unlikely source of petrol-based high-performers in the ST/RS moulds, though hybrid or all-electric versions might emerge. Stranger things have happened…

FOCUS ROAD-CAR DEBUT DRIVES IN FRANCE

The European media launch for the Focus took place in France on 14 and 15 September 1998. Journalists were flown from Stansted to Deauville, and I had the opportunity to drive two examples of the Focus, in 1.6 LX and 2.0 Ghia trim levels, within a 24-hour trip. Both cars had petrol engines: a 1.6 Zetec 16v of 100bhp, and a Duratec 2.0-litre producing 128bhp. Turbo-diesels weren't scheduled for the UK until 1999. The 1.6 Zetec packed 16-valve technology, along with a formidable 11:1 compression ratio, and was expected to deliver 115mph, with 0–62mph occupying 10.9sec, and an urban economy claim of 30.1mpg. The 2-litre was naturally thirstier (an unimpressive 23.9 urban mpg), but offered 0–62mph in 9.2sec, coupled to a 125mph max.

The standard equipment lists recalled some of Ford's past meanness. Only the 2-litre Ghias had ESP stability control and ABS as standard, but all had power steering and central locking, and front-seat occupants benefitted from two airbags and electrical assistance for the front windows.

The weather for the test was poor, but that highlighted what an excellent experience laid in store for Focus owners. Without doubt, what impressed most was the quality of the beautifully balanced chassis attributes, from accurate steering on Pirelli P600 tyres to a compliant ride. Other assets included plenty of space for baggage and occupants, and an airy and comfortable interior with straightforward and quickly comprehended controls.

Driving characteristics included an easily adjustable mid-corner attitude on a small test track (EIA at Pont L'Évêque) that swept the newcomer to the forefront of mass-production motoring pleasure. These impressive abilities would be widely acclaimed, and were partially owed to class-leading low weight (1,070kg for a 1.6 Zetec-SE) and enhanced rigidity of the basic bodies, although the drag-coefficient was hardly impressive at 0.32Cd.

There were some early production glitches, such as pronounced resonance from the 1.6-litre engine at around 4,000rpm, but it was a good start, coupled to an equally good platform, available in an estate five-door derivative. Yet, in the bitterly contested global category for three- and five-door hatchbacks, the Focus had to deliver, which it did, but Toyota, Volkswagen Group and General Motors would continue to contest every sale.

ST170 LEADS THE WARM-HATCH WAY

Another trip to France, but this time the warmer southern region, on 22 March 2002, for the debut of the Focus ST170, alongside the Mondeo ST220. Again, I shared the experience with excellent peddler and my last Editor at *CCC*, Steve Bennett. From memory, and media comment 20 years onwards, the ST170 wasn't quite the hot-hatch you'd hope with 171bhp on call. It was really quite a subdued performance hatchback, partly because the development team had civilised it so that there weren't the immediate visual or performance thrills the UK faster-Ford sector often demanded.

It was also four years since the debut of the Focus, and there were plenty of quick rivals from GM, and Volkswagen-Audi-Seat group, plus Honda's legendary R-type Civic. Also, BMW had acquired the MINI brand and had immediately taken commercial advantage of the Mini Cooper and S legends to ensure a warm nostalgia glow, backed by a decent product. Ford didn't nominate the admittedly smaller MINI as a rival, but I have – and I think the comparison stands up when shopping nowadays for a modern classic with a bit of zip.

My 2002 driving notes commented first on the ST170's extensively reworked Zetec 2-litres. The engine felt less impressive than expected: 'Lethargic as if weighted and padded heavily, especially in the 3,000–3,500rpm band. Between 4,000–7,250rpm felt best, but no ball of fire, especially remembering the Honda Civic R's rev-happy sophistication.'

▲ The initial performance derivative of the Focus, the ST170, was a neat but unobtrusive performer. It was so well engineered and balanced, with a full quota of weighty safety aids, that it didn't feel as exhilarating as some demanded. Underrated. *Ford*

Other aspects were welcome, including the Getrag six-speed box (shared by the MINI Cooper S) and the handling/road-holding balance was rated as: 'Super ride, knobblier than less powerful Focus only at lower speeds on poor surfaces. On the limit, without traction control, you can overcome the natural understeer with the brakes.' We ran without Ford's traction-control system because it gave a smoother flow of grip on dry and tighter (hairpin) French surfaces, but used it on less-challenging roads. The bigger brakes, versus a kerb weight of 2,829lb (1,283kg) – and French twisty tarmac sections – ensured that we reached the endurance limits of the ST's retardation capabilities.

The interior earned some brickbats: 'Mundane Ford in the basic-price spec, without optional Recaros. The substitute cheaper seats have passable bolstering for harder cornering and the ST-embossed dials look tacky and too small to deliver best clarity.' Such cluttered, multi-colour and smallish dials, subsidiary and main, remain one of my main dislikes in later ST and RS variants. Yet old-school Escorts and Capris were hardly exemplary on this score either, so just old-age grumpiness I guess…

'THE HANDLING/ROADHOLDING BALANCE WAS RATED AS: "SUPER RIDE, KNOBBLIER THAN LESS POWERFUL FOCUS ONLY AT LOWER SPEEDS ON POOR SURFACES. ON THE LIMIT, WITHOUT TRACTION CONTROL, YOU CAN OVERCOME THE NATURAL UNDERSTEER WITH THE BRAKES."'

Looking back in hindsight, and with the benefit of contemporary notes, I thought I maybe had my BMW Group rose-tinted glasses on. My handwritten comparison sheet survives, one that records plus and minus scores against the contemporary MINI Cooper S. The MINI range-leader was supercharged to 163bhp and 155lb ft, from 1.6-litres, and rather than displaying the higher-rpm power peaks of the normally aspirated 2-litre Ford ST170, the MINI offered 10lb ft more torque at lower revs (4,000rpm) than the Ford's peaky 5,500 maximum. Also taken into account was a 7,000rpm delivery summit for the Ford's 8bhp power bonus, versus 6,000 peak-power revs for the MINI.

The lighter (2,513lb/1,140kg) and supercharged MINI naturally felt more responsive to mid-range and initial throttle inputs. Official performance figures for both put the ST170 at a maximum 134mph and 0–60mph in 7.9sec, versus 135mph and 7.2sec for the MINI.

Another MINI benefit was pricing, at £14,545 versus the ST170's £15,995 without options. Today, early BMW MINIs are beginning to grow in value, and the ST Fords have a strong following, especially the later five-cylinder reprobates.

There were also the handling and driving-pleasure factors. The Mini legend was kept alive by BMW deliberately biasing all the range toward 'darty', instant responses, with many media references to 'go-kart' handling. That may be a good thing for a kart, but there are oncoming-traffic limits on public roads, and a minimal level of comfort is required, even of a hot-hatchback. Here, I think the Ford had the more honest balance of safe and enjoyable handling, coupled to far higher levels of comfort and accommodation.

My conclusion on the BMW MINI versus ST170 comparison was harsh: 'I'd take the MINI Cooper S because of its sparkling performance, price and equipment. Yes, the MINI is more cramped, but it is also more fun.'

GGR-TUNED FIRST-GENERATION FOCUS RS

A rare opportunity came my way on 19 May 2004 to take to the road and test track with the first-edition Focus RS, which proved very rewarding. Graham Goode Racing (GGR) had taken the raw 212bhp RS package and thoroughly reworked it into a 285bhp enjoyable and manageable package – one conscientiously balanced by bigger brakes and effective lowered ride-heights that satisfied me on and off the Bruntingthorpe test-track.

Recalling that 285bhp was about the same power provided by the production-racing Janspeed Sierra RS Cosworth I was privileged to share in 1988, I was surprised what a docile and fine-handling machine this Focus had become. It also had the distinct advantage over those '80s Sierra Cosworth legends of not attracting so much unwelcome public or police attention.

Under the skin: a comprehensive upgrade

What went into this desirable RS upgrade? In 2021, Alastair Mayne, still a loyal and skilled technical manager at GGR, gave me chapter and verse: 'Our GGR 285 conversion featured a stainless-steel high-flow exhaust downpipe with far less restrictive sports catalytic converter that remained legal and MoT friendly. That was complemented by a higher-flow-rate cold-air induction system in spun aluminium, replacing the original plastic air-filter box. A stainless-steel exhaust manifold, with polished stainless-steel heat-shields, was used instead of the original standard cast-iron manifolding.

'The cylinder head was gas-flowed and ported, with the valve seats cut at three modified angles to Group A type standards, although there was no racing plan – or homologation – for the Focus RS then. Additionally, the engine-management

▲◀ This pair of pictures records the typically careful and thorough upgrades from Graham Goode Racing (GGR) for the first RS Focus. *Author*

system contained a unique programme of fuel, ignition and boost-control parameters to suit these mechanical modifications, downloaded via a Superchips Bluefin handset.'

Alastair recalled that GGR tested the conversion's effectiveness – as I did that day – and it truly transformed the performance. Aside from offering better throttle response, improved low-down torque and an increase in higher-rpm-range horsepower, all accompanied by improved drivability, it also delivered the following statistics (standard RS figures in brackets) 0–60mph in 5.5sec (6.4sec), 0–100mph in 13.28sec (15.5sec), and a top speed of 155mph (144mph).

Other GGR uprates on this RS included enhanced braking, with 343mm x 32mm (13.5in x 1.3in) AP Racing ventilated front brake discs, coupled to AP Racing six-piston callipers. The rear disc brakes were standard Ford, carrying uprated pads. Equally important were H&R uprated and lowered road springs, which lowered the ride height by 25mm (1in), plus 19in OZ Superleggera 12-spoke wheels, carrying Goodyear Eagle tyres.

Behind the wheel: a deceptively quick performer

In action, the road manners remained intact, with just a slightly stiffer ride on low-speed lanes and completely civil city throttle pick-up. It felt so completely original and well balanced – the chassis changes had damped down most of the front-drive raggedness the media chuntered about extensively for the showroom RS. On the perimeter track, with a combination quick, quick, slow through turnaround areas, it just went where pointed.

The Ford-Quaife ATB limited-slip differential contributed much to the snatch-free power transmission, and it wasn't until we had covered a couple of laps that I noticed that a decent BMW that had started off as company for our exploratory laps had now faded into the distance. Amazing what can be achieved with front-wheel-drive…

The interior covered the Ford omission of water- and oil-temperature dials with a SPA combined analogue monitor, exclusive to GGR, and detailed right down to the blue facing to match the blue themes of the showroom RS.

Externally, the demonstrator often wore a Mach 5 twin-scoop bonnet with aluminium heat-resistant cladding beneath, but as Alastair commented: 'I also liked that the car was quite subtle [especially without the Mach 5 bonnet!]. The car could be parked in a Little Chef car park often without getting a second glance from other visitors. This wasn't something you could do with an Escort Cosworth fitted with the huge whale-tail rear spoiler. Or the later bright orange Focus ST, or high-vis green Focus RS MkII…'

FUELLISH ADVENTURES FOR SECOND-GENERATION FOCUS RS

When it came to the now-revered five-cylinder, 300bhp second edition of the still front-drive Focus RS, I did the anticipated UK 2009 media driving sessions, but really got to know this RS type properly when a scheme was gradually cooked up in those preliminary sessions to boldly go where no right-minded RS would ever go: an officially monitored two-day economy run.

With the help of www.wheelsalive.com proprietor and supreme classic-car guru Kim Henson, I took a Ford press-fleet RS out on a warm day and established some basics on main and country roads. Pushing the air-con button firmly off, windows shut tight, boost-gauge never provoked unless vital – and instant mpg readout selected – we returned over 35mpg. Since the official Ford urban figure was 21mpg, we thought the RS could surprise the results sheets – not on an overall figure, because tiny economy cars could nuzzle 100mpg back then, but as an improvement percentage over official test figures. So, a formal proposal was put to Ford UK via press-garage manager Paul Wilson (my 1987 Snetterton 24 Hours Sierra Cosworth trackside saviour) and accepted, albeit with some strict instructions about preserving the RS Focus showroom specification.

As it transpired, over an official 360-mile

route out from a Coventry hotel HQ, the statistic we had to improve on was much harder: the EU figure translated to 30.1mpg. Also, more than 21 manufacturers and importers contested the 30 September–1 October 2009 events, with varying degrees of professionalism. A particularly hard-fought sector had become the light-van overall consumption, where the leading drivers and navigators were proven economy-run performers, but our overall fight for percentage improvements within the car category was taken pretty seriously, particularly by Mazda with a 230bhp engine.

I had the eye-catching Focus RS, registered CAR 1, well before the event, as the Ford-owned vehicle stayed with me from 16 September to 4 October. I had to promise not to get up to any old saloon-car racing tricks with CAR 1, like pulling out all the sound-deadening, or any other unobtrusive weight or friction-saving manoeuvres. I did fill it with the highest-octane pump fuel I could find within 10 miles of the hotel and scrutineering base, and pumped up the tyre pressures madly.

The scrutineering before and after the event, supervised and implemented with AA service-van back-up, was as thorough as most race meetings. Fairly obviously, the fuel-tank cap was sealed after checking that it had enough for the route, and any sign of tampering/removal meant disqualification.

Alyson Marlow became my professional navigator. A resourceful journalist and special-stage co-driver, Alyson ensured some sensible basic discipline. No pre-event alcohol, and an early night prior to the start. I'm not a naturally disciplined or team-player person, but I had to obey my co-driver in my RS2000 rally adventures. As it happened Alyson and I also covered a Peugeot Paris–Geneva economy competition the following year. I watched as she correctly predicted where participants would miss an early and vital right turn after the start: sorry to say all

▲◀ The five-cylinder turbo ST Focus was a great character, including traces of an Audi quattro soundtrack. It was the first Focus claimed to reach 150mph, but also an enthusiastic drinker! *Author/Ford*

the British entries missed the turn and none of them figured in the final top 10.

I'd say we had a pretty unremarkable event, but that does mask some essential motorway slip-streaming tactics behind 60–65mph lorries, preferably with the large and squared-off rear bodywork. Downhill descents also needed careful speed to exploit RS grip, ensuring a clean, and no-boost, run up the following slopes. The only fun I had was trying to out-psych improvement-class leader Iain Robertson in that performance Mazda 3 MPS, lurking in his wake on dual carriageways, or occasionally arriving alongside at roundabouts. All to no avail, as he was too professional to lose his focus to a Focus!

Now, I quote excerpts from *Driving Spirit,* one of many media post-event reports.

'Neither the Ford Focus RS nor the Mazda 3 MPS are the most obvious choices for a competition that's all about fuel economy, and yet they were two of the competitors in the 2009 MPG Marathon. The idea of the MPG Marathon is to improve on the official fuel-consumption figures, as provided by the EU, demonstrating that even the less fuel-efficient vehicles can return an improved fuel performance when driven smartly. The cars with the highest overall MPG and biggest percentage improvement are crowned the winners.

▲ My first experience of the 300bhp second-generation Focus RS was at an October 2009 Ford west-country media day that included a fun run out to Castle Combe race track. *Author*

➤ I did get some time off from the economy leash to appreciate the second-generation RS Focus, both at this Millbrook test track in 2010, plus an epic 2009 night run home from the event. Mighty! *SMMT*

'This year, the car that achieved the best improvement was a Mazda 3 MPS, powered by a 2.3-litre, four-cylinder engine with 230bhp on tap. The drivers, Iain Robertson and Rob Marshall, covered the route with just 7.61 gallons of petrol, an average consumption of 43.34 miles per gallon, and an astonishing improvement of 48.95 per cent on the Mazda's official combined figure of 29.1mpg. Another top performer was a Ford Focus RS, a car with a not-very-green 2.5-litre turbocharged petrol engine under the bonnet. The RS's official fuel consumption is 30.1mpg, and yet it managed to cover the route at an average of 38.53mpg. Not bad for a 300bhp hot-hatch that can reach 163mph. You see, now you can afford to run a Focus RS!

'The best overall consumption of the event was 96.81mpg, recorded by a diesel Smart ForTwo Coupe.'

Incidentally, former Peugeot UK motorsport manager Mick Linford drove that Smart, who was similarly effective in 2010. Overall, it had been an educational experience, but not one I was keen to repeat. Ford was well rewarded for its support in terms of editorial preview and reportage. In fact, some years later, winning rally crew Andrews Dawson and Marriott put our performance in the shade, winning the improvement division with a 5-litre Mustang GT!

▲ A selection of memories from the MPG Marathon, shared with experienced rally co-driver Alyson Marlow (right). It wasn't a popular event, with some adverse comment from hardcore media, but it demonstrated that the turbo Ford could deliver 38.4mpg on public roads, so long as you didn't let that turbo dial twitch! *Innes Marlow/Author*

◀ A couple of months before the 2009 MPG Marathon, coincidentally I had taken the chance of driving a BP Ultimate-backed urban-based economy test in a five-door Econetic 1.6 Focus. Driving with instructor Anthony Sale (pictured right chatting with me outside Reading FC stadium), I returned nearly 48mpg versus Sale's 55.69mpg. *Author*

FOCUS WRC WRITES A TALE OF WINS AND LOSSES IN THE WORLD RALLY CHAMPIONSHIP

On 25 November 1998, it was time to trek back round to the eastern loop of the M25 to attend another Essex media debut. Our destination this time was the Dunton Engineering HQ. Here, M-Sport's redevelopment of the Focus for the World Rally Championship (WRC), made its debut, having appeared as a static exhibit at the Paris motor show. The Dunton occasion marked the end of a 30-year factory Ford Escort championship rallying era, one I had celebrated on 19 November with a grand reunion of key players at Sudeley Castle, within striking distance of Britain's 1998 WRC qualifier, based at Cheltenham racecourse that year.

Just two days after the Castle farewell banquet, we began to follow the last Escort entries around the Network Q Rally of Great Britain. Escort RS Cosworths in WRC-homologated format were present, tended by M-Sport personnel, as they had been since 1997. Scrutineering on Saturday at the prestigious Cheltenham (horse) racecourse saw corporate blue-and-white Escorts for Juha Kankkunen and Bruno Thiry to drive, entered by Ford Motor Company, for the last time. They would finish a fine second and third, but it was time for radical changes, as Richard Burns and his Mitsubishi Ralliart Team entry finished over three minutes ahead.

That British round of the championship was now firmly about English stages and (primarily) Welsh forest terrain. All done in a couple of days for me, with just a B&B overnight needed, rather than the night watches and telephone-box reporting of the rear-drive Escort era, back when Scotland was a critical part of the 2,000-mile-plus route. This time I was back in a Berkshire office before the rally finished on Tuesday 24 November, turning editorial around before leaving for Dunton on the Wednesday.

I rated the Dunton Focus outdoor display 'excellent', but didn't fully appreciate the depth of detail engineering that the specialists at M-Sport had applied – especially in distribution of weight, placed as low within the Focus as possible. Also truly significant was the replacement of the inline transverse five-speed gearbox with a longitudinal Xtrac six-speed. Other clever weight-bias moves included a near-central fuel tank mounted under the area occupied by the rear seats on the production car, crafted to allow space for the rear drive components passing beneath. In another move to improve weight distribution, the engine was inclined towards the front bulkhead and moved back 20mm (0.79in) on an elongated wheelbase.

That engine relocation, and a lengthy sequential-shift gear lever within a hand-span of the driver, were inspired by the example set

▼ Wednesday, 25 November 1998, and the media debut of the radically reworked WRC (World Rally Championship) Focus weapon. Created by M-Sport in Cumbria, the competition Focus won its third event, the tough East African Safari rally. *Author*

by Super Touring circuit racers, particularly the Williams Engineering Renault Lagunas. Until this was written, and I filled in the accompanying specification comparison between first-generation showroom Focus RS and the M-Sport WRC variety, I naively hadn't appreciated how fundamentally redesigned that WRC Focus was.

The personnel to emerge into the Focus limelight included 1995 World Champion (for Subaru-Prodrive) Colin McRae. At 30 years old, the now revered Scot was billed as 'the fastest man in rallying' by the Ford press material: an unsurprising claim, since the company had paid a reported record £3 million for Colin's services. Another high-profile team figure was Guenther Steiner, who at the time of writing had risen to fame as a popular TV interview subject while managing the Haas F1 team. Guenther had been with M-Sport from the start of the direct Ford liaison in 1997, and then oversaw the 317 days devoted to re-engineering of the Focus into WRC format at the temporary Millbrook project engineering base. M-Sport boss and former Ford-contracted driver Malcolm Wilson drove the Focus WRC for the first time on 22 October 1998.

What we could not have gathered from that 1998 Focus preview was just how tough the WRC version had become. The basic body shell torsional strength was rated at 30 per cent up on the Escort predecessor. The Focus WRC's long-travel strut suspension, front and rear, received its ultimate accolade when notoriously hard driver McRae and his co-driver Nicky Grist won the punishing Safari rally. They would score another WRC win in Portugal that debut year, but otherwise the Focus suffered some first-year mechanical problems, making the Safari win a welcome but baffling durability exception.

The 2000 and 2001 seasons didn't bring either Ford or McRae the expected championship rewards, although the McRae/Grist combination won nine WRC events during their 1999–2002 Ford partnership. The striped Martini-Valvoline sponsorship colours remain a powerful Focus memory today.

For 2003, Ford cut back some areas of financial support for the M-Sport team, and both expensive and established winners McRae and

M-SPORT YESTERDAY AND TODAY

Chosen to mastermind Ford's 1999 assault on the FIA World Rally Championship, the Cumbrian-based M-Sport team has secured seven FIA World Rally Championships and many other victories, making it one of the sport's most successful outfits. Yet it has not all been about Ford, or just rallying, although the company does report as: 'The biggest manufacturer of Ford racing vehicles in the world.'

In 2013, M-Sport led the design and development of Bentley's Continental GT3 international race car, and in 2018 the company entered the world of electric motorsport as Jaguar's technical partner in the IPACE eTROPHY. Running alongside the FIA Formula E Championship, the zero-emission series completed two seasons and was M-Sport's first step into sustainable motorsport.

As well as its imposing Dovenby Hall site, where ancient and modern buildings co-exist on the edge of the Lake District National Park, M-Sport now has a European base, with a well-resourced facility in Krakow, Poland. The UK base has developed with an M-Sport Evaluation Centre, which includes a 2.5km test-track.

In 2023, M-Sport continued to represent Ford in the WRC, with the Puma Rally1.

▼ Sparkling action from Colin McRae/Nicky Grist in the Martini-backed V6 FMC on their ninth-placed run in the 2001 Swedish Rally. A similar 2001 McRae WRC Focus (Y4 FMC) – albeit credited with an outright victory on the Acropolis WRC round – sold for £423,300 at auction in 2022. *Ford*

▲ ➤ The author poses alongside the WRC Focus crew credits. Although it looked like a showroom Focus at first glance, every dynamic aspect, from powertrain location to enhanced structural integrity, had been investigated. *Author*

Carlos Sainz were dropped in favour of cheaper, but promising conductors, including a winner in Estonian Markko Märtin. A fresh chief engineer (Christian Loriaux) created a respected 2003 WRC Focus.

Ford's corporate lust for world titles remained unsatisfied until 2006–07, when it won the World Rally Championship as constructors twice. Ford and M-Sport had lured Peugeot star and double World Champion driver, Finland's Marcus Grönholm, into the Focus fold. This combination broke a Ford World Championship drought of 27 years, albeit Grönholm finished as runner-up to Sébastien Loeb in the drivers' championship both years.

The Focus, with yearly updates, would be M-Sport's primary weapon in the WRC from 1999 to 2010, when the Fiesta took over, with a fundamentally smaller body and a wide appeal to customer teams. That latter point was important for the now expanding M-Sport, as support from Ford could vary wildly, with budgets unconfirmed until late in the preparation schedule for each season. More recently, M-Sport and Ford were enjoying a second honeymoon, as the Puma ST hybrid won the most famous of all World Championship rallies, Monte Carlo, at the beginning of the 2022 season, and took victory on Rally Sweden in early 2023.

In 176 WRC rallies from 1999 (Monte Carlo) to the 2010 Wales Rally GB, the M-Sport-prepared and entered Focus WRC designs won 44 rounds outright, secured 142 podium places, plus that pair of WRC manufacturers' titles. The first win came on that epic 1999 Safari, and the last was 2010 Rally Finland, seized by home-country pairing Jari-Matti Latvala/Miika Attila, who accrued four Focus wins in the closing seasons of the car's WRC life.

Footnote *In March 2022, Silverstone Auctions sold McRae's 2001 Focus WRC, as used on the 2001 Acropolis Rally and rolled on Rally GB, costing McRae the 2001 WRC title. It fetched £423,300 against a guide price of £350,000. Colin's contemporary race suit sold for an equally astonishing £16,650, underlining the financial power of his legacy as a Scottish rally-driving legend.*

1998 FORD FOCUS RS ROAD CAR vs 1999 FOCUS WRC RALLY CAR

M-Sport WRC statistics in brackets

Engine

Configuration: Transverse, inline four-cylinder, 16-valve, forged pistons and connecting-rods, with under-piston oil jets. Garrett GT2560L turbocharger, 1.2 bar max boost. Multi-point fuel-injection electronically managed by Ford EEC-V (Engine tilted and moved rearward at 25° and 20mm (0.79in) respectively. Relocated competition Garrett turbo, 34mm (1.3in) air-intake restrictor, air intercooling and additional water sprays. Reprogrammed engine management) **Type:** Zetec rebranded Duratec RS (Mountune redeveloped Zetec E) **Capacity:** 1,998cc **Compression ratio:** 8:1 (Unknown) **Maximum power:** 212bhp at 6,500rpm (300bhp at 6,500rpm) **Maximum torque:** 229lb ft at 4,000rpm (406lb ft at 4,000rpm)

Transmission

Type: Front-wheel-drive, uprated Focus components, Quaife torque-bias limited-slip differential, AP uprated clutch, quick-shift with Sparco aluminium gear lever. (Permanent four-wheel-drive, Ford co-designed front, centre and rear electronically monitored active differentials, competition limited-slip differentials) **Gearbox:** Transverse, Ford MTX five-speed (Longitudinal Xtrac 240 six-speed sequential)

Suspension

Front: Independent, via MacPherson struts, Sachs dampers, uprated geometry (Specially fabricated MacPherson struts, adjustable damping and links) **Rear:** Multi-link Control Blade, Sachs damping (Specially fabricated MacPherson struts, adjustable damping and links)

Steering

Type: Power-assisted rack-and-pinion, 2.3 turns lock-to-lock (Power assisted rack-and-pinion, 2.0 turns lock-to-lock, 12:1 high-ratio rack)

Brakes

Type: Servo-assisted, Bosch anti-lock braking system (Cockpit-adjustable bias front to rear) **Front:** 324mm/12.8in ventilated discs, Brembo four-piston callipers (Asphalt and gravel Brembo disc choices from 300mm to 380mm/11.8 to 15in, eight-piston callipers for asphalt, otherwise four-piston) **Rear:** 280mm/11in solid discs, Brembo two-piston callipers (Asphalt and gravel Brembo disc choices from 300mm to 315mm (11.8 to 12.4in), four-piston callipers)

Wheels and tyres

Wheels: OZ alloy 8x18in, 5-spoke (Asphalt, gravel and snow OZ choices: 5in to 7in widths, 15, 16 and 18in diameters) **Tyres:** 225/40 18 Michelin Pilot Sport (Michelin surface choices)

Body and chassis

Type: Three-door, pressed-steel monocoque. (M-Sport reinforced body welding, limited use of production Focus panels, plus integrated steel roll cage and extensions)

Weight

Kerb weight: 1,275kg, distributed 60/40 per cent front/rear (Competition weight, 1,230kg distributed 58/42 per cent front/rear, 'near' 50/50 balance when larger rear spoiler intervened at 'higher speeds')

Dimensions

Length: 4,152mm/163.5in (4,174mm/164.3in)
Width: 1,770mm/69.7in without mirrors (1,998mm/78.7in)
Height: 1,430mm/56.3in (Adjustable from 1,420mm/55.9in)
Wheelbase: 2,615mm/103.0in (2,635mm/103.7in)

Performance

Maximum speed: 144mph (Unknown) **0–60mph:** 6.2 sec (Unknown)

▼ **This extreme WRC look was seen at a Hendy Ford dealer gathering.** *Author*

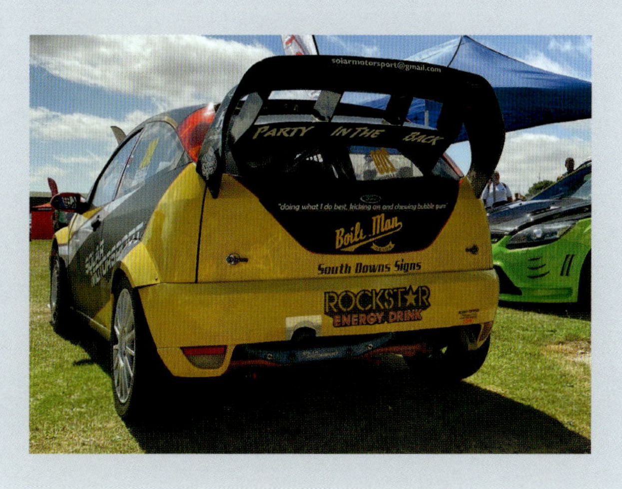

CHAPTER 17

POWERED BY FORD

Ford engines in road and competition cars of other makes

The 'Powered by Ford' corporate slogan and engine-bay badge has been used on many road and competition cars produced outside of the Ford company, from Formula Ford single-seaters to specialist road cars such as the iconic Cobra, plus cars from manufacturers such as Lotus, Morgan, TVR and Marcos to name but a few.

I tend to associate 'Powered by Ford' with specialist sports or competition cars, because so many British-based constructors carried Ford motors in their engine bays, and some still do. When I came to look into it, I saw that Ford actually deployed the slogan officially way before my time of employment, as it was associated with trucks, as in 'Powered by Ford, Built to Last,' a spin-off from the company's better-known 'Built Ford Tough' tag line.

◀ Ford V8s have powered many hot-rods based on pre-war US hardware. My favourites are the 289/302 competition units from the 1960s used in the GT40, Mustang and Falcon. *Author*

▲ Lotus Elan, road or race, uprated or period perfect, coupé or convertible, remains an author favourite. This is David Brodie's ex-Jeff Goodliff 2-litre racer with 178bhp that I enjoyed around Silverstone Club circuit. *Motorsport Images*

THE GO-TO ENGINE SUPPLIER

What mattered to me for this chapter was the sheer variety of machinery that could be driven, which complied with that advertising strapline. Most obvious in Britain were the early Formula Ford single-seaters, and I had my first track instruction in Lotus 51s pushed along by converted Cortina 1500 GT power – cars that Brands Hatch race school used with instructors such as John 'Lotus' Miles and Tony Lanfranchi. I tested more Formula Fords over the decades, the last in 1996 (a Swift car at Silverstone Racing School), but never had any ambition to race one. However, I did enjoy the chance to drive a more responsive, purpose-built racer, rather than the less exhilarating feedback of a compromised sports or saloon car.

Closer to my capabilities were the enclosed Sports 2000 racers of 1977–81. I steered examples including Tiga's 77SC and Lola T490/492 series at various tracks, including Brands Hatch and Snetterton, and set lap-times that were less embarrassing than later Formula Ford tests. I did write a full-length feature about S2000 for *Motor Sport*, published in May 1978. The S2000s had 126–130bhp from blueprinted and uprated Pinto 2-litre motors that, at the time, cost £900–£1,200 from Burton, Titan or Sam Nelson. The driving experience was outstanding for an affordable formula (the Lola cost £4,750 to £5,250). Handling was divinely direct compared with much of my production-saloon-car fare, as were prompt formula-car brakes and constantly chatty communication from tarmac to conductor, via unassisted and high-geared steering.

I never made it to race a Sports 2000, but the Brands club circuit times were OK and a test day at Snetterton with factory technician and Lola T492 hardware tempted me to try, and fail, to raise the necessary sponsorship between 1987 and 1989. Incidentally, the closest I ever got to doing a major sponsorship deal was via major American car-rental company Hertz – but the sponsorship was for the Tyrrell F1 team. My cut would have bought an ex-Ford press-fleet Granada V6 estate.

I include the Lotus-Ford Twin Cam motors in this section and the Cosworth BD series, as production foundations (including the main cylinder block) were supplied by Ford factories. Ford's contributions to outside new applications for its showroom motors, including motorsports, usually demanded specialist subcontractors. However, Ford usually ended up paying the *big* bills…

Although the most valuable and powerful Ford-powered road and race cars I drove were of the venomous Cobra breed, my more affordable passion was directed at Lotus, particularly its Seven/7 and Elan models. In later decades, the Caterham-built updates on a Seven theme, including the 2-litre, 16-valve, 175bhp, HPC variant, came my way, along with the chance to race a Caterham S3-based 1600 GT. The 1991 HPC I tested for a colour feature packed a GM-Vauxhall motor, so not relevant here, although it was an excellent balance between high-rpm thrills and amiable traffic manners. The original Lotus Sevens were mine to steer in 1600 GT or Lotus-Ford twin-cam form. The Ford pushrod motors were right in the spirit of this chapter, combining reasonable purchase cost, economy and startling acceleration through flyweight construction.

Lotus delivered variety, from sophisticated Elans to flyweight Seven satisfaction. I raced a 1600 GT for Caterham Cars on a one-off Silverstone GP track appearance to prove its road and race capabilities. Elans came in many formats, including my *Motoring News* editor's beautiful blue BRM-uprated twin-cam coupé (sadly rear-ended to oblivion when stationary), plus a maniacal Broadspeed-tuned roadster with Cosworth BDA power. Similarly powerful, I also loved the David Brodie/Richard Lloyd 2.0 twin-cam racer.

I experienced Ford four-cylinder or V6 power in the British trio of Morgan, Marcos and TVR, but I sadly missed out on the TVR Griffith with Ford 4.7 V8 motivation: the closest I got was writing about some of Gerry Marshall's exploits in the Barnet Motor Company's versatile road, race and hill-climb example. The more powerful versions of Morgans and TVRs I drove featured Rover (*née* Buick) aluminium V8s, from soft ex-Range Rover 165bhp ratings to a 400 race-horsepower TVR Tuscan, and a successful Bryant-family Morgan competitor.

I did some Reliant 3-litre Scimitar steering time on a Marcos factory visit to see how the two low-volume performers compared, as both used the lazy British 'Essex' V6 engine. The Scimitar belonged to *CCC*'s MD, Julian Berrisford, well known in VSCC circles, so this hairy Associate Editor wasn't going to thrash it mercilessly. The marked contrast between the Reliant and Marcos characters showed how useful the Ford V6 was to British specialists. The Scimitar was a practical and comfortable companion, fit for British royalty (HRH Princess Anne), while the Marcos was a sharply dressed strict two-seater with loud twin exhausts and race-car-style driver posture.

Also powered by Ford, but Ford of Germany's compact V4, was a striking yellow road test Matra 530A, with distinctly French glass-fibre panels over a steel chassis. It had but 70bhp from 1.7 short-stroke litres, and less than 100lb ft of torque to propel 864kg (1,905lb) to less than 100mph, coupled to a 0–60mph time of a less-than-sparkling 16sec. The upsides were driving joy in twisty motion, and 30mpg. I remember that Matra mostly for excellent ride and handling, spoiled by a sloppy and elongated gear-change mechanism, not dissimilar to that of another mid-engine machine, VW-Porsche's 914.

The mid-engine layout was fashionable back then, but my experience pointed to Lotus doing it best in the Europa, albeit the gear-change linkage actually fell apart on two of three Renault-powered examples I tested. The Lotus-Ford twin-cam version, with Austin Maxi five-speed gears installed in a Lotus casing, was probably the best, but I cannot find any evidence that I drove that more desirable package. However, for less than 24 hours, I did drive Formula 5000 race-car constructor Tony Kitchener's much more dramatic Europa, packing a 4.7-litre/271bhp ex-Mustang V8, hitched to a ZF five-speed transaxle, all in association with F5000 rear-end components and a reinforced backbone chassis. This was really a development drive; yet the surprisingly good road impressions were published in *Motoring News* on 19 March 1970.

Another Ford powered one-off appeared in my 1969 regular *CCC* column 'Walton's Wanderings', alongside a very un-PC lady wearing a rally jacket that couldn't be zipped up, as her twinned assets required an airing. I was just as interested to discover that this Ipswich oval-track single-seater was an aristocrat amongst Spedeworth Superstox competitors. In my 53-year-old words, the 'brightly hand-painted, fun-loving 1,800cc tractor' comprised an ancient Ford four-cylinder, tweaked to 70bhp within a £75 capped budget. The effective work on a mandated pre-1956 engine highlighted the craftsmanship of former stock-car World Champion Doug Wardropper, proprietor of Scholar Conversions. Scholar became legendary providers of Formula Ford power in the following decades.

The vehicle was a little out of my comfort zone – as was the oil-slicked track. A centrally mounted gear lever for the four-speed gearbox ensured that my legs were splayed. I commented: 'Into third gear, from which I never shift until my return some 30 laps later: with no windows and wearing goggles, it feels like a biplane. Especially on one corner, which has a nasty bump on the clipping point, which throws it all off-line. I'm sawing at the steering wheel, (apparently) going in three directions at once down the next straight. There's loads of torque, so that you can boot the tail out a fair way on this greasy surface. On this occasion, I didn't manage to spin the car – which just goes to show that beam-axles do enable a moron to get a feel of what is going on.'

Perhaps the craziest Ford-powered device I drove was the DAF Marathon 55 rallycross car from Jan and Harry de Rooy, where I sat atop an automatic continuously variable transmission, sharing the cramped cabin with a Lotus twin-cam-derived 175bhp. The mechanically cacophonous cabin demanded a large blister in the roof to accommodate the elevated, centrally seated, driver. I expect this appeared in *Motoring News,* since I was the rallycross reporter for the 1969–72 televised seasons, amongst other duties. Yet, the only published text I can see was from *Motor Sport* in 1972.

By 1976, British race-goers were treated to some amazing automotive concoctions brewed for the Special Saloons series. Basically high-end club racing, but because General Motors took it seriously through its Dealer Team Vauxhall (DTV) participation, so did Ford. GM/Vauxhall/DTV set the benchmarks with the 'Big Bertha' and 'Baby Bertha' V8-powered cartoon creations – driven by flamboyant racing entertainment genius Gerry Marshall – which vaguely resembled showroom Ventora six-cylinder and 2.3-litre Firenza coupés.

Ford and others took notice, and the construction of some amazing devices resulted. We've covered the ineffective major Ford effort behind David Brodie's Capri II in Chapter 8, but privateers achieved excellent results for the Blue Oval, particularly former Post Office engineer Mick Hill in his Boss Capri V8s.

I had a soft spot for the immaculately turned-out creations of Superspeed employee Colin Hawker, working in the company's traditional Romford, Essex base. He made the most of a scrap F1 Cosworth DFV V8 in his sparkling Capri, but the better results, and my personal experience, was founded in a VW Type 3 Fastback body (originally designed around redeveloped VW Beetle rear-engine hardware) over a monocoque race chassis. I tested the ingenious racer, dubbed 'DFVW', at Silverstone for *Motor Sport* to publish in December 1976. Its 425bhp allowed 'over 150mph', whereas the original 44bhp Wolfsburg showroom item kissed the *autobahn* at a peak velocity of 78mph.

Excerpts from that DFVW period piece – and my driving memories of the assorted Cobras, Lotuses, TVRs and Morgans, appear later in this chapter, along with details of a rare Ford Zodiac (Ford of Britain's '60s range-leading 3-litre, V6-engine pioneer) equipped by Willment with Ford's American 4.7-litre V8.

'PERHAPS THE CRAZIEST FORD-POWERED DEVICE I DROVE WAS THE DAF MARATHON 55 RALLY CAR FROM JAN AND HARRY DE ROOY.'

COBRAS: RACE, ROAD AND CONTINUATION

The AC Cobra cocktail of aged but effective British body and accessible US V8 power is the most charismatic and costly of the machines in this chapter, a proper performance idol. A descendant of the '50s AC Ace sports cars, with straight-six-cylinder power, the Cobra was the 1961 creation of Carroll Shelby, mating a light, but strong AC rolling chassis with Ford V8 power. There was a bewildering variety of Cobras through the '60s, including 365 production Cobra 427s (durable power from 7-litres), plus the rare (six-off) Cobra Daytona Coupe, born to run in the international GT class against Ferrari. Whilst the early GT40s struggled for reliability and outright victories, the Daytona Coupe won the 1965 FIA World Championship for GT cars.

The Cobra is a reminder to both sides of the Atlantic that we can build great things together, especially under pressure. Although the GT40 became another winner via transatlantic cooperation, its existence depended on motorsport, rather than the Cobra's dual-purpose race and recreation roles. Yes, the Cobra can also fulfil lazy cruises, or amiable touring tasks, more credibly than that viciously reptilian name implies.

More than 20 years of Cobra outings inspired my enthusiasm for such a fraught alliance of skinny anatomy and accessible power, testing cars at British tracks and equally demanding British byways. Since the original cars were built, the AC company itself and a variety of others have produced continuations of the classic '60s Cobra: at the time of writing there was another brand licence holder, AC Cars Ltd, based in Norfolk.

Judging by the prices achieved by the originals today – and the still-significant costs associated with purchasing Cobra clones – there are plenty out there who remain bitten by Cobra fever.

DRIVEN

Cobra-Ford 4.7 V8 Track test on Brands Hatch Club circuit in September 1971, published in *Motoring News*, 1 October 1971; race driver and co-owner Shaun Jackson with Bill Harding; driven in 4.7-litre, 340bhp hardtop format; 10 Chevron Oils/STP Modsports race wins during 1970 season. **Cobra-Ford 5.0 V8** Track test on Silverstone club circuit, 1985, published in *Sporting Cars International*, December 1985, 13 COB, Bell & Colvill Ltd (owner Martin Colvill); simply 452bhp fabulous! **Cobra-Ford 4.7 V8** Road weekend and Silverstone Club circuit, June and winter 1985, published in *Sporting Cars International*, December 1985, JA11; owner John Atkins, multiple winner (30 victories) of mainly HSCC events; 375bhp. **Cobra-Ford 5.0 V8 1992 continuation** Road test, appeared *Motor Sport*, April 1993, J955 SPH; continuation car priced at £94,842 in 1992, built at Vickers Drive, Brooklands Industrial Park, via Brian Angliss, then owner of the AC Cars Ltd trademark; c345bhp, 0–60mph in 4.35sec.

Cobra-Ford 4.7 V8 hardtop race car: a legend in UK club racing

By the autumn of 1971, the combination of *Motor Sport* and *Motoring News* as my joint employers had allowed me access to cars and the booming motorsport scene that were only in my Mini teenage dreams. A supreme example was a Brands Hatch opportunity to drive an AC Cobra that had won 12 Modsports races that season, co-owned by sales wizard Bill Harding, and driven by promising Corbeau Seats co-owner Shaun Jackson. Even when I tested it, more than 50 years ago, this Cobra had a long and pedigree history: I'm told it is still seen at classic events today, but I haven't traced it.

Built in 1964, the car's competition preparation was credited to C.T. 'Tommy' Atkins for leading British-based Grand Prix and sports-car exponent Roy Salvadori. The results for that former Aston Martin ace were disappointing, and the legendary Chequered Flag concern received the car during the winter of 1964–65. The 1965 results – with a welter of fiercely ambitious 'coming men' drivers – brought five top-level international victories. The Chiswick sports-car specialists retained the Cobra during 1966 without notable results, but Roger Mac did set a March 1966 lap record with it at Brands Hatch (56.6sec/78.87mph) when leading a GT race, where he retired after five laps.

The car passed into the 1967–69 private ownership of Wendy and Keith Hamblin, who both established class victories at the Brighton Speed Trials over 1km, recording a best of 24.11sec in 1968. Messrs Harding and Jackson bought it in 1969 with the incentive that the Cobra 4.7 V8 was now allowed into production-based sports-car races. This point marked a low in Cobra values, and Bill Harding commented it was bought for: 'A very reasonable sum.' It was up for sale in the winter of 1971–72 at an asking price of £2,000 (the equivalent of £31,000 at the time of writing), unbelievably cheap by today's standards.

Under the skin: muscular simplicity

In 1971, the car appeared in a low state of tune and chassis development compared with the front runners I watched at Goodwood or Silverstone Classics in the 2020s. The engine was rated at 340bhp, with an amicable working range of 3,000–6,500rpm, despite the ferocious bank of downdraught 48mm (1.9in) IDAs, mounted between factory-fresh looking 'Cobra by Ford' rocker covers. This 4.7-litre/289cu in unit was prepared by John Roberts and Dick Webster at their premises in Bromley, Kent, and could survive multiple outings without expensive rebuilds. Much of that durability owed thanks to a low 9.6:1 compression ratio from TRW pistons, and a 'nice, strong, bottom half' extracted from the Paul Hawkins GT40. A small-bore exhaust system and inlet manifolding were the original Shelby-specification items.

Outside the engine bay, my concerns as the 24-year-old survivor of just a quartet of novice circuit races centred on the suspension: so many bar-room tales emphasised how viciously this snake could bite back under duress. My apprehension caused great merriment at the time, as I was told the system was certainly 'all independent, and that means each wheel!' In reality, both the front and rear of the car were transverse leaf sprung, bearing on lower wishbones and telescopic shock absorbers.

Before I crouched beneath the hardtop and stared over curvy bonnet bulges with a red-and-gold colour scheme, I did take stock of other hardware needed for some fortunately dry laps. The gearbox was a straightforward four-speed BorgWarner T10, the clutch came from a contemporary Boss 302 Mustang, and the Salisbury limited-slip differential ran a 3.55:1 final drive. That meant a top speed at the rev limit of 144mph, and they expected 0–100mph to occupy around 9sec. A massive 35-gallon fuel tank remained in residence as a legacy of long-distance events, when it gobbled at a rate of 3–4mpg.

The brakes – solid discs with no servo – would be judged inadequate today. Dry race tyres were early low-profile ZB Firestones, used and scrubbed on a Ferrari 512S, resting on 18x9in JA Pearce front rims and 18x10in rears grafted on from a Felday 4x4 racing sports-car.

Behind the wheel: one of my most memorable drives

This example was a little more difficult to enter than three other roadster-spec Cobras I drove, because it had adopted a simple bubble hardtop in aluminium, along with Alan Mann-inspired red livery with gold stripe. Once installed, you inhabited a world I described as: 'Bare of all carpets as one would expect, but it is also a mixture of riveted plates, matt-black paint and a host of instruments. A tall roll-bar, together with a fire-extinguisher and full Britax safety harness, were reassuring.'

Not so comforting was that my cautious opening brace of laps was curtailed by a red flag and a Mini flattening itself, though fortunately not the owner/driver too. There was a 20-minute wait before we could tackle another half-hour/75p

▼ A serial winner of 1970 UK club races for Shaun Jackson and owner Bill Harding, this 340bhp Cobra introduced me to the addictive nature of 4.7-litre snake bites. A full account of my Brands Hatch foray appeared in *Motoring News*. *Motorsport Images*

session. Regular driver Shaun had established a 55.6s/80.29mph lap-record for the short Brands layout, but his times were off that pace for our test, especially after the Mini incident, which left a trail of oil on the approach to Clearways.

Nevertheless, the bold owners let me out for another 20 laps, so I had a real chance to feel Cobra charisma, rather than the usual teaser laps allowed in a car for sale or auction. My report stated: 'I arrived at the big hummock before Paddock with 5,800–6,000rpm indicated, somewhere in the 125mph bracket, judging by the time it took to slow! I braked for much too long, but that did at least allow the power to be progressively applied in third, so that the Cobra dived into the dip like an aircraft. It swooped up towards the bridge with 5,500 indicated [on a large and flickering US-sourced tachometer]. For me, this was the best part of the whole test, because the brakes could be let off at any point and the car pointed into Druids with all that power ready for use in second gear. 'There was very strong (initial) understeer at this point and it needed nearly full throttle to kick the back wheels into a power slide on the exit.'

Next on the menu was a dive down to the left-hander that led along South Bank: 'A most critical part of the circuit, as third gear and a gradual increase to flat-out revs are needed: judgement is so important because the Cobra does not like full power on the outside of the track.' In hindsight, the feeling was of an edgy weaving that could lead to disaster on the original suspension.

The following left, opening the line into Clearways, featured a temporary oil slick. Even so, novice and thunderous ride arrived at the tricky rising-contour right of Clearways to: 'Lift off the brakes, squealing into the corner, resulting in the nose sliding outwards until a hefty dose of third-gear urge made the front rear up and the back move round to a neutral or oversteer stance. Still in third, the Cobra gobbled up space until, just before the pits, fourth went home and – with a resounding bellow – it streaks up to Paddock again.'

It was a fantastic driving experience, right amongst the most memorable in this book… But there were more race Cobras to come and none were dull!

Cobra-Ford 5.0 V8 HSCC and Inter-Marque competitor: hot laps on a cold day

This was a story that really deserved a larger audience, as the magazine that carried it wasn't an extended survivor. However, *Sporting Cars International* employed an excellent photographer that day at Silverstone day, plus it was published by loyal supporters of my work at CCC, so I was just grateful for another chance of a Cobra drive.

Not just any old Cobras either, this car had history. Registered 13 COB, it was the well-used, but beautifully presented, property of Bell & Colvill co-owner, Martin Colvill, and when I drove it in December 1985, had completed 107 races in his hands, with 102 finishes. The car was raced by Martin with such consistency that he won the HSCC Classic Sports Car Championship in 1978, 1980 and 1981: he also became the highest-scoring individual within the 1981 and 1985 Inter-Marque series. Winning outright was tough, as later competition cars arrived, but this Cobra had achieved that feat 12 times, plus 44 class victories, prior to December 1985.

The car had been transformed from its skeletal form as part of an exchange deal at the multi-franchise Bell & Colvill dealership (an original Lotus outlet) into a thing of beauty, via the craftsmanship of Emilio Garcia. Initially, Garcia created a very nice 271bhp road-runner. Consummate motor-trade professional Martin, complete with immaculate manners and height that meant his other competition car (a pedigree GT40, what else?) needed a bump in the roof, wanted more.

On that wintry but dry day at Silverstone Club circuit, Martin explained that a year after the restoration was finished, in 1976, he had slipped a set of race Dunlops onto the shiny AC V8, and tackled a few sprints. Still unsatisfied because, as he said with a wide grin: 'Inevitably, sprints became rather frustrating. You have to stop, just when you've *really* got motoring!'

Under the skin: sympathetic modifications

Martin certainly didn't stop, this Cobra consistently redeveloped under the care of respected former BRM employees Rick Hall and Rob Fowler. Some nine years after it

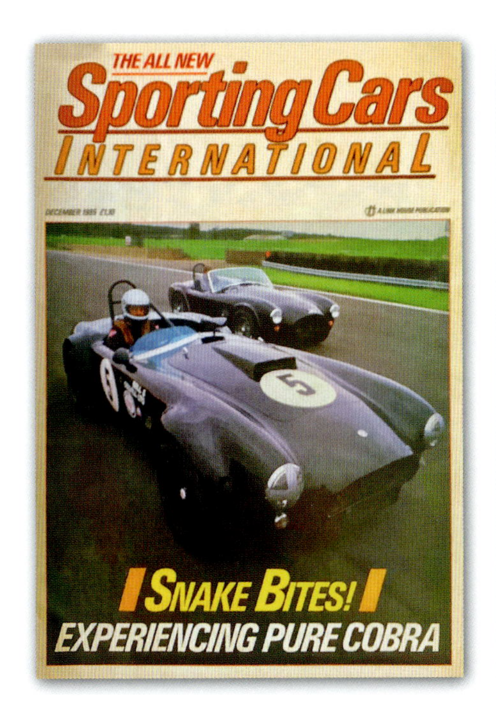

▲ These are the best records I have of a memorable pair of Cobras tested at Silverstone in 1985. The Bell & Colvill pure racer of Martin Colvill is closest the camera on the cover picture, whilst the public-road warrior was owned and steered by bearded John Atkins. *Author's archive*

had emerged from restoration, the mild (by Cobra/GT40 standards) engine had expanded from 4.7 litres/289cu in to a full 5.0 litres/302cu in, and had gained 181bhp, some of those equines owed to Gurney-Weslake alloy heads, fed by a bank of downdraught Webers. Yet more debt was owed to the consistent V8 expertise of specialist Stuart Mathieson at Mathwall Engineering. Mathieson revealed after my Silverstone session that the 452bhp peak was achieved at 6,500–6,700rpm, backed by a 391lb ft torque summit at 5,000rpm. Stuart also disclosed that the V8 within Colvill's GT40 retained a rather more original specification and components than the 'from scratch' Cobra unit.

Outside the engine bay, 13 COB – a plate uncovered by John Atkins, owner of the other Cobra we drove that Silverstone day – also progressed over the decades. The leaf-sprung suspension was retained, but careful repositioning of key parts such as fuel tank and battery had brought surprisingly even 52/48 (front/rear) distribution of 2,240lb (1,016kg). Another critical change was the switch from wire wheels to Dymag-manufactured versions of the classic Halibrand, adopted for regular-racing safety. They also allowed the use of effective Michelin TB15 wets from 1984 onwards, replacing the period Dunlop CR65s. The brakes proved magnificent, the result of a 1978 rethink by Bob Green's BG Developments, which implanted a system featuring enlarged callipers and discs, with ventilated units at the front.

Behind the wheel: raw power

The cockpit was functional in the old biplane aviation way, and demanded I borrow Martin's helmet with a full visor. A single aero-screen replaced the conventional windscreen, and within the cockpit was a simple leather-rimmed three-spoke Mota-Lita wheel and capacious seat, taped strategically at the sides and leading edge of the seat cushion to suit Martin's elongated frame. Right down to the gridded foot pedals, this was a purposeful racer, but that 'was' is a deliberate reference, as 13 COB was scheduled to revert to a quieter and fully roadworthy two-seat, 400bhp specification soon after my test..

A heavyweight clutch and heavy-duty four-speed gearbox were accompanied by the blessing of a long-travel, but sweetly progressive, throttle action. In usual Cobra character was a central bank of minor instruments, clock and speedometer inert, whilst the large Jones rev counter was hyperactive. The red tell-tale marker remained at 7,200rpm; I used 6,500 regularly, but needed a raucous 7,000 in second through the Becketts U-turn.

That Silverstone Club layout was then simply triangulated, ignoring the flat-out, slight left-hand kink of Maggotts that eased into Becketts.

The generous straights linked to three right-handers, from the third-gear quickness of Copse, to the slowest at Becketts. Also towards rapid was the final Woodcote curve, which often saw desperate last-lap lunges for premier places on race days. Martin had seized the lap record at 61.2sec/94.59mph. My 15 laps allowed 63.77sec/90.78mph, underlining what a properly redeveloped machine that Colvill Cobra became.

Rousing the V8 to life repeatedly was too much for the race battery, so many easy second-gear bump-starts were required. Thereafter, it was hyper-action all the way: 'With all that throttle travel exploited, Silverstone's simple corners arrive at stunning speed. Copse demands third gear and firm braking, but the Cobra seems to revel in this fast sweeper and tracks for the apex with flattering accuracy. You emerge with the V8 now rampantly exploring that 5,000rpm peak torque. The Cobra then bolts for the easy left-hander at Maggotts with an exuberance that sees it in top gear *very* quickly. Again, the speed of the corner helped, and it was easier to take the bend flat-out than in some of the V6 TVRs I have driven.

'However, at 120mph, Becketts now loomed large and the brakes got a chance to haul us back into subsonic reality. Most of my mistakes were made at the hairpin, and tremendous understeer is the reward of the timid Cobra pilot. Of course, there is sufficient power to sweep the back out into lurid oversteer at any time. Once you have sufficient courage and faith, this proved the safest approach. However, we did manage to launch ourselves onto the main straight in a series of second- and third-gear flurries.' There was a limit to how much grunt the Cobra could transmit, even with the assistance of a Salisbury limited-slip differential and a tall 3.54:1 final-drive, allied to 15in-diameter wheels. Yet the transmission of so much power in a comparatively light body was just as well-managed as at my Brands Hatch Cobra debut.

There followed a slight breather as: 'The long ride down the straight to Woodcote was but a brief pleasure, the worried apprentice sat up, out of the slipstream, to plot a braking point with clarity.' It transpired that pushing the pedal a gasp after the 300m board returned the best lap times. Yet there were some obstacles: that tall axle meant lugging through below peak torque at 4,000rpm in third, or putting on a spectacle as you emerged into view, deploying second gear's bass-to-shrill rev range, before slipping into third for a final thunderous flypast – meanwhile hoping the owner and associates within the pits weren't laughing too loudly.

Back in 1985, we insured 13 COB for £30,000 (£101,033 in 2023 money). Even then it was too little: Martin Colvill was gentle with the magazine's skinny finances, as Cobras were more commonly valued in the £40,000 region at that time.

'BACK IN 1985, WE INSURED 13 COB FOR £30,000 (£101,033 IN 2023 MONEY). EVEN THEN IT WAS TOO LITTLE: MARTIN COLVILL WAS GENTLE WITH THE MAGAZINE'S SKINNY FINANCES, AS COBRAS WERE MORE COMMONLY VALUED IN THE £40,000 REGION AT THAT TIME.'

Cobra-Ford 4.7 V8 road-legal racer

Back in the '80s, I had frequent contact with John Atkins and Gordon Bruce, both Cobra fanatics and owners, although I knew Gordon better because he was a former staffer at *Motor* and proprietor of GBA (Gordon Bruce Associates), a PR company with an automotive bias. I think it must have been Gordon who ensured that I drove this Atkins Cobra (registered JA 11), not just as a double-act with the previously described Colvill Cobra, but also for a blissful weekend on public roads, proving its dual personality. Few write that the Cobra can be a good touring companion when fitted with the Ford production V8 in 271bhp showroom trim. It is actually quite amiable, but with that reserve of legendary Cobra venom when the driver becomes bored of being saintly…

Emilio Garcia had restored this 1965 MkII example, in conjunction with Autokraft (Restorations). By 1981 it had hit the concours scene as a winner, in metallic purple, riding on glittering wire wheels. When I experienced it in 1985, it was still smartly presented, particularly when compared with typical British club racers. The specification embraced a much less radical approach than the Colvill Cobra, but it still had a hearty 104bhp bonus over a production 4.7-litre example.

Then AC Owners' Club registrar for the Cobra breed, Atkins staunchly supported the HSCC's Post Historic Road Sports championship (PHRS), which demanded complete road legality, so John drove his Cobra to and from race meetings. Race prep then primarily consisted of swapping over to a set of race wheels, with tyres pressured up, and removing the windscreen, behind which there was a fixed shallow aero screen. Results were far from shabby, with two PHRS championships (1982 and 1985) and 30 victories recorded over a five-year span.

Behind the wheel: road friendly, but still a racer at heart

I tested the car in civilised windscreen-fitted guise, and was cosseted by smart black leather seats as a reminder of its concours pedigree. Yet, there was the large roll-over hoop above my head and a three-spoke, leather-rim, race wheel, along with the red plastic on/off master-switch for the electrical systems.

The central six minor dials remained functional, as did the larger speedometer alongside a matching rev counter. There was even a swivelling interior torch. As ever, the steering was towards heavy at low and manoeuvring speeds, and the race-biased 375bhp beneath a giant four-choke Holley carburettor needed concentration to keep it alert below 2,000rpm. After that: 'At 4,000rpm, the tone hardens and the chassis seems to tauten under the massive torque. By 6,500rpm the rush for power seems unabated, and stays to 7,000 revs.'

Other items in this well-used Cobra's armoury were showing their hard-earned middle-age. The clutch and driveline were a trifle snappy, but the four-speed gear-change earned the comment: 'Bliss, the stubby lever almost falling from ratio to ratio with very little effort.'

Dating back to the '50s, the transverse-leaf-sprung chassis (a subsequent third generation was equipped with coil springs) had limitations, as I had discovered at Brands Hatch. This time at a smoother and less-contoured Silverstone, I reported: 'At best, progress is harsh in the traditional sports-car manner. At worst, the car has a tendency to be deflected from its line by mid-corner bumps, as the wheels and tyres suffer some interesting changes in camber.' All was forgiven as I welcomed: 'The sheer euphoria of rushing from one corner to the next… The sheer exhilaration engendered by such massive acceleration, by the powerful brakes (discs all round) and the apparently near-neutral, highly responsive handling, is immeasurable.'

It was fun all right: I even became a John Atkins customer for a registration plate from his specialist business. However, there were some side issues to that transaction – and my road miles in another reborn Cobra that still cannot bear the illumination of publication!

As observed in my Cobra introduction, the AC name, and the classic vehicle itself, have been through many iterations since the '60s originals. Nowadays, Norfolk-based AC Cars Ltd displays the UK trademark, and Shelby American Inc is the parent to six Shelby-associated Las Vegas enterprises, following the death of the larger-than-life Texan legend Carroll Shelby on 11 May 2012. Shelby divisions still manufacture Cobra continuations in the classic mould, and a variety of spin-off Fords, from trucks by the thousand, to contemporary Mustangs.

Cobra-Ford 5.0 V8 1992 continuation: a continuing snake bite

Back in 1992, I had my last Cobra outing for publication, appearing in *Motor Sport* in April 1993. That Cobra was an official continuation car, and it was an excellent rendition, carefully brought up to more modern roadworthy standards by Brian Angliss, 30 years after its debut. It was branded officially by AC Cars Ltd, the company residing within spacious Brooklands Industrial Park premises in Surrey. Its Cobra continuation lived up to the promise of its creation by pedigree craftsmen. Yet so much has dated so quickly in another 30 years of the Cobra extended production story, official or otherwise, that I have just used that magazine article as a factual base, with a few direct quotes where the joy of steering a Cobra again burst through…

The enormous number of fakes and/or replicas

that abound assaults the authenticity of even the most genuine of Cobras. Even our demonstrator, then fresh from the factory in a lightweight trim that demanded £94,842 from a British buyer (£80,623 export), was accosted by a knowledgeable motor trader who enquired: 'What sort of Rover V8 runs in that?'

We winced for the 1992 Cobra's pride, for it was every bit as enjoyable as the original, and rather better made. Logical when the price for the exquisite endeavours of AC's skilful 60-strong workforce nudged toward six figures. The premises were home to AC Cars and Autokraft from 1988 onwards. Autokraft proprietor Brian Angliss also acquired total ownership of AC Cars Ltd in 1992, purchasing the remaining interests of Ford Motor Co.

That capacious main factory floor was packed with up to 170 historic motorcycles. Scotts, Rudges and Velocettes stood mute alongside MV Agustas and Nortons. Yet even that array of polished metal and seductively painted frames and tanks was overshadowed by the aeronautical restoration work. A completed Hurricane, but the real rarities were a pair of Hawker Tempests, reassembled spar by spar, panel by panel. Yet the Cobra's construction sat easily alongside those aviation heroes. The 16-gauge aluminium panels draped over the 4in steel tubes of the main chassis members, and the high standards of trim, made this one of the most eagerly awaited test drives.

Behind the wheel: a worthy successor to 1960s Cobras

As I rumbled along Vickers Drive, away from the bustle of Brooklands Industrial Park, an invigorating breeze of fresh February air across the cockpit blew in the magic that pervades a Cobra waiting to strike. The amiable eight-cylinder rumbled up 2,000rpm (only TVRs provided a better 'rolling thunder' soundtrack), the comfortable embrace of leather seats, the outstanding ride on coil springs and Koni dampers. All lulled you into thinking this was just a tame touring car. The boot was large enough to make the Cobra a suitable longer-distance companion, although carrying around a full set of tonneau equipment *and* a full-size (front) spare diminished practicality.

On the move, the AC proved far more comfortable than a Morgan and less heated in the cockpit than a TVR Griffith. I sampled one customer Cobra with the optional hardtop and I found it unpleasantly claustrophobic: neither draughtproof nor waterproof. Still, on a wet day that could be a better option than *not* using your Cobra.

The demonstrator was beautifully carburated. I use the phrase more because of the car's manners than for its impressive underbonnet polished-chrome decor. Only the sharp drop in rpm whilst repeatedly turning around during our photographic shoot betrayed that this was a not a fuel-injected version of the then current Ford 5.0 V8.

As with any original Cobra that has not been viciously modified, this one drove and rode with an easy charm that contradicted expectations from the moment that the clutch was fully home. I used a conservative 6,000rpm in first and second.

The first serious depression of the long-travel throttle told you that the 'Lightweight'-spec Cobra had every right to its name. No horsepower figure was quoted for J955 SPH, but the performance certainly felt on a par with the 345bhp claimed for the fuel-injected, catalysed Lightweight. Company performance claims (obtained by the same electronically monitored methods as used by *Motor Sport*) were credible. Are you sitting comfortably? Absorb 0–62mph at an *average* 4.35sec over two runs. Perhaps a trip from 0–100mph? Can you spare around 10 seconds of your time, sir? Actually, 10.06, to be pedantic.

There was long-legged performance potential, generated via a BorgWarner five-speed gearbox (T50D, a relative of the T5 that lurked within earlier Sierra Cosworths) and the 3.31:1 final drive. The Cobra whipped to 45mph in first gear, and 70mph in second. In top, non-standard 15in wheels, shod with Pirelli P7s, aided acceleration rather than cruising. Plumper 275/55 VR rears and the 225/65 (rather than the usual 225/60) fronts enhanced the handling. This still allowed us to lope along at 70mph, at little more than 2,000rpm in the overdriven (0.68:1) fifth.

More ambitious use of the throttle, however, showed that the Cobra's venomous sting remained intact. From 2,000 to 6,000rpm, approximately 2,500lb (1,134kg) of Cobra was flung forward. The Pirellis and Salisbury limited-slip differential restrained any wild hip-shakes once the initial take-off was completed, but the Cobra still quivered with vibrant energy under full throttle. One of the very few cars that has the overtaking-distance requirements of a motorcycle, an asset made all the more accessible by the broad power band.

Legend has it that a provoked Cobra is very hard to handle. Yet the examples I have tried since that original first 1970 assignment have done 95 per cent of what I asked of them. The 1992 Cobra remained composed, even when provoked on damp tarmac. The wooden-rimmed Nardi wheel, attached to rack-and-pinion steering, swung more easily through the 3.6 turns from lock to lock than anticipated. One expected 225 tyres to be harder to park without power-assistance than was the case, and I was rewarded with excellent feedback, as well as moderate loads on the move. Under open-throttle provocation in the first three gears, unsticking the rear 275 Pirellis was likely, yet they would carry on relaying precise feedback. At the same time, the loyalty of the front wheels was assured beyond the bounds of public-road reason.

My conclusion does bear direct quotes, for I felt the same about the Anglo-Saxon brute in 2022: 'That I enjoyed driving the Cobra may be evident. Any enthusiast would be alert to such pleasures. What isn't so obvious is that I – and the people who bought these aluminium-panelled artefacts – gain almost as much pleasure from simply looking over the Cobra. It isn't beautiful in the manner of a '60s V12 Ferrari, but that continuation was immaculately constructed, painted and assembled. Old or new, the Cobra seems to attract the same price premium. It is motoring's equivalent of the biggest, roughest diamond you can find; one you should collect yourself from its creators, savour… And never, ever, sell!'

LOTUS/CATERHAM SEVEN

The Lotus Seven and its licensed Caterham 7 successor ('Seven' originally, '7' post-Caterham) were my passion, and remain desirable today. My back-to-front chronological experience stretches from a pair of '50s Lotus VIs tasted more recently on road and track, back to 1968 and a wonderfully uncharacteristic show-standard Seven-Holbay (TNG 7G) from the factory. That car gave me a week of unmitigated pleasure for a full-colour feature test.

Naturally, I did drive Caterham versions with powerplants other than Ford – notably the flyweight Austin Rover K-series – and in Westfield SEight format with Rover V8, plus an example with Cosworth-Ford Sierra turbo motivation. My favourites for everyday use and purchase today would be the original Lotus S3 (too much cash), or the Caterham 1600 GT updates from the later '70s onwards.

My big regret, as a driver and Seven fan, was missing out on a current (at the time of writing) 420 Cup format, with 210bhp Ford Duratec power, six-speed sequential gearbox, limited-slip differential and all the goodies. So effective was the Cup 7 that it made a worthy *Evo* magazine feature in July 2022 as it set faster Anglesey track lap times than a V12/631bhp Lamborghini STO supercar, costing approximately five times as much as the flagship Caterham's £55,000 retail price.

▼ Caterham Cars continued to provide steadily improved twists on the Lotus Seven theme through the 1970s and into the 1980s. This is a 1978 test car I had for *Motoring News*. *Motorsport Images*

DRIVEN

Lotus-Holbay Seven S Three-page colour feature for CCC, published in June 1969, TNG 7G; £1,600 built; metallic burgundy paint/white (carpeted!) interior, tan weather equipment; kerb weight 1,204lb (546kg); rebuilt Ford 1600 GT, twin 40 DCOE Webers; 120bhp and 110lb ft torque at 6,800rpm limit; 0–60mph in 7.3sec, 0–100mph in 19.5sec, 114mph max, 22–26mpg as tested. **Lotus Seven S3 kit** Build story in parts for CCC, July 1969; owner Derek Carmichael; cost £824.75p with optional heater, roll-over bar and workshop manual; constructed in a driveway, I became a 'terrified' passenger... **Lotus Seven TC** Road test in winter 1969 for *Motoring News*, VVF 7H; expensive as £1,225 kit; Holbay 125bhp twin-cam; heavier and no proven performance bonus over 109–110bhp 1600 GT, our test figures too 'disappointing' to publish. **Lotus Seven S4 kit build** Construction and driving article, in parts, for *Motoring News*, published 20 August 1970; constructed with Michael Black and Peter Osborne from kit with red exterior, XNG 117H; was Lotus-loaned for 6 months; better than reputation; green-colour and still DVLA roadworthy in 2022. **Caterham 7 1600 GT** Road test for *Motoring News*, 1978, PPH 633R. **Caterham 7 1600 GT** Raced at Silverstone GP circuit, Production Sports Cars, 7 September 1980, published in *Motor Sport*, January 1981, GPL 39V; drove car on road to circuit; standard 88bhp 1600 GT motor; 6th in practice, 5th overall in race; clutch slip, and boiled at 110°C. **Caterham 7 development 1600 GT Sprint** Road test, October 1980, published in *Motor Sport* January 1981; development 7 with 110bhp Sprint-spec 1600 GT engine; 0–60mph in 'under 7sec', 110mph, 25mpg; Caterham 7 prices began at £5,046, but £21 for seat belts(!), £56 for a heater, £28 for a roll-over bar and £109 for preferred Goodyear 185x13 tyres and alloy wheels. **Caterham 7 de Dion** Privately assessed de Dion rear-suspension layout for road 7s on local Surrey roads, 1985; better ride, good location of (BL) axle, Ford outboard drum rear brakes. **Westfield Cosworth Turbo** Drive on very wet French race track, Friday 13 October 1995, purely for fun and not published; from memory, former World Rally Championship co-driver and subsequent Ford Sweden PR Gunnar Palm was the unlucky passenger. **Lotus VI Ford** Various outings 2014–19, usually as a passenger/push-starter, plus passenger ride at Castle Combe Lotus Day on 29 May 2020, SPW 990; owner Andrew Morland; tuned period MG XPAG side-valve engine.

1969 Lotus-Holbay Seven S: my first Seven experience

Let's get started, in the early summer of 1969 with that outrageous (in period) TNG 7G. For me, this encounter stood the test of time. The impact of my first Seven – and such a gloriously dotty factory car at that – ensured I remained a lifelong fan of the starkest Lotus. I still recall it with the pleasure that so many (not all) Lotus types have imprinted over the decades. I should balance that gushing recall, underlining that I'm now aware that longer periods in Lotus company – such as my 2015–18 ownership of an Elise 135 Sport – can inflict pain thanks to durability issues. That familiarity with glaring Lotus snags sits alongside the finest track and public-road miles of my 57 years' driving.

Back to 1969, and three pages of colour coverage have been awarded to the Seven S, that metallic machine dubbed 'The Ultimate Lotus Seven' by then sole concessionaires Caterham Cars, in a supporting Lotus Cars (Sales) advertisement to accompany my feature. That striking paintwork and carpeted (!) interior (it even had a radio) spoke of a commercial show life. All as pictured by Jasper Spencer-Smith portraits from Chobham Common, back when the M3 motorway had yet to be completed. Yet this special Seven still did the exhilarating performance business. Under the headline 'The Go & Show Machine', I got straight to the point in the first paragraph: 'Lying inside the Seven, our driver takes in the first chance he's had to try the delights of Lotus-Holbay power on a fine day. With the hood down and side-screens up, it felt remarkably like a single-seater formula car. There was a hard – but not harsh – ride, slot-car-racer cornering, and that fascinating heave in the back at anything from 3,000 to 6,800rpm. The fuel gauge read full, so we decide to detour from the cluttered main road and into the country.

'Amongst the twists and hills, the car seemed ever more in its element, with 120-horsepower and 110lb ft torque hurling the spaceframe Lotus ever faster through the scenery… well, the tarmac parts of it anyway! However many hairpins we desperately slow for, the brakes stay right with us – no pulling to one side, no fade, nothing to worry at all. The [brake pedal] pressure maybe a trifle hard at first, but we find after a time we like its unvarying response that makes it so easy to judge

lotus holbay – 7S

THE GO & SHOW MACHINE

LYING INSIDE the Seven our driver takes in the first chance he's had to try the delights of Lotus-Holbay power on a fine day. With the hood down and sidescreens up it feels remarkably like a single seater formula car There's a hard but not harsh ride, slot racer cornering and that fascinating heave in the back at anything from 3-8000 r.p.m. The fuel gauge reads full so we decide to detour from the cluttered main road and into the country Once in among the twists and hills the car seems ever more in its element with 120 horsepower and 110 lb. ft. torque hurling the space frame Lotus ever faster through the scenery well the tarmac parts of it anyway! However many hairpins we desperately slow for the brakes stay right with us: no pulling to one side, no fade, nothing to worry at. The anchor pressure maybe a trifle hard at first, but we find after a time we like its unvarying response that makes it so easy to judge the lock-up point. The light and dead accurate steering is fully up to racing standards, which we're very glad of because with this much power you can power out of a corner at any attitude you care to name. The acceleration is feenominal with the ton coming up in under 20 seconds and 105 m.p.h. appearing on the magic dial in the most unlikely places! The efficiently located rear axle (radius rods, coil springs, shockers and an "A" bracket) give tyre burning traction out of every curve the wiff of Mr Dunlop's incinerated 165 by 13 radials on the same gent's natty 5½ inch rims. providing solace for those masochists among us.

The tubular frame chassis with aluminium panels, stressed undertray résin bonded glassfibre nose and wings are standard. So are the independent wishbones and coil spring front suspension incorporating an anti-roll bar The secret isn't among that lot so we turn to the rear axle which is basically an Escort unit with a modded diff casing to take the triangulated bracket and radius arms: a 3.77 final drive is used giving about four thousand revs at 70 m.p.h. using a standard Ford box one of the car's nasty points. Rack and pinion steering controlled by a dished 14 in. wheel. Front 9 in. discs, rear drums and a 7¼ in. Borg and Beck clutch:— all these are normally supplied, so it's to the engine we look for some of that extra loot (standard Seven costs £777 in kit form).

With Holbay's name on the smart rocker cover twin 40 DCOEs gurgling away on the end of some flexibly mounted cast alloy inlet manifolding and some very large bore free flow exhaust piping, things look interesting. And so they are for this is the CFR 1600 GT crossflow engine with a power output guaranteed to within ± 2 per cent. of 120 neddies. That peak output is at 6200 r.p.m. while peak torque is at 5000 revs. Cost for this little lot is £325 and this includes all that we've mentioned plus such goodies as an R120 cam which seems quite content to idle at 900 r.p.m. and deliver the goods up to valve bounce at 6800. The head has larger valves, stronger springs and a 10:1 c.r This and the high lift cam has meant using recessed pistons of die-cast alloy with three rings. Oil pressure has been considerably boosted by using a re-designed pump, this gives you about 80 lb.sq. in. at anything over 4000 revs. Compare that with the normal 40 lbs. and nothing on tickover and you'll see someone has done a bit more than muck about with the spring! Con-rods are forged steel and like everything else in the engine, these have been balanced. Holbay assure us that all the vital components have been subjected to flaw detection. They also seem to have done a lot of polishing, including the rod flanks. There

ROAD TEST

Photos by: Spencer Smith

592

593

▲ One of the earliest colour features I did for *Cars & Car Conversions* featured this very special Holbay-powered Lotus Seven. Its rare features included metallic plum paint, leather cockpit trim and fancy alloy wheels. *Author's archive*

the lock-up point. The light and dead-accurate steering is fully up to racing standards, which we are very glad of because with this much power you can power out of a corner at any attitude you care to name.'

The factory-built Seven S was quoted at £1,600 with all taxes, versus a standard Seven kit at £777. So what did you get for the extra money? Much went on the highly developed Holbay engine at £325, for extracting a warranted 120 or so horsepower at 6,200rpm, coupled to a 6,800rpm limit, wasn't routine in 1969. Such power, and maximum torque at 5,000rpm, all from a mass-production pushrod motor, demanded skilled staff and some special kit. Items such as twin double-choke 40 DCOE Weber carburettors featured, alongside replacement pistons giving a 10.5:1 compression ratio, an R120-coded camshaft, allowing a sustained but lumpy 900rpm idle, steel connecting rods, and a great deal of hand finishing, plus weight-balanced reciprocating parts. The only further refinement would have been a £30 dry-sump lubrication layout, but they had already taken the precaution of doubling oil pressure to 80psi beyond 4,000rpm.

Much of the basic Seven engineering remained, including a production Ford gearbox (with notably wide ratios) that we disliked. There were good performance reasons for our public displeasure, as the acceleration runs were dogged by the stiff change and ratios that allowed 35mph, 54mph and 90mph in the first three gears respectively, thus the benchmark 0–60mph times averaged 7.3sec, rather than sinking below the 7sec mark, which would have been outstanding in those days.

Listening to, and reading back to, *CCC* Editor Martyn Watkins's earlier test experience with a 1500 Cosworth-Ford Seven, I realised that the earlier Seven on 95bhp had actually beaten the Holbay newcomer to 60mph, running 7sec… And that Cosworth flyweight kept ahead to 70mph (8.8sec versus 9.2sec). The Cosworth 1.5-litre had a close-ratio 'box that gave 42mph, 63mph and 83mph, with a change quality rated 'as fast as the hand can move'. Just as significantly, my beloved 'Go & Show' Seven tipped the scales at 1,204lb (546kg), whereas the earlier and less-powerful Cosworth model had to haul but 1,064lb (483kg). It pays to tailor any Seven carefully to your priorities; certainly the factory scored an own goal there.

▲ My first Ford rally jacket saw action at the 1969 Chobham Common Lotus Seven test, before the M25 was complete! By November 2023 the pedigree 7S-Holbay was for sale again via a specialist Kent classic-car dealer at £54,995. *Author's archive*

However, an Escort 3.77:1 axle with A-bracket, coil springs and location rods gained approval for traction. Those components cooperated with 165-section radials, sitting on Fort Dunlop company-branded 5.5x13in multi-spoke alloys. The chassis, alloy panels, glass-fibre nose-cone, disc/drum braking and Borg & Beck clutch, were all standard Seven items beneath that show glamour.

I praised the handling and road-holding, but back then few, if any, road cars would have given the agility sensations that a Lotus could generate. In fact, the other quick car around our Chobham test track was a Lotus, a 1968 Europa with Hermes-modified Renault power, but the acceleration wasn't in the same bracket as the Seven. I drove a Caterham 7 during a 1975 media day at the same Chobham track, and it simply passed all-comers, inside, outside, probably via my Lady's Chamber. All the while, the driver wondered why all other track vehicles looked so cumbersomely slow and tall!

More relevantly for this book, I'm a believer in Ford/Lotus-Ford power for these earlier models, specifically believing that an uprated 1600 GT still offers a fine cost-to-fun ratio in a Lotus or Caterham version. I have also occupied a Westfield running that Ford unit, and my endorsement holds. Yes, I know there are plenty of higher-power alternatives for Westfields or Seven-lookalike chassis, from car and motorcycle sources. I would like to try Ford Zetec or Duratec power, but have no great desire to repeat the mad, turbocharged Cosworth Pinto punch, as sampled in 1995, via a Westfield on a wet French race track.

Such a simple formula, from the 1953 Lotus VI/Six to today's supercar-baiting Sevens – surely the Sixes and Sevens in all forms must be the most loyal providers of automotive thrills?

1970 Lotus Seven S4 build and test: an abbreviated privilege

The Series 4 Lotus Seven – often referred to as a bathtub for its squared-off glass-fibre panels – wasn't a commercial success. Caterham only sold the S4 briefly when it took over all sales and manufacturing rights in June 1973. Under the shrewd leadership of Graham Nearn, Caterham concentrated on the classic alloy-body S3 and Seven-themed redevelopments. Ironically, I saw a press clip in autumn 2022 of Caterham's heavily revised 7s for the future and they had more than a hint of the S4's boxier lines.

Back in 1970, I had the unexpected pleasure of receiving a brand-new S4 kit with 84bhp Ford 1600 GT motivation, associated four-speed gearbox and heavy-duty Escort back axle. The revamped tubular-frame chassis was now clothed in Lotus glass-fibre. Suspension featured double wishbones up front, with coil-spring/damper units. As used on showroom RS Escorts, a Ford live axle was restrained by rods, forming a Watt's linkage, and sprung by vertical coil spring/damper units. A mixed disc/drum braking system needed no servo assistance and worked well by period standards, tasked with slowing just 1,288lb (584kg) – fat by alloy-bodied Seven standards.

I would co-build that car at friend Michael

Black's house, with my now business partner Peter Osborne. The loan deal was to write about it for *Motoring News* and *Motor Sport*, and that deal followed staff visits from both publications to Lotus Components. It was with that subsidiary – used primarily to sell customer race-cars – that our editorial staff dealt until the red S4, registered XNG 177H, was unexpectedly claimed back by Lotus, less than six months and 5,000 miles later.

That snatch-back was a disappointment, as we had planned to do some low-level racing with it, although some preliminary laps at Hethel – when our build was checked out and a few updates installed – told us in no uncertain terms that was a forlorn hope: especially as Sevens weren't admitted to production-based sports-

▲ There were bigger plans for the S4 'bathtub' final Lotus version of the Seven, both at the factory and for my loaned 1970 self-build. Peter Osborne (draped within the engine bay) and Michael Black (photographer) did most of the heavy lifting and I wrote progress reports for *Motoring News* and *Motor Sport*. *Michael Black*

1970 LOTUS SEVEN S4 PERFORMANCE

Figures from Motoring News, 12 August 1970

Acceleration

0–30mph: 3.6sec **0–40mph:** 5.1sec **0–50mph:** 7.0sec **0–60mph:** 9.7sec **0–70mph:** 12.8sec **0–80mph:** 17.4sec

Gear speeds at 6,500rpm First: 39mph; Second: 58mph; Third: 81mph; Fourth: 100mph

Overall economy

Average: 24.3mpg

Price

Kit: £991

car formulae until 1980, so the only categories we could contest featured radically modified, flyweight Clubmans cars.

During the spring of 1970, we started to assemble the shiny kit. We set no spare-time construction records, and Messrs Black and Osborne tackled most of the labour. I did what I was told, or took notes for four *Motoring News* part-work features and a spin-off to *Motor Sport*. I did get the honour of an initial drive, curtailed by a leaking fuel-filler cap, but by then I had already had a young girl pedestrian wanting to hop in as I waited at Great West Road traffic lights.

By August 1970, Lotus had sold over 100 examples of its last factory Seven, and we had racked-up around 4,000 miles. We had driven the car up to Hethel and Lotus Components with just under 1,000 initial miles completed. A thorough safety check was carried out before I drove those unpromising Hethel circuit laps to see if there was any potential in racing an S4.

On our return to the workshops, a new axle bracket was installed, and more attempts made to leak-proof the fuel-filler cap, which had already caused the failure of one rear lamp. The S4 proved pretty reliable up to the 4,000-mile mark, but needed a constant spanner-and-screwdriver watch for faults that could bite. The exhaust kept working loose, a nut on the clutch arm followed suit with some accompanying clutch slip, and the throttle stop/cable adjustment fell out of range too, needing readjustment to regain full performance.

Overall, my memories are of an abbreviated privilege. Now that classic and new prices are so elevated, the S4 may be the only way of affordably enjoying Lotus Seven standards of steering, handling and outstanding, accelerative, agility.

I still miss it, 52 years on!

LOTUS ELAN

Elan +2: sophisticated lightweight pleasures

I supposed I had achieved a future with my employers, as *CCC* circulation rocketed once its youth formula secured a readership with the spare cash to tune their cars. May 1969 bloomed, and I was arrogant enough to think my contributions made a difference. Actually, I got a *CCC* management offer, which I refused, opting out in a huff, rather than sharing associate editorship with a new recruit that *CCC* bosses had hired while I was overseas.

Thanks to Teesdale Publishing's Andrew Marriott and Editor Michael Cotton, I joined *Motoring News* and *Motor Sport* by September 1969, responsible for TV rallycross, plus tuned/competition car tests and features in a newly created 'Plus Performance' page for *Motoring News*. Looking back to *CCC*, there was an August 1968 Tuesday penance to pay before my first Lotus performance test session in an Elan +2. It involved towing a broken Ginetta G15-Imp out to Essex from Ealing behind a 1275 Mini Cooper S.

Not unnaturally, the police intervened, unhappy that our trailer was wider than the Mini, but we completed the mission after a prolonged roadside chat – one of many police interviews I participated in when towing tuned cars away from mechanical incontinence at that Surrey test track. When I got to Ford in 1972, I renewed regular roadside contact with Essex patrol vehicles, discussions/fines centring on velocities apparently incompatible with British law. Oh, and a kerbside exhaust repair of Roger Clark's legendary 'Esso Blue' Escort. In order to work off some of my Naughty Blue Oval Boy points, I was 'volunteered' to provide confessional entertainment at the annual Chigwell Christmas Police Feast in an exclusive part of Essex.

So, come an August 1968 Wednesday, I took +2-seat stretch of an Elan to Chobham track, after my Editor/road test writer Martyn Watkins granted the privilege. I had also been allowed a weekend at Brands Hatch, ensuring that young women could fit into the +2 element of the Elan, another mission enjoyably accomplished. Although I had enough experience to know that the Elan was a benchmark performer in every respect, the Elan +2 wasn't my first or most-exhilarating Lotus drive. I had good reason to be grateful to the original Renault-powered Europas, which arrived with *CCC* in tuned format. They delivered an outstanding driver education when the military guardians of the Chobham circuit turned a blind eye to my lunch-time activities, shutting the access gates to allow unrestricted speed over their rapid outer track, or testing the blind brows and sharp curves of the infield 'Snake' layout.

This Elan +2 was the 118bhp first edition – a +2S version was announced as our test was published – and carried obvious drawbacks. Most obstructive, and shared with the two-seat Elan – was the dreaded rear drive-shaft doughnuts winding up and surging during stop-start driving in traffic. Since that sort of tedious going was a feature of our test routes (West London suburbs and Essex commutes) that got a mention in the review, along with very slow two-speed wipers and a very shallow boot.

However, for me, driving any Lotus was a privilege at the age of 22. The delicate steering and fine ride/handling balance was complemented in the +2 by bonus grip on wider tyres, and even more predictable extreme-cornering reactions, thanks to a longer wheelbase. I always was hot-headed, not overly impressed with a 0–60mph sprint occupying 9.4sec in the elongated Elan, when *CCC* had

▲ This yellow 1968 road Elan, seen at a pub classic gathering in 2019, tempted me to buy it, boasting a fine finish, healthy hardware and that classy cockpit at an affordable price. *Author*

DRIVEN

Elan +2 Road test and performance figures at Chobham track, 28 August 1968, published in CCC, November 1968, RPW 444G; £2,197 built (kit £1,718); Lotus-Ford 1,558cc twin-cam, 118bhp at 6,250rpm; 0–60mph in 9.4sec, 0–90mph in 20.8sec, 123mph max, 24–28mpg. **Elan 2.0 coupé race car** Track test at Silverstone Club circuit, October 1971, published in *Motor Sport* January 1971; David Brodie/Richard Lloyd club racer, ex- Jeff Goodliff hill-climb; Chevron front suspension, Racing Services 178bhp, full wood dash, electric windows (!); 15 laps of Silverstone Club, with best 1m 6.4sec/87.18mph, 1.8sec slower than class record. **Broadspeed Elan Sprint convertible BDA, tuned** Road test, 5–9 March 1972, published in *Motor Sport*, May 1972, DUE 886K; Broadspeed 1,601cc/170bhp BDA, Group 2 race four-speed gearbox, Salisbury limited-slip differential, uprated chassis, extended wheel arches; camshaft seized, no published acceleration figures, 17–22mpg running up to 7,500rpm.

➤ ▼ These two Lotus good-lookers were at a Club Lotus day at Castle Combe in 2017. They reminded me of my late-1960s days with *Cars & Car Conversions* magazine. The +2S was perhaps less exciting but more reliable than the Europas I experienced (to be fair two of the three were modified). *Author*

tuned examples of the smaller Elan twin-cam, or the Seven, turning 0–60 times in the 7sec bracket, along with sharper handling responses, which I preferred.

The factory-built Elan +2, at £2,197, also cost a fair amount, equivalent to £35,828 in 2022. Back then, versus contemporaries it was no sporting snip: the +2 had risen above the £2,000 barrier swiftly after debut. The flagship Lotus cost up to £500 more than contemporary pricey imports from Alfa Romeo (1750 GTV) and BMW (2002), never mind Jaguar's bargain-basement 4.2 E-type at £2,273. However, in hindsight I'd be kinder today and would agree that the Elan +2 was a fine Lotus in its own shapely right, and featured a more-practical body.

Elan 2.0 twin-cam coupé race car: highly effective club racer

There were a lot of Elans to drive during my formative working life, from production examples to tuned road cars (YYY 830G was a BRM-tuned favourite) and circuit racers. The Elan was, and remains, a wonderful combination of accessible pace and rewarding handling, and no wonder that there have been imitations of such individual style and agile format. Frustratingly, and sadly for their economy, the British never learned the lessons of reliable sporting performance and high-volume production that the Japanese understood, from Datsun 240Z to the world's biggest-selling sports-car MX-5 line.

I've chosen excerpts from an article published in January 1972 that I wrote for *Motor Sport* to relay the depth of my feelings for the charismatic Lotus, when the Elan established a versatile road and race quality before Colin Chapman decided to take Lotus upmarket, with a second

Elite and Esprit. I explained: 'Mr Chapman and his team called it the Elan, and it was welcomed by many enthusiasts for features such as the light and rust-proof glass-fibre body [oops, I forgot to mention the poorly rustproofed sheet-steel chassis] and the famous independent suspension, giving incredibly accurate handling, without the traditional spine-jarring ride of small British sports cars. It had other factors in its favour as well: terrific acceleration without the normal penalties of bulky sheet metal and fuel gobbling 6-, 8- or 12-cylinder engines, plus an easily tuned twin-overhead-camshaft engine that could be developed for even better road performance, with no fear of handling or braking deficiencies spoiling the job.'

Under the skin: a highly developed package

I then introduced: 'A currently successful Elan Sprint derivative that uses a massively enlarged 2.0-litre Twin Cam engine and Chevron B15 Formula Three front suspension. Driven by David Brodie (the most successful Escort driver at club level during 1970), and owned by Richard Lloyd, the public relations consultant, it has won six club races for modified sports cars, and holds the Brands Hatch Club circuit lap record. The writer was able to try this Elan recently and its track manners were so astonishingly refined that it seemed a good idea to explore, in print, the ways in which some of the lessons from this racing car can be applied to Lotus Elans acquired by our readers.'

That advice meant re-tracing some of the Elan experience that backed my comments then and now: 'The current [1971–72] Elan Sprint has the latest Tony Rudd development of the 82.6mm by 72.8mm (1,558cc) Twin Cam engine and is no mean performer in standard form. The press road test car we enjoyed *did* employ an rpm limiter, unlike some earlier Lotus demonstrators. Yet 126bhp at 6,500rpm was enough to pull just less than 14cwt [1,568lb/711kg] of red-and-white Sprint from rest to 60mph in little over 7sec: a time that even the V12 E-type has to breathe hard to beat. The big-valve Lotus engine has a torque curve which peaks at 5,500rpm and 113lb ft of torque. It incorporates some of the (deep-breathing) ingredients that the specialist engine builders have used in the past to extract more power from the Twin Cam unit.'

I talked to Ian Walker, who had sold both showroom and circuit Elans from its debut onwards. Walker sold modifications through his North London Ian Walker Racing (IWR) company. I asked him for an overview of tuning priorities: 'Walker commenced by pointing out that his company do six stages of engine tune for the car, at prices ranging from £40 to over £400, whilst braking, handling and transmission changes can be ignored, at least until the fourth stage of tune is applied.' Because the factory stopped quoting gross figures in favour of more realistic net/DIN results in 1971, Walker told us that customers could expect from a 20 to 50 per cent increase in brake-horsepower. IWR fitted replacement camshafts, and inserted new carburettor jets, before modifying a cylinder head, which featured revised inlet and exhaust porting, bigger valves, and a machined face to increase compression.

At Stage 4 level – approximately 150bhp – Walker recommended changing the front shock absorbers for the adjustable type. He reckoned the axle ratio could be revised to suit a particular purpose. For example, IWR put a higher-ratio differential into the rear of its Elan SE coupé, to find that the little car was accelerating to 100mph in under 20sec. During 1971–2, my comments recorded: 'A standard Sprint occupied 21sec to repeat that exercise, which would probably suit the dedicated traffic-light fiend down to the tarmac.'

On behalf of *Motor Sport* I then looked at circuit-racing options. I concluded there were two major options: 'Either shop around for a well-used ex-factory competition (or lightweight 26R) Elan, or rebuild it completely. Or build up a completely new car using a production backbone and glass-fibre body.' I added: 'The latter method was used by the builder of the Gold Seal Elan, back in 1969. The gentleman who did the original work was Jeff Goodliff, a director of British Vita Racing Team (BVRT) in Littleborough, Lancs. Just how successful he was can be seen from the fact that he won the 1970 Castrol BARC Hillclimb Championship in the car. Goodliff used a smaller-capacity Twin Cam engine than the present unit, which he planned to supercharge, but never did. Goodliff also carried out the redesign of the front end to carry the wide-based wishbones of the Chevron formula car. That operation involved the manufacture of a small spaceframe cradle attached to the front of the standard backbone, and enabled him to fit the uprights and disc brakes of the Chevron. At the rear, the suspension was based on Elan components, as were the brakes, the entire stopping system utilising aluminium Girling callipers. Koni shock absorbers were fitted at the front, and Armstrong adjustable units at the rear, using suitably-higher-poundage racing coil springs.'

When I drove the Brodie Elan Sprint in late 1971 for January 1972 publication, the Elan's exterior was distinguished by black-and-gold pinstripe paintwork, as made famous by John Player Special Lotus Formula 1 winners, livery also utilised by Brodie for his 2.1-litre club-racing Escort. Underneath extended wheel arches, genuine Minilite wheels of 8.5x13in dimensions carried Dunlop 350 intermediate-compound 200/550 tyres. Charles Beattie, a regular West London-based Brodie chassis sub-contractor, worked to produce the instant responses at circuit-racing speeds that his customer demanded. Luckily, the car was built up to incorporate easily adjustable suspension, so camber, ride and wheel alignment (toe-in and out) could be altered without a major fuss, but no rear anti-roll bar was installed when I drove.

The most obvious performance boost under Lloyd/Brodie parentage was to install one of the big-capacity twin-cams like those used by Brodie in his Escort. UK sports-car rules mandated less than 2-litres, so this Elan ran at 1,974cc instead of the legendary 'Run Baby Run' Escort's 2,150cc. This meant that a bore and stroke of 90mm x 77.62mm was needed, achieved by use of a new cylinder block, pistons, long-throw Gordon Allen steel crankshaft, and BRM steel connecting-rods. Hillthorne Engineering in West London carried out machining work, followed by assembly via Racing Services of Twickenham.

Incidentally, the pistons were forged from blanks provided by Brodie to run in the line-bored block. The credit for the big twin-cam concept (2-litres and more, rather than a production 1.6-litres) was attributed to Norman Abbott, who ran a specialist engineering business in Ilford, Essex. Abbott constructed some very clever and properly finished racing cars, including a dainty Formula Three car and an Escort with a spaceframe and independent rear suspension, hauled along by a Formula 2 Ford-Cosworth FVA engine. I wrote of the engine modification results: 'When Racing Services had completed the Elan unit, it was found to boast 178bhp at 6,800rpm and a very flat torque curve, culminating in 165lb ft at 5,800rpm. One of the few original BVRT engine parts was the 12.5:1 compression cylinder head with all the right cams and valves already installed and ready to go.'

The transmission was next on my list to detail: 'This strong torque output was entrusted to the Ford-developed Bullet gearbox (based on the Cortina GT unit, long ago) with close ratios, giving of their best when mated to a Borg & Beck clutch and 3.9:1 rear axle, which incorporates a Salisbury limited-slip differential. The unlikely combination of Brabham and Zodiac driveshaft couplings relay the power via Brabham F2 shafts outboard of the differential.'

The interior trim and some power-assisted luxuries were a total surprise, features I never saw repeated in a successful club racer: 'Left magnificently standard, even down to the electric windows, wooden dash and operable ventilation fan. Secured at all four points by safety harness, the writer hardly felt as if he was in a racing car when he looked around the interior. One could amuse oneself by flicking the electric window idly up and down, or popping the headlamps up and down, all whilst the owners were trying to relay important instructions.' I childishly observed that: 'It was a shame the radio had to go in favour of a proper chronometric rev-counter, even though all the standard instruments were connected up as well. Below the dashboard, a combined oil- and water-temperature gauge was suspended, its function being to indicate 60psi of oil pressure from the dry-sump system, and preferably less than 90°C on the section covering H_2O – otherwise there would be trouble brewing.'

Behind the wheel: few words needed to describe an easy drive

The track test résumé for such an effective racer was abbreviated, considering it was actually one of the best-known and now fondly remembered British competition cars of the period. When I put a faded picture of this Elan sliding round Becketts hairpin at Silverstone on Facebook in 2022, it proved one of the most popular pictures I had posted in a decade.

Here's what I reported rather curtly at the time: 'The car had already been fully warmed up by Brodie, so it was no trouble to start the engine and move quite smoothly away on to the Club track at Silverstone. No more than three or four laps were needed to find that the car was exceptionally easy to handle, whilst circulating in roughly the same times as we had managed after considerably more exertion in the Blydenstein Viva, track tested earlier this year. The brakes seemed to lack feel, but a second session with the Elan toward the cooler evening revealed that, though they could be improved for a proper racing driver, they were quite adequate for this tester. Our fastest lap occupied 1m 6.4s (87.18mph average), compared to the class record of 1m 4.6s (89.61mph) set up by Norman Cuthbert in a lightweight Elan with a smaller Twin Cam engine. The best lap came up after 15 laps or so, which serves better than any long-winded account to prove how easy it was to drive. Only snag that I know of is that the asking price would be £2,000 plus.'

That £2,000 in 1972 equated to £24,780 when I wrote this, actually ridiculously cheap for a UK race winner in 2023, or even an immaculate classic Elan.

Broadspeed Elan Sprint convertible BDA: thrills and failures

I'm at a damp and misty Brands Hatch on 5 March 1972 for my first race in the Janspeed Escort Sport. I was thrilled to win the class, but almost equally excited to pick up Broadspeed customer David Pannel's freshly converted (the engine had run 350 miles) Lotus Elan for a full *Motor Sport* road test. At least that was the plan. In the exhilarating days that followed, I had some fabulously exciting B-road miles in an Elan that officially had only 8bhp less than that the Brodie racing 2-litre recalled previously. Sadly, like so many tuned cars of the period, it's mechanical heart did not survive subsequent acceleration runs.

Judging from the fact that the 16-valve Cosworth never became a popular Elan transplant, in the way some subsequent Ford Zetec engine implants have been adopted, the roadgoing BDA motor was too tricky an ownership experience. Rare indeed is the Escort RS1600 or RS1800 in regular, unmolested, road use today.

Under the skin: radical revisions

Anyway, back in March 1972 I was just happy to have another Elan to steer, albeit the bright-orange finish and extended wheel arches meant it would attract a lot of often unwelcome public-road attention. The professionals at Broadspeed made the conversion sound a simple job, but admitted that a reshaped cross-member was required to accommodate the BDA (wet) sump. Plus, the radiator had to be moved forward, a Serck oil cooler installed, and fresh plumbing, including a four-branch tubular manifold, led out to a single large-bore tailpipe.

The BDA engine remained at the homologated 1,601cc, but was radically modified to incorporate much of the Formula Atlantic formula-car technology Broadspeed was supplying to paying customers outside its contracted Ford Escort racing programme. The cylinder head carried oversize valves and Cosworth BD2 camshafts, laboriously reworked around a 10.5:1 compression ratio. It gargled generous quantities of petrol at 17–22mpg via over-size Weber 45

'JUDGING FROM THE FACT THAT THE 16-VALVE COSWORTH NEVER BECAME A POPULAR ELAN TRANSPLANT, IN THE WAY SOME SUBSEQUENT FORD ZETEC ENGINE IMPLANTS HAVE BEEN ADOPTED, THE ROADGOING BDA MOTOR WAS TOO TRICKY AN OWNERSHIP EXPERIENCE.'

DCOE sidedraught carburettors. Basic engine-build durability precautions covered a Tuftrided crankshaft, whirring alongside balanced and lightened reciprocating components.

There were some radical revisions outside the engine bay as well. The four-speed gearbox featured much of the Escort's 1971 Group 2 technology, running needle-roller bearings and circuit ratios, including an elongated first gear. A production 3.77:1 final-drive was allied to a Salisbury limited-slip differential. Using a safe, sustained 6,500rpm we recorded gear speeds of 52, 68 and 90mph in the first three ratios. We didn't record an accurate top speed, as the hood blew off the screen mountings three times, so I garaged the car and used it hood-down only. The fastest I saw in fourth was 120mph, with 110mph coming up in places where you would be arrested on the spot today.

The chassis was thoroughly reworked, featuring replacement steering arms for the repositioned and stiffly mounted rack-and-pinion gear. Bilstein shock absorbers complimented shortened coil springs all round, and the Elan ran notably lower than any showroom contemporaries, on wide, but still 13in, Minilites carrying Goodyear 195-section GP-branded rubber. A vestigial front spoiler was added, but there was no mention of any brake uprates!

The results of such engine work unharnessed much of the 200bhp potential the Cosworth BDA series contained, particularly at higher revs. Torque was notably weak, at 125.8lb ft by an elevated 5,500rpm, and brake-horsepower didn't reach that 170bhp summit until 6,200rpm. So it wasn't such a surprise when I was told on handover that the rev limit was: '7,000 – but you can use 7,500 *if* you really must.' There was a pause and snort of laughter from the Broadspeed technician: 'Slide the clutch out slowly, keep the revs above 3,000 – and I bet you'll stall it!'

Behind the wheel: blindingly fast on the road

Elan and Walton made it out of the uphill wet grass exit to Brands that day. I survived commuting the very orange wonder into London and out to our Essex printers, because I learned you could get the plot rolling with the clutch fully out from 1,200rpm on flat ground, with practice… In such traffic conditions the water temperature would climb close to 90°C, but the electric fan could haul it back to 85°C safely.

The strongest memories came from a mid-week solo evening outing I made from West London down to my Mum's home in West Sussex. My diary simply records of the trip: 'Blindingly fast along the lanes with the roof down.' *Motor Sport*'s published version was more detailed and ignored the advice I'd been given by a legendary motoring writer: 'Never write about anything you enjoy – somebody somewhere will put a stop to whatever it is!'

Remembering this was written 50 years ago, when Lotus and reliability were unfamiliar travelling companions, here are excerpts from what appeared in that report: 'As the Elan swooped along country lanes, I reflected on how much we miss in an ordinary saloon. Would that it was practical for everyone to enjoy the thrills of extracting perfect unison from engine, chassis, brakes and gearbox, as is possible with this Elan…. It is the sheer verve with which the car responds that endears it to the driver, especially on the sort of switchback represented by an 80mph crest, followed by a downhill 60mph left-hander, 70mph right, bumpy 110mph straight and 30mph downhill hairpin. To follow that sort of terrain you need the Elan's red-hot-needle performance package. This is a sophisticated Lotus Seven, for it is unbelievably rapid in the sort of situations a skilled driver could enjoy, whilst offering creature comforts when required. Even the electric windows worked throughout our tenure!'

Sadly, the driven bliss ended abruptly after five days. We took the Elan-BDA to Chobham, clamped the fifth wheel over the rear number plate via a convenient plank, and managed 12 runs. Practically, that meant just six statistics could have been published, as we followed industry practice in searching for the best two-way-average acceleration times. The results were poor – in other words inferior to the factory showroom demonstrators – and then a camshaft seized in its carrier, and that was that.

We never got that Elan-BDA back after publishing our findings.

Vixen S2 1600 GT: Blackpool rocks!

My acquaintance with steering a TVR with Ford power began with a full road test of a pretty Vixen 1600 GT. Outwardly, I was still credited as an Associate Editor of a now flourishing magazine, but it was all change within. For a start, the magazine moved offices from handy Ealing to premises adjacent to East Acton tube station: *CCC* was then soon sold to the larger Link House Group in Croydon.

I was also about to leave *CCC* with three months' salary after refusing to work alongside a second Associate Editor, recruited during my absence, driving to Sicily and back for a Targa Florio road-race feature. So, I spent the dodgy summer of my 23rd year sweating about maintaining a mortgage and supporting a wife and two kids. I tried selling cars on commission, and writing job-application letters to such varied potential employers as Castrol and the *Financial Times*, before *Motoring News* sports editor Andrew Marriott ensured I was interviewed by the owner of that weekly paper and *Motor Sport* magazine.

I became a feature writer/reporter at the Teesdale Group, owned by the Tee family of strongly individual Victorian Christian characters: they loved musical and operatic daily business outbursts. They published *Motoring News* and *Motor Sport*, amongst other titles on subjects such as guns and motorcycling. As with Ford, I worked as a freelance for *CCC* and the Teesdale magazines into the '90s, long after I left employment with both organisations.

Ah yes, the attractive TVR Vixen S2. My article was published in the May 1969 issue of *CCC*. The magazine became more sophisticated during 1969: although the TVR test was just two pages of grey text, a third sheet was devoted to colour photography, offering a side action shot that emphasised the TVR's distinctive style marriage between sleek sports-car nose and abbreviated tail carrying large wrap-round tinted rear-window glass. All this was an echo of an earlier Griffith V8/Tuscan V6 coupé, sadly minus the Griffith's beefy Ford V8, or the Cortina tail-lights. However, there was a connection under the glass-fibre panelling, as a later and reportedly much stronger Tuscan tubular chassis debuted at the 1968 British motor show, subsequently shared by the S2 Vixen.

The test Vixen was hauled by the more affordable and frugal 1600 GT crossflow, which looked pretty standard, but had the benefit of a tubular four-branch exhaust manifold tailored within the spacious engine bay. It was rated at 92bhp, rather than the usual Ford quotes in the 84–88bhp bracket. You could pay extra and have power uprated by double World Champion Graham Hill's Speedwell tuning company.

CCC road tested a pretty basic specification, albeit it did sport wire 5Jx15in wheels and the weight was skimpy by today's standards at 1,680lb (762kg). The Ford inline four-cylinder was hitched to a contemporary Cortina four-speed manual gearbox and a 3.89:1 final-drive, allowing a sustained 75mph at 4,000rpm. Since the tank generously held 15 gallons of leaded fuel, you could travel for miles at 25–30mpg.

We were asked to observe a 6,000rpm limit – the car was a 10,000-mile demonstrator from

DRIVEN

TVR Vixen S2 1600 GT Colour road test for CCC, May 1969, CTW 932G; Burnt Yellow dealer demonstrator, £1,212; standard Ford 88bhp 1.6-litre crossflow; weight, 1,680lb (762kg); 0–60mph in 11.1sec, 0–90mph in 27.5sec, 104mph max; exhaust centre box cracked, Ford at Boreham fixed; nimble. **1969 TVR Tuscan and Vixen** Double road/track test, August 1970, published in *Motoring News*, 10 September 1970; showroom cars from Barnet Motor Co; Tuscan V6 used London–Nürburgring, 139bhp, 0–60mph in 8.2sec, 124mph max, 24mpg, £1,678; Vixen, DYF 2J, 88bhp, 0–60mph in 10.2sec, 110mph max, £1,342 including £100.50p sunroof. **TVR 1600/3000 race cars** Track test of two TVR production racers with Clive Richardson, 1976, written by Clive, published in *Motor Sport*, October 1976. **TVR 3000S race car** Track test, March 1981, published in *Motor Sport*; 1980 BRDC/DB Motors Champion Colin Blower's TVR 3000, prepared by Blower/Peter Butterworth; Racing Services 200bhp/7,000rpm race V6; quirky and quick!

Bridge Motors at Bocking in Essex – which allowed 40mph in first, but only 55 in second. That meant the 0–60mph benchmark statistic was never going to be as good as it could be, entailing an extra change into an 80mph third. However, in period it wasn't embarrassing at 11.1sec average, and 90mph came up in less than half a minute. The best top speed we measured was 106mph, about 10mph short of the theoretical maximum on this gearing, *if* it could have pulled the full 6,000rpm in fourth.

I liked the small-capacity TVR. The handling was rewarding, and traction outstanding. So much so that I had to use over 4,000rpm and a clutch-balancing act for worthwhile standing-start acceleration runs. The steering was quick enough to catch the plot tidily over wet tarmac under full power. It was also a lot more comfortable to drive than pretty well any British sports car at the price, and quite baggage-practical, with a good-sized carpeted area beneath that large rear window.

However, we had this dealer demo TVR for longer than usual (10 days) and – inevitably – there were faults. As delivered, abused brakes wilted, along with the Dunlop SP tyres. We also needed the centre-section exhaust box welded up at Ford of Boreham. Those aristocrats of rough-road resourcefulness complimented the TVR chassis and wishbone front suspension. Other snags included a wickedly out-of-balance propshaft and the lack of an engine fan, which led to one boiling incident. I suspected the demonstrator had been kept short of water: when refilled it never repeated the exercise.

Rather rare and undervalued today, I think a private owner could made a better job of the Vixen S2 than the example we tried – and should obtain a great deal of pleasure from the TVR's practicality.

1969 Tuscan and Vixen: a pair of TVRs and a Nürburgring adventure

August 1970 brought a second chance to test and enjoy a 1.6 Vixen. The long-distance adventures arrived over the 20–24 August packaged alongside a 3.0 TVR Tuscan, carrying Ford Essex V6 power and four-speed Ford Zodiac transmission, plus an electric overdrive on third and fourth gears. Whilst the dainty Vixen flaunted her optional £100 Webasto sunroof and danced for *Motoring News* staff and me at home, I picked up the 3-litre Tuscan from fabled Barnet Motor Co. That leading TVR dealership was earlier the backer and employer of the legendary Gerry Marshall, who twirled the rare, original Ford V8 Tuscan to race and hill-climb wins, before his Dealer Team Vauxhall decade.

I don't think I have ever worked with TVRs without dramas from the start. Despite operational failures, that Tuscan satisfied me all the way to the Nürburgring and back. It was an exciting prospect, as the largely unmodified V6 had been lobbed into the tubular spaceframe to mate with running gear that originally resisted 124bhp more from the Mustang's 289 V8. The V6 carried tubular exhaust manifolding and a simple sports air cleaner, and was now rated at 147 *very* gross horsepower. The Tuscan name remained for the V8-to-V6 transition – and would surface again in '90s TVRs with exhilarating Rover V8s for road and track.

Back in late summer 1970, I zoomed off from the dealership in a pretty foul mood, as I had been forced to wait for the Tuscan to be fettled; 22 years later at another TVR dealership, the handover delay was longer, but the driving experience also totally rewarding. I headed for a sleepless four-hour overnight ferry crossing to Ostend. The minor setbacks continued in Belgium, frequently going off route, not assisted in those pre-sat-nav days by the Tuscan's lights blinking out on main beam, the horn intermittently parping as the steering lock was applied, and the winkers staying resolutely lit *if* dipped beam was selected for the one headlight that stayed with us the longest!

Fortunately, the well-lit E42 motorway carried the long-legged TVR across Belgium at a regular 105–110mph and allowed an exit just over the German border for my hotel at Blankenheim. Knocking on the door of the proprietor at 3am sounds outrageous, but it had been friendly and affordable accommodation for *Motoring News* and *Motor Sport* staff for years.

I slept until about 8.30am and then enjoyed a short run out to the circuit, where the 86-hour Marathon de la Route completed its apparently endless competitive mileage. In 1960 the event

TVR VIXEN ROAD TEST

IF YOU'RE on the lookout for a pretty wild looking form of locomotion which is exciting to drive, more reliable than average with cheap spares — then come along with us as we try the Vixen Series 2. This 1600c.c. machine can hold its own with Porsches and Elans on twisty roads and the price tag shouldn't leave the bank manager gibbering on the counter. On paper it doesn't look like a ball of fire — but try it on B class roads and you'll find that superbly spaced cogs, a terrific adhesion limit that stays with you even when all seems to be lost, coupled with inspiring "hurlability" — make driving a fun happening for even the most jaded enthusiast. There's also a true top speed of just over 105 m.p.h. and easy to bear fuel consumption from the standard Ford engine. For more detail come peek with us at the eye-boggling machine you see strewn all over these pages. If there aren't any pictures use your initiative and imagine them!

Life, we've often been moved to remark, is not a bowl of cherries or apples or even grapefruit for the smaller specialist car makers. Most of 'em have beautifully bodied machinery to offer using as near standard running gear as possible in the interests of long life and cheap spares. It also helps a small maker if he can use plenty of miscellaneous bits that the mass producer uses for his big sellers: things like windscreens, tailamps and instruments can cost a bomb if they've got to be made specially for a car which isn't going to sell more than five hundred units a year. So the specialist manufacturer plumps for as many standard parts as can be used. Even then his problems have just started: delivery of the bits may lag behind — especially if there's a strike such as Ford had recently. Even when the men go back to work there's bound to be a shortage of parts until they've caught up with orders.

One of these specialists who relies on a Ford engine and gearbox is TVR Engineering of Fielding's Industrial Estate, Bispham Road, Blackpool. They market three versions of the Vixen S2, all of which use the Cortina 1600 GT engine and box: we tested a standard model, but the other two versions offer a bit more go using Speedwell engine mods. Using the same chassis and similar suspension, TVR also sell a V6 Zodiac engined device called the Tuscan — a name that was used for the V8 powered TVR in the past.

Martin Lilley took over as the boss in November 1965, which was a few months after the firm had announced the Mk 4 1800 S engined Vixen with the large rear window and Mk 1 Cortina tailamps: the 1800 engine, gearbox and final drive were all taken from the MG B. The type we tested started life as the Series 1 Vixen back in 1967. For the 1968 Motor Show they produced a very much improved model, still with Ford's 1600 GT crossflow power, but having the V8 Tuscan tubular chassis — which is longer and much stronger. Another change for the better was to bolt the body onto the all-independently suspended chassis, instead of the previous bonding process. Since then TVR have reduced the price by a hundred quid and changed the interior trim to comply with American safety regs; late model cars also have a Tuscan V6 styled bonnet. Although ours had a G registration plate, it didn't have any of the above features.

The test car is owned by Bridge Motors of Bocking, Nr Braintree, Essex, and is regularly used as a customer's demo car. Which we all appreciate means its had a hard life — and then some! So what had gone wrong with the car during at least 10,000 miles of vigorous use? The answer seemed to be nothing much: the brakes and one tyre had seen better days, but were still within safe limits; the exhaust system had been grounded, cracking the centre silencer box — the mechanics at Ford Boreham fixed this in 20 minutes, after they'd all had a crawl round the underneath muttering favourablly about the chassis design. In spite of the fact that the basic Vixen design has been around for some time, most of its features strike us as being up-to-the-minute; while some of 'em are ahead of the opposition.

Among the Goodies we count independent suspension for all corners, managed by upper and lower wishbones (looking at the top front W. bones a Boreham mechanic sighed, "baby that's really the wildest shape!"), which have coil spring and damper units attached to the lower arms, at the top they're bolted onto the chassis. TVR don't stop worrying about the roadholding at that point, no they carry on and provide five inch rims shod with Dunlop SP radials. Add the fact that the front track is 53 inches and the rear 54, plus a low centre of gravity, and anyone can see that this is going to be fun around the curves.

We've covered the engine and gearbox by saying it's as used in the current Cortina GT; which is one of the nicest combinations we've come across. The motor is oversquare and gives claimed 92 b.h.p. at five-five with enough torque to pull 15 cwt. of TVR around very nicely, thank you. The gearbox is all-beautiful with four forward slots and what feels like synchro reverse! The final drive differential comes from a Vitesse and has a ratio of 3.89. Coupled to 15 inch wheels this gives a relaxed 18.8 m.p.h. per thousand revs in fourth: in uvver wuurrds 112 point something miles an hour at 6,000 r.p.m. in top is the theoretical top speed — the rev limit being six. This gearing gives really useful top speeds in the lower gears: at the red line we got nearly 40 m.p.h. in first, 55 in second and 80 in third. That high first gear was a blessing to us, but a curse to TVR, because it

464

465

▲ In the summer of 1969 I had my first extended experience of a TVR with Ford power for *Cars & Car Conversions*. The pretty Vixen, with 1.6 litre ex-Cortina/Capri four-cylinder power has become a rare classic today, yet it proved agile, affordable and practical 106mph fun. *Speed Sport Publications/ Author's archive*

had been run over the old Liège–Rome–Liège rough-road and stage-rally route, and Pat Moss had won, beating all the top blokes and conquering the cockpit heat of her factory Austin-Healey 3000. Partly in homage to that victory, British Leyland sent over a rapid but fragile contingent of works machines: the pioneering 4.3-litre Rover P6 V8 saloon initially led for 16 hours, then a Mini Clubman from Abingdon took over and lasted nearly to the end, but by that time, factory-crewed and prepared Porsche 914/6s had swept into 1–2–3 winning formation. There were interesting drivers within those works 914s including today's high-profile Red Bull Formula 1's manipulator Helmut Marko in the winning car.

In the absence of Abingdon in the paying positions, I talked to hardy British privateers. I encountered the resourceful Holman Blackburn, for many years *the* expert on competing in private and semi-works Capris, culminating in the 1975 purchase of the ex-Boreham Capri IIs on behalf of Hermetite (Chapter 8). At Nürburgring in 1970, Holman ran a 3-litre Capri with a modified version of the V6 I had in the Tuscan, which created a lot of interest in that vast and cobbled German paddock. Together with his team-mates, including a serving army officer, Holman finished an honourable 12th overall of 24 survivors in a massive entry. Naturally, I shared their traditional British breakfast fry-ups – and rewarded that worthy achievement by misspelling their names in the race report!

At the conclusion of the Nürburgring competition, the competitors swept back into Belgium to Liège for the prize-giving. Creditably, the TVR managed to stay in the high-speed convoy, most machines still wearing their competition wheels and tyres. I returned via the Jabbeke Belgian motorway, one that had seen so many British sports-car speed records set in the '50s, including a 132mph run from a Jaguar XK120. The TVR recorded a kilometre-post-checked 124mph/130mph-indicated on that section. I was back at home, typing up a lengthy race report and TVR notes by Sunday. The TVR's stubby 12ft 1in length flitted back into London and then overnights in Essex to complete the long

TVR TUSCAN V6 PERFORMANCE

Figures taken from Motoring News, August 1970.

Acceleration

Vixen 1600 GT figures in brackets

0–30mph 2.5sec (3.6sec) **0–40mph** 3.7sec (5.5sec) **0–50mph** 6.0sec (7.4sec) **0–60mph** 8.2sec (10.2sec) **0–70mph** 11.1sec (13.1sec) **0–80mph** 14.0sec (17.0sec) **0–90mph** 18.3sec (22.1sec) **0–100mph** 24.8sec **Gear speeds at 6,000rpm** First: 38mph; second: 55mph; third: 86mph; third overdrive: 105mph; fourth: 121mph; fourth overdrive: 124mph

Overall economy

Average 24mpg

Price

Kit £1,558 plus £120 of options

race weekend on Tuesday afternoon, *Motoring News* appearing on Thursday news-stands.

I think I gave that 3.0 Tuscan and myself a thorough workout. Amongst my conclusions for *Motoring News* (16 September 1970) were these remarks: 'We came back from the trip very pleased with all important aspects of this enthusiast's car, such as braking, steering (despite the strong kickback) and straight line go… The V6 offers exceptional value for money at £1,558 in component form… Quick enough for most people and yet easy to maintain.'

1600/3000 production sports-car racers

It was 1976 before I turned laps in a track TVR, and then it was as a support act to text by Clive Richardson in *Motor Sport*. We had a troubled day around the Silverstone Club circuit with a pair of representatives from a then booming Production sports car racing category: similar to my 1972 Production saloon car season in that the class divisions were decided by price, rather than the usual cubic-capacity splits. There was just as much shouting and controversy about permitted modifications and eligible vehicles within both series, championship titles sometimes being decided in off-track tribunals, plus plenty of back-chat about the performance of Chris Meek in a Ford twin-cam Lotus Europa, the only mid-engine front-runner seeing off conventional higher-power, front-engine, rear-drive machinery.

That day in 1976 we had a 3-litre TVR Tuscan in 200bhp Racing Services V6 format from self-described 'Hinckley knicker manufacturer' Rod Gretton. However, the more durable choice was a sister TVR-Ford 1600 GT with an engine provided by David Minister of Formula Ford power fame. The smaller-engine TVR, for experienced 1975 champion Chris Alford, had been a consistent 1976 winner: 17 of 18 starts can't be bad!

The Alford 1.6-litre TVR was credited with 102bhp and was great, agile fun. Sadly, the Gretton V6 TVR suffered a broken throttle linkage within two laps of me getting behind the wheel, and I only got an exciting taste of the bellowing V6. However, in 1981 I would be rewarded for some patience, having another day at Silverstone with a different Hinckley-based TVR racer, the successful Colin Blower 3000S convertible.

Meanwhile, it was technically interesting in 1976 to compare four- and six-cylinder TVRs… In Clive's words they were astonishingly similar outside the engine bays: 'Not only do the 1600 and 3000M TVRs look alike; they really are almost identical apart from engines and transmissions. The bodies, tubular chassis and basic all-round double-wishbone suspension are the same. In the case of these racers, the 1600 was fitted with coil springs from the Triumph 2.5-engined 2500M, the 3-litre has standard springs, and all eight corners wear Spax adjustable shock absorbers. Even the 6in-wide, 14in-diameter alloy wheels are shared, along with TR6-type Girling disc/drum brakes, lined in both cases with DS11 pads and identical standard rear linings.

'Understandably, therefore, the 16cwt (1,792lb/813kg) 1600 stops better than the 18cwt (2,016lb/914kg) 3000M. The 1600 has a Spitfire differential with 4.1:1 final-drive, the 3000M a TR6 differential with 3.45:1 final-drive; neither has a limited-slip differential, banned unless standard, which was applicable only to the Morgan Plus 8. A standard 3-litre Capri gearbox is housed in the 3000M, the standard small Ford gearbox, which Alford believes to be of Escort Sport type, sitting in the 1600 GT.

'Power is certainly not the answer to the 1600's success: 102bhp at 6,000rpm recorded at the flywheel. Alford claims careful blueprinting and the right choice of standard camshaft is the limit of the work carried out on the 1600 GT engine by Dartford-based David Minister. The result is impeccable reliability, and no maintenance is required between races. It was also a cheap engine to prepare initially: just £120-worth of work on top of the brand-new engine originally delivered. The end-of-season rebuild cost £100. Both prices included dynamometer testing.'

Those were the racing days!

3000S: on track for '80s success

In the '80s and '90s I had a lot of contact with the meticulous preparation and driving skills of Leicestershire-based Colin Blower at Hinckley and his local Mallory Park circuit. We had a common competition-tyre contract at BF Goodrich (one Colin held for a decade), but his path to championship success was always calculated in a different direction from the track herd. In the later '80s, when Production saloon car racing was overwhelmed with Sierra Cosworths at the sharp end, Colin opted for Mitsubishi's Starion turbo. So quick, we even tried to qualify for a full Group A European Championship TT race at Silverstone. Despite the charitable best efforts of fellow motoring writer/race ace Pierre Dieudonné to tow me along fractions faster in the wake of his TWR Jaguar XJ-S, we failed.

Initially, and for more than a decade, Colin was a Production sports car star, right from a 1973 road-car MGB and then a 1974 Lotus Europa season that saw some heroic tussles with Jaguar engineering employee Peter Taylor's rapid Jaguar V12, and nearly netted a championship in just his second race season. From 1975 onwards, 'CB' hooked up with TVR, and by the time I tested his fourth such Blackpool production sports racer, he had both commercial sponsorship and additional employees, required, as he prepared many other road and track TVRs. I particularly liked his amiable sense of humour and his ability to go motor racing effectively without ruining his embryonic business. That mischievous nature and commercial common-sense was typified by the Indestructible Socks local sponsorship on the 3000S I tested on a blustery and typically wet/dry Silverstone track. Nobody forgot that sponsor, emblazoned in big red capitals on a winning TVR!

That was in December 1980 for 1981 publication as two pages in *Motor Sport* with just one grainy black-and-white image of TVR #14 sliding out of Becketts. It hadn't been an easy route to Colin becoming 1980 BRDC/DB Motors Champion, as I explained, having raced in the category with the Caterham 7: 'It is worth making a point about the way Blower won his title. While Meek ran a class contender (either Panther Lima 2.3 or TVR 1600) versus largely Ginetta G15 1-litre opposition, Blower had to

▼ Colin Blower's championship-winning TVR 3000S convertible proved another exciting drive when tested for *Motor Sport* in 1981. I always enjoyed the Blackpool machines in Ford 1.6 or 3-litre formats, but I have equally warm road and race memories of the Rover V8 Griffiths, Tuscans and Chimaeras in the 1990s. *Author's archive*

TRACK TEST

TVR's champion convertible

J. W. steers into a drift with Colin Blower's production (well, nearly) TVR convertible on a wet Silverstone corner.

FOR the majority of seasons since the 1973 start of production sports car racing, it has been easy to answer readers' queries over who was champion in a particular year. You spelt the name of Leeds entrepreneur Chris Meek. Then added the brand of car he drove that particular season. A Lotus Europa being the most effective of all, though Meek seems to have driven virtually every machine from MG Midget upward, with great success in this category. When readers ask about 1980, it will be different. For the established BRDC series, the DB Motors of Leicester Championship, went to Hinckley, Leics, garage owner Colin Blower, in a TVR convertible.

Colin was also declared champion in the BRSCC-organised CAV series for similar cars and competitors, but Meek protested the points awarded at an Aintree round. An acrimonious RAC tribunal followed and Meek was subsequently declared champion, even though the BRSCC's Peter Browning was practically foaming at the mouth after a display of "club sportsmanship" that made the FISA/FOCA row look like a good-natured discussion.

Bravely pushing aside the unacceptable face of motor racing, we decided to talk to Colin Blower about his long-standing interest in the category. He agreed to meet us at Silverstone one blustery pre-Christmas day, bringing with him the 200 + horsepower TVR that netted him most of the 22 wins in 24 outings that he scored in 1980. We tried this exciting convertible, officially designated the 3000S, listening later to Blower's equally interesting 1981 project to run the 2.8 fuel injected TVR Tasmin. This will be the first car to utilise a racing version of the 2.8i Ford engine, to our knowledge, though that V6 has a cylinder block that is a direct result of the 1970s racing programme.

A mechanic by trade, Colin Blower has stuck with production sports car racing — a category where modifications are theoretically quite limited, like Group 1 racing of saloon cars, but which now offers rather more scope for tuning than originally intended! Blower first saw the category when spectating at his local Mallory Park circuit in 1973. "I had a brand new MGB at the time, and I was hooked! My first race was soon afterwards at Llandow in the B. I came last, but what an eye-opener after spectating! Everyone else looked like lunatics on the track, but I soon learned to join in . . ." By the end of the year his self-prepared B was capable of showing reasonable class results, winning the division at the last Silverstone of the season.

For 1974 Blower bought a Europa and "enjoyed a fabulous year dicing for first and second overall. We nearly won the title that time, but it was a fascinating season because our main rival was Peter Taylor from Jaguar." Amongst other duties Peter conscientiously ensured the health of press road test vehicles that usually have more mileage displayed than those of any rival manufacturer, yet run superbly.

"It was so interesting because Peter's E-type V12 was fast along the straight and our Europa was nippier around the twiddly bits. The result was some very close racing that I will never forget."

Even during 1974 Blower was talking to TVR management, a logical development for a garage owner with a high proportion of TVR clientele. "Since 1975 I have run TVRs, this convertible being the fourth one. They've been very good to me, but as with most others in club racing today I couldn't afford to run without additional sponsorship." As with many effective sponsorship arrangements Colin Blower's backing came from a local company and a casual acquaintanceship. The result is the legend across the red convertible's flanks to the effect that Indestructible Socks are the greatest: nobody forgets Colin's sponsor in a hurry!

From our own earlier participation in prodsports with the Caterham Seven it is worth making a point about the way Blower won his title. While Meek ran a class contender (either Panther Lima 2.3 or TVR 1600) versus largely Ginetta G154 1-litre opposition, Blower had to fight three quick Morgan 3½-litre V8s. Colin knows the Morgan car and Charles Morgan well, for he raced against ITN cameraman Charles closely in 1979. At the end of that season, instead of courtroom rancour, the two rivals swopped cars and tried the merits of each for themselves. That's more like the club racing spirit . . .

The Car

Built by Blower in what spare time he could find, the convertible we tried was actually the second such racing TVR that Blower and local enthusiasts headed by Peter Butterworth had constructed. The first, apparently lightweight model, was destroyed in a 110 m.p.h. crash at Gerrards Bend, Mallory Park, in April 1980.

"A driveshaft broke and it rolled, and rolled, and rolled!" said Blower, his eyes still wide at the memory, "it seemed to dismantle itself around me. Doors, everything seemed to fly off until I ended up clutching the steering wheel with a seat and a roll cage for company: I let go of the steering wheel and it flopped in my lap! Incredibly I had just two cracked ribs." A seat bolted down to the chassis and proper seat belts were the keys to his survival.

They missed only two Championship races. The result was the TVR we drove, of which the heart is a Ford 3-litre V6 with a single production twin-choke Weber carburetter. The V6 was built by Racing Services, now at Hitchin rather than the Twickenham base we once featured.

Blower conservatively quotes 200 b.h.p. at 7,000 r.p.m. for the V6. Similar units in Group I saloon cars gave 220 b.h.p. and the 1980 mid-season figure for such V6s, in a slightly more developed form for the carburation, was in excess of 250. I was quoted about £2,000 as the cost of the modified engine, which features a 9.8:1 c.r. (a lot less than those saloons I was talking of), a camshaft to the measurements Ford specify (which allows a profile akin to the Weslake inspired 649 of BMC Mini fame, a profile still used by Mini racers today) and a thorough balance of standard crankshaft and associated reciprocating components. The engine is the smoothest and most satisfying of the many Ford V6s we have encountered over the years.

The gearbox is a production Ford 3-litre Capri unit (close ratio gears from Hewland are available in the production casing, but not permitted in this category), and the clutch is also from a production Capri.

TVR's independent wishbone rear suspension allows the use of a Jaguar Salisbury differential. The limited slip is torqued to 90 lb. and the ratio will vary between 4.5 and 4.7:1, according to circuit.

TVR suspension is ideal for the production racer because the small volume manufacturer provides quick adjustment facilities for all important aspects, including castor and camber, in much the same way as a formula racing car. Selecting a spring is eased too, for TVR say they have tried just about everything in production at one time or another, so the competitor can pick from a wide choice of ratings — unlike a TR7 owner, for example.

Blower commented, "we just said to TVR what we wanted to do and they sent down some suitable hard springs, which also provide a lower ride height than usual. We use Spax adjustable damping. Konis would probably be better, but the Spax adjustable feature is valuable to us. Basically we keep the front hard and the rear soft with a hefty front anti-roll bar also included. We tried a rear anti-roll bar, but traction was not so good. Now it is smashing off the line, just blows the others off every time!" For our test it was necessary to run a softer front bar as it rained

302 MOTOR SPORT, MARCH 1981

fight three quick Morgan 3.5-litre V8s. Colin knew the Morgan car and Charles Morgan well, for he raced against ITN cameraman/former Morgan heir Charles closely in 1979. At the end of that season, instead of courtroom rancour, the two rivals swopped cars and tried the merits of each for themselves.'

What went into a winning Production sports car 42 years ago? Sat inside the winning TVR, I appreciated a rather smarter interior than most club racers of the era, partly Colin's preparation standards and also the vehicle was a lot fresher than most, barely a season old. There was an eye-rolling reason for that. The convertible we tried was actually the second such racing TVR that Blower and local enthusiasts, headed by Peter Butterworth, had constructed. Colin explained that the first, apparently lightweight, model was destroyed in a 110mph crash at Gerard's, Mallory Park, in April 1980.

Colin reported: 'A driveshaft broke and it rolled, and rolled, and rolled. It seemed to dismantle itself around me. Doors, everything seemed to fly off until I ended up clutching the steering wheel with a seat and a roll cage for company: I let go of the steering wheel and it flopped in my lap! Incredibly I had just two cracked ribs.' That was said with wide eyes and a chuckle. Apparently, the race seat bolted down to the chassis and proper safety belts were the keys to his survival.

Under the skin: an ideal production base

They missed only two championship races before the TVR I drove was up and ready for battle. The heart was a Ford 3-litre V6 with a single production twin-choke Weber carburettor. Racing Services had rebuilt that V6; Blower conservatively quoted 200bhp at 7,000rpm for the engine, which gave no more than 138bhp in the showroom. Similar units in Group 1 saloon cars gave 220–250bhp in 1980. I was quoted about £2,000 as the cost of the modified engine, which featured a modest racing 9.8:1 compression ratio, a competition-profiled camshaft to the measurements Ford specified, and a thorough balance of standard crankshaft and associated reciprocating components. That engine was amongst the smoothest and most satisfying of the many Ford V6s I encountered.

The gearbox was a production Ford 3-litre Capri unit (close-ratio gears not permitted), and the clutch was also from a contemporary production Capri. TVR's independent wishbone rear suspension allowed the use of a Jaguar Salisbury limited-slip differential. The final-drive ratio varied between 4.5 and 4.7:1, according to circuit.

The TVR's suspension was ideal for the production racer, because the small-volume manufacturer provided quick-adjustment facilities for critical aspects, including castor and camber, as for a formula racing car. Selecting a spring rate was eased too, as TVR had tried just about everything in production at one time or another, so competitors could pick from a wide choice. Blower commented in 1980: 'We just said to TVR what we wanted to do and they sent down some suitable hard springs, which also provide a lower height than usual. We use Spax adjustable damping. Konis would probably be better, but the Spax adjustable feature is valuable to us. Basically, we keep the front hard and the rear soft with a hefty front anti-roll bar also included. We tried a rear anti-roll bar, but traction wasn't so good. Now it is smashing off the line, just blows the others off every time!' For our test it was necessary to run a softer front bar, as it rained pretty much throughout.

A production brake layout of discs at the front and drums at the rear featured, deploying anti-fade friction materials, with the brake bias being towards the front. Wheels and tyres to make this 2,100lb (953kg) TVR trustworthy were Kleber V12 GTS in 195/70 HR (front) and 205/70 HR (rear). We used wets, which in this case simply means they have more tread than the worn-down road tyres turned into semi-slicks by the production racers. Colin said of the Klebers: 'They suit the car well. The soft walls like some negative camber to give their best, and the adjustable suspension allows us to get the best from them. I have tried Michelin, but didn't find them so predictable.' The alloy TVR wheels were a generous 14in-diameter, which allowed a good-size brake disc to be properly ventilated at the front.

Behind the wheel: a slippery customer

Technicalities absorbed, it was time to clamber within for a pretty adventurous session for a visitor faced with plenty of wet track sections and ample power to embarrass us. The cloth finish of the bucket seat, and a Britax full harness, combined with a sensible fire extinguisher to lull the new driver into thinking a straightforward task awaited. Not so.

Colin Blower didn't practise traditional heel-and-toe, brake-to-accelerator pedal ballet. Always the individual, Colin reversed the process and toe'd and heeled! Thus the throttle pedal was almost impossible to blip during hard braking and I settled for separating each action. The Formula sports steering wheel exhibited a bit of unwelcome wobble and, the *pièce de résistance*, ventilation was boosted by a driver's side window that wanted to sit on my lap at speed.

Instruments were straightforward and largely of TVR production origins. Oil pressure was indicated at a steady 60psi and water at 90°C. Clearances inside the V6 tend to be generous, in a racing application anyway. This thrice-rebuilt unit had even larger piston-to-bore and valve-guide tolerances than normal, to judge by the blue smoke wreathed around the twin rear pipes whenever the engine was started.

As soon as the motor was running, detail thoughts disappeared. The uneven lope that passed for a 2,000rpm idle yielded to a solid beat of power as 3,000rpm arrived in first gear. As Blower remarked, the rear-end traction and the V6's sheer 'get-up-and-go' spirit were remarkable. The first three gears were used up quickly, which meant this TVR travelled at approximately 110mph when you selected top. Along the Silverstone Club straight, the rpm built, before Woodcote demanded that 6,000rpm in fourth became unwise, emphasised by an impending foot-pedal tap dance, cumbersomely required to change down from approximately 128mph, working through Blower's unique brake and accelerator spacing.

For the opening laps, I concentrated on changing gear with the gear lever, instead of mistaking the equally stubby and convenient handbrake for the gear-selection control. I added the refinement of a change in routine from my usual heel-and-toe, switching to brake, clutch in, blip throttle, and change down in separate movements. This cumbersome sequence occupied so much space that I soon settled for what little throttle-pedal action I could find during heavy braking.

A slippery approach to Becketts hairpin, where decelerating from 120mph to 50mph and slotting into second gear in the wet were on the agenda, provided some intensely interesting motoring, right down to the full Timo Mäkinen 'put-it-sideways' arrival – the opposite way to the direction you intend to go. This shocked the photographer sufficiently to ensure that one fuzzy picture was filed of the most exhilarating part of the test.

Coming out of Becketts was *the* challenge. Blower convinced me that the throttle should be left applied and that the resulting long spells of sideways motoring could be deftly handled by the steering. The theory was to arrive slightly below the best speed and just blast through. I quickly discovered that any chicken-heartedness at this point with the throttle led to stubborn understeer. Sometimes I just could not believe that the TVR, which accelerated very rapidly from 50 to 65mph in second gear, could carry on sliding without spinning. Thus a long slide could end with the throttle sharply eased, my equally sharp intake of breath released, and the car lurching into third gracelessly. However, on the rare occasions when I did get it right, the resultant slide onto the main straight was utterly satisfying.

Quicker Woodcote and Copse right-handers were wet too. I settled for third gear and constantly-increasing throttle for both. Copse became even faster than its easier curvature suggested in this car. For Woodcote, which

'THIS THRICE-REBUILT UNIT HAD EVEN LARGER PISTON-TO-BORE AND VALVE-GUIDE TOLERANCES THAN NORMAL, TO JUDGE BY THE BLUE SMOKE WREATHED AROUND THE TWIN REAR PIPES WHENEVER THE ENGINE WAS STARTED.'

was traditionally the scene for desperate last-lap overtaking manoeuvres, it was better to brake earlier in the TVR, turn in gently over the glistening tarmac and apply the considerable V6 torque with respect.

There was hardly any need for fourth gear on the short pits straight, but I felt the owner might be happier if there was some evidence of gear-changing going on when the engine was obviously going to touch 7,000rpm. Barely had the pedestrian bridge been passed, than I started fumbling for third again. Provided the car had been settled into Copse easily, it came out with fine poise, third gear nearly used up and eager to gobble up the space before the theoretically flat-out Maggotts curve. Theoretically, because in the wet conditions the TVR tended to slide so far that the line into Becketts hairpin was all wrong.

Easing the throttle in fourth gear seemed a better way on that treacherous day. Slick surface conditions laid the emphasis on neat braking ability. Blower commented that the use of Aeroquip lines had cut out the spongy showroom feel that the brakes would normally exhibit in circuit use. Certainly they performed well for this class of car, but there was still enough pedal travel to allow contact with that elusive throttle pedal.

It would be difficult to imagine a better combination of grip and predictability from a production road tyre. The special Kleber rubber made it possible for an unfamiliar driver to get in and return laps in the 1m 13s region, giving high 70mph average speeds.

You still see TVR-Fords out giving their drivers circuit pleasure, and I certainly enjoyed my outings between 1969 and 1991. The capability of the chassis is largely forgotten, but if racing isn't your thing, I believe a classic TVR to be an underestimated blend of stylish and practical pace, whatever the engine… But go BIG if you can, it's just fiendish fun!

Footnote *Researching for this book, I found some text in the November 1969 edition of* Motor Sport *in which I assessed the merits of various Ford V6 implants, from conversions of mass-production Capris, Escorts and Transits to the specialist sports-car makers. Here's a prophetic concluding paragraph relevant to TVRs: 'Finally, if I were picking a V6-propelled machine, my choice would fall on the TVR Tuscan. It may not have the luggage space of the Ford-manufactured saloon/GT models, but it does have just enough of everything to satisfy those who want an individual sports machine and who have to make some concessions to a family. The Marcos is out for me because of the strictly twosome seating, while the Scimitar's looks are old-fashioned, but their GTE model could be just the job when the family rebel at being compressed into the TVR!'*

MARCOS

3.0 V6 GT: plenty of power, sharp looks

The Marcos factory at beautiful canal-straddling Bradford-on-Avon, and founder/boss-engineer Jem Marsh, were a shock to me in the summer of 1969. An established specialist builder of machines with (partial) wooden hearts that punched well above their weight, I really felt out of my depth. Marsh was a tall and forbidding figure, attired in country-gentleman manner. Jem, hard racer and resourceful constructor, expected maximum respect for Marcos achievements, not some long-haired Jesse who might have just escaped his teens without changing his Rolling Stones fan-boy appearance. Marcos products took a bit of learning when your main steering diet was 75 per cent tuned saloon cars, for nobody had – at that stage – let me into a Jaguar E-type's long-bonnet world. I had come to drive a then-new Marcos V6, which featured floor-level seating to peer over an elongated, cowled-headlamp, bonnet protuberance.

The most popular of the three models being built at that time, and a very good starter racer or road car, was the Mini Marcos, which could accommodate any Mini running gear from mild to wild. More relevant here were two Ford engine choices to haul a notably sleek glass-fibre outline, owing some influence to celebrated aerodynamicist Frank Costin, although the overall style was credited to Dennis and Peter Adams. This body was fitted over the unique Marcos chassis, one then constructed in marine plywood from floor to outriggers. A steel frame was required at the front to install adapted Triumph GT6 suspension and brakes: the rear carried a live axle, well restrained by links and a Panhard rod, and sporting coil-spring/damper units.

Steering was by rack-and-pinion, and the heaviest model I drove hit the weighbridge at 1,905lb (864kg). The 1600 GT had a very similar spec, but naturally weighed less, at 1,660lb (753kg). This was the period when Ford's ex-Zodiac/Capri V6 was quoted by the company at 144 gross horsepower, when it was actually below 130bhp, but the Marcos demonstrator I drove had been uprated: there were Janspeed manifolds to go with the unique twin-pipe exhaust system, and the Weber compound twin-choke carburettor had been rejetted and asked to breathe a little deeper via a wire-mesh air cleaner. No power claims were made for the modified V6, but performance was quoted as a 130mph maximum, coupled to 0–60mph in some 7sec, versus the independently measured 109mph maximum and 0–60mph in 11.4sec reported for the 1600 GT Marcos. You could expect over 25mpg from the smaller Ford engine and hope for slightly over 20mpg from the bigger V6 in mixed urban and country mileage.

DRIVEN

Marcos 1650 GT-Lawrencetune Track test day at Silverstone (8 laps), July 1968, published in CCC; Cortina 1500 GT bored to 1,650cc, high-lift camshaft, 10.5:1 compression ratio, Tecalemit Jackson fuel-injection. **Marcos 3.0 V6 GT (two examples)** Factory visit and two-hour Wiltshire drive, May 1969, published in CCC; car priced at £1,770 on debut; quoted 0–60mph in 7sec, 130mph max; extraordinary, laid-back, two-seater. (I also drove a 3.0 Marcos for *Motor Sport* colour feature, but retain no details.)

Footnote *Long after my 1969 visit, during the '90s, I visited Marcos again, this time as V8 endurance racing constructors under the helm of Jem's son, Chris Marsh. It was a more relaxed occasion, in Nissen-hut premises not far from my present home. I subsequently drove the bellowing racer at Snetterton. It was fun, and I was sorry to see a reborn Marcos concern implode a few seasons later, but the V8 was from the outer regions of hell, beyond the Blue Oval, so only a passing mention here.*

◄ I live locally to a couple of sites where Marcos cars were built. The Mini Marcos was probably the company's biggest seller and certainly the most affordable of the breed, but my favourite was the red coupé shown here (that's a '90s Mantis on the right) with 1600 GT or Ford V6 power. *Author*

MORGAN

Amidst the miscellany of Morgans produced over the years using power units from numerous sources, most of my experiences were in Ford-powered 4/4 models, the second '4' referring to the presence of a useful rear seat. However, I tasted examples with engines from Standard Vanguard (big four-cylinder, similar to TR2/3) on a '60s outing to Thruxton, plus the inevitable Ford Cortina/Capri/Escort Mexico 1600 GT units, on road and track. As a track car, I did enjoy Chris Alford's 1980 championship-winning Mog with 1.6 crossflow motor lightly modified by Minister, as smooth tracks suited the primitive suspension and low weight of this supremely traditional Brit sports car. That session was simply as a support to a *Motor Sport* double track test written by Clive Richardson.

Informally, whilst on an '80s factory visit, I drove a 1.6 EFi CVH example of the 4/4 that was credited with 105bhp, but seemed a wheezy disappointment in a flip around Worcestershire lanes. Morgan must have felt the same, as they had graduated to the 1.8-litre Zetec R in the '90s, in 120bhp alliance with a five-speed gearbox.

I carried on driving Morgans for the media, but none with Ford power. There was a well-balanced Plus 4 model with Rover 2-litre for *Fast Lane*, pitted against a 2.3 Panther. During a 2001 pre-production debut, I sampled the vastly more expensive BMW V8-engined Plus 8 model at Oulton Park in 4.4-litre format, and I featured a much better sorted 2014 example, carrying smooth and powerful 4.8-litre clout, for an American client.

Feeling that I should refresh my Morgan memories with a shot of later Ford power, I took advantage of Morgan and Lotus Dealership Williams Automobiles Ltd, outside Bristol, offering June 2021 factory-funded national test drives. The result was a solo outing in a 2012 used machine, registered AA12MOG. This one had the 110bhp/1.6-litre Sigma-branded Ford four-cylinder beneath traditional butterfly-wing, vented panels, which allowed engine access within the signature extended snout.

The car was well presented, and the

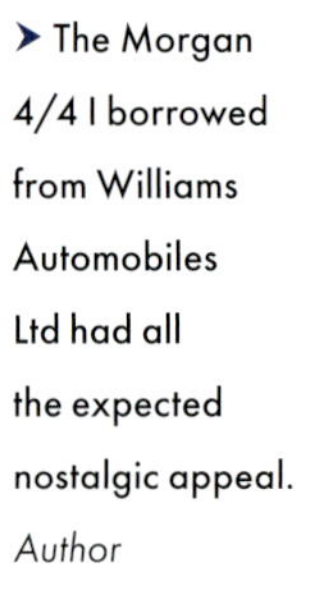

➤ The Morgan 4/4 I borrowed from Williams Automobiles Ltd had all the expected nostalgic appeal. *Author*

appearance and finish – especially if accompanied by a factory tour – clinched many a Morgan sale. I drove it only with side-screens up, hood down, and found many of the traditional characteristics: steering that goes strangely over-centre, suspension that just doesn't absorb much, and the Ford four-cylinder was just as unconvincing as the CVH had been. Plenty of torque to pull a still-low kerb weight, but no higher-rpm joys above 4,500.

Despite the gripes, the red roadster planted a silly grin on my face, and I could understand once more why the company attracted so many loyal, repeat owners, even when you had to wait more than 10 years for a new example. It's not like that now, and the company is no longer family-owned, for the last Morgan family member involved with the business ended more than 110 years of majority ownership when Italian group Investindustrial acquired the company in March 2019. At the time of writing, Morgan prices were set from £39,309 to £94,665, and waiting lists aren't the conversation piece they were back in the day.

I thank Williams for that driving opportunity, but would still opt for a Lotus Seven or Elise when buying in that sector.

DRIVEN

Morgan 4/4 and Plus 4 (multiple examples) Many cars driven (models from 1958 to 2012), with engines from Standard Vanguard (TR2/3) and Ford (1600 GT and 1.6 EFi CVH). **2012 Morgan 4/4** Driven in June 2021, courtesy Williams Automobiles Ltd, AA12 MOG; 1.6-litre, four-cylinder Sigma (Zetec), 110bhp; memory refresher.

▲ ◀ Country lanes suit Morgans in the 21st century, just as they always did. The demonstrator does not appeal quite so much with a wheezy 110bhp from the Ford Sigma motor hitched up to a slick six-speed gearbox, partnered by tall motorway ratios beyond third. However, smiles are still guaranteed. *Author*

OTHER FORD-POWERED CARS

Over the years, a wide variety of Ford power units have been transplanted into a fantastic array of individual vehicles, from humble 1.6-litre crossflow to period Grand Prix-dominating Cosworth Ford DFV V8. The resulting creations have been used for all kinds of motor sports, and on the roads, in the USA, Europe, South Africa, Australia and New Zealand. Here are some I drove in period.

V6-powered variety

Ford 2.5- or 3-litre V6-powered machines in late 1969 could be classified into three groups. The first comprised Ford products in the Zephyr/Zodiac large saloon range, and the Capri 3000 GT coupé. Next, a group of specialist, low-volume producers, who made a small number of sporting vehicles each year, available in kit form, or ready-built, with a glass-fibre body becoming a common feature that all these specialists shared. For example, TVR, revived when Martin Lilley took over, had success via the Vixen 1600 GT, plus the Tuscan, powered by a Ford V6. Gilbern's revamped 3-litre (renamed Invader) also fell within this specialist group. The third category arguably comprised not a group, but a single company – Reliant. Its Scimitar and GTE were produced on a larger scale. Reliant was exceptional, as it didn't offer kit-form assembly, and sold more cars than most specialists. Reliant was also the first of the outside firms to realise the V6 Zodiac engine's potential, and success with this and the 2.5-litre Ford V6 Scimitars took the company closer to Lotus production volumes.

Many British Ford 3-litre V6 and occasional American 5.0 or 4.7-litre V8 engine implants featured in my working life during 1968–72. They included upwards of a dozen firms transplanting the compact, but heavy, iron V6s into the unsuspecting engine compartments of Escorts, Cortinas, Transit vans, and – before the September 1969 availability of a factory Ford Capri – I was deluged with converted V6 Capris. Even back then, I explained that there were a couple of advantages to owning a V6 Capri, rather than a converted Cortina or Escort. The reason was that the Capri was designed with the Zodiac engine in mind, so that it felt better-balanced in V6 format. Another plus point was one of quieter progress, attributed to the Capri's cleaner aerodynamics.

Otherwise, I also steered V6 conversions in the form of Crayford Automobile Developments' excellent Escort V6 Eliminator, plus Race Proved Ltd's Savage Cortina automatic Estate – and the same company's Transit-based Sprite Motorhome. The latter was assessed with two other families. It was a riot as a mobile fun wagon, rather than a driving machine: V6 torque was a bonus, but not the raging fuel consumption.

V6 Sprite motorhome

I think some pearls from my November 1969-published review of the Jeff Uren/Race Proved 3.0 V6 (replacing Ford's unloved 2.0 V4) Transit-based, Sprite-bodied motorhome might cheer us. Bear in mind that the writer was then

DRIVEN

Lotus Europa V8 Brief drive in 1970, published as an aside in *Motor Sport*, November 1970 (also published 19 March 1970, *Motoring News*) within 'V8 Rumblings' that also featured SuperSpeed Boss 302 Capri and Escort-Rover V8; F5000 race-car constructor Tony Kitchener inserted formula-car hardware and Ford 4.7 V8 into the Europa. **Willment Zodiac 4.7 V8** Road test, 1971, published in *Motor Sport*, March 1971, XMT 912G; Willment/Mike Crabtree-loaned Ford Zodiac road car; Ford 4.7 V8 and Mustang three-speed automatic transmission; 0–60mph in 10.8sec, 98mph max, 14–17mpg. **DAF 55 rallycross Twin Cam 4x4** Drive at Lydden Hill, 28 August 1972, published in *Motoring News* and *Motor Sport*, November 1972, 70-77-MF; factory/de Rooy rallycross DAF, BRM 175bhp modified Lotus-Ford twin-cam, 4x4 with Variomatic central transmission, two foot pedals; driver straddled centre tunnel, effective 'bitza'. **'DFVW' Special Saloon** Tested on Silverstone GP track, 29 October 1976, published *Motor Sport* colour feature, December 1976; Colin Hawker's follow-up to race Capri with Ford Cosworth DFV V8; rear-engine layout of VW Fastback, and ex-race monocoque chassis suitable to accommodate 425bhp plus Hewland transaxle; fifth-wheel acceleration measured on Hangar Straight.

23 years old, and bred on tuned, small sports or saloon cars, with racing ambitions: 'The motorhome seems enormous when looking from the outside, and the thought of the vehicle's apparent skyscraper proportions certainly dampened this enthusiast's leaning towards hard cornering. Normal straight-line travel is easy, so long as one remembers that the 'home' section of the caravan is a couple of inches wider than the front; judging from the dents along the sides, other drivers need a reminder of this!

'The bulky body, with its external aluminium panelling, really demands something more than the manufacturer's V4 to pull it along, for even with the V6 installed, acceleration to 60mph isn't that much faster than an 848cc Mini. Top speed depends on how strong the crosswind is, and the bravery of the driver. The flat, tall sides make the Sprite-Uren machine a suitable training ground for all-in wrestlers when trying to correct the steering on a windy motorway. Under such conditions we found the safe cruising speed to be about 65mph, while less-exposed tarmac can be negotiated at over 85mph. A 4.6:1 final drive, combined with the standard gearbox, provided reasonable acceleration, but didn't aid pleasant motorway cruising, as the engine transmits 'I am working hard' messages, while turning at 4,000rpm to return 60mph. To be fair, that ratio aided the motorhome's very useful top-gear flexibility. So, one can accelerate fridge, cooker, kitchen sink, lavatory, shower, generous cupboards and five-berth sleeping arrangements from 8mph in top gear: at 25mph in top, the acceleration is quite adequate to keep up with Sunday drivers.'

Willment Zodiac 4.7 V8

I later graduated to commenting on a wider range of V6 and V8 conversions. Under the heading 'A thrusting executive with American V8 power' for the March 1971 edition of *Motor Sport*, I explored a Ford V8 conversion that had been tailored to an individual, rather than offered for multiple sales. That large Zodiac belonged to company director Peter Mahne, and had a second-hand 4.7-litre Ford V8 implanted alongside a factory-fresh three-speed Mustang automatic transmission. All work was carried out by respected Ford dealership Willment, under the care of racer/employee Mike Crabtree.

▲ The 1966–72 Zodiac MkIV marked Ford's fourth-generation crack at the six-cylinder UK market, and adopted independent rear suspension. In the popular Executive trim it carried a 126bhp version of the 3.0 Essex V6, but I also tested a 4.7-litre Ford V8 implant for *Motor Sport* magazine, which cheered it up considerably, despite a lazy automatic transmission. *Alamy*

Although the Zodiac was immaculate, it had covered a total of 20,000 miles, approximately half that distance in V8 form. Prior to the insertion of the V8, a mildly tuned production 3-litre V6 had been used in conjunction with modified suspension. Significantly, Willment found that, when they lowered the V8 into the space once occupied by a V6, there was no change in ride height. So if there was a difference in weight between Brit V6 and US V8, it was marginal and didn't require a ride-height reset.

Earlier suspension and braking uprates allowed this large gin palace on wheels to drift across Britain at a creditable pace, while the driver pondered idly behind the power-assisted steering. The plush seats offered little in the way of support, and recirculating-ball steering, linked up to the padded 16.5in steering wheel, needed much twirling for any real deviation in course.

The production cross-member was suitably braced, and the V6 engine mountings were adapted to a new role in the cause of supporting the V8. An engine-bay brace was also installed to fight the engine's flabby surroundings. Early tests showed that the engine had a tendency to overheat, and a bigger radiator and separated twin exhaust systems were installed. The bigger radiator's temperature was controlled by a pair of thermostatically operated Kenlowe electric fans. Their cut-in occurred between 70–80°C on the Smiths oil-temperature gauge, neatly fitted into

WILLMENT ZODIAC 4.7 V8 PERFORMANCE

Acceleration

0–30mph 4.1sec **0–40mph** 6.2sec **0–50mph** 8.0sec
0–60mph 10.8sec **0–70mph** 14.5sec **0–80mph** 9.6sec
0–90mph 26.9sec **50–70mph (in Hold 2)** 6.5sec

Gear speeds (automatic transmission)

Hold 1 58mph (25mph in Drive) **Hold 2** 68mph (45mph in Drive)

Maximum speed

In Drive 98mph at 5,000rpm

Overall economy

Average 14–17mpg

the wooden hole vacated by the showroom water-temperature dial.

Apart from a discreet V8 badge on the boot, substantial silencers were the only giveaway to this Q-battleship, unless a side view of 6in-wide Minilites was accessible. Generous in period, 185x14in Goodyear G800 tyres coped nobly with the demands of test-track and road use. As the Zodiac was the range leader in Britain, Ford had broken its usual period equipment rules and allowed disc brakes all round, 9.63in-diameter at the front and 10in at the back. Willment fitted Ferodo DS11 pads in a successful effort to provide enhanced retardation. The brakes were light and superbly graduated, so the lock-up point on the powerful servo-assistance could be easily judged.

The MacPherson-strut front suspension on the Willment car was substantially uprated by modifying the struts' valves. The rear suspension featured a pair of Armstrong adjustable shock absorbers, as the coil-spring independent system provided for a heavy load.

From a standing start, the automatic Mustang three-speed transmission was reluctant to accept a loading of more than 2,000rpm before accelerating away, so the initial take-off was leisurely until 50mph, when the Ford Executive started strutting along at a pace towards 90mph. The standard rear-transmission components were retained, including a 3.7:1 final-drive. The ex-Mustang eight developed maximum power at 4,400rpm, so top speed was limited by valve float at 5,000rpm.

This soft V8 was rated at 200bhp gross, developed at 4,400rpm when new, coupled to a more-impressive torque figure of 282lb ft at a leisurely 2,400 revolutions. That test V8 was an automotive tuning-recipe staple: the performance 4.7-litre/289cu in unit offered 271bhp at 6,000rpm and served as the basis for many a competitive power unit in Falcons, Mustangs and GT40s.

The characteristic rumble of a V8 came through clearly. Trickling through London suburbs, the flagship Ford felt slightly like a Mercedes in one respect: you aimed it by the bonnet mascot! Outside town, the twin-headlamp system picked up the contours of a long, straight B-road leading towards Box Hill. The inertia of the automatic in the 'Drive' position was overcome, and the palace on wheels burst forward in an impressive manner.

Slower corners were deliberately circumnavigated with respect, and at a velocity appropriate to an unmodified Mustang running on the same rim widths. A vigorous session at our test-track confirmed that the modified Zodiac tended to plough straight off course on understeer lock if liberties were taken at low speed. It had far more resistance to roll than the author had observed in standard Zodiacs at lower speeds. So the conversion had paid off, without sacrificing the American soft-ride characteristics which 150,000 customers, who had bought the MkIV Zephyr/Zodiac models, apparently preferred.

This Executive by Willment was suitable for Britain's speed-limited roads, continuous speeds of an honest 90mph+ accompanied by a temperature of 90°C, and 40psi oil pressure, instead of the usual 50psi. So good was the soundproofing – and so gallant the V8 – that you could flog it to death without knowing. No roughness betrayed the engine's distress at operating flat out, just the wavering rev-counter needle and other instrument readings told that harsh story. We found the lack of a throttle kick-down below 60mph a handicap, but anticipation allowed a Drive-selected 50–70mph performance in 6.5sec (close to a contemporary Triumph TR6 in third gear).

We managed something over 200 miles in

the 24-hour loan period. How long one stayed ensconced without visiting a petrol station depended, naturally, on how twitchy one's right foot became. A range of 210–260 miles proved within the car's 14–17mpg capabilities; the lower figure including performance testing.

DAF 55 rallycross twin-cam 4x4: improvised and extraordinary Dutch technology

The 26–28 August 1972 Bank Holiday weekend was crammed with personal motorsport missions. I had a lairy yellow Capri 2.6 V6/RS2600 mongrel by then. It hummed to Leicester Forest East on the Friday night to stay close to the M1 motorway for an early Saturday morning run up to Oulton Park to race a Janspeed Escort. That was followed by an Escort-racing Sunday at Mallory Park, then a plunge south to Kent and the insalubrious facilities of Lydden Hill rallycross circuit on the Bank Holiday Monday. Oulton and Mallory were amongst my favourite tracks, long Oulton in particular with the challenge of a rapid right called Knickerbrook!

Sadly, my affection for those Cheshire and Leicestershire tracks was only rewarded with third places for the Janspeed Escort. I also scraped the small red Ford repeatedly, deleting an exterior mirror in frustration as I tried to get by a larger-class car between me and our class leaders at Oulton. These were amongst my worst results – rare DNFs aside – in the 1.3 Escort Sport, so I went along to Lydden looking for a better story. That I certainly obtained, in the hand-built Dutch DAF concern's front-running rallycross mutation of its unique Variomatic-transmission 55 Marathon coupé.

Normally thrifty, small two-doors, with inoffensive performance but great charm, these DAF 55s carried a variety of engines in the search for competitive power to fulfil the company's desire to promote the Variomatic belt-and-pulleys stepless substitute for a conventional manual or automatic gearbox. The extraordinary 55s were operated and driven by heavy-transport company owners Jan and Harry de Rooy, on behalf of the compact but efficient DAF sports department. The modest coupés managed top-ten places even on less than 130bhp with rear-wheel-drive, versus grids of factory V6 Fords and British Leyland offerings. When Ford and BL fielded 4x4 variants of the Capri, Mini and Triumph 1300, DAF and the de Rooys decided that could be just the boost they needed to raise their game and seize places on the podium during 1971–72.

Their three-week transformation of the first 4x4 Marathon resulted in the coupé I drove. My experience behind the wheel followed the car's two years of harsh but successful use in rallycross events across continental Europe, with the Netherlands and Belgium especially keen on that form of bump-and-grind motorsport. My test, in the lunch hour at a full Lydden international meeting, centred on that 1971 incarnation, complete with Lotus-Ford twin-cam power, a bump in the roof, and the chance to share cabin-quarters with the transverse engine and Variomatic drive.

The Eindhoven factory and its de Rooy representatives had attracted very yellow Camel cigarettes sponsorship colours. That finance enabled an upgrade to lighter construction for two more coupés (no head-clearance bumps in their roofs), albeit these carried Ford-Cosworth 16-valve motors offering at least 25bhp more than the twin-cam vehicle I drove.

Under the skin: daring to be different

Here are some of my reported technical details involved in the rallycross test DAF, which I had also seen under construction at Eindhoven in 1971: 'The Variomatic has no reduction gears on the original DAF 4WD, transmitting power via standard secondary pulleys to separate prop shafts running fore and aft to BMW 2002 differentials, with limited-slip control. Front drive-shafts, MacPherson struts, and hubs were from a Taunus 12M, while the all-disc braking came from a Renault Gordini. The rear suspension is based on that used by the company's prototype rallying 55 coupé, a de Dion axle sprung by leaves. Both front and rear suspension is restrained by anti-roll bars. Minilite wheels normally wear dry-weather racing tyres, but for our non-competitive runs heavily grooved wet-weather Michelins of 13in diameter rested on the 7in rims.'

The first of the DAF rallycross trio sported a Renault engine, modified to give 128bhp by Bernard Collomb of Nice. The twin-cam I handled replaced the Renault unit in 1971. This former race engine was also tuned by BRM to Phase 4 specification. It had been sold, via Ford-backed John Taylor, to DAF. In its new home, the twin-cam, mounted transversely where the front seat would normally be, drove a Variomatic DAF-patented transmission, also arranged transversely in the front-seat area. Peak power was estimated at 175bhp, reached at 7,500rpm, elevated revs that the Variomatic tended to hang on to under full-throttle conditions – yet the engine performed very reliably over many 1971–72 events.

Behind the wheel: a unique experience

You would expect the DAF drive to be different, and so it proved, but first a bit of my rallycross-driving perspective. I had competed at Lydden rallycross with a notable lack of Imp-mounted success, but decades later I would also test Will Gollop's imposing 400bhp+ Metro 6R4 at this venue. I had another Ford-linked rallycross outing, tasting 'terrible twins' Rod Chapman and John Taylor's pushrod 1,850cc/140bhp Escorts – fitted with Ford factory Mexico World Cup engines – at Mike Cannon's forested Kent farm.

All of these machines were entertaining and impressive in their own ways. 'My' *CCC* Imp had 70 or 90bhp, depending which motor was installed, and was pretty tricky to balance between the engine's need for high rpm and understeer/oversteer on ice and snow. The Escorts were easy for a stranger to handle, but both Chapman and Taylor attained cosmic results, way beyond expectations. The Metro was the most enthralling, a normally aspirated V6 having a wonderful soundtrack and responses, so that I could enjoy it at a pace that talented Gollop regarded as suitable for warm-up laps.

Against that backdrop of other experience, the DAF was still a mighty machine, frightening to watch or steer under full power, but a monument to small-factory and team ingenuity. Here's how it felt to me at the time in a revised excerpt from *Motor Sport*'s two-page feature: 'Once installed, another obvious individual trait of the car is apparent: the steering wheel is in the middle, so one has to straddle the old propshaft tunnel to operate the brake, with a throttle on the right-hand (engine) side. We were sitting above the transmission! To connect the steering column up to the standard rack-and-pinion, the constructors took the easy way out – a chain, though one would never know from the steering action, which betrays no sign of this unorthodox link.

'The car was fairly hot from a previous thrashing, so I was advised to do no more than five laps. That DAF normally covered only three tours in competition, so I had to keep a close eye on the water and oil temperatures. Gauges were also provided for engine rpm, ammeter, oil pressure and fuel. I found that I needed to slow up on straights to read any of them, underlining that in this form of competition you just get on with it, leaving engineers to read the dials before and after…

'The getaway from a standing start was certainly very different to anything I've experienced before: just a feeling that one was encased within a catapult, which had just been released. Once the throttle is pressed, the engine note just goes up slightly with hills and whilst combating loose surfaces. For my trial, the normal Lydden summer rallycross course was used, in a very dry and dusty state. The speed at which the car could enter the notorious chalk section was quite astounding, this smooth, loose surface allowing one to set the car up just on the throttle.

'The flat-out ride – across the bumpy meadow – was accomplished in limousine comfort. Choking clouds of dust promptly caught us up at the Devil's Elbow hairpin, just after the tarmac track section unravelled. After the ease of control on grass and cinders, it was a rude shock to fight the heavy steering on tarmac – definitely a Tarzan task! At Lydden, there is another sharp right-hand U-bend at the top of the hill, then a dreadful plunge onto comparatively smooth track into what used to be called the Cinders/Mabbs Bank; nowadays this section is much smoother and seems predominantly covered in chalk as well. The right-hand hairpin calls for a lot of braking and understeer lock on the way in, but a quick flick out to the grassy kerb seems to set the car up straight for the downhill run. Whereas the car has probably been under 70mph previously, down that hill it attained 90 quickly. The cinders/chalk right-hander loomed forbiddingly. With brakes and some power applied simultaneously, the yellow DAF rocketed onto the rough with joy. Even full throttle only produced mild tail-out at my pace, and the left and right flick to complete a lap on the rough merely reinforced my enthusiasm for 4x4.'

Incidentally, DAF engineers were able to switch the Marathons rapidly from 4x4 to rear-drive, and measured the results thoroughly. On the dry, mixed surface track of the test-day, 4x4 turned laps averaging almost 3sec faster than rear-drive only.

My *Motor Sport* text continued: 'When I take the car back to make way for the proper racers, the way to 4WD round corners in comfort has become clear – apply dabs of power to lighten the steering. Naturally that is no advice for a (loose surface) competition driver. That may be why Roger Clark so obviously disliked the Capri 4WD, for any such layout can give you stubborn understeer.'

Prophetic words, as Hannu Mikkola was initially prejudiced against the Audi quattro's rallying potential after experiencing the 4x4 Capri's equally recalcitrant power-understeer traits: one loose-surface Audi test drive was enough to convince the Finnish Ford star, but it has to be said that even the massive Audi Sport World Championship rallying effort never saw success on pure tarmac trails such as Corsica.

▲ My August Bank Holiday 1972 wild ride. The de Rooy brothers' factory-backed DAF 55 Marathon, with Ford twin-cam power and Variomatic gearbox. Both were snugly installed in the cabin, so the driver was elevated on a perch that required that bump in the roof! *Peter J. Osborne*

'DFVW' Super Saloon: the VW with a Grand Prix heart

This vehicle involved perhaps the most resourceful and effective use of a 'scrap' DFV F1 engine I ever tried. Colin Hawker, a private individual, and star technician at the SuperSpeed Ford specialists, created at least three shimmering special-saloon competition cars using Ford motivation, but not all with Ford outer skins… Hawker's DFV-engined, VW 1600TL-bodied weapon became his best-know creation.

Under the heading 'The epitome of Super Saloons: Colin Hawker's VW-Cosworth V8', I introduced readers of the December 1976 issue of *Motor Sport* to the device, unusually granted the privilege of some colour illustrations: 'Colin Hawker is a mild-mannered young man of mechanical disposition, who earns a modest living as a mechanic in and around Ilford and Romford in Essex. Colin has a shock of grey hair, precocious for his years, and a two-tone blue VW, which has added a fair share to those greying locks. Now completing its second racing season, Hawker's DFVW is the latest in a long line of mongrels that Colin has constructed and raced with equal enthusiasm in the Special Saloon and Super Saloon categories. This year's tally of 12 wins and a couple of lap records has been Hawker's best with this interesting mid-engine cocktail of F1 (engine and gearbox), sports-car racing (monocoque and suspension were originally from the Alain de Cadenet Le Mans special) and that original choice of body shape from VW's obsolete 1600 fastback.'

I remember that both monocoque chassis and race-car suspension had been significantly modified when I drove the VW with an F1 V8 heart. All helpful for access during installation of our regular fifth-wheel and electronic timing gear. I commented that allowed us: 'To actually sample the phenomenal acceleration (from rest to 120mph averaged just 12sec) first-hand. Silverstone's Hangar Straight provided the base for the acceleration runs, while the track session was conducted on the Club circuit. Throughout the majority of the day, the circuit was either wet… or drying out from the spluttering of the leaden sky.' At the time, these were the quickest figures we had measured, made more impressive by hostile weather and a far-from-flat Hangar Straight.

These ingenious Super Saloon cocktails were usually concocted around V8 power implants to purpose-built formula-race-car underpinnings. They were often the products of people who built them privately, thus finances were limited, but ingenuity was unbounded. The exception that ruled this roost was Gerry Marshall and his 5-litre Dealer Team Vauxhall Firenza V8: Gerry simply dominated proceedings, taking a second successive championship win in 1976. The combination was so much faster than many privateers, that Marshall-DTV wins were a foregone conclusion, and entries withered away, as did the category, eventually.

Hawker started work on the DFVW in 1972 when, for £3,000, he bought a second-hand Cosworth V8 from Ken Tyrrell. That compact V8 had served the Tyrrell Formula 1 Jackie Stewart/François Cevert partnership well, and had history, dating back well before 1972. What Colin wanted was the kind of power those iron-construction Chevrolet V8s gave, but without a hefty mass.

Hawker was pleased with the choice, which served him in an earlier ('much too heavy') Capri, as well as the VW. He commented: 'It has all the F1 cams and valves for the 1975 season, but we don't have the super single-seater exhaust system and high-rpm inlet trumpets. I reckon that all costs at least 45bhp: put it this way, with me rebuilding the engine I think we would be happy to know we had a real 425bhp. No doubt about it, it's a fantastic engine. My Cosworth was totally scrap by their standards, just the bore-wear alone is five or six times what Cosworth recommend for F1, but it still doesn't puff out oil. Since I have had it, the Cosworth has required a set of liners, replacement pistons and three sets of bearings. The company recommends scrapping everything at 500 miles/five hours for F1 use, but by using 10,500rpm, instead of over 11,000, and taking part in these short races, I have had excellent service from the V8. It once chewed up a valve spring, passed it through the oil pumps and spewed it out in the sump-pan without misfiring during the race!'

Under the skin: an eclectic mix of engineering ingenuity

Although the engine was transferred from Hawker's Capri, precious little else made the trip. Colin acquired the VW's monocoque in 1975. It had been damaged when John Nicholson spun the de Cadenet at Le Mans 1974. By May 1975, Colin, brother Ray, and Brian Grove stripped the remains of the de Cadenet down. They created a mock-up VW body (just as a major manufacturer would from clay, wood and wire) and used that mock-up to make their own glass-fibre moulds for the VW 412 body. When the car was readied to race, the steel windscreen surround and roof from a VW were grafted onto their glass-fibre rear and front sections, which were in turn superimposed over de Cadenet's hardware, which used a large number of Brabham parts around the unique monocoque.

Subsequently, both the front and rear suspension was totally redesigned, following a rose-joint failure after just three races, one that sent the car into the wall at Brands Hatch shortly after exiting Clearways. A fresh parallel-link rear suspension and fabricated aluminium rear-upright design was substituted, but there were problems with the uprights, which failed until it was found that they hadn't been properly heat-treated. At the front, solid-magnesium uprights were installed. The Len Terry design for the double-wishbone front end and parallel-link/radius-arm rear proved itself. Originally, Hawker was struggling with 550lb/in front springs, suspecting there was no way his glass-fibre/alloy creation should be using spring rates normally used in a steel-bodied V6 race Capri. A phone call to Len Terry set them on the path to a suitable rating. Another major change to the original specification came in May 1976, with the adoption of a single front radiator, actually a former pair of side radiators amalgamated as one. Now oil radiators for the engine sat behind NACA ducts.

Wheels, brakes and drive-shafts were totally revised after the car made its debut in May 1975.

◀ As for his previous Capri with recycled Ford Cosworth DFV eight-cylinder F1 power, Colin Hawker's 'DFVW' was beautifully crafted and presented to professional standards. It was a privilege to drive at Silverstone for the December 1976 issue of *Motor Sport* magazine. *Alamy*

For dry use, Shadow F1 17in-wide rears were combined with Brabham F1 fronts of 11in spread. For wet tarmac, the same-diameter fronts were utilised, but the backs were 1in narrower, the wet-weather wheels all ex-Brabham. Goodyear rubber, from its Formula 1 stock, was used, but definitely not at F1 prices! Hawker expected 30 races from the bulbous rear covers, but the fronts could be worn out in four events.

Hawker acquired a Hewland five-speed DG400 transaxle from March at exceptionally low cost, but a minimum of £1,000, including a few gear-ratio sets, was the true value. Hawker paid £30–35 for his ratio sets, required to gear the car for each circuit with a DFV installed. Another £18 was regularly spent on gear-selector engagement dogs, which wore out after three or four races: that metallic appetite proved our undoing during track testing.

Colin Hawker felt his DFVW needed to overcome the lack of accessible torque in the DFV conversion, and confessed to 'having to overtake half the field after a poor start, before I can start motor racing!'

Other regular competitors that 1976 season lined up in machinery as diverse as Vince Woodman's ex-Broadspeed Capri RS with a Cosworth Ford 24-valve V6 of 3.4-litres; Tony Strawson's Capri-Chevrolet 7-litre; Mick Hill's Beetle-shaped machine with Chevrolet V8 power (following his Capri success); and Tony Hazelwood's Jaguar with a Chevrolet V8 unit placed in the cabin with its intrepid pilot. There were a variety of Escorts, notably Nick Whiting's screaming 2-litre Ford-Cosworth-engine example and Doug Niven's machinery, which in the past had included both Capri and Escort with American V8 power-units.

At Silverstone that Friday 29 October 1976 morning, the 'seen 'em all' staff looked at the author with their 'now we know you need locking up' expressions. Puddles abounded and I kept recalling the more vivid details of the component failures this car had suffered. Despite a tightly scheduled trip to the Jochen Rindt show in Vienna, followed by a complete strip-down, Hawker calmly accepted that we went ahead.

'DFVW' SPECIAL SALOON PERFORMANCE

Acceleration

0–60mph 4.45sec **0–100mph** 9.8sec **0–120mph** 12.0sec

Gear speeds

First 60mph **Second** 85mph **Third** 108mph **Fourth** 137mph **Fifth** 'over 150mph'

Behind the wheel (and in the footwell!): surprisingly refined power

Complete with 425bhp powertrain and gleaming monocoque finish (Solvol Autosol's finest hour), the DFVW weighed in at over 1,790lb (812kg). Preparing the car for our test procedures dissipated my apprehension. First the pictures were taken on the shiny slick-shod wheels, then we changed rubber for business, and installed blue-spoked devices that had cracked in hard, dry use before (gulp!). Cursed with an efficient photographer for the day – usually John Dunbar or Maurice Selden on my test sessions – it wasn't long before I was tremulously admiring the neat bracket the boys had fabricated for our fifth-wheel. We ran the electronic cable back through the engine compartment, via the bulkhead, as the engine was neatly partitioned off from the driver's compartment by a sheet of 'chauffeur glass'. The broad sills of the monocoque allowed plenty of room for the speedometer head and data board to be accommodated, while I read off the times and speeds perched, without a formal seat, towards the left of the cockpit during timed runs. My right thigh rested carefully against the on/off switchgear and cable, and the press-button starter.

Hawker settled into the luxury of a Corbeau seat and Willans belts, mounted behind a sturdy steering wheel of unfashionably large diameter. 'Easier to handle,' he commented. I gauged that the centre dashboard was a well-finished fabrication that wouldn't look out of place in a production car. Nothing showroom about the volumes of information relayed, pertaining to oil pressure and temperature, water temperature, fuel pressure and engine revs, the latter electronically limited to 10,600rpm. The heavy-duty Formula 1 non-synchromesh gear-change to the driver's right was laid out with first/third/fifth in the right-hand plane and reverse/second/fourth on the left.

It was a tight fit inside, especially when the surprisingly flush doors were pegged shut from within. Having gone through the rigmarole of starting the DFV up (it had been fired up that morning in the workshop), then stopping it to change onto a weak mixture on the metering unit, I wanted to get on with it, after imploring a lightweight battery under my feet to retain a charge. The latticework of the integral steel roll-cage allowed a handy brace for the feet. As the digital watch reported for duty, I gave our chauffeur the thumbs up.

Everything happened at once. As the plump Goodyears searched amongst the crevices of Silverstone's former runways for grip, the DFVW shot swiftly right. Hawker barely moved the wheel to twitch the car straight, but by the time he had done just that, we were well on the way to 50mph. We could just reach 60mph in first at 10,500rpm. A shift to second meant just over 5sec to reach this speed, while our best first-gear-only run occupied 4.36sec.

The effect on one's stomach as the car rushed from 30 to 60mph was awesome, but exhilarating nonetheless. There was a little bit of slip as second gear went home, but the car tracked straight on to 85mph. The 108mph third gear really brought a sense of speed, and it was worth looking up to see the tarmac rushing madly by from our humble position. Since the sprint from 0–100mph sizzled by in less than 10sec – about the same as an (Alfa Romeo) Alfetta then took to reach 60mph – we spent little time in third. With fourth engaged, things did slow a little, but the temporary two-seater leapt forward at a formidable rate, so that the electronic speedometer needle just clicked onto 137mph before we slowed.

The gearbox had taken quite a thrashing during the acceleration runs, and a race that previous weekend at Mallory Park, where the hairpin required first, and that meant an awkward dog-leg gear-change. Those abused selectors spoiled our test-session a little, but the thrill of steering a car with this performance and handling overcame all. The triple-plate Borg & Beck clutch had to be eased home delicately to get a clean start, but its manners were really little worse than the contemporary Ferrari 308 GTB I had road tested. The transparent section to the rear wing enabled some rear vision, but the overall rear view remained restricted.

The steering felt surprisingly heavy for a rear-engine machine, but the response was all that you would expect: virtually as if attracted by magnets the car swooped from left to right, those massive rears only broken free under hard, second gear, duress at Silverstone Club circuit's hairpin. The Club straight became enjoyable – deploying 10,000rpm in the first four gears, I found the rise in the middle shot towards one like a motorised mountain, instantly followed by the appearance of the grandstands and Woodcote.

The brakes, using a four-pot calliper system on ventilated Lockheed discs mounted inboard at the rear, were outstanding. Hawker reported: 'This is one of the car's strengths, there doesn't seem to be anything better under braking than this car amongst our opposition.' For our use, they just provided reassurance, in what was really a very easy car to feel confident in.

When we had completed only a handful of laps, the poor gearshift deteriorated – it kept falling out of second and third, highlighting the flexibility of that obsolete race V8. I was astonished to find that Keith Duckworth's masterpiece was quite capable of pulling the car smoothly forward from 5,000rpm in third at the hairpin, which still equated to 9,000rpm in fifth well before Woodcote. Building the revs up to 6,500 or 7,000rpm in fifth and accelerating hard delivered an apparently instant 9,000rpm and an eagerness to hit 10,000rpm in top.

I was privileged to drive that previously mentioned showroom Ferrari 308 GTB at our test track, and I think anybody would have been amazed at how close the Cosworth full-race engine came to Ferrari's standards of road-car docility. Abundant torque – albeit not enough racing against large-capacity American V8s – and clean starting, struck a welcome civilised note amongst the memorable sensations of raw speed.

▼ The future – does the electric Mustang Mach-E mean the end of all fossil-fuel performance Fords? It looked that way in 2023. The electric route can quietly deliver stunning acceleration, in conjunction with four-wheel drive, as for this Mustang. *Ford*

APPENDIX 1
GATHERING PERFORMANCE DATA

Performance figures are mentioned regularly throughout this book, so I must comment on how such figures and model facts were gathered over my 50+ years of experience. I was employed by, or a freelancer/sub-contractor for, seven British weekly and monthly motoring magazines, and between 1980 and 2017 had experience working for Italian, French and American print media. I must admit that the methods of assessing acceleration and maximum speed for my first magazine job at CCC were hardly the authentic engineering-orientated stuff of British weeklies *Autocar* or *Motor*.

Back in 1967–72, the proper method involved using a trailing-frame bicycle wheel towed behind the car, with a cable feed to a corrected independent speedometer within the car. A passenger monitored and recorded speeds via multiple stopwatches on a clipboard, filling out on paper all minutiae of performance data that could be gathered. The most popular in-gear statistic was 50–70mph as a measure of top-gear flexibility. Not only 0–60mph figures were recorded, but in-gear acceleration times from 20mph to whatever the test track could accommodate. Sometimes we just ran out of space on runs above 0–90mph. I recall a very senior technical journalist of considerable engineering intelligence simply running into the barriers at the close of a 0–120mph run in a Jaguar XJ-S. He was distracted by the oh-so-slow final 115–120mph increments, failing to respect a diminishing supply of tarmac and impending crash barriers. The damage was mainly to his pride…

The two weeklies would also measure weight on a public weighbridge, fuel consumption using flow-rate monitors, and every interior dimension, particularly rear-seat space and headroom, plus boot capacity. These were proper and thorough road tests to the standards of the motor industry, usually carried out at MIRA (Motor Industry Research Association), where there was a banked track and interconnecting straights, with ride-and-handling and figure-of-eight test tracks, and many other facilities, including a skid pan. The published results made dreary reading, and accompanying grey pictures in those largely pre-colour inside pages added to the feeling of a rather stodgy establishment bulletin. Yet, if you wanted accurate measurements, it seemed this was the only way.

However, there were successful alternatives to this expensive and labour-intensive business and these appealed to magazines like CCC with only three full-time editorial employees. Our ex-*Autosport* editor, Martyn Watkins, took his secretary (Brenda Slough), or me, along for the test car ride. Out from our West London offices, along the M4 past Heathrow airport, until we got to the Colnbrook exit. After negotiating a large roundabout, you arrived on a reasonably flat strip of A-rated road running westwards parallel to the M4.

By then, we would have cumbersomely cross-checked speedometer mileage readings at speed readings between 30 and 90mph, utilising a stopwatch against the motorway mileage posts to check the calibration of the vehicle's production speedometer. Rough and ready, but marginally better than believing the production speedometers of the period, particularly as most of CCC's fare comprised tuned cars on replacement wheel and tyre sizes.

At the time I was personally ultra-keen on the performance stats, so would keep an eye on the figures reported by the weeklies with their accurate timing methods. Although these magazines did a thorough job, some British-based manufacturers often provided them with road-test specials that were capable of better-than-standard figures. Occasionally, we'd get one of these cars in at CCC, and you could usually tell because they wouldn't run on anything but five-star high-octane petrol. Not unexpectedly, they would often pink if you didn't drive them like any low-grade competition car with frequent gear-changes.

The worst and most memorable offenders? Mainly the culprits were British manufacturers of sports machinery, notably MG and Triumph, whereas the cars provided by BMW, Alfa Romeo and other imported brands were more representative of what the public could buy.

Fortunately for the preservation of my driving licence, and insurance documents referring to 'speed testing' of any sort as totally excluded, our methods changed within months of my involvement with testing. *CCC*'s rapid escalation in circulation – and the arrival of an enforced 70mph speed limit – quickly prompted us to move from public tarmac to what was then Chobham FVRDE (Fighting Vehicle Research Development Establishment) army premises in Surrey.

I would spend more time at that banked track, Snake Pass handling course, skid pan and infield loose (forest) roads than any other test resource. It's still used for filming, albeit with civilian personnel and management, and no longer littered with army tanks and other military hardware – solid obstructions that you would usually encounter as you tried some illicit speed measurement (there was a frequently ignored 70mph limit). Sections like that Snake Pass selection of humps and twists often feature in TV dramas that involve mishaps, from punctures to fatal accidents.

When I arrived at *Motor Sport* and *Motoring News* as an employee, I found they were doing a more professional job with a tatty, but effective, fifth wheel – a gadget that we tied onto most things. For instance, the Brian Muir/Malcolm Gartlan 1971 Chevrolet Camaro racer, which threw our abused wheel over our heads at the end of Silverstone's Hangar Straight. By then the snorty beast had provided the exciting data we needed.

I used timing equipment regularly at the Millbrook two-mile bowl and associated infield tracks created for General Motors in Bedfordshire, plus the Ford sites at Boreham in Essex and Lommel in Belgium. We also used many racing circuits, either as pictorial backdrops (Brands Hatch was particularly helpful in period) or for competition cars. For assessments that were more about emotion than measuring statistics, Mallory Park and Snetterton proved particularly useful on many occasions.

Performance measuring kit progressed to electronic Correvit read-outs and associated automatic printer updates when I returned to *Motor Sport* and *Motoring News* as a contracted freelancer. I also tested cars for *Performance Car* (very demanding test-track data masters) through most of the '80s, and *Fast Lane*. Mark Hales actually did most heavy lifting; I just attended track sessions and, after the most rigorous tests, obeyed cooling-down orders aimed at preserving turbo life. As a freelancer, I delivered properly timed outings for competition machines such as factory Sierra and Escort RS Cosworths for *Motor* and *Autocar*, using their equipment and personnel.

When dealing with factory Fords, I usually had the regular competition driver undertake the acceleration runs while I operated the timing equipment. John Fitzpatrick completed the sprints in the Group 2 Broadspeed Escort (Chapter 7), John Welch the Xtrac rallycross Escort (Chapter 10), Boreham chief engineer John Wheeler the Corsica-winning Sierra RS Cosworth (Chapter 13) and John Taylor the Corsican 1993 Escort RS Cosworth (Chapter 14). The trickiest timed standing-start runs I ever did were for Colin Hawker's DFVW-Cosworth V8 Super Saloon (Chapter 17) , as it had so much traction versus a Formula 1 power band.

The task of satisfying younger editors who wanted the best recorded 0–60mph times in high-grip turbo cars (Audi's quattro, Porsche's 911 turbo) was a pretty protracted and destructive regular routine, involving repeated tests that demanded intense concentration and robust test vehicles on quality tyres.

APPENDIX 2
AUTHOR'S PREVIOUS 'FAST FORD' BOOKS

All the titles listed are long out of print, the oldest (Capri) dating back to the early '80s and most recent (Cosworth) published in the '90s.

I do have Facebook and Amazon author pages, and some of my shorter editorial/news appears on the website www.fromthedrivingseat.com.

Capri

- *Capri: The Development and Competition History of Ford's European GT Car*, Foulis/Haynes Publishing (1981, 1987 and 1990).
- *A Collector's Guide: The Sporting Fords Volume 3 – Capris, Including RS2600, RS3100, 2.8i and 280*, Motor Racing Publications (1983 and 1990).

Escort

- *Escort Mk1, 2 & 3: The Development & Competition History*, Foulis/Haynes Publishing, 1985 (also updated book-club edition that year, incorporating significant extra data, then 1987 and 1989 reprints). Some key Ford of Britain people risked their jobs to give insider talk.
- *Escort Mk1, 2, 3 & 4: The Development & Competition History*, Haynes Publishing (1990). Updated edition of the previously listed book, with MkIV added, plus other new material, and corrections/enhancements. Also includes photographs/captions of MkV, announced just before book went to press.
- *A Collector's Guide: The Sporting Fords Volume 5 – Front-drive Escorts, From XR3 to RS Cosworth 4x4*, Motor Racing Publications (1994).

Multiple Fords

- *XR: The Performance Fords*, Motor Racing Publications (1985). Covered Escorts, Fiestas and Sierras, with independent performance figures and specifications.
- *Ford Escort XR3 & XR3i: The Enthusiast's Companion*, Motor Racing Publications (2000). Six authors contributed nine chapters. I did a piece about the mechanical modifications then available, particularly for suspension. Peter Newton added a useful round-up of all sporting applications.
- *RS: The Faster Fords*, Motor Racing Publications (1987 and 1990). Covered Escort, Capri, RS200, Sierra and Fiesta. Extended factory and competition specifications and performance figures in 1990 edition.
- *The Cosworth Fords: A production and competition history*, PSL/Haynes Publishing (1994). Coverage from Lotus 18 to Mondeo. Particularly strong on Escort and Sierra competition cars, plus RS Cosworth rear-drive and 4x4. Ford Motorsport chief engineer John Wheeler gave me the most open access and information I had enjoyed at Ford, including driving factory winners and pre-production Escort RS Cosworths. For me it is the most informed Ford book I have written. I can only compare it with my experience of Lotus 'open-door' access for my work on the Esprit and Elise.

BIBLIOGRAPHY

Excluding author's books, listed separately

Websites

www.touringcarracing.net
Motor Sport (online and hard copy)
Multiple auction houses (for prices)

Magazines

Autocar
Autocar & Motor
Autosport
Cars & Car Conversions
Classic & Sports Car
Evo
Ford Heritage
Motor
Motoring News/Motorsport News
Motor Sport
Octane
Performance Car

Books

A Collector's Guide: The Sporting Fords Volume 4 – Sierras Including XR4i, Merkur XR4Ti, XR4x4, RS Cosworth, RS500 and RS Cosworth 4x4, by Graham Robson, Motor Racing Publications (2000)

Fitz: My Life at The Wheel, by John Fitzpatrick, Autosports Marketing Associates (2016)

Ford Cortina: The Complete History, by Russell Hayes, Haynes Publishing (2012)

Ford in Touring Car Racing, by Graham Robson, Haynes Publishing (2001)

Niki Lauda: His Competition History by Jon Saltinstall, Evro Publishing (2019)

Rallye Sport Fords: The Inside Story, by Mike Moreton, Veloce Publishing (2007)

The Complete Catalogue of the Ford Escort Mk3, Mk4, Mk5 & Mk6, by Dan Williamson, Herridge & Sons (2020)

INDEX